Current Topics in Neurotoxicity

More information about this series at http://www.springer.com/series/8791

Norbert Müller • Aye-Mu Myint
Markus J. Schwarz
Editors

Immunology and Psychiatry

From Basic Research to Therapeutic Interventions

Springer

www.wpanet.org
WORLD PSYCHIATRIC ASSOCIATION

Editors
Norbert Müller
Department of Psychiatry and Psychotherapy
Ludwig-Maximilians-University Munich
Munich, Germany

Aye-Mu Myint
Department of Psychiatry and Psychotherapy
PsychoNeuroImmunology Research Group
Ludwig-Maximilians-University Munich
Munich, Germany

Markus J. Schwarz
Institut fuer Laboratoriumsmedizin
Klinikum der Universitaet Muenchen
Muenchen, Germany

ISSN 2363-9563 ISSN 2363-9571 (eBook)
Current Topics in Neurotoxicity
ISBN 978-3-319-13601-1 ISBN 978-3-319-13602-8 (eBook)
DOI 10.1007/978-3-319-13602-8

Library of Congress Control Number: 2015934820

Springer Cham Heidelberg New York Dordrecht London

Printed on acid-free paper

Springer International Publishing AG Switzerland is part of Springer Science+Business Media (www.springer.com)

Preface

Although Emil Kraepelin, the founder of modern psychiatric classification, described the influence of infections on psychiatric disorders as early as 1890 in his manuscript "Ueber Psychosen nach Influenza" ("On psychoses after influenza"), he was by far not the first one to perform research in the field of psychoneuroimmunology. In 1887, the later Nobel Laureate Julius Ritter Wagner von Jauregg published a sort of ancient meta-analysis on the therapeutic influence of typhus infections on patients with psychiatric disorders. He merged data from observations in Austrian, German and Swiss asylums during typhus epidemics, showing that psychiatric symptoms improved in about half of the patients and that about one-third of them were cured after the infection had subsided. These observations served as the basis for Wagner von Jauregg's fever therapy, which was adopted in some European countries during the 1920s and 1930s. One extremely intriguing example of the effect of "immunological" research is the identification of *Treponema pallidum* as the causative agent of neurosyphilis and the discovery of Salvarsan, the first chemotherapeutic drug against *Treponema*. These findings resulted in the causative treatment of about one-third (!) of all psychiatric patients at that time. He won the Nobel Prize in 1927 for his treatment of paralysis using malaria inoculation.

After World War II and the triumphal procession of the neuroleptics in the 1950s and 1960s, research in biological psychiatry focused on neurotransmitter disturbances and their influence on psychiatric disorders. Without doubt, those neurotransmitter disturbances play a key role in disorders such as schizophrenia, major depression, autism and anxiety, but—despite huge amounts of research—the causes of the neurotransmitter changes are widely unknown. Unfortunately, the spectacular progress in psychopharmacotherapy had a negative impact on immunological research in psychiatry. In the 1980s, investigations of autoantibody titres in patients with schizophrenia were a kind of revival of immunological research in psychiatry. Only in recent years was it proven, for example, that autoantibodies directed against the NMDA receptor can cause schizophrenia (or rather a schizophrenia-like syndrome). This example highlights the extremely difficult path from a hypothesis to a proven finding in the field of immunological research in psychiatry.

The growing, fascinating and future-oriented field of psychoneuroimmunology claims to close the gap between the neurotransmitter disturbances and the underlying processes such as infections or other body/environmental processes. The immune system, consisting of an innate and adaptive part, shows an extremely high multiplicity and variability of cellular and humoral components and in addition underlies multiple influences that hinder especially clinical research with patients.

Thanks to encouraging scientific results, interest in the field of psychoneuroimmunology has grown over the last few years with regard to basic and clinical research, including therapeutic studies. The growing number of scientific groups in the field reflects this growing interest.

Therefore, the editors decided to prepare a more or less representative overview of the current activities in the field by asking internationally established scientists and junior researchers from experienced groups to contribute a chapter reflecting their current research. Practically, all scientists we addressed agreed to write a chapter.

One of the scientific platforms of psychoneuroimmunological research in the field of psychiatry is the section "Immunology in Psychiatry" of the World Psychiatric Association (WPA), which was founded in the early 1990s by Manfred Ackenheil from Germany, Oakley Rey from the USA and Costas Stefanis from Greece. Many of the contributors to this book are active members in the section, which regularly organizes workshops, symposia and lectures at WPA congresses and co-organizes many additional meetings and research activities. The editors of this book, Aye-Mu Myint (secretary of the section), Markus J. Schwarz (past secretary) and Norbert Müller (chair), therefore dedicate their contributions to the WPA section "Immunology in Psychiatry."

The editors cordially thank all authors of this book for the tremendous work they have invested and the publisher Springer for their help and support. Furthermore, we thank Richard Kostrzewa, the editor of the series, for the inspiration for this book and Karin Koelbert for her help with handling the manuscripts. We think we have succeeded in assembling a representative overview of the current state of psychoneuroimmunological research in the field of psychiatry and related topics.

Munich, Germany

Norbert Müller
Aye-Mu Myint
Markus J. Schwarz

Contents

Contributors

Karl Bechter Clinic for Psychiatry and Psychotherapy II, Ulm University, Bezirkskrankenhaus Günzburg, Günzburg, Germany

Department Psychosomatics/Psychotherapy, Günzburg, Germany

Michael Eriksen Benrós Faculty of Health Sciences, Mental Health Centre Copenhagen, Copenhagen University Hospital, Bispebjerg Hospital, Copenhagen NV, Denmark

National Centre for Register-Based Research, Aarhus University, Aarhus, Denmark

Elisa Brietzke Interdisciplinary Laboratory of Clinical Neuroscience (LINC), Department of Psychiatry, Universidade Federal de São Paulo, São Paulo, Brazil

Erik F.J. de Vries University of Groningen, University Medical Centre Groningen, Nuclear Medicine Molecular Imaging, Groningen, The Netherlands

Lot de Witte Department of Psychiatry, University of Utrecht, Utrecht, The Netherlands

Janine Doorduin University of Groningen, University Medical Centre Groningen, Nuclear Medicine, Molecular Imaging, Groningen, The Netherlands

Harald Engler Institute of Medical Psychology & Behavioral Immunobiology, University Hospital Essen, University of Duisburg-Essen, Essen, Germany

Tatiana Falcone Departments of Psychiatry and Neurology, Neurologic Institute, Cleveland Clinic, Cleveland, OH, USA

Epilepsy Center, Neurologic Institute, Cleveland Clinic, Cleveland, OH, USA

Casara Jean Ferretti Autism and Obsessive Compulsive Spectrum Program, Montefiore Medical Center, Albert Einstein College of Medicine, Bronx, NY, USA

Kathleen Franco Cleveland Clinic Lerner College of Medicine, Cleveland, OH, USA

Steven J. Garlow Department of Psychiatry and Behavioral Science, Emory University School of Medicine, Emory University, Atlanta, GA, USA

Jan-Sebastian Grigoleit Institute of Medical Psychology & Behavioral Immunobiology, University Hospital Essen, University of Duisburg-Essen, Essen, Germany

Laboratory of Neuronal Structure and Function, The Salk Institute for Biological Studies, La Jolla, CA, USA

Angelos Halaris Department of Psychiatry, Loyola University Chicago Stritch School of Medicine, Maywood, IL, USA

Andrew Harkin Trinity College Institute of Neuroscience, Trinity College Dublin, Dublin 2, Ireland

The School of Pharmacy and Pharmaceutical Sciences, Trinity College Dublin, Dublin 2, Ireland

Pieter J. Hoekstra Department of Child and Adolescent Psychiatry, University of Groningen, University Medical Center Groningen, Groningen, The Netherlands

Eric Hollander Department of Psychiatry and Biobehavioral Sciences, Montefiore Medical Center, Albert Einstein College of Medicine, Bronx, NY, USA

Autism and Obsessive Compulsive Spectrum Program, Montefiore Medical Center, Albert Einstein College of Medicine, Bronx, NY, USA

Georg Juckel Department of Psychiatry, Ruhr University, LWL University Hospital, Bochum, Germany

Hans C. Klein Department of Psychiatry, University Medical Centre Groningen, Groningen, The Netherlands

Daniela L. Krause Department of Psychiatry and Psychotherapy, Ludwig-Maximilians University Munich, Munich, Germany

Brian E. Leonard National University of Ireland, Galway, Ireland

Barry McGuinness Trinity College Institute of Neuroscience, Trinity College Dublin, Dublin 2, Ireland

Preben B. Mortensen National Centre for Register-based Research, Aarhus University, Aarhus, Denmark

Michael Maes IMPACT Strategic Research Center, Deakin University, Geelong, VIC, Australia

Faculty of Medicine, Department of Psychiatry, Chulalongkorn University, Bangkok, Thailand

Health Sciences Graduate Program, Health Sciences Center, State University of Londrina, Londrina, Paraná, Brazil

Rodrigo Mansur Interdisciplinary Laboratory of Clinical Neuroscience (LINC), Department of Psychiatry, Universidade Federal de São Paulo, São Paulo, Brazil

Mood Disorders Psychopharmacology Unit (MDPU), University Health Network, University of Toronto, Toronto, ON, Canada

Urs Meyer Physiology and Behaviour Laboratory, ETH Zurich, Schwerzenbach, Switzerland

Norbert Müller Department of Psychiatry and Psychotherapy, Ludwig-Maximilians University Munich, Munich, Germany

Aye-Mu Myint Department of Psychiatry and Psychotherapy, PsychoNeuro-Immunology Research Group, Ludwig-Maximilians-University Munich, Munich, Germany

Otto von Guericke University, Magdeburg, Germany

School for Mental Health and Neuroscience, Faculty of Medicine, Maastricht University, Maastricht, The Netherlands

Cristiano Noto Interdisciplinary Laboratory of Clinical Neuroscience (LINC), Department of Psychiatry, Universidade Federal de São Paulo, São Paulo, Brazil

Mark Hyman Rapaport Department of Psychiatry and Behavioral Science, Emory University School of Medicine, Emory University, Atlanta, GA, USA

Veronika M. Reinisch Department of Psychiatry and Psychotherapy, Ludwig-Maximilians University Munich, Munich, Germany

Lucas B. Rizzo Interdisciplinary Laboratory of Clinical Neuroscience (LINC), Department of Psychiatry, Universidade Federal de São Paulo, São Paulo, Brazil

Tammy Saah Department of Psychiatry and Behavioral Science, Emory University School of Medicine, Emory University, Atlanta, GA, USA

Manfred Schedlowski Institute of Medical Psychology & Behavioral Immunobiology, University Hospital Essen, University of Duisburg-Essen, Essen, Germany

Markus J. Schwarz Institut fuer Laboratoriumsmedizin, Klinikum der Universitaet Muenchen, Muenchen, Germany

E. Fuller Torrey The Stanley Medical Research Institute, Chevy Chase, MD, USA

Elif Weidinger Department of Psychiatry and Psychotherapy, Ludwig-Maximilians University Munich, Munich, Germany

Robert H. Yolken Stanley Division of Developmental Neurovirology, Department of Pediatrics, Johns Hopkins University School of Medicine, Baltimore, MD, USA

About the Editors

Norbert Müller After studies of psychology and medicine, Dr. Müller was trained in psychiatry, psychotherapy and neurology at the University Hospital of the Ludwig-Maximilians-University in Munich. He did research in the field of psychoneuroimmunology since 1983, primarily in schizophrenia, affective disorders and Tourette's syndrome. The research focused on pathophysiological aspects and on therapy with anti-inflammatory compounds. Since 2000, he is Professor of Psychiatry, at the department of Psychiatry and Psychotherapy, LMU Munich. He was president of the German Society of Biological Psychiatry and member of the executive committee and treasurer World Federation of Societies of Biological Psychiatry (WFSBP), since 2006 he was chair of the section "Immunology in Psychiatry" of the World Psychiatric Association (WPA). He won several honours and scientific awards including the Emil-Kraepelin Research Award.

Dr. Aye-Mu Myint is a Medical Doctor and obtained her Ph.D. in Neuroscience from the University of Maastricht, The Netherlands and has done Habilitation in Experimental Psychiatry at Magdeburg University, Germany. She is working as visiting scientist at Ludwig-Maximilian University Munich, Germany since 2007 as well as senior research scientist at Advanced Practical Diagnostics bvba (apDia), Belgium since 2006. She is also an honorary assistant professor at the School for Mental Health and Neuroscience from Maastricht University. In 2003, she proposed the "neurodegeneration" hypothesis explaining the neurotoxic changes induced through the involvement of immune system imbalance and imbalance of the kynurenine metabolites beyond the activity of tryptophan pathway. She is one of the leading scientists in the field of neuroscience and psychiatry, and is working on major psychiatric disorders, depression-dementia link, psychoneuroimmunology, kynurenine pathway, and related neuroendocrinology in clinical settings as well as animal and in vitro models of depression, schizophrenia, and neurodegenerative disease. She is also involved in antibodies and immunoassay developments through EU consortia. Outside the EU consortium, she has collaborations with several universities including the University of New South Wales, Australia, and the Universities of Chicago, Illinois, and John Hopkins of the United States.

Markus J. Schwarz studied Medicine in Munich from 1990 to 1996. After finishing his dissertation on autoimmune mechanisms in schizophrenia, he started his research career and received his habilitation (venia legendi) in Experimental Psychiatry in 2005. From 2004 to 2012, he was head of the laboratory section "PsychoNeuroImmunology and Therapeutic Drug Monitoring" at the Psychiatric Hospital of Munich University. Since 2012, he is head of the research group on Neurobiochemistry at the Institute for Laboratory Medicine of Munich University. His main research interests are the impact of the two main metabolism pathways of tryptophan (serotonin and kynurenine) in psychiatric disorders, immunological investigations to identify a distinct subgroup of schizophrenia with immune-related pathogenesis, basic psychoneuroimmunologic research on the crosstalk between neurotransmitter and cytokine system and role of therapeutic drug monitoring for enhancing efficacy and safety in psychopharmacotherapy. He published more than 130 research articles in international peer-reviewed journals.

Part I
Basic Science

Chapter 1
Animal Models Based on Immune Challenge: The Link to Brain Changes and Schizophrenia

Georg Juckel

Abstract Within the pathophysiology of schizophrenia, microglial cells seem to play the most important role. Early changes within the embryonal phase and neurodegenerative processes lead to activation of microglia cells, which—via the neurotoxic activities of these cells—induces a rarification of synaptic connections in frontal and temporal brain regions, i.e. reduction of the neuropil. Promising inflammational models in rodents for schizophrenia with high validity can be today used to mimic behavioral as well as neurobiological findings in patients, e.g. the well-known neurochemical alterations within the dopaminergic, glutamatergic, serotonergic, and other neurotransmitter systems. The microglial activation can also be well modelled within one of these models, i.e. the inflammational PolyI:C animal model of schizophrenia, having a time peak in late adolescence/early adulthood. The exact mechanism, by which activated microglia cells then trigger neurodegeneration, must be now investigated further on in broad detail. Thus, these animal models can be used to understand the pathophysiology of schizophrenia especially concerning the interaction of immune activation, inflammation, and neurodegeneration. Furthermore, this should lead to the development of better treatment options and of preventive interventions.

Keywords Inflammational animal models • Schizophrenia • Dopamin • Glutamate • Microglia • Neurodegeneration

Introduction

A leading hypothesis today about the pathophysiology of schizophrenia is the so-called neurodevelopmental hypothesis. Impairment of important brain developmental genes in the 2. trimenon of pregnancy leads to disturbances of the normal brain maturation with the result of neurotransmitter imbalance and dysfunction of

---

G. Juckel (✉)
Department of Psychiatry, Ruhr University Bochum, LWL University Hospital, Alexandrinenstraße 1-3, 44791 Bochum, Germany
e-mail: georg.juckel@rub.de

N. Müller et al. (eds.), *Immunology and Psychiatry*, Current Topics in Neurotoxicity 8, DOI 10.1007/978-3-319-13602-8_1

the limbic system, accompanied by structural changes in the areas of hippocampus, regio entorhinalis, and the ventricular system (Juckel et al. 1994, 1996, 2003a, b; Weinberger 1995; Heinz et al. 1999; Falkai et al. 2003). This neural dysgenesis becomes clinically relevant with considerable postnatal delay. It is supposed that further brain development processes or additional pathogenic factors are likely to elicit the illness finally (Heinz and Weinberger 2000; Heinz et al. 2003). In many patients, progressive cortical volume reduction is found mainly in frontotemporal regions (gyrus cinguli, prefrontal cortex) (Mathalon et al. 2001; Cahn et al. 2002; Pantelis et al. 2003). These findings may be caused by a reduction of neuropil with relatively increased number of cells (Scherk et al. 2003).

Inflammatoric and immunological reactions are discussed as being main reasons for the impairment of brain development in the 2. trimenon. Big cohort studies have shown the result that children of mothers who suffered from, e.g., influenza in the 2. trimenon developed schizophrenic disorders more often than others (Mednick et al. 1988; Takei et al. 1996; Munk-Jørgensen and Ewald 2001; Limosin et al. 2003). Serologic evidence also points to multiple prenatal exposures to various viruses as causative factors in rise of schizophrenic births with a sevenfold increase of risk (Brown et al. 2004). This could be the reason for the activation of the immune system and immune competent cells in schizophrenia, which was often found in patients with schizophrenic disorders since years. Thus, the unspecific immune system shows clear signs of overactivation in unmedicated patients with schizophrenic disorders with an increased number of gamma/delta cells and monocytes/macrophages, and, as consequence of the activation of these cells, to an increase of interleukin 6 (Müller et al. 1997, 1998; Gaughran 2002). The specific immune system shows an imbalance of cytokines in favor to an increased activity of T-helper cells; in addition, there were changes concerning the immunglobulines, antibodies, and the complement system (Müller and Ackenheil 1998; Rothermundt et al. 2001). Interestingly, a 5-week add-on therapy with the antiinflammatory substance celecoxib, a COX-2 inhibitor, showed better clinical improvement in acute patients with schizophrenic disorders than the treatment with the atypical neuroleptic risperidone alone (Müller et al. 2002).

Activation of microglia/macrophages is a key event in the reaction of the central neuro-immune system as answer to any pathological changes (Kreutzberg 1996; Monji et al. 2009). Two post-mortem studies found activated microglia (stained by HLA-DR) in older patients with chronic schizophrenic disorders in particular in the dorsolateral prefrontal cortex (DLPFC), in the gyrus temporalis superior (GTS) and in the anterior gyrus cinguli (ACC), which are known as the main areas of schizophrenic pathophysiology (Bayer et al. 1999; Radewicz et al. 2000). Concerning astroglia, no differences were found in these two first studies. Activated microglia could be either a reaction of the brain immune system to a constant disturbance of neuroplasticity due to, e.g., the degenerative changes that are found in patient suffering from schizophrenia (increasing number of microglia cells with age) or could be stored as an early answer to any inflammatory or non-inflammatory events during pregnancy staying up to the adult age (relatively stable cell number) (Radewicz et al. 2000; Munn 2000; Rothermundt et al. 2001). Thus, activated microglia could lead to neuropil reduction with rarification of synapses, dendrites, and spines in patients with schizophrenic disorders due to their neurotoxic properties (e.g., release of cytokines or free radicals), either in the early stage of brain development or progressively with the course of illness (Radewicz et al. 2000). Many questions concerning the role of activated microglia for schizophrenia are still unexplained. Furthermore, there is still no proof of activated microglia in vivo in patients suffering from schizophrenia. Using PK11195, a PET ligand for the peripheral benzodiazepine receptor, which is expressed in activated microglia, the Dutch working group around R. Kahn was able to find first hints that there seem to be more activated microglial cells in living patients with schizophrenia (van Berckel et al. 2008), which was replicated by Doorduin et al. (2009).

A promising animal model of the inflammatory genesis of schizophrenia is the model of the motherly influenza infection in mice (BALB/c) developed by H. Fatemi in Minneapolis (Fatemi et al. 1998a, b, 1999, 2000; Shi et al. 2003). Descendants of female mice, who were exposed to a mouse-adapted influenza virus in the middle of pregnancy and had gone through a respiratory infection, show several changes in brain morphology, physiology, and behavior after puberty (fertility) comparably to those of schizophrenic patients. Influenza infection at E9 of pregnancy in mice leads to abnormal corticogenesis, pyramidal cell atrophy, and alterations in levels of several neuroregulatory proteins, such as Reelin, and GFAP, in the exposed mouse progeny: volume reduction of neocortex and hippocampus with relatively raised cell number, pyramid cell dystrophy, wide ventricles, reduced Reelin and immune reactivity, changes from nNOS and SNAP-25, deficits in prepulse inhibition (PPI) within the acoustic startle response, in the open field test, in the novel object test and in social interactions (Fatemi et al. 1999; Shi et al. 2003; see Table 1.1).

A further promising animal model is the administration of viral mimetic polyriboinosinic–polyribocytidilic acid (PolyI:C) in vulnerable periods during pregnancy: Descendants of female mice, who were exposed to PolyI:C at day 9, show changes in brain morphology, physiology, and chemistry as well as in behavior after puberty that are partly reminiscent to changes observed in human schizophrenia (e.g., Meyer and Feldon 2010; Winter et al. 2009). Especially, the following findings were described (Meyer 2014):

- Impairments in several behavioral measurements incl. PPI, LI, startle response, sensory gating etc.
- Changes of interleukin-10 and -1 beta, and other immune reactions

Table 1.1 Similar chances of behavioral and neurobiological markers in influenza mouse model and in patients with schizophrenia (H. Fatemi)

	Brain markers			Brain structure		Brain genes				Behavior
Group	nNOS	Reelin	GFAP	Pyramidal cell atrophy	Brain atrophy	MBP	PLP	Net-nd P	HSCP70	Prepulse inhibition
Influenza mouse model										
Adulthood	↓	↓	↑	+	+	↓	↓	↑	↑	Abnormal
Schizophrenia										
Adulthood	↓	↓	↓/–	+	+	↓	↓	↑	↑	Abnormal

- Changes of GABA-A receptors in limbic areas
- Dopamine- and glutamate related pharmacological and neuroanatomical
- Disturbances
- Reduced D1 receptors in PFC and reduced hippocampal NMDA receptor
- Increased number of mesencephalic dopamine neurons in the fetal brain
- (middle/late gestation) (accompanied by specific gene changes)
- Reduction of Reelin- and parvalbumin expressing PFC neurons

Evidence has shown that the time of prenatal insult may provide distinct changes in the exposed offspring. In a recent series of experiments by Meyer et al. (2006) using the viral mimic polyribocytidilic acid (PolyI:C) at E9 (which corresponds to midpregnancy) and E17 (which corresponds to late pregnancy) there were distinct behavioral deficits, neuropathological differences, and acute cytokine responses (Meyer et al. 2006). Adult mice that were exposed on E9 displayed reduced exploratory behavior while those exposed on E17 displayed perseverative behavior (Meyer et al. 2006). At P24, mice that were exposed on E9 displayed a more pronounced reduction of Reelin immunoreactivity in hippocampus than mice exposed at E17 (Meyer et al. 2006). In contrast, mice exposed at E17 displayed an increase in apoptosis as visualized by immunoreactivity of caspase-3, a key enzyme involved in apoptosis (Rami 2003), in the dorsal dentate gyrus (Meyer et al. 2006). Finally, Meyer et al. (2006) found that late gestational immune challenge uniquely stimulated increased IL-10 and TNF-α in fetal brain (Meyer et al. 2006). Taken together, these results provide evidence that the time of prenatal insult results in important differences that are persistent through adulthood.

Neurochemical Findings in the Influenza and PolyI:C Models

Since neurochemical alterations such as in the dopaminergic, glutamatergic, or serotonergic system are highly characteristic for the neurobiology of schizophrenia, some of such findings in the animal models should be reported here in greater detail. The efficacy of dopamine D2 receptor blocking drugs in the treatment of schizophrenia, as well as SPECT studies on neuroleptic naïve patients, suggests that dopamine hyperfunction in the ventral striatum and dopamine hypofunction in the prefrontal cortex may be responsible for the positive symptomology of schizophrenia (Abi-Dargham et al. 2000; Abi-Dargham 2002). Additionally, electrophysiological studies have suggested increased serotonergic function in schizophrenia (Juckel et al. 2003a, b, 2008).

As mentioned above, viral infection causes deleterious effects on brain structure and function in mouse offspring following late first trimester (E9) and late second trimester (E18) administration of influenza virus. Neurochemical analysis following infection on E18 using this model has revealed significantly altered levels of serotonin, 5-hydroxyindoleacetic acid, taurine, but not dopamine. In order to monitor these different patterns of monoamine expression in exposed offspring in more detail and to see if there are changes in the dopamine system at another time point,

C57BL6J mice were infected with a sublethal dose of human influenza virus or sham-infected using vehicle solution on E16. Male offspring of the infected mice were collected at P0, P14, and P56, their brains removed and cerebellum dissected and flash frozen. Dopamine and serotonin levels were then measured using HPLC-ED technique. When compared to controls, there was a significant decrease in serotonin levels in the cerebella of virally exposed mice at P14. No differences in levels of dopamine were observed in exposed and control mice although there was a significant decrease in dopamine at P14 and P56 when compared to P0. This study (Winter et al. 2008) has shown that the serotonergic system is disrupted following prenatal viral infection potentially modelling disruptions that occur in patients with schizophrenia, but there was no significant effect on the dopaminergic system. In an additional study (Fatemi et al. 2008), microarray, qRT-PCR, DTI and MRI scanning, western blotting and neurochemical analysis were performed to detect differences in gene expression and brain atrophy. Expression of several genes associated with schizophrenia or autism including Sema3a, Trfr2 and Vldlr was found to be altered as were protein levels of Foxp2. E18 infection of C57BL6J mice with a sublethal dose of human influenza virus led to significant gene alterations in frontal, hippocampal, and cerebellar cortices of developing mouse progeny. Brain imaging revealed significant atrophy in several brain areas and white matter thinning in corpus callosum. Finally, neurochemical analysis revealed significantly altered levels of serotonin (P14, P35), 5-hydroxyindoleacetic acid (P14), and taurine (P35), but again not concerning the dopaminergic or the glutamatergic system.

In the established mouse model mimicked viral-like infection with PolyI:C, we were able to show much clearer neurochemical effects in one study (Winter et al. 2008) as in the influenza model. Prenatal infection-induced changes in brain and behavioral functions are found to be associated with multiple changes at the neurochemical level. Pregnant dams on gestation day 9 were exposed to viral mimetic polyriboinosinic–polyribocytidilic acid (PolyI:C, 5 mg/kg i.v.) or vehicle treatment, and basal neurotransmitter levels were then compared in the adult brains of animals born to PolyI:C- or vehicle-treated mothers by high-performance liquid chromatography on post-mortem tissue. We found that prenatal immune activation significantly increased the levels of dopamine and its major metabolites in the lateral globus pallidus and prefrontal cortex, whilst at the same time it decreased serotonin and its metabolite in the hippocampus, nucleus accumbens, and lateral globus pallidus. In addition, a specific reduction of the inhibitory amino acid taurine in the hippocampus was noted in prenatally PolyI:C-exposed offspring relative to controls, whereas central glutamate and c-aminobutyric acid (GABA) content was largely unaffected by prenatal immune activation. These results thus confirm that maternal immunological stimulation during early/middle pregnancy is sufficient to induce long-term changes in multiple neurotransmitter levels in the brains of adult offspring. This further supports the possibility that infection-mediated interference with early fetal brain development may predispose the developing organism to the emergence of neurochemical imbalances in adulthood, which may be critically involved in the precipitation of adult behavioral and pharmacological abnormalities after prenatal immune challenge (Fig. 1.1, Table 1.2).

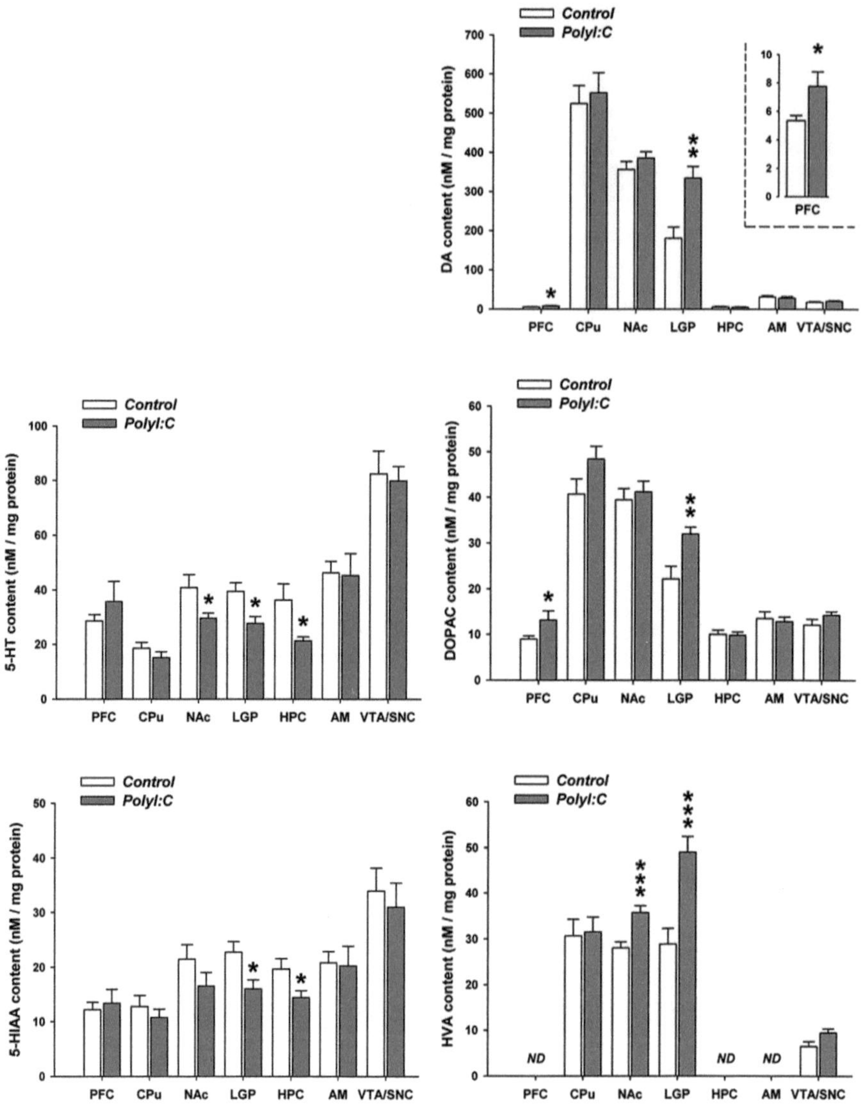

Fig. 1.1 PolyI:C administration at day E9 and measurement of neurotransmitters at day P84 (from Winter et al. 2008)

Table 1.2 PolyI:C administration at day E9 and measurement of neurotransmitters at day P84 (from Winter et al. 2008)

	GABA	GLU	GABA	GLU
	NaCl		PolyI:C	
PFC	11.82±1.24	89.33±8.28	12.83±1.12	88.53±8.72
CPu	11.84±1.13	70.05±6.85	10.83±0.86	64.65±5.64
NAc	29.20±3.14	60.53±5.14	24.52±2.64	61.82±6.15
LGP	27.93±3.36	52.43±6.82	32.04±4.12	52.85±4.97
HPC	14.54±2.47	64.82±11.31	14.82±1.45	64.54±6.35
AM	26.51±2.99	78.04±10.14	25.22±2.75	69.52±10.21
VTA/SN	39.70±6.03	46.43±6.72	35.97±4.77	41.21±7.15

Microglial Activation

According to the hypothesis above declared, a first own study (Juckel et al. 2011) was aimed to explore microglial activation in PolyI:C exposed mice at postnatal day 30 quantitatively and qualitatively in comparison with NaCl exposed control animals in brain regions involved in the pathogenesis of schizophrenia (frontal cortex, hippocampus, and striatum). Female and male BALB/c mice were maintained under standard laboratory conditions by breeding. Mice were mated overnight, and the presence of vaginal plug marked that day as embryonic day 0. At embryonic day (ED) 9 pregnant mice were treated with intraperitoneal injections of PolyI:C at a single dose of 20 mg/kg. Control mice were given injections of sterile sodium chloride (0.9 %). After this procedure pregnant mice were single housed. Thirty days after birth, descendants were deeply anesthetized and transcardially perfused. Subsequently, 40-μm-thick sagittal sections were cut by using a vibratome. Fluorescence immunostaining was performed according to standard protocols (e.g., Ohsawa et al. 2000). Brain sections were incubated with rabbit polyclonal anti-Iba1 antibody. In addition, a combined CD11b/Iba1 staining was performed to reveal the activation and immune competence of microglial cells. Immunostaining was performed in offspring from PolyI:C treated mice and respective controls. Averaged values from two sagittal sections were taken at the medio-lateral position. Quantitative evaluation of Iba1-positive microglia cells was done in four brain areas: hippocampus, frontal cortex, striatum, and as a control region the occipital cortex. Furthermore, microglial branches and processes were analyzed by using a confocal microscope. PolyI:C treatment caused a significant increase in hippocampus and a significant increase in striatum. There were no significant effects of treatment in the frontal cortex as well as in the control region of the occipital cortex. The density of the Iba1-positive processes of the microglial cells was reduced in the brains of the descendants of the injected mothers as compared to the controls. In the control mice, Iba1-positive cells showed round or oval cell bodies with highly branched processes that formed a dense meshwork. In descendants of the mothers injected with PolyI:C, fine branches were largely diminished, particularly in hippocampus. Microglial cells of descendants from PolyI:C infected mothers exhibited significantly less branches and processes as compared to the control mice, which was supported by confocal microscopy showing reduced surface of processes in PolyI:C mice (Fig. 1.2), suggesting that PolyI:C treatment of mothers caused a higher activation status in the offspring generation. Activated and immune competent status of microglial cells could be also demonstrated by combined CD11b/Iba1 staining. This study was able to demonstrate a higher number of activated microglia in the hippocampus and striatum of PolyI:C exposed mice descendants at day 30 which is comparable to adolescence/early adulthood in man, a time window, in which first symptoms of schizophrenia appear or its first manifestation can be seen. The branches and processes of the microglial cells derived from PolyI:C exposed mice were different to those of the control group. While microglial cells from control animals showed more highly branched arborization, indicative for a non-inflammatory state of microglia, while offspring from PolyI:C treated mice were characterized by not or less branched cells, characteristic for an activated and

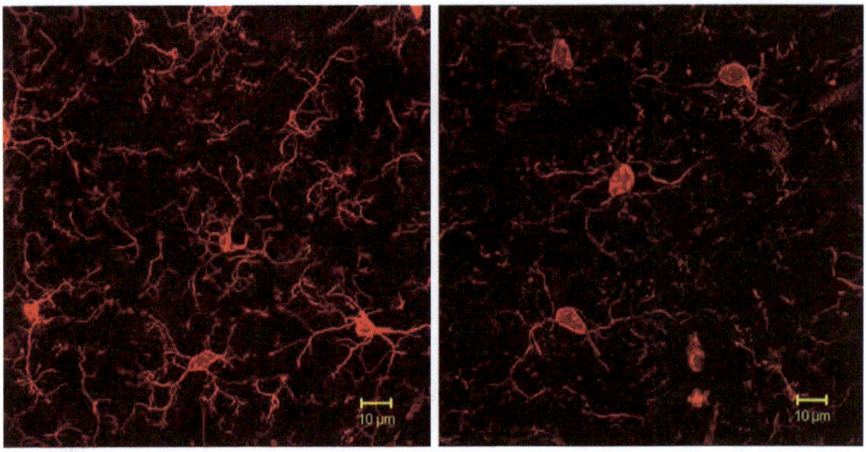

Fig. 1.2 Confocal imaging of Iba1-stained microglial cell: resting microglia in control versus activated microglial cell in PolyI:C animals (Data from own laboratory)

inflammatory or phagocytic state of microglia (e.g., Shapiro et al. 2009; Stence et al. 2001). Activated microglia could lead to neuropil reduction with rarification of synapses, dendrites, and spines in patients with schizophrenic disorders due to a direct neurotoxic effect (e.g., NO, release of cytokines or free radicals) causing the well-known neuroanatomical changes in the early phases of this disease (e.g., Stehen et al. 2006; Witthaus et al. 2009) as well as the psychopathological symptoms and cognitive symptoms. Alternatively, microglial derived mediators could led to neurophysiological changes that cause changes in the neuronal morphology. Given the duration of this activation, one may also consider negative effects on neural stem cell proliferation and integration into the respective brain areas.

In a second study (Manitz et al. 2013), we have performed longitudinal observation across the life span concerning microglial activation in the PolyI:C model and have analyzed also data from PND10 and PND100 offspring. There was no statistical difference between treatment groups concerning number of microglial cells per mm at PND10. At PND30, there was a significant higher number of microglial cells in all regions studied in descendants from Poly(I:C) treated mothers: FrA ($p=0.009$), CPu ($p=0.005$), DG ($p=0.009$), V1 ($p=0.019$), with the exception of the M1. Interestingly, at PND100 a significant increase ($p=0.023$) of microglial cell number was found only in the FrA of the offspring from Poly(I:C) injected dams. Thus, we were able to show a developmental dynamics of microglial activation in different brain regions in the inflammatory Poly(I:C) mouse model of schizophrenia. While mice pups displayed no microglial alterations in response to prenatal exposure to Poly(I:C), adolescent mice showed a significant increase in the number of microglial cells in frontal association cortex, ventral striatum, dentate gyrus of hippocampus, and secondary visual cortex. In contrast, in adult mice the immunological activation remains only within the frontal association cortex. These developmental dynamics of microglial activation in our animal model fit very well to the well-known findings about course and pathogenesis of schizophrenia. In mice pups

(PND10), the process leading to schizophrenia related pathophysiological and behavioral changes is still "sleeping." However, in adolescent mice (PND30), a huge immuno logical reaction with activation of microglial cells occurs. Adolescence in humans is the time of fully manifested disease, which slows down with further adulthood. Actually, in adult descendants of Poly(I:C) treated mice the immunological process seems to remain active only within the frontal association cortex. This could be related to the known cognitive deficits and negative symptoms remaining during the late course of schizophrenia (Howes et al. 2012). Thus, these developmental dynamics of microglial activation in our animal model point to an important role of microglia during life span in schizophrenia.

Conclusion

Maternal infection during pregnancy is an environmental risk factor for descendants to develop severe neuropsychiatric disorders such as schizophrenia. The course of schizophrenia with lifelong remissions and relapses has been widely debated. Immunological processes interfering with brain development are discussed as one cause of schizophrenia. There are excellent and well-validated immune animal models of this disease with an early inflammational signal such as influenza virus or PolyI:C during pregnancy. They mimic a huge range of behavioral and neurobiological characteristics similarly to patients with schizophrenia. In own studies within the PolyI:C model, we were able to demonstrate an increased number of microglial cells during the life span, especially during adolescence. It can be hypothesized that these cells contribute to disease pathogenesis and may actively be involved in the well-known gray matter loss in frontal and temporal brain areas as well as other related neurobiological changes such as neurochemical alteration mainly in the dopaminergic system. Thus, these animal models can be used to understand the pathophysiology of schizophrenia especially concerning the interaction of immune activation, inflammation, and neurodegeneration. Furthermore, this should lead to the development of better treatment options and of preventive interventions. In the following years, the molecular mechanism should be studied how microglial cells exhibit neurotoxic effects on the neuronal substrate. Hereby, the role of cytokines such as interleukin 6 (Ye and Johnson 1999) or of free radicals such as NO (Fatemi et al. 2000) might become clearer.

References

Abi-Dargham A. Recent evidence for dopamine abnormalities in schizophrenia. Eur Psychiatry. 2002;17 Suppl 4:341s–7.

Abi-Dargham A, Rodenhiser J, Printz D, Zea-Ponce Y, Gil R, Kegeles LS, Weiss R, Cooper TB, Mann JJ, Van Heertum RL, Gorman JM, Laruelle M. Increased baseline occupancy of D2 receptors by dopamine in schizophrenia. Proc Natl Acad Sci U S A. 2000;97(14):8104–9.

Bayer TA, Buslei R, Havas L, Falkai P. Evidence for activation of microglia in patients with psychiatric illnesses. Neurosci Lett. 1999;271(2):126–8.

Brown AS, Begg MD, Gravenstein S, Schäfer CA, Wyatt RJ, Bensahan M, Babulas VP, Susser ES. Serologic evidence of prenatal influenza in the etiology of schizophrenia. Arch Gen Psychiatry. 2004;61(8):774–80.

Cahn W, Hulshoff Pol HE, Lems EB, van Haren NE, Schnack HG, van der Linden JA, Schothorst PF, van Engeland H, Kahn RS. Brain volume changes in first-episode schizophrenia: a 1-year follow-up study. Arch Gen Psychiatry. 2002;59(11):1002–10.

Doorduin J, de Vries EF, Willemsen AT, de Groot JC, Dierckx RA, Klein HC. Neuroinflammation in schizophrenia-related psychosis: a PET study. J Nucl Med. 2009;50(11):1801–7. doi:10.2967/jnumed.109.066647.

Falkai P, Schneider-Axmann T, Honer WG, Vogeley K, Schönell H, Pfeiffer U, Scherk H, Block W, Träber F, Schild HH, Maier W, Tepest R. Influence of genetic loading, obstetric complications and premorbid adjustment on brain morphology in schizophrenia: a MRI study. Eur Arch Psychiatry Clin Neurosci. 2003;253(2):92–9.

Fatemi SH, Sidwell R, Akhter P, Sedgewick J, Thuras P, Bailey K, Kist D. Human influenza viral infection in utero increases nNOS expression in hippocampi of neonatal mice. Synapse. 1998a;29(1):84–8.

Fatemi SH, Sidwell R, Kist D, Akhter P, Meltzer HY, Bailey K, Thuras P, Sedgwick J. Differential expression of synaptosome-associated protein 25 kDa [SNAP-25] in hippocampi of neonatal mice following exposure to human influenza virus in utero. Brain Res. 1998b;800(1):1–9.

Fatemi SH, Emamian ES, Kist D, Sidwell RW, Nakajima K, Akhter P, Shier A, Sheikh S, Bailey K. Defective corticogenesis and reduction in Reelin immunoreactivity in cortex and hippocampus of prenatally infected neonatal mice. Mol Psychiatry. 1999;4(2):145–54.

Fatemi SH, Cuadra AE, El-Fakahany EE, Sidwell RW, Thuras P. Prenatal viral infection causes alterations in nNOS expression in developing mouse brains. Neuroreport. 2000;11(7):1493–6.

Fatemi S, Reutiman TJ, Folsom TD, Huang H, Oishi K, Mori S, Smee DF, Pearce DA, Winter C, Sohr R, Juckel G. Maternal infection leads to abnormal gene regulation and brain atrophy in mouse offspring: implications for genesis of neurodevelopmental disorders. Schizophr Res. 2008;99:56–70.

Gaughran F. Immunity and schizophrenia: autoimmunity, cytokines, and immune responses. Int Rev Neurobiol. 2002;52:275–302. Review.

Heinz A, Saunders RC, Kolachana BS, Jones DW, Gorey JG, Bachevalier J, Weinberger DR. Striatal dopamine receptors and transporters in monkeys with neonatal temporal limbic damage. Synapse. 1999;32(2):71–9.

Heinz A, Weinberger D. Schizophrenie: Die neurobiologische Entwicklungshypothese. In: Helmchen et al., editors. Psychiatrie der Gegenwart 5. Berlin: Springer; 2000. p. 89–104.

Heinz A, Romero B, Gallinat J, Juckel G, Weinberger DR. Molecular brain imaging and the neurobiology and genetics of schizophrenia. Pharmacopsychiatry. 2003;36 Suppl 3:S152–7. Review.

Howes OD, Fusar-Poli P, Bloomfield M, Selvaraj S, McGuire P. From the prodrome to chronic schizophrenia: the neurobiology underlying psychotic symptoms and cognitive impairments. Curr Pharm Des. 2012;18:459–65.

Juckel G, Reischies FM, Müller-Schubert A, Vogel AC, Gaebel W, Hegerl U. Ventricle size and P300 in schizophrenia. Eur Arch Psychiatry Clin Neurosci. 1994;243(6):352–4.

Juckel G, Müller-Schubert A, Gaebel W, Hegerl U. Residual symptoms and P300 in schizophrenic outpatients. Psychiatry Res. 1996;65(1):23–32.

Juckel G, Gallinat J, Riedel M, Sokullu S, Schulz C, Möller HJ, Müller N, Hegerl U. Serotonergic dysfunction in schizophrenia assessed by the loudness dependence measure of primary auditory cortex evoked activity. Schizophr Res. 2003a;64(2–3):115–24.

Juckel G, Sass L, Heinz A. Anhedonia, self-experience in schizophrenia, and implications for treatment. Pharmacopsychiatry. 2003b;36 Suppl 3:S176–80.

Juckel G, Gudlowski Y, Müller D, Ozgürdal S, Brüne M, Gallinat J, Frodl T, Witthaus H, Uhl I, Wutzler A, Pogarell O, Mulert C, Hegerl U, Meisenzahl EM. Loudness dependence of the auditory evoked N1/P2 component as an indicator of serotonergic dysfunction in patients with schizophrenia—a replication study. Psychiatry Res. 2008;158(1):79–82.

Juckel G, Manitz MP, Brüne M, Friebe A, Heneka MT, Wolf RJ. Microglial activation in a neuroinflammational animal model of schizophrenia—a pilot study. Schizophr Res. 2011;131: 96–100.

Kreutzberg GW. Microglia: a sensor for pathological events in the CNS. Trends Neurosci. 1996;19(8):312–8.

Limosin F, Rouillon F, Payan C, Cohen JM, Strub N. Prenatal exposure to influenza as a risk factor for adult schizophrenia. Acta Psychiatr Scand. 2003;107(5):331–5.

Manitz MP, Esslinger M, Wachholz S, Plümper J, Friebe A, Juckel G, Wolf R. The role of microglia during life span in neuropsychiatric disease—an animal study. Schizophr Res. 2013;143(1):221–2. doi:10.1016/j.schres.2012.10.028.

Mathalon DH, Sullivan EV, Lim KO, Pfefferbaum A. Progressive brain volume changes and the clinical course of schizophrenia in men: a longitudinal magnetic resonance imaging study. Arch Gen Psychiatry. 2001;58(2):148–57.

Mednick SA, Machon RA, Huttunen MO, Bonett D. Adult schizophrenia following prenatal exposure to an influenza epidemic. Arch Gen Psychiatry. 1988;45(2):189–92.

Meyer U. Prenatal poly(I:C) exposure and other developmental immune activation models in rodent systems. Biol Psychiatry. 2014;75(4):307–15.

Meyer U, Feldon J. Epidemiology-driven neurodevelopmental animal models of schizophrenia. Prog Neurobiol. 2010;90(3):285–326. doi:10.1016/j.pneurobio.2009.10.018.

Meyer U, Nyffeler M, Engler A, Urwyler A, Schedlowski M, Knuesel I, Yee BK, Feldon J. The time of prenatal immune challenge determines the specificity of inflammation-mediated brain and behavioral pathology. J Neurosci. 2006;26(18):4752–62.

Monji A, Kato T, Kanba S. Cytokines and schizophrenia: microglia hypothesis of schizophrenia. Psychiatry Clin Neurosci. 2009;63(3):257–65.

Müller M, Ackenheil M. Psychoneuroimmunology and the cytokine action in the CNS: implications for psychiatric disorders. Prog Neuropsychopharmacol Biol Psychiatry. 1998;22(1):1–33.

Müller N, Dobmeier P, Empl M, Riedel M, Schwarz M, Ackenheil M. Soluble IL-6 receptors in the serum and cerebrospinal fluid of paranoid schizophrenic patients. Eur Psychiatry. 1997;12(6):294–9. doi:10.1016/S0924-9338(97)84789-X.

Müller N, Schlesinger BC, Hadjamu M, Riedel M, Schwarz M, Ackenheil M, Wank R, Gruber R. Increased frequency of CD8 positive gamma/delta T-lymphocytes (CD8+ gamma/delta+) in unmedicated schizophrenic patients: relation to impairment of the blood-brain barrier and HLA-DPA*02011. Schizophr Res. 1998;32(1):69–71.

Müller N, Riedel M, Scheppach C, Brandstätter B, Sokullu S, Krampe K, Ulmschneider M, Engel RR, Möller HJ, Schwarz MJ. Beneficial antipsychotic effects of celecoxib add on therapy compared to risperidone alone in schizophrenia. Am J Psychiatry. 2002;159(6):1029–34.

Munk-Jørgensen P, Ewald H. Epidemiology in neurobiological research: exemplified by the influenza-schizophrenia theory. Br J Psychiatry Suppl. 2001;40:s30–2. Review.

Munn NA. Microglia dysfunction in schizophrenia: an integrative theory. Med Hypotheses. 2000;54(2):198–202.

Ohsawa K, Imai Y, Kanasawa H, Sasaki Y, Kohsaka S. Involvement of Iba1 in membrane ruffing and phagocytosis of macrophages/microglia. J Cell Sci. 2000;113(Pt 17):3073–84.

Pantelis C, Velakoulis D, McGorry PD, Wood SJ, Suckling J, Phillips LJ, Yung AR, Bullmore ET, Brewer W, Soulsby B, Desmond P, McGuire PK. Neuroanatomical abnormalities before and after onset of psychosis: a cross-sectional and longitudinal MRI comparison. Lancet. 2003;361(9354):281–8.

Radewicz K, Garey LJ, Gentleman SM, Reynolds R. Increase in HLA-DR immunoreactive microglia in frontal and temporal cortex of chronic schizophrenics. J Neuropathol Exp Neurol. 2000;59(2):137–50.

Rami A. Ischemic neuronal death in the rat hippocampus: the calpain-calpastatin-caspase hypothesis. Neurobiol Dis. 2003;13(2):75–88.

Rothermundt M, Arolt V, Bayer TA. Review of immunological and immunopathological findings in schizophrenia. Brain Behav Immun. 2001;15(4):319–39.

Scherk H, Vogeley K, Falkai P. The importance of interneurons in schizophrenic and affective disorders. Fortschr Neurol Psychiatr. 2003;Suppl 1:S27–32. Review. German.

Shapiro LA, Perez ZD, Foresti ML, Arisi GM, Ribak CE. Morphological and ultrastructural features of Iba1-immunolabeled microglial cells in the hippocampal dentate gyrus. Brain Res. 2009;1266:29–36.

Shi L, Fatemi SH, Sidwell RW, Patterson PH. A mouse model of mental illness: maternal influenza infection causes behavioural and pharmacological abnormalities in the offspring. J Neurosci. 2003;23(1):297–302.

Stehen RG, Mull C, McClure R, Hamer RM, Lieberman JA. Brain volume in first-episode schizophrenia: systematic review and meta-analysis of magnetic resonance imaging studies. Br J Psychiatry. 2006;188(6):510–8.

Stence N, Waite M, Dailey ME. Dynamics of microglial activation: a confocal time-lapse analysis in hippocampal slices. Glia. 2001;33(3):256–66.

Takei N, Lewis S, Jones P, Harvey I, Murray RM. Prenatal exposure to influenza and increased cerebrospinal fluid spaces in schizophrenia. Schizophr Bull. 1996;22(3):521–34.

van Berckel BN, Bossong MG, Boellaard R, Kloet R, Schuitemaker A, Caspers E, Luurtsema G, Windhorst AD, Cahn W, Lammertsma AA, Kahn RS. Microglia activation in recent-onset schizophrenia: a quantitative (R)-[11C]PK11195 positron emission tomography study. Biol Psychiatry. 2008;64(9):820–2.

Weinberger DR. From neuropathology to neurodevelopment. Lancet. 1995;346(8974):552–7. Review.

Winter C, Reutiman TJ, Folsom TD, Sohr R, Wolf RJ, Juckel G, Fatemi SH. Dopamine and serotonin levels following prenatal viral infection in mouse-implications for psychiatric disorders such as schizophrenia and autism. Eur Neuropsychopharmacol. 2008;18(10):712–6.

Winter C, Djodari-Irani A, Sohr R, Morgenstern R, Feldon J, Juckel G, Meyer U. Prenatal immune activation leads to multiple changes in basal neurotransmitter levels in the adult brain: implications for brain disorders of neurodevelopmental origin such as schizophrenia. Int J Neuropsychopharmacol. 2009;12(4):513–24.

Witthaus H, Kaufmann C, Bohner G, Özgürdal S, Gudlowski Y, Gallinat J, Ruhrmann S, Brüne M, Heinz A, Klingebiel R, Juckel G. Gray matter abnormalities in subjects at ultra-high risk for schizophrenia and first-episode schizophrenic patients. Psychiatry Res. 2009;173(3):163–9. doi:10.1016/j.pscychresns.2008.08.002.

Ye SM, Johnson RW. Increased interleukin-6 expression by microglia from brain of aged mice. J Neuroimmunol. 1999;93(1–2):139–48.

Chapter 2
Developmental Immune Activation Models with Relevance to Schizophrenia

Urs Meyer

Abstract It is increasingly appreciated that altered neuroimmune mechanisms might play a role in the development of schizophrenia and related psychotic illnesses. On the basis of human epidemiological findings, a number of translational rodent models have been established to explore the consequences of prenatal immune activation on brain and behavioral development. The currently existing models are based on maternal gestational exposure to human influenza virus, the viral mimic polyriboinosinic–polyribocytidilic acid [Poly(I:C)], the bacterial endotoxin lipopolysaccharide, the locally acting inflammatory agent turpentine, or selected inflammatory cytokines. These models are pivotal for establishing causal relationships and for identifying cellular and molecular mechanisms that affect normal brain development in the event of early-life immune exposures. An important aspect of developmental immune activation models is that they allow a multifaceted, longitudinal monitoring of the disease process as it unfolds during the course of neurodevelopment from prenatal to adult stages of life. An important recent refinement of these models is the incorporation of multiple etiologically relevant risk factors by combining prenatal immune challenges with specific genetic manipulations or additional environmental adversities. Converging findings from such recent experimental attempts suggest that prenatal infection can act as a "neurodevelopmental disease primer" that is likely relevant for a number of chronic mental illnesses. Hence, the adverse effects induced by prenatal infection might reflect an early entry into the neuropsychiatric route, but the specificity of subsequent disease or symptoms is likely to be strongly influenced by the genetic and environmental context in which the prenatal infectious process occurs.

Keywords Animal models • Cytokines • Infection • Maternal immune activation • Schizophrenia

U. Meyer (✉)
Physiology and Behaviour Laboratory, ETH Zurich, Schorenstrasse 16,
8603 Schwerzenbach, Switzerland
e-mail: urmeyer@ethz.ch

N. Müller et al. (eds.), *Immunology and Psychiatry*, Current Topics in Neurotoxicity 8, DOI 10.1007/978-3-319-13602-8_2

Introduction

The idea that altered neuroimmune mechanisms might play a role in the development of schizophrenia and related psychotic illnesses attained increasing popularity in recent years. A first line of evidence supporting such mechanisms might even date back more than 100 years. Karl A. Menninger was one of the first to reveal an association between influenza exposure and subsequent psychotic disease in patients who were admitted to the Boston Psychopathic Hospital subsequent to the outbreak of the 1918 influenza pandemic (Menninger 1919). In his early observational study, Menninger documented the clinical courses of 80 patients admitted to the psychiatric hospital with “mental disturbances” associated with influenza, of whom 16 were diagnosed with delirium, 25 with dementia praecox, 23 with “other types of psychosis,” and 16 were left unclassified (Menninger 1919). After it had largely sunken into oblivion, this early neuroimmune hypothesis of psychotic disease was reanimated by the seminal work of Torrey and Peterson (1973) in the 1970s suggesting that latent viruses might be involved in the development of schizophrenia. The field has since greatly expanded, and various infectious agents are now being considered to play an etiopathological role in schizophrenia and related disorders (Torrey et al. 2012).

Within this neuroimmune framework of psychotic disease, a great deal of interest has been centered upon the possible contributions of infections in prenatal life. The prenatal period seems highly sensitive to the damaging effects induced by environmental insults such as infections (Meyer et al. 2007). Indeed, infection-induced disturbances directed at the maternal host can lead to pathophysiological changes in the fetal environment and negatively affect the normal course of early brain development of the offspring. This can have long-lasting consequences for the emergence of postnatal brain dysfunctions, in which relevant cerebral insults or pathological processes occur during early brain development (Meyer et al. 2007). Such processes seem particularly relevant for schizophrenia and other disorders with developmental etiologies, including autism and bipolar disorder, which are all believed to be associated with aberrations in early neurodevelopmental processes (Fatemi and Folsom 2009; Bale et al. 2010).

The prenatal infection hypothesis of schizophrenia received a strong boost in the 1980s after Mednick et al. (1988) reported an increased risk of schizophrenia after prenatal maternal exposure to an influenza epidemic in greater Helsinki. Prenatal exposure to a number of other viral agents have since been associated with schizophrenia risk, including rubella (Brown et al. 2001), measles (Torrey et al. 1988), polio (Suvisaari et al. 1999), herpes simplex (Buka et al. 2001a), as well as bacterial pathogens causing sinusitis, tonsillitis, and pneumonia (Sørensen et al. 2009), genital and/or reproductive infections (Babulas et al. 2006), and the protozoan parasite Toxoplasma gondii (Mortensen et al. 2007). Importantly, prospective epidemiological research has provided serologic evidence for at least some of the pathogens implicated in the prenatal infectious etiology of schizophrenia (Brown and Derkits 2010). For example, two epidemiological studies using prospectively collected and quantifiable measurements have provided serologic evidence that maternal influenza infection during pregnancy increases the risk of schizophrenia of the offspring (Brown et al. 2004a, 2009). Epidemiological studies involving clinical examination and serological testing have also confirmed a higher risk of schizophrenia and other psychosis-related disorders after prenatal exposure to rubella virus (Brown et al. 2001) and Toxoplasma gondii (Mortensen et al. 2007; Brown et al. 2005), whereas they have thus far provided equivocal results with respect to the role of herpes simplex virus (Buka et al. 2008; Brown et al. 2006) and specific cytokines (Buka et al. 2001b; Brown et al. 2004b).

The prospective nature of such epidemiological studies, in which a specific infectious pathogen or inflammatory marker in prenatal life is accessible to quantitative measurements, is arguably a very powerful approach to link developmental immune abnormalities with the neuropsychiatric disease risk. For ethical and technical reasons, however, human epidemiological research cannot directly establish causality for such associations and is often limited in its capacity to unravel the downstream cellular and molecular mechanisms affecting normal brain development. Experimental research in animals provides a unique opportunity to overcome these limitations, and this is perhaps the best reason "why schizophrenia epidemiology needs neurobiology," as pointed out by McGrath and Richards (2009). An increasing number of experimental studies in rodents (Meyer et al. 2009a; Meyer and Feldon 2010) and more recently in monkeys (Short et al. 2010) now provide robust evidence for the emergence of long-term functional and structural brain abnormalities after prenatal exposure to specific infectious or inflammatory agents. This review provides a concise overview of existing developmental immune activation models in rodent systems and discusses their value to the identification of developmental neuroimmune factors relevant to schizophrenia and beyond.

The Beginning: Models of Viral Infections

Kneeland and Fatemi (2013) have pioneered, on the basis of the reported association between prenatal influenza infection and adult schizophrenia (Mednick et al. 1988; Brown et al. 2004a, 2009), an experimental mouse model of prenatal exposure to human influenza virus in mice. In this model, pregnant mice on gestation day 9 receive intranasal infusion with a sublethal dose of a mouse-adapted human influenza strain, and the long-term brain and behavioral effects are then evaluated in the resulting offspring relative to control offspring born to mock-infected mothers. By exposing pregnant dams to influenza virus at distinct gestational stages, the prenatal influenza model has also been used to explore the impact of the precise prenatal timing (Kneeland and Fatemi 2013). As extensively reviewed elsewhere (Meyer et al. 2009a; Meyer and Feldon 2010; Kneeland and Fatemi 2013), maternal influenza infection in mice leads to a variety of neuropathological signs in the brains of the offspring postnatally, some of which are dependent on the precise timing of influenza exposure. These neuropathological signs include deficient corticogenesis and brain atrophy, impaired development of the corpus callosum, reduced hippocampal volumes, and decreased expression of γ-aminobutyric acid (GABA) markers such as Reelin (Kneeland and Fatemi 2013; Fatemi et al. 1999, 2008; Moreno et al. 2011). Furthermore, long-term deficiency in serotonin (but not dopamine) production is present after prenatal exposure to influenza virus in mice (Winter et al. 2008). Prenatal exposure to influenza virus in mice also induces a set of behavioral abnormalities in adulthood (Meyer and Feldon 2012; Shi et al. 2003), some of which are highly relevant to schizophrenia and related psychotic illnesses (Table 2.1).

Table 2.1 A sample of long-term behavioral and cognitive dysfunctions as identified in developmental immune activation models in rats and mice

Immunogen	Principal mode of action	Behavioral and cognitive abnormalities in adult offspring born to immune-challenged mothers							Methodological advantages	Methodological disadvantages
		Senso-rimotor gating	Selective attention	Social behavior	Exploratory behavior	Working memory	Cognitive flexibility	Sensitivity to psychoto-mimetic drugs		
Influenza	Broad innate and adaptive antiviral immune response, including production of cytokines and antibodies, and activation of B- and T- cells following maternal intranasal application	↓	ND	↓	↓↓	ND	ND	↑	Full spectrum of immune response normally induced by infections, including antibody production	Stringent biosafety precautions necessary No easy control of immune response intensity and duration
Poly(I:C)	Recognition by TLR3 and induction of cytokine-associated viral-like acute phase response after systemic maternal administration	↓↓↓	↓↓↓	↓↓	↓↓	↓↓↓	↓↓↓	↑↑↑	No stringent biosafety precautions necessary Strictly limited duration of immune response Easy control of (cytokine-associated) immune response intensity	Limited immune response spectrum
LPS	Recognition by TLR4 and induction of cytokine-associated bacterial-like acute phase response after systemic maternal administration	↓↓	ND	↓↓	↓	ND	ND	↑	No stringent biosafety precautions necessary Strictly limited duration of immune response Easy control of (cytokine-associated) immune response intensity	Limited immune response spectrum Marked fetal losses due to spontaneous abortion (high doses)

Turpentine	Local tissue damage, recruitment and activation of immune cells, and secretion of pro-inflammatory cytokines after maternal intramuscular injection	↓	ND	ND	ND	ND	ND	↑	No stringent biosafety precautions necessary Strictly limited duration of inflammatory response Maternal localization of maternal inflammatory response	Limited immune response spectrum Muscular injury/trauma
IL-6	Induction of fever and acute phase responses after systemic maternal administration	↓	↓	↓	↓	ND	ND	ND	No stringent biosafety precautions necessary Strictly limited duration of immune response Easy control of immune stimulus intensity	Limited immune response spectrum

The models are based on prenatal exposure to human influenza virus, the viral mimic polyriboinosinic–polyribocytidilic acid (poly[I:C]), the bacterial endotoxin LPS, the locally acting inflammatory agent turpentine, and the pro-inflammatory cytokine interleukin (IL)-6. Downward and upward arrows indicate an impairment or enhancement of the particular phenotype, respectively. The number of arrows reflects the number of replications by independent research institutes, with three arrows signifying three or more independent replications. Data from Meyer et al. (2009a) and Meyer and Feldon (2010, 2012). Note that the precise timing and dosing of the prenatal immunogens might significantly influence the nature and/or severity of the outlined functional changes [not indicated in the table; for a detailed discussion, see (Meyer et al. 2007, 2009a, 2006)]. Some of the methodological advantages and disadvantages of each model are also summarized. *ND* not determined, *TLR* transmembrane protein toll-like receptor

At least some of the prenatal influenza-induced behavioral deficits can be normalized by acute administration of typical or atypical antipsychotic drugs (Moreno et al. 2011; Shi et al. 2003), suggesting that they are sensitive to treatments used in the symptomatic pharmacotherapy of psychotic illnesses.

The mouse prenatal influenza model has recently extended to experimental investigations in rhesus monkeys, demonstrating the emergence of reduced gray and white matter in distinct cortical and parieto-cortical brain regions of neonates born to influenza-infected mothers (Short et al. 2010). The extension of such translational research to rhesus monkeys is especially relevant in the present context, because prenatal corticogenesis is more advanced in primate as compared with rodent species, and therefore, primate models help to verify the relevance of findings in animal models to the human condition. Taken together, the experimental data obtained in mouse and primate prenatal viral infection models can be taken as experimental evidence to support causal effects of prenatal influenza infection in the development of long-term brain abnormalities.

The Present, Part 1: Models of Viral-Like Immune Activation by Polyriboinosinic–Polyribocytidilic Acid

Another class of animal models of prenatal immune challenge makes use of immune-activating agents that evoke cytokine-associated immune responses in the mother without using live viral or bacterial pathogens (Meyer et al. 2009a; Meyer and Feldon 2010). These models were initially developed to test whether altered expression of maternal and/or fetal cytokines might assume a key role in mediating the link between maternal infection during pregnancy and abnormal brain development in the offspring (Gilmore and Jarskog 1997; Meyer et al. 2009b). One of the most popular and widely used methods nowadays is maternal administration of polyriboinosinic–polyribocytidilic acid (poly[I:C]) (Meyer and Feldon 2012). Since its initial application in mouse developmental biology, the prenatal poly(I:C) model has exerted an appreciable impact on researchers concentrating on the neurodevelopmental and neuroimmunological basis of complex human brain disorders such as schizophrenia (Meyer and Feldon 2012).

In the mouse prenatal poly(I:C) model, pregnant dams are exposed to the immunological manipulation at a specific gestational stage, and the brain and behavioral consequences of the prenatal immunological manipulation are then compared in the resulting offspring relative to offspring born to vehicle-treated control mothers (Meyer and Feldon 2012). Poly(I:C) is a commercially available synthetic analog of double-stranded RNA. Double-stranded RNA is generated during viral infection as a replication intermediate for single-stranded RNA or as a by-product of symmetrical transcription in DNA viruses (Akira and Takeda 2004). It is recognized as foreign by the mammalian immune system primarily through the transmembrane protein toll-like receptor 3 (Akira and Takeda 2004). Transmembrane protein toll-like receptors are a class of pathogen recognition receptors that recognize invariant structures present on virulent pathogens. Upon binding to toll-like receptor 3, double-stranded RNA or its synthetic analog poly(I:C) leads to the expression of an extensive collection of innate immune response genes and proteins. The array of these responses involves the production and release of many pro-inflammatory cytokines, including interleukin (IL)-1β, IL-6, and tumor necrosis factor (TNF)-α (Cunningham et al. 2007). In addition, poly(I:C) is a potent inducer of the type I interferons (IFNs): IFN-α and IFN-β (Kimura et al. 1994). Administration of poly(I:C) can therefore efficiently mimic the acute phase response to viral infection (Kimura et al. 1994) and leads to significant inflammatory processes in the fetal brain when given systemically to pregnant dams (Meyer et al. 2006; Abazyan et al. 2010).

As extensively reviewed elsewhere (Meyer et al. 2009a; Meyer and Feldon 2010, 2012), numerous neurochemical and brain morphological abnormalities have been detected in adult mice and rats after maternal gestational exposure to poly(I:C). An increasing amount of rodent studies further provides robust evidence for the emergence of a multitude of behavioral, cognitive, and pharmacological dysfunctions after prenatal poly(I:C)-induced immune activation (Table 2.1). One intriguing feature of the prenatal poly(I:C) model is that the full spectrum of behavioral, cognitive, and pharmacological abnormalities only emerges after the offspring have reached late adolescence or early adulthood (Zuckerman et al. 2003; Ozawa et al. 2006; Piontkewitz et al. 2001; Vuillermot et al. 2010). This maturational delay is indicative of a progression of pathological symptoms from pubescence to adulthood, which is consistent with the post-pubertal onset of full-blown psychotic behavior in schizophrenia and related disorders (Tandon et al. 2009).

Numerous theories have been put forward to explain why most subjects at risk for schizophrenia develop full-blown psychotic disease only after reaching late adolescence or early adulthood. One prevalent hypothesis suggests that this delayed clinical course might be related to the functional maturation of intracortical connectivity, especially within prefrontal-temporolimbic cortical pathways (Weinberger and Lipska 1995). Other theoretical accounts of this maturational delay focus on the influence of (sex-dependent) hormonal refinements occurring during the periadolescent stage of life (Halbreich and Kahn 2003) and/or interactions with exposure to stressful situations and associated changes in the stress-response system (Corcoran et al. 2003). The prenatal poly(I:C) model provides a unique opportunity to identify the developmental character and molecular processes of neuropathological changes across postnatal brain maturation (Meyer and Feldon 2012; Piontkewitz et al. 2012). Such efforts are highly desirable, because therapeutic interventions during the periadolescent life span might represent an effective strategy to reduce the incidence of or even prevent the emergence of multiple brain dysfunctions in high-risk subjects with a (immune-mediated) neurodevelopmental predisposition to adult mental illness. Experimental efforts toward this direction are underway, and the prenatal poly(I:C) model has already provided initial promising data suggesting that periadolescent treatment with reference antipsychotic or antidepressant drugs can successfully block at least some of the behavioral and brain abnormalities in prenatally poly(I:C)-exposed offspring (Meyer et al. 2010; Piontkewitz et al. 2009, 2011; Richtand et al. 2011).

Another important feature of the prenatal poly(I:C) model is that the severity of the long-term brain and behavioral changes is dependent on the intensity of the cytokine-associated immune reaction (Shi et al. 2003; Meyer et al. 2005). The apparent impact of the precise immune stimulus intensity suggests that there is a threshold of (viral-like) immune activation that is required to induce long-term brain and behavioral pathology in the offspring. The dose-dependent nature of the prenatal poly(I:C)-induced effects parallels findings indicating that more severe forms of prenatal immune abnormalities are associated with more intense structural brain abnormalities in offspring affected with schizophrenia (Ellman et al. 2010). Another feasible assumption derived from such dose-response effects is that prenatal (viral-like) infection enhances the risk of schizophrenia only if the infectious process is associated with relatively strong immunological reactions in the maternal/fetal compartments and/or if this early-life immunological insult takes place in conjunction with additional genetic or environmental risk factors (see following).

Finally, the prenatal poly(I:C) model is also very useful to test the hypothesis that the vulnerability to infection-induced neurodevelopmental abnormalities differs between distinct stages of fetal development. This can be achieved by comparing the effects of prenatal poly(I:C) at distinct gestational stages relative to corresponding prenatal control treatment (Meyer et al. 2006, 2008a; Li et al. 2009; Fortier et al. 2007). The poly(I:C)-induced immune reactions in the maternal host are time-limited, ranging from 24 to 48 h, depending on the precise dose used (Cunningham et al. 2007), thus allowing the experimenter to precisely time the maternal immune response according to specific periods of fetal development. Such experimental investigations seem highly relevant to address some of the ongoing debates as to whether the strength of the association between prenatal maternal infection and enhanced schizophrenia risk is dependent on the precise prenatal timing (Meyer et al. 2007).

The Present, Part 2: Models of Bacterial-Like Immune Activation by Lipopolysaccharide

Maternal administration of the bacterial endotoxin lipopolysaccharide (LPS) is a widely used model system to mimic an innate acute phase response to bacterial infection in the absence of live bacteria exposure. LPS is an inherent cell wall component of gram-negative bacteria, which is recognized mainly by the pathogen recognition receptor transmembrane protein toll-like receptor 4 (Akira and Takeda 2004). Upon binding to toll-like receptor 4, LPS stimulates the expression of a wide array of innate immune responses that include the synthesis and release of various pro-inflammatory cytokines (Akira and Takeda 2004). There are some notable similarities between the cytokine-associated inflammatory responses triggered by LPS and poly(I:C) (Akira and Takeda 2004; Kimura et al. 1994). Therefore, it might not seem surprising that prenatal LPS treatment, similarly to poly(I:C), precipitates a number of behavioral and neurochemical changes relevant to schizophrenia and related psychotic illnesses (Table 2.1).

Despite the apparent similarities between the LPS- and poly(I:C)-induced effects, there are also some noticeable differences between the two models with respect to the nature of brain and behavioral changes. For example, early prenatal poly(I:C) treatment in mice has been shown to increase midbrain dopamine cells (Vuillermot et al. 2010), whereas prenatal LPS exposure leads to the opposite effect (Carvey et al. 2003). Prenatal LPS exposure in rhesus monkey has also been found to cause a significant increase in global white matter volume (Willette et al. 2011), whereas an opposite pattern (i.e., decreased white matter volume) has been noted in rhesus monkey offspring born to influenza-infected mothers (Short et al. 2010). Such differences in the long-term outcomes between prenatal exposures to bacterial-like and viral/viral-like immunogens might be taken to support the idea that different pathogens can induce a distinct set of neuroimmune abnormalities across brain development.

It remains to be determined whether such differences can be explained by the differential immune signatures and pathophysiological responses triggered by different immunogens. Unlike LPS, for example, poly(I:C) is a potent inducer of type I IFNs that stimulate antiviral immune responses (Kimura et al. 1994). By contrast, LPS is more effective in stimulating the production and secretion of TNF-α from innate immune cells such as macrophages (Reimer et al. 2008). The differential effects on TNF-α might be a reason why LPS is more potent than poly(I:C) in triggering anorexia, lethargy, and febrile responses (Hopwood et al. 2009). In view of the strong apoptotic activity of TNF-α (ClarkI 2007), maternal LPS exposure might also be more effective in inducing neurotoxic effects in the fetal brain compared with maternal poly(I:C) treatment (Hagberg et al. 2012). Indeed, maternal or intrauterine exposure to LPS has been widely used as an experimental model system for the induction of perinatal white and gray matter damage in relation to cerebral palsy and other developmentally acquired neurological conditions (Hagberg et al. 2012; Burd et al. 2012; Harvey and Boksa 2012).

The Present, Part 3: Models of Local Inflammation and Exposure to Individual Cytokines

There have been recent attempts to explore whether local maternal inflammation during pregnancy is sufficient to induce long-term brain and behavioral changes in the offspring. One promising model is based on maternal intramuscular injection of turpentine oil (Aguilar-Valles and Luheshi 2011; Aguilar-Valles et al. 2010, 2012). After its intramuscular injection, turpentine remains confined at the site of administration and locally causes tissue damage, recruitment and activation of immune cells, and secretion of pro-inflammatory cytokines (Aguilar-Valles and Luheshi 2011; Aguilar-Valles et al. 2010, 2012). This experimental approach offers the opportunity to study the effects of circulating inflammatory mediators that are solely produced by the maternal immune system. Hence, in contrast to poly(I:C)- or LPS-based models of systemic immune activation (Ashdown et al. 2006; Hsiao and Patterson 2011), placental secretion of inflammatory markers is minimal, and this readily facilitates the delineation of the relative contribution of maternally produced versus placenta-derived inflammatory factors in the link between prenatal inflammation and abnormal brain and behavioral development (Aguilar-Valles and Luheshi 2011; Aguilar-Valles et al. 2010, 2012). Intriguingly, maternal turpentine treatment is effective in inducing long-term behavioral and pharmacological changes in the offspring, some of which are highly similar to the pathologies found after prenatal exposure to other immune-activating agents (Table 2.1). These findings provide further support for the hypothesis that induction of maternal inflammatory responses might be a key factor mediating the association between maternal infection during pregnancy and altered behavioral development in the offspring. Moreover, given that intramuscular injection of turpentine causes localized tissue damage, the findings from this model might be taken to encourage human epidemiological studies exploring potential associations between maternal physical trauma and increased risk of neurodevelopmental disorders in the offspring.

Another valuable approach to delineate the role of cytokine-associated mechanisms is to treat pregnant animals with specific cytokines. This approach has been successfully implemented both in rats and mice (Smith et al. 2007; Samuelsson et al. 2006). The existing data indicate that the inflammatory cytokine IL-6 might be a crucial immunological mediator of the link between maternal immune activation and altered brain development. Indeed, administration of exogenous IL-6 to pregnant animals is sufficient to induce long-lasting structural and functional abnormalities in the adult offspring, some of which are highly comparable to those induced by prenatal exposure to other immune-activating agents such as poly(I:C) (Smith et al. 2007; Samuelsson et al. 2006). Moreover, when IL-6 is eliminated from the maternal immune response by genetic interventions or with IL-6 blocking antibodies, maternal immune challenge by poly(I:C) is no longer efficient in inducing behavioral maldevelopment in the resulting offspring (Smith et al. 2007). Prenatal exposure to other pro-inflammatory cytokines alone, including IL-1β, IFN-γ, or TNF-α, seems to be insufficient to precipitate similar behavioral deficits in the adult animals; and co-administration of soluble IL-1β or IFN-γ receptor antagonist to pregnant dams does not prevent the behavioral deficits caused by prenatal poly(I:C) exposure (Smith et al. 2007).

It remains to be further elucidated why IL-6 might be a key cytokine in mediating the link between prenatal immune challenge and altered brain development. One parsimonious explanation might be that IL-6 is readily capable of crossing the placental barrier, whereas other inflammatory mediators such as IL-1β and TNF-α display only minimal transplacental transfer (Zaretsky et al. 2004). Interestingly, IL-6 exerts some noticeable anti-inflammatory effects in addition to its known pro-inflammatory functions (Xing et al. 1998). The disruption of the balance between pro- and anti-inflammatory signaling in prenatal life might thus represent an important mechanism precipitating changes in brain and behavioral development (Meyer et al. 2008b). Taken together, it seems that the nature and/or severity of neuropathological outcomes after prenatal infection and/or inflammation might be, at least in part, influenced by the specificity of cytokine-associated immunological reactions. Additional research toward this direction is highly warranted, because the identification of specific key mechanisms might offer an effective strategy to attenuate or even prevent the neurodevelopmental sequelae associated with prenatal infection.

The Future: Deconstructing the Phenotypes and Broadening the Concepts

It seems no longer a matter of debate that the etiology of multifaceted neuropsychiatric disorders such as schizophrenia is multifactorial. With the establishment of more sophisticated genetic techniques and epidemiological approaches, the list of potential candidate genetic and environmental risk factors for schizophrenia is constantly rising (Brown 2011; Owen 2012a). Furthermore, it is becoming increasingly evident that seemingly remote disorders such as schizophrenia, autism, attention-deficit/hyperactivity disorder, and major depression share considerable amounts of risk factors and brain dysfunctions (Cheung et al. 2010; Meyer et al. 2011; Smoller et al. 2013; Moreno-De-Luca et al. 2013). The presence of shared genetic and environmental risk between those illnesses has led to the proposal that they might lie along a continuum of genetically and environmentally induced neurodevelopmental causality (Moreno-De-Luca et al. 2013; Owen 2012b). This concept has a number of implications for those of us who are attempting to study disease-relevant neurobiological and behavioral correlates in animal models. First and foremost, animal researchers should not aim for an animal model that could possibly recapitulate the full spectrum of behavioral symptoms associated with a complex brain disorder such as schizophrenia. This is especially true when the model is based on one particular genetic or environmental manipulation only. Such single risk-factor models should rather be expected to induce a restricted set of brain and behavioral changes, and those changes should not be canalized or fitted into one specific neuropsychiatric disease entity (O'Donnell 2011).

Some of these considerations have encouraged animal researchers to develop translational models that incorporate multiple etiologically relevant risk factors (Ayhan et al. 2009; Kas et al. 2011). On the basis of the role of prenatal infection in schizophrenia, there have been recent attempts to explore the cumulative impact of immune-related environmental challenges and distinct genetic abnormalities in the disruption of brain and behavioral development. The rationale for these attempts is given by epidemiological findings suggesting that prenatal infections seem to have rather modest effects in large populations in spite of their relatively frequent occurrence (Selten et al. 2010). It has therefore been proposed that early-life immune challenge might unfold its neuropathological impact primarily in genetically predisposed subjects (Clarke et al. 2009). Another feasible scenario is that initial exposure to a prenatal environmental insult, such as infection, can render the offspring more vulnerable to the pathological effects of a second postnatal stimulus, such as exposure to traumatizing experiences or chronic consumption of drugs of abuse (Maynard et al. 2001).

Rodent prenatal immune activation models seem highly suitable for these kinds of gene–environment or environment–environment interaction studies. With the maternal poly(I:C) administration model in mice, it has recently been shown that the brain and behavioral consequences of (mild and physiologically relevant) prenatal immune activation are markedly exacerbated in offspring with genetic predisposition to neurodevelopmental abnormalities induced by mutant expression of disrupted-in-schizophrenia 1 (Abazyan et al. 2010; Lipina et al. 2013) or by mutations in the dopamine-related transcription factor Nurr1 (Vuillermot et al. 2012). Intriguingly, the combination of such genetic manipulations and prenatal immune activation does not only produce additive effects but further leads to behavioral disturbances that are not manifest after exposure to either manipulation alone (Abazyan et al. 2010; Lipina et al. 2013; Vuillermot et al. 2012). Our laboratory has recently developed an environmental "two-hit" model, in which exposure to physiologically relevant dose of maternal poly(I:C) treatment served as the first hit and exposure to sub-chronic stress in pubescence served as the second hit (Giovanoli et al. 2013). In this model, mild prenatal immune activation and peripubertal stress caused synergistic effects in the development of specific behavioral abnormalities such as sensorimotor gating deficiency and enhanced sensitivity to psychotomimetic drugs (Giovanoli et al. 2013). Neither immune activation alone nor stress alone affected these behavioral functions in adulthood, so that abnormalities in these domains became evident only after combined exposure to the two environmental factors (Giovanoli et al. 2013). Hence, prenatal infection can act as a "disease primer" that increases the vulnerability of the offspring to the detrimental neuropathological effects of other environmental insults such as peripubertal stress.

Taken together, animal models might critically help to identify complex interactions between discrete genetic and environmental risk factors in the development of chronic mental illnesses. These novel approaches hold promise especially for psychiatric disorders with neurodevelopmental components, considering how little we understand about the disruption of brain development induced by combined effects of multiple genetic or environmental adversities.

Concluding Remarks

Modeling the epidemiological association between prenatal immune challenge and altered brain and behavioral development in rodent systems has produced an astonishing amount of experimental data supporting a role of immune-mediated neurodevelopmental abnormalities in major psychiatric illnesses. Many of the models can mimic a broad spectrum of behavioral, cognitive, and pharmacological abnormalities relevant to schizophrenia and beyond. Perhaps one of the most important features of prenatal immune action models is that they are "neurodevelopmental disruption models": they allow a multifaceted, longitudinal monitoring of the disease process as it unfolds during the course of neurodevelopment from prenatal to adult stages of life. Moreover, they do not rely on any presumption of the neuronal substrates of a specific disorder, and therefore they offer an unbiased way to identify etiopathological processes underlying the changes in neurodevelopmental trajectories and behavioral functions after exposure to prenatal adversities such as infection. Quite surprisingly, however, prenatal immune activation models have thus far largely ignored the (pathological) processes occurring during the early postnatal period. Indeed, the investigation of physiological, behavioral, and neuroanatomical functions at neonatal or early juvenile stages of life is relatively rare in the existing models (Baharnoori et al. 2012; Escobar et al. 2011). It would seem highly important to gain knowledge about possible disturbances in early neurobehavioral functions, because subjects who later go on to develop chronic mental illness such as schizophrenia often show (subtle) cognitive, motor, and social disabilities during the premorbid phase of the disease (Fatemi and Folsom 2009; Tandon et al. 2009).

Related to this, it is highly conceivable that prenatal immune activation models are likely to be relevant for a number of brain disorders with neurodevelopmental components (Harvey and Boksa 2012). Epidemiological support for this notion is manifold: prenatal maternal infection and/or inflammatory processes have also been linked to—besides schizophrenia—increased risk of autism (Atladóttir et al. 2010; Brown et al. 2013), attention-deficit/hyperactivity disorder (Pineda et al. 2007), and cerebral palsy (Dammann and Leviton 1997, 2000). As pointed out by Harvey and Boksa (2012), one needs to consider that prenatal exposure to infection (in animal models) is a general vulnerability factor for neurodevelopmental disorders rather than a disease-specific risk factor. The findings derived from the recently established models, in which prenatal poly(I:C) exposure is combined with a specific genetic (Abazyan et al. 2010; Lipina et al. 2013; Vuillermot et al. 2012) or additional environmental (Giovanoli et al. 2013) factor, strongly support this concept. Hence, the adverse effects induced by prenatal infection should be generally considered as an early entry into the neuropsychiatric route, but the specificity of subsequent disease or symptoms is likely to be strongly influenced by the genetic and environmental context in which the prenatal infectious process occurs.

The epidemiological literature reporting enhanced risk of chronic mental illnesses after early-life exposure to infection and/or inflammation is still evolving, and so are the attempts to model these associations in experimental animals. McGrath and Richards (2009) correctly point out that there is a "need to build shared discovery platforms that encourage greater cross-fertilization between schizophrenia epidemiology and basic neuroscience research." The experimental models discussed in the present chapter represent an important step toward this direction. It is the continual integration of epidemiological and experimental work that will truly further our understanding of how prenatal infection and inflammation increases the risk of neurodevelopmental brain disorders.

References

Abazyan B, Nomura J, Kannan G, Ishizuka K, Tamashiro KL, Nucifora F, et al. Prenatal interaction of mutant DISC1 and immune activation produces adult psychopathology. Biol Psychiatry. 2010;68:1172–11781.

Aguilar-Valles A, Luheshi GN. Alterations in cognitive function and behavioral response to amphetamine induced by prenatal inflammation are dependent on the stage of pregnancy. Psychoneuroendocrinology. 2011;36:634–48.

Aguilar-Valles A, Flores C, Luheshi GN. Prenatal inflammation-induced hypoferremia alters dopamine function in the adult offspring in rat: relevance for schizophrenia. PLoS One. 2010;5:e10967.

Aguilar-Valles A, Jung S, Poole S, Flores C, Luheshi GN. Leptin and interleukin-6 alter the function of mesolimbic dopamine neurons in a rodent model of prenatal inflammation. Psychoneuroendocrinology. 2012;37:956–69.

Akira S, Takeda K. Toll-like receptor signaling. Nat Rev Immunol. 2004;4:499–511.

Ashdown H, Dumont Y, Ng M, Poole S, Boksa P, Luheshi GN. The role of cytokines in mediating effects of prenatal infection on the fetus: implications for schizophrenia. Mol Psychiatry. 2006;11:47–55.

Atladóttir HO, Thorsen P, Østergaard L, Schendel DE, Lemcke S, Abdallah M, Parner ET. Maternal infection requiring hospitalization during pregnancy and autism spectrum disorders. J Autism Dev Disord. 2010;40:1423–30.

Ayhan Y, Sawa A, Ross CA, Pletnikov MV. Animal models of gene-environment interactions in schizophrenia. Behav Brain Res. 2009;204:274–81.

Babulas V, Factor-Litvak P, Goetz R, Schaefer CA, Brown AS. Prenatal exposure to maternal genital and reproductive infections and adult schizophrenia. Am J Psychiatry. 2006; 163:927–9.

Baharnoori M, Bhardwaj SK, Srivastava LK. Neonatal behavioral changes in rats with gestational exposure to lipopolysaccharide: a prenatal infection model for developmental neuropsychiatric disorders. Schizophr Bull. 2012;38:444–56.

Bale TL, Baram TZ, Brown AS, Goldstein JM, Insel TR, McCarthy MM, et al. Early life programming and neurodevelopmental disorders. Biol Psychiatry. 2010;68:314–9.

Brown AS. The environment and susceptibility to schizophrenia. Prog Neurobiol. 2011;93:23–58.

Brown AS, Derkits EJ. Prenatal infection and schizophrenia: a review of epidemiologic and translational studies. Am J Psychiatry. 2010;167:261–80.

Brown AS, Cohen P, Harkavy-Friedman J, Babulas V, Malaspina D, Gorman JM, Susser ES. Prenatal rubella, premorbid abnormalities, and adult schizophrenia. Biol Psychiatry. 2001;49:473–86.

Brown AS, Begg MD, Gravenstein S, Schaefer CA, Wyatt RJ, Bresnahan M, et al. Serologic evidence of prenatal influenza in the etiology of schizophrenia. Arch Gen Psychiatry. 2004a;61:774–80.

Brown AS, Hooton J, Schaefer CA, Zhang H, Petkova E, Babulas V, et al. Elevated maternal interleukin-8 levels and risk of schizophrenia in adult offspring. Am J Psychiatry. 2004b;161:889–95.

Brown AS, Schaefer CA, Quesenberry Jr CP, Liu L, Babulas VP, Susser ES. Maternal exposure to toxoplasmosis and risk of schizophrenia in adult offspring. Am J Psychiatry. 2005;162:767–73.

Brown AS, Schaefer CA, Quesenberry Jr CP, Shen L, Susser ES. No evidence of relation between maternal exposure to herpes simplex virus type 2 and risk of schizophrenia? Am J Psychiatry. 2006;163:2178–80.

Brown AS, Vinogradov S, Kremen WS, Poole JH, Deicken RF, Penner JD, et al. Prenatal exposure to maternal infection and executive dysfunction in adult schizophrenia. Am J Psychiatry. 2009;166:683–90.

Brown AS, Sourander A, Hinkka-Yli-Salomäki S, McKeague IW, Sundvall J, Surcel HM. Elevated maternal C-reactive protein and autism in a national birth cohort. Mol Psychiatry. 2013;9:259–64 (published online ahead of print January 22).

Buka SL, Tsuang MT, Torrey EF, Klebanoff MA, Bernstein D, Yolken RH. Maternal infections and subsequent psychosis among offspring. Arch Gen Psychiatry. 2001a;58:1032–7.
Buka SL, Tsuang MT, Torrey EF, Klebanoff MA, Wagner RL, Yolken RH. Maternal cytokine levels during pregnancy and adult psychosis. Brain Behav Immun. 2001b;15:411–20.
Buka SL, Cannon TD, Torrey EF, Yolken RH, Collaborative Study Group on the Perinatal Origins of Severe Psychiatric Disorders. Maternal exposure to herpes simplex virus and risk of psychosis among adult offspring. Biol Psychiatry. 2008;63:809–15.
Burd I, Balakrishnan B, Kannan S. Models of fetal brain injury, intrauterine inflammation, and preterm birth. Am J Reprod Immunol. 2012;67:287–94.
Carvey PM, Chang Q, Lipton JW, Ling Z. Prenatal exposure to the bacteriotoxin lipopolysaccharide leads to long-term losses of dopamine neurons in offspring: a potential, new model of Parkinson's disease. Front Biosci. 2003;8:s826–37.
Cheung C, Yu K, Fung G, Leung M, Wong C, Li Q, et al. Autistic disorders and schizophrenia: Related or remote? An anatomical likelihood estimation. PLoS One. 2010;5:e12233.
Clarke MC, Tanskanen A, Huttunen M, Whittaker JC, Cannon M. Evidence for an interaction between familial liability and prenatal exposure to infection in the causation of schizophrenia. Am J Psychiatry. 2009;166:1025–30.
ClarkI A. How TNF, was recognized as a key mechanism of disease. Cytokine Growth Factor Rev. 2007;18:335–43.
Corcoran C, Walker E, Huot R, Mittal V, Tessner K, Kestler L, Malaspina D. The stress cascade and schizophrenia: etiology and onset. Schizophr Bull. 2003;29:671–92.
Cunningham C, Campion S, Teeling J, Felton L, Perry VH. The sickness behaviour and CNS inflammatory mediator profile induced by systemic challenge of mice with synthetic double-stranded RNA (poly I:C). Brain Behav Immun. 2007;21:490–502.
Dammann O, Leviton A. Maternal intrauterine infection, cytokines, and brain damage in the preterm newborn. Pediatr Res. 1997;42:1–8.
Dammann O, Leviton A. Role of the fetus in perinatal infection and neonatal brain damage. Curr Opin Pediatr. 2000;12:99–104.
Ellman LM, Deicken RF, Vinogradov S, Kremen WS, Poole JH, Kern DM, et al. Structural brain alterations in schizophrenia following fetal exposure to the inflammatory cytokine interleukin-8. Schizophr Res. 2010;121:46–54.
Escobar M, Crouzin N, Cavalier M, Quentin J, Roussel J, Lanté F, et al. Early, time-dependent disturbances of hippocampal synaptic transmission and plasticity after in utero immune challenge. Biol Psychiatry. 2011;70:992–9.
Fatemi SH, Folsom TD. The neurodevelopmental hypothesis of schizophrenia, revisited. Schizophr Bull. 2009;35:528–48.
Fatemi SH, Emamian ES, Kist D, Sidwell RW, Nakajima K, Akhter P, et al. Defective corticogenesis and reduction in Reelin immunoreactivity in cortex and hippocampus of prenatally infected neonatal mice. Mol Psychiatry. 1999;4:145–54.
Fatemi SH, Reutiman TJ, Folsom TD, Huang H, Oishi K, Mori S, et al. Maternal infection leads to abnormal gene regulation and brain atrophy in mouse offspring: implications for genesis of neurodevelopmental disorders. Schizophr Res. 2008;99:56–70.
Fortier ME, Luheshi GN, Boksa P. Effects of prenatal infection on prepulse inhibition in the rat depend on the nature of the infectious agent and the stage of pregnancy. Behav Brain Res. 2007;181:270–7.
Gilmore JH, Jarskog LF. Exposure to infection and brain development: cytokines in the pathogenesis of schizophrenia. Schizophr Res. 1997;24:365–7.
Giovanoli S, Engler H, Engler A, Richetto J, Voget M, Willi R, et al. Stress in puberty unmasks latent neuropathological consequences of prenatal immune activation in mice. Science. 2013;339:1095–9.
Hagberg H, Gressens P, Mallard C. Inflammation during fetal and neonatal life: implications for neurologic and neuropsychiatric disease in children and adults. Ann Neurol. 2012;71:444–57.
Halbreich U, Kahn LS. Hormonal aspects of schizophrenias: an overview. Psychoneuroendocrinology. 2003;28 Suppl 2:1–16.

Harvey L, Boksa P. Prenatal and postnatal animal models of immune activation: relevance to a range of neurodevelopmental disorders. Dev Neurobiol. 2012;72:1335–48.

Hopwood N, Maswanganyi T, Harden LM. Comparison of anorexia, lethargy, and fever induced by bacterial and viral mimetics in rats. Can J Physiol Pharmacol. 2009;87:211–20.

Hsiao EY, Patterson PH. Activation of the maternal immune system induces endocrine changes in the placenta via IL-6. Brain Behav Immun. 2011;25:604–15.

Kas MJ, Kahn RS, Collier DA, Waddington JL, Ekelund J, Porteous DJ, et al. Translational neuroscience of schizophrenia: seeking a meeting of minds between mouse and man. Sci Transl Med. 2011;3:102mr3.

Kimura M, Toth LA, Agostini H, Cady AB, Majde JA, Krueger JM. Comparison of acute phase responses induced in rabbits by lipopolysaccharide and double-stranded RNA. Am J Physiol. 1994;267:R1596–605.

Kneeland RE, Fatemi SH. Viral infection, inflammation and schizophrenia. Prog Neuropsychopharmacol Biol Psychiatry. 2013;42:35–48.

Li Q, Cheung C, Wei R, Hui ES, Feldon J, Meyer U, et al. Prenatal immune challenge is an environmental risk factor for brain and behavior change relevant to schizophrenia: evidence from MRI in a mouse model. PLoS One. 2009;4:e6354.

Lipina TV, Zai C, Hlousek D, Roder JC, Wong AH. Maternal immune activation during gestation interacts with Disc1 point mutation to exacerbate schizophrenia-related behaviors in mice. J Neurosci. 2013;33:7654–66.

Maynard TM, Sikich L, Lieberman JA, LaMantia AS. Neural development, cell-cell signaling, and the "two-hit" hypothesis of schizophrenia. Schizophr Bull. 2001;27:457–76.

McGrath JJ, Richards LJ. Why schizophrenia epidemiology needs neurobiology—and vice versa. Schizophr Bull. 2009;35:577–81.

Mednick SA, Machon RA, Huttunen MO, Bonett D. Adult schizophrenia following prenatal exposure to an influenza epidemic. Arch Gen Psychiatry. 1988;45:189–92.

Menninger KA. Psychoses associated with influenza, I: general data: statistical analysis. JAMA. 1919;72:235–41.

Meyer U, Feldon J. Epidemiology-driven neurodevelopmental animal models of schizophrenia. Prog Neurobiol. 2010;90:285–326.

Meyer U, Feldon J. To poly(I:C) or not to poly(I:C): advancing preclinical schizophrenia research through the use of prenatal immune activation models. Neuropharmacology. 2012;62:1308–21.

Meyer U, Feldon J, Schedlowski M, Yee BK. Towards an immuno-precipitated neurodevelopmental animal model of schizophrenia. Neurosci Biobehav Rev. 2005;29:913–47.

Meyer U, Nyffeler M, Engler A, Urwyler A, Schedlowski M, Knuesel I, et al. The time of prenatal immune challenge determines the specificity of inflammation-mediated brain and behavioral pathology. J Neurosci. 2006;26:4752–62.

Meyer U, Yee BK, Feldon J. The neurodevelopmental impact of prenatal infections at different times of pregnancy: the earlier the worse? Neuroscientist. 2007;13:241–56.

Meyer U, Nyffeler M, Yee BK, Knuesel I, Feldon J. Adult brain and behavioral pathological markers of prenatal immune challenge during early/middle and late fetal development in mice. Brain Behav Immun. 2008a;22:469–86.

Meyer U, Murray PJ, Urwyler A, Yee BK, Schedlowski M, Feldon J. Adult behavioral and pharmacological dysfunctions following disruption of the fetal brain balance between pro-inflammatory and IL-10-mediated anti-inflammatory signaling. Mol Psychiatry. 2008b;13:208–21.

Meyer U, Feldon J, Fatemi SH. In-vivo rodent models for the experimental investigation of prenatal immune activation effects in neurodevelopmental brain disorders. Neurosci Biobehav Rev. 2009a;33:1061–79.

Meyer U, Feldon J, Yee BK. A review of the fetal brain cytokine imbalance hypothesis of schizophrenia. Schizophr Bull. 2009b;35:959–72.

Meyer U, Spoerri E, Yee BK, Schwarz MJ, Feldon J. Evaluating early preventive antipsychotic and antidepressant drug treatment in an infection-based neurodevelopmental mouse model of schizophrenia. Schizophr Bull. 2010;36:607–23.

Meyer U, Feldon J, Dammann O. Schizophrenia and autism: both shared and disorder-specific pathogenesis via perinatal inflammation? Pediatr Res. 2011;69:26R–33.
Moreno JL, Kurita M, Holloway T, López J, Cadagan R, Martínez-Sobrido L, et al. Maternal influenza viral infection causes schizophrenia-like alterations of 5-HT2A and mGlu2 receptors in the adult offspring. J Neurosci. 2011;31:1863–72.
Moreno-De-Luca A, Myers SM, Challman TD, Moreno-De-Luca D, Evans DW, Ledbetter DH. Developmental brain dysfunction: Revival and expansion of old concepts based on new genetic evidence. Lancet Neurol. 2013;12:406–14.
Mortensen PB, Nørgaard-Pedersen B, Waltoft BL, Sørensen TL, Hougaard D, Yolken RH. Early infections of Toxoplasma gondii and the later development of schizophrenia. Schizophr Bull. 2007;33:741–4.
O'Donnell P, editor. Animal models of schizophrenia and related disorders. New York: Humana Press; 2011.
Owen MJ. Implications of genetic findings for understanding schizophrenia. Schizophr Bull. 2012a;38:904–7.
Owen MJ. Intellectual disability and major psychiatric disorders: a continuum of neurodevelopmental causality. Br J Psychiatry. 2012b;200:268–9.
Ozawa K, Hashimoto K, Kishimoto T, Shimizu E, Ishikura H, Iyo M. Immune activation during pregnancy in mice leads to dopaminergic hyperfunction and cognitive impairment in the offspring: a neurodevelopmental animal model of schizophrenia. Biol Psychiatry. 2006;59:546–54.
Pineda DA, Palacio LG, Puerta IC, Merchán V, Arango CP, Galvis AY, et al. Environmental influences that affect attention deficit/hyperactivity disorder: study of a genetic isolate. Eur Child Adolesc Psychiatry. 2007;16:337–46.
Piontkewitz Y, Arad M, Weiner I. Abnormal trajectories of neurodevelopment and behavior following in utero insult in the rat. Biol Psychiatry. 2001;70:842–51.
Piontkewitz Y, Assaf Y, Weiner I. Clozapine administration in adolescence prevents postpubertal emergence of brain structural pathology in an animal model of schizophrenia. Biol Psychiatry. 2009;66:1038–46.
Piontkewitz Y, Arad M, Weiner I. Risperidone administered during asymptomatic period of adolescence prevents the emergence of brain structural pathology and behavioral abnormalities in an animal model of schizophrenia. Schizophr Bull. 2011;37:1257–69.
Piontkewitz Y, Arad M, Weiner I. Tracing the development of psychosis and its prevention: what can be learned from animal models. Neuropharmacology. 2012;62:1273–89.
Reimer T, Brcic M, Schweizer M, Jungi TW. Poly(I:C) and LPS induce distinct IRF3 and NF-kappaB signaling during type-I IFN and TNF responses in human macrophages. J Leukoc Biol. 2008;83:1249–57.
Richtand NM, Ahlbrand R, Horn P, Stanford K, Bronson SL, McNamara RK. Effects of risperidone and paliperidone pre-treatment on locomotor response following prenatal immune activation. J Psychiatr Res. 2011;45:1194–201.
Samuelsson AM, Jennische E, Hansson HA, Holmäng A. Prenatal exposure to interleukin-6 results in inflammatory neurodegeneration in hippocampus with NMDA/GABA(A) dysregulation and impaired spatial learning. Am J Physiol Regul Integr Comp Physiol. 2006;290:R1345–56.
Selten JP, Frissen A, Lensvelt-Mulders G, Morgan VA. Schizophrenia and 1957 pandemic of influenza: meta-analysis. Schizophr Bull. 2010;36:219–28.
Shi L, Fatemi SH, Sidwell RW, Patterson PH. Maternal influenza infection causes marked behavioral and pharmacological changes in the offspring. J Neurosci. 2003;23:297–302.
Short SJ, Lubach GR, Karasin AI, Olsen CW, Styner M, Knickmeyer RC, et al. Maternal influenza infection during pregnancy impacts postnatal brain development in the rhesus monkey. Biol Psychiatry. 2010;67:965–73.
Smith SE, Li J, Garbett K, Mirnics K, Patterson PH. Maternal immune activation alters fetal brain development through interleukin-6. J Neurosci. 2007;27:10695–702.
Smoller JW, Craddock N, Kendler K, Lee PH, Neale BM, Cross-Disorder Group of the Psychiatric Genomics Consortium, et al. Identification of risk loci with shared effects on five major psychiatric disorders: a genome-wide analysis. Lancet. 2013;381:1371–9.

Sørensen HJ, Mortensen EL, Reinisch JM, Mednick SA. Association between prenatal exposure to bacterial infection and risk of schizophrenia. Schizophr Bull. 2009;35:631–7.

Suvisaari J, Haukka J, Tanskanen A, Hovi T, Lönnqvist J. Association between prenatal exposure to poliovirus infection and adult schizophrenia. Am J Psychiatry. 1999;156:1100–2.

Tandon R, Nasrallah HA, Keshavan MS. Schizophrenia "just the facts" 4. Clinical features and conceptualization. Schizophr Res. 2009;110:1–23.

Torrey EF, Peterson MR. Slow and latent viruses in schizophrenia. Lancet. 1973;2:22–4.

Torrey EF, Rawlings R, Waldman IN. Schizophrenic births and viral diseases in two states. Schizophr Res. 1988;1:73–7.

Torrey EF, Bartko JJ, Yolken RH. Toxoplasma gondii and other risk factors for schizophrenia: an update. Schizophr Bull. 2012;38:642–7.

Vuillermot S, Weber L, Feldon J, Meyer U. A longitudinal examination of the neurodevelopmental impact of prenatal immune activation in mice reveals primary defects in dopaminergic development relevant to schizophrenia. J Neurosci. 2010;30:1270–87.

Vuillermot S, Joodmardi E, Perlmann T, Ögren SO, Feldon J, Meyer U. Prenatal immune activation interacts with genetic Nurr1 deficiency in the development of attentional impairments. J Neurosci. 2012;32:436–51.

Weinberger DR, Lipska BK. Cortical maldevelopment, anti-psychotic drugs, and schizophrenia: a search for common ground. Schizophr Res. 1995;16:87–110.

Willette AA, Lubach GR, Knickmeyer RC, Short SJ, Styner M, Gilmore JH, Coe CL. Brain enlargement and increased behavioral and cytokine reactivity in infant monkeys following acute prenatal endotoxemia. Behav Brain Res. 2011;219:108–15.

Winter C, Reutiman TJ, Folsom TD, Sohr R, Wolf RJ, Juckel G, Fatemi SH. Dopamine and serotonin levels following prenatal viral infection in mouse—implications for psychiatric disorders such as schizophrenia and autism. Eur Neuropsychopharmacol. 2008;18:712–6.

Xing Z, Gauldie J, Cox G, Baumann H, Jordana M, Lei XF, Achong MK. IL-6 is an antiinflammatory cytokine required for controlling local or systemic acute inflammatory responses. J Clin Invest. 1998;101:311–20.

Zaretsky MV, Alexander JM, Byrd W, Bawdon RE. Transfer of inflammatory cytokines across the placenta. Obstet Gynecol. 2004;103:546–50.

Zuckerman L, Rehavi M, Nachman R, Weiner I. Immune activation during pregnancy in rats leads to a postpubertal emergence of disrupted latent inhibition, dopaminergic hyperfunction, and altered limbic morphology in the offspring: a novel neurodevelopmental model of schizophrenia. Neuropsychopharmacology. 2003;28:1778–89.

Chapter 3
Rodent Models of Stress-Induced Depression: The Link Between Stress and Immune System Related Changes

Barry McGuinness and Andrew Harkin

Abstract Bi-directional communication between the brain and immune system is a research area gaining in prominence in the search for biological factors associated with the development of a number of psychiatric disorders. In this regard, animal models make substantial contributions to our understanding of these interactions. For instance, long-term adaptive changes to the immune system, in response to stress, have been documented and include stress-induced desensitisation of immune cell glucocorticoid receptors (GR) and β_2-adrenoceptors (AR) which in turn can impact on the ability of these factors to elicit their natural anti-inflammatory actions. It is of interest that psychological stress has been reported to induce microglial activation and pro-inflammatory cytokine expression in the central nervous system (CNS) in animal studies, and whilst there is little evidence of microglial activation or an increase in CNS cytokines in depressed humans to date, animal studies are providing important insights into the potential role of central inflammatory events in the pathophysiology of stress related psychiatric disorders. Stress can be linked to the development of a pro-inflammatory state, and it is of interest to consider the reciprocal impact of peripheral inflammatory mediators on the brain. In this regard, there is accumulating evidence that inflammation plays a role in the pathogenesis of major depressive illness. To further explore the behavioural effects associated with immune system activation, a number of animal models are currently in use and include a range of agents used to initiate immune activation including bacterial endotoxin lipopolysaccharide (LPS), the inflammatory cytokine interferon (IFN) alpha, inoculation with Bacille Calmette–Guerin (BCG), experimental autoimmune encephalitis (EAE) and administration of synthetic double stranded ribonucleic acid

B. McGuinness
Trinity College Institute of Neuroscience, Trinity College Dublin, Dublin 2, Ireland

A. Harkin (✉)
Trinity College Institute of Neuroscience, Trinity College Dublin, Dublin 2, Ireland

The School of Pharmacy and Pharmaceutical Sciences,
Trinity College Dublin, Dublin 2, Ireland
e-mail: aharkin@tcd.ie

N. Müller et al. (eds.), *Immunology and Psychiatry*, Current Topics in Neurotoxicity 8, DOI 10.1007/978-3-319-13602-8_3

(dsRNA) polyinosinic:polycytidylic acid (poly-IC), a viral mimetic. Immune activation in these models provokes signs of depression which may be ameliorated following treatment with antidepressant drugs. Despite these findings the question as to how an immunological basis for a disorder can be reconciled with more traditional hypotheses including impairment of monoamine transmission and the functional impairment of neurotrophins remains to be addressed. In this regard, the activation of indoleamine 2,3 dioxygenase (IDO) and activation of the kynurenine pathway or the ability of cytokines to influence monoamine transport, receptor activation or neurotrophic factor expression are possibilities. The impact of inflammation on such systems is providing new avenues to develop a better understanding of the role of stress and related immune changes in the pathophysiology of depression and to find new and improved therapies to address current unmet clinical needs.

Keywords Immune system • Stress • Depression • Cytokine • Animal model • Glucocorticoid • Catecholamine • Neurotrophic factor • Kynurenine pathway

Immune System Activation Is Associated with Anxiety and Depression Related Behaviours

Activation of the immune system in vivo occurs when a foreign or non-self antigen is detected by cells of the immune system. Various pathogen mimics exist which can induce a range of immune responses based on different receptors that recognise pathogens including toll-like receptors (TLRs). LPS is a component of the cell wall of gram negative bacteria and is recognised by innate immune cells as a foreign antigen by binding to TLR4 leading to downstream intracellular signalling. LPS challenge is routinely used as an experimental tool to act as an immunostimulus and mimics the early stages of a bacterial infection. TLR4 receptors are also present on the cell surface of astrocytes and microglia. The physiological changes following administration of LPS have been well characterised in animal models including peripheral and central immune activation (Olson and Miller 2004), elevations in circulating and central pro-inflammatory cytokines (Connor et al. 2005; Chung and Benveniste 1990), behavioural changes (Frenois et al. 2007) and impairments in synaptic plasticity including reduced neurogenesis (Guan and Fang 2006). Although usually considered to be immune privileged, receptors for numerous cytokines, soluble protein mediators of immune cells, are expressed throughout the brain. Moreover there are mechanisms by which cytokines communicate with the brain via the circumventricular organs, via active transport or the activation of cells lining the cerebrovasculature. Cytokines are also capable of activating peripheral nerves such as the vagus nerve which transmit signals to the brain (Miller et al. 2009; Quan and Banks 2007). Poly I:C is a viral mimetic, bearing analogy to double stranded viral RNA, that can also be used to activate the immune system. Whereas LPS activates TLR4 receptors on innate immune cells, poly I:C binds to TLR3 receptors. Like LPS it is widely used as an experimental immune stimulus and has been shown to reliably induce sickness behaviour in animals, inducing fever and an inflammatory response characterised by an increase in the expression and release of pro-inflammatory cytokines both peripherally and centrally. BCG is an attenuated strain of mycobacterium bovis, which is commonly used to immunise against mycobacterium tuberculosis. Administration of this vaccine to mice has been shown to induce a sustained pro-inflammatory response accompanied by sickness behaviour (O'Connor et al. 2009a).

The demonstration of depression-like behaviours in animals following challenge with immunostimulants is frequently confounded by the overlap of cytokine-induced sickness behaviours such as fever, hypoactivity, weight loss and the suppression of feeding. Reports that antidepressant responsive depression-like behaviours in immune stimulated animals can develop independently of sickness related performance impairments, however, have enhanced the validity of these models as animal analogues of clinical depression. For instance, peripheral administration of LPS to laboratory mice or rats induces sickness behaviour, including hyperpyrexia, hypoactivity, suppressed feeding and bodyweight loss, which peaks 2–6 h later and then gradually subsides. Depression-like behaviours as measured by increased immobility, regarded as behavioural despair, in the forced swimming test or tail suspension test, reduced voluntary wheel running and decreased preference for a sweet solution, indicative of anhedonic behaviour, emerge later. Such behaviours are believed to require the activation of pro-inflammatory cytokine signalling in the brain (Dantzer et al. 2008; Dunn et al. 2005).

In response to the viral mimetic poly I:C, Sprague-Dawley rats showed significant body weight loss, a reduction in saccharin preference (anhedonia), decreased locomotor activity in the home-cage activity test and increased anxiety-related behaviours in the open field test. This altered behavioural phenotype was associated with poly I:C-induced temporal increases in the expression of the cytokines IL-1β, IL-6 and TNFα and the microglial activation marker CD11b in the frontal cortex and hippocampus. Decreased expression of the neurotrophic marker BDNF and its receptor TrkB was observed in response to poly I:C administration. An increase in the expression of the tryptophan degrading enzyme IDO was also found in the brain which was associated with increased central concentrations of the tryptophan metabolite, kynurenine. These findings suggested that the depressive-like behaviour elicited by poly I:C administration may be associated with cytokine-induced alterations in growth factor expression and/or tryptophan metabolism. Activation of the kynurenine metabolic pathway may be of significance in relation to the development of viable therapeutic strategies for inflammation-related depression and will be discussed later (Gibney et al. 2013).

Immune stimuli including LPS and poly I:C increase circulating concentrations of pro-inflammatory cytokines both peripherally and centrally, which can induce depressive behaviour (Dunn et al. 2005; Burton et al. 2011; Fortier et al. 2004; Gibb et al. 2011; Konat et al. 2009). The role that individual cytokines play in the pathogenesis of depressive symptoms, and in particular IL-6, is not well characterised in general. IL-6 may be induced following exposure to psychological stressors (Butterweck et al. 2003; Jankord et al. 2010; Maes et al. 1999; Voorhees et al. 2013).

It is of particular interest as it has been the cytokine which has been reported to be most reliably increased in patients with depression (Dowlati et al. 2010; Howren et al. 2009). Central administration of IL-6 in mice induced depressive behaviours, such as increased immobility in the forced swim and tail suspension tests which were sustained for 48 h coupled with reduced interactions in a social paradigm which were blocked by co-administration with anti-IL-6 antibody (Sukoff Rizzo et al. 2012).

However, IL-1β and TNFα are considered the main cytokines responsible, as there are numerous reports that systemic or central administration of these cytokines to laboratory rats and mice induces sickness behaviour in a dose- and time-dependent manner. Systemic administration of IL-1β to mice has been reported to mediate endotoxin induced sickness behaviour (Bluthe et al. 2000) and to increase immobility (behavioural despair) in the forced swimming and tail suspension tests (Dunn and Swiergiel 2005). Skelly et al. (2013) have reported that systemic administration of TNFα and IL-1β to mice provoked a reduction in activity in the home-cage. It has been shown that IL-1β can provoke anhedonia in laboratory rodents independent of effects on anorexia (Larson et al. 2002; Merali et al. 2003). Other cytokines may play a less direct role including IL-6 which contributes to the expression of brain IL-1β and TNFα in response to immune challenge and may play a role in some behaviours such as LPS induced mediated, hippocampal related, cognitive impairments (Sparkman et al. 2006). A recent study by Karson et al. (2013) reported that treatment of rats with the TNFα inhibitor infliximab reduced depressive-like behaviours induced by a chronic mild stress (CMS) exposure. In other reports to date, IL-2, but not IL-1β or IL-6, has been reported to provoke long-lasting anhedonic effects in mice and rats (Dunn et al. 2005; Anisman et al. 2002).

In humans, cytokine immunotherapy for viral hepatitis or various forms of cancer provokes severe psychological disturbances including depression. Numerous studies have reported that previously psychiatrically healthy individuals, treated with high doses of exogenous cytokine IL-2 or IFNα, develop depressive-like symptoms such as depressed mood, increased somatic symptoms and stress reactions and cognitive impairment (Capuron et al. 2000, 2003, 2004; Valentine and Meyers 2005). Investigations into IFNα immunotherapy in patients have reported an increase in the circulating and CSF concentrations of kynurenine which may contribute to the appearance of depression related symptoms (Raison et al. 2010). A role for tryptophan metabolism and the kynurenine pathway in depression related behaviour is reviewed in detail below.

It is not entirely clear why only a proportion of individuals, typically less than 50 %, who receive cytokine immunotherapy become depressed. It has been suggested that psychological stress may sensitise to the neurochemical and behavioural actions of cytokines and therefore individuals with an anxious/stress-prone phenotype may be more susceptible to developing psychiatric sequelae in response to cytokine administration (Anisman et al. 2002). In support of this idea, it was reported that patients with a higher cortisol response (stress response) to the initial injection of IFNα displayed a greater propensity to develop depression following treatment with IFNα (Capuron et al. 2003). Further support for this notion stems from a recent pre-clinical study demonstrating that mice exposed to psychosocial stress showed exaggerated central monoamine changes, HPA axis reactivity and sickness behaviours to IFNα treatment (Anisman et al. 2007). A systemic injection of IFNα was reported to induce long-term behavioural effects including reduced locomotor activity and sickness behaviour in rats (Kentner et al. 2006).

When one is considering the literature on cytokines as a trigger for depressive illness, it is indisputable that treatment with IL-2 or IFNα can induce depression. However one must remember that the doses of cytokines administered to patients in these studies far exceed physiological concentrations. This is a factor that should be considered when implicating elevated endogenous cytokine secretion as a causal factor in the development of depressive symptoms. Thus one cannot simply regard the psychiatric and biological sequelae that occur following exogenous administration of cytokines such as IFNα or IL-2 as being akin to the biology of idiopathic depression where no obvious exogenous agent is driving changes in the psychiatric state.

A connection between inflammation and depression may also be supported given the high levels of co-morbidity of depression reported in patients suffering with inflammatory disorders. Rheumatoid arthritis (RA) patients are twice as likely to suffer from depression than the general population (Isik et al. 2007; Katon and Sullivan 1990), and a meta-analysis by Anderson et al. (2001) reported that people suffering with diabetes were also twice as likely to develop depressive disorders than non-diabetic controls. Prevalence of depression is also greater in patients with multiple sclerosis (MS) than in healthy controls (Minden and Schiffer 1990). Sadovnick reported that MS patients had a lifetime prevalence for depression of 50 % (Sadovnick et al. 1996). Indeed, even people with a history of allergies were more likely to be diagnosed with major depression (Hurwitz and Morgenstern 1999). The prevalence of depression is three times higher in patients with inflammatory bowel diseases (IBD) when compared to the general population (Graff et al. 2009), and the incidence of depression and anxiety in patients with Crohn's disease reported to be fivefold higher when compared to healthy controls (Walker et al. 2011).

Biological Mechanisms Implicated in Mediating the Depressive Effects of Cytokines

Currently, elucidation of mechanisms mediating the ability of immune activation to induce depression-like behaviours is an area of considerable interest and includes cytokine effects on neurotransmitter metabolism, neuroendocrine function and neuronal plasticity. Such mechanisms have been reviewed extensively elsewhere (Miller et al. 2009; Anisman et al. 2008; Anisman 2009).

Of these mechanisms, considerable attention has recently been focused on the ability of pro-inflammatory cytokines such as IL-1β, TNFα and IFNγ to enhance the activity of IDO. IDO breaks down tryptophan, the primary amino acid precursor of serotonin, into kynurenine. The breakdown of tryptophan is believed to contribute to reduced serotonin availability and increased kynurenine. In mice, IDO is induced following an immune challenge in both the periphery and brain in a time-dependent fashion and peaks at 24 h following LPS administration (Lestage et al. 2002). In an alternative model to LPS administration in mice, chronic stimulation of the immune system by inoculation with BCG induces a sustained elevation in circulating levels of IFNγ and a chronic activation of IDO (O'Connor et al. 2009a). Activation of IDO results in decreased circulating tryptophan concentrations and increased production of kynurenine and other kynurenine derived metabolites.

Activation of IDO is also associated with late onset depression-like behaviours following BCG inoculation that may be blocked in IFNγ knockout and IDO knockout mice (O'Connor et al. 2009a). Pre-treatment with the second generation tetracycline minocycline, which has potent anti-inflammatory effects both peripherally and in the brain, blocks both LPS-induced sickness and depression-like behaviours. By contrast, treatment with the competitive IDO inhibitor 1-methyl tryptophan (1-MT), without influencing IL-1β or IFNγ expression or LPS-induced sickness behaviour, attenuates late onset depression-like behaviours in tandem with a reduction in the plasma kynurenine/tryptophan ratio. Similar mechanisms were reported in the BCG inoculation model in mice (Dantzer et al. 2008; O'Connor et al. 2009b). Although the activation of IDO by cytokines provokes a reduction in circulating levels of tryptophan, this may not translate into alterations in the synthesis of serotonin as the attenuation of LPS induced depression-like behaviours in mice by the inhibition of IDO are not associated with central changes in serotonin turnover. Rather degradation of tryptophan into kynurenine and related metabolites such as quinolinic (QUIN) and kynurenic acid (KYNA) with effects at the *N*-methyl-D-aspartate (NMDA) receptor could account for changes in glutamatergic transmission which promote the development of depression related behaviours (Muller and Schwarz 2007). In support of this, administration of kynurenine to mice dose dependently induces depression-like behaviour (O'Connor et al. 2009c). Such findings support a role for the kynurenine pathway in mediating late onset depression-like behaviours induced following immune challenge.

Alternative mechanisms to kynurenine pathway activation have also been proposed in the development of depression associated with inflammation including the ability of cytokines to influence tryptophan uptake in the brain (Dunn et al. 2005), serotonin uptake in nerve terminals (Morikawa et al. 1998; Tsao et al. 2008; Zhu et al. 2006, 2010) and the ability of cytokines to influence the expression of serotonin 5-HT$_{1A}$ receptors in cells in culture (Cai et al. 2005) which when taken together are suggestive that cytokines modulate serotonergic neurotransmission by mechanisms other than an IDO related decrease in tryptophan and the generation of kynurenine related metabolites. Increased serotonin transporter expression has been associated with elevated IFNα (Morikawa et al. 1998) and following immune stimulus (Zhu et al. 2010). This could mediate the depressive effects of cytokines by removing serotonin from the synapse and reducing serotonin related synaptic transmission.

It is also of interest that auto-antibodies, antibodies produced by the body which respond to resident "self" protein signals, implicated in autoimmune diseases such as rheumatoid arthritis, lupus erythematosus and Grave's disease (Eggert et al. 2010), have been found against serotonin (Maes et al. 2012; Schott et al. 2003) and the 5-HT$_{1A}$ receptor (Tanaka et al. 2003) in circulating serum of depressed patients.

Non-serotonergic mechanisms such as hyperactivity of the hypothalamic pituitary adrenal (HPA) axis may also be significant in mediating depression associated with inflammation on account of the ability of cytokines to activate the HPA axis (Pace et al. 2007). In addition, pro-inflammatory cytokines promote the expression of the inactive β isoform of the GR that can bind to response elements in the nucleus, while decreasing expression of the active α isoform (Pace and Miller 2009). In vitro experiments have shown greater increases in GRβ expression following treatment with TNFα and IL-1 than GRα (Webster et al. 2001).

A recent study has shown that depressed patients have lower expression of GRα which is strongly correlated with increases in pro-inflammatory cytokines (Carvalho et al. 2014). Such mechanisms promote resistance to glucocorticoids and may account for cytokine-induced dysregulation of the HPA axis on account of the reduced ability of glucocorticoids to feedback to the hypothalamus. Depressed patients show elevated levels of corticotrophin releasing hormone (CRH), a hypothalamic neuropeptide under the regulation of glucocorticoids that drives HPA axis activation and plays an important central role in behavioural, endocrine and immune responses to stress (Pace et al. 2007; Irwin 2008). It has been proposed that at the level of peripheral and central immune cells, inhibitory effects of glucocorticoids on cytokine production may no longer be operative, facilitating an unrepressed increase in the production of pro-inflammatory cytokines. Such a mechanism consolidates a stress driven inflammatory loop in tandem with an over-production of CRH (Dantzer et al. 2008). A model of CNS stress circuitry activation involving CRH activation and ultimately sympathetic nervous system outflow are also proposed to link the brain and stress to altered immune responses. Further consideration of sympathetic effector mechanisms in stress and depression and the critical role of sympathetic output in mediating immune response, competence and inflammation is reviewed elsewhere (Irwin 2008).

Mapping of neuronal activation in response to LPS administration has been carried out and the peak of sickness behaviour is associated with increased expression of the cellular activation marker and immediate early gene product c-fos in the paraventricular nucleus and the bed nucleus of the stria terminalis (BNST), areas involved in endocrine and autonomic components of LPS-induced sickness. By contrast, immunostaining for FosB and its truncated splice variant ΔfosB, both of which have a longer half life than c-fos and accumulate during repeated or long-lasting stimulation, was increased in several hypothalamic nuclei and had extended to the amygdala and hippocampus. Such activation patterns point towards the involvement of these structures in cytokine-induced depression (Frenois et al. 2007; Dantzer et al. 2008).

It is also of significance that immunological stimulation with agents such as LPS provokes a reduction in the expression of brain derived neurotrophic factor (BDNF) and neurogenesis in the hippocampus (Fujioka and Akema 2010; Ormerod et al. 2013). In rodent models IFNα treatment was reported to reduce the expression of BDNF within the hippocampus (Kaneko et al. 2006). BDNF expression is also reduced following immune challenge with LPS (Tanaka et al. 2006) which is antagonised by inhibition of IL-1β signalling following intrahippocampal administration of IL-1 receptor antagonist (IL-1ra). Moreover IL-1ra prevents stress related reductions in BDNF expression suggesting a role for that pro-inflammatory cytokine in stress related mechanisms (Barrientos et al. 2003). Such reports of reduced hippocampal growth factor expression and neurogenesis may contribute to reductions in hippocampal volume as indicated in numerous neuroimaging studies in depressed patients (Chen et al. 2001; Sheline et al. 2003). It is proposed that reduced hippocampal volume may reflect neuronal atrophy in the region associated with chronic stress or inflammatory coupled mechanisms.

Effects of Stress on the Immune System of Relevance to Clinical Depression

The role of stress-induced changes in the immune system has been more routinely considered in disease vulnerability, wound healing and peripheral or central inflammatory conditions (Glaser and Kiecolt-Glaser 2005), e.g. IBD (Mawdsley and Rampton 2005) or multiple sclerosis (Gold et al. 2005). Accumulating data also indicate that stress related effects on the immune system and inflammation may also play a role in neuropsychiatric disorders including major depression (Miller et al. 2009) and analyses of stressor and immune activation effects in animal models have been useful in defining some of the alterations that may subserve pathophysiology.

The most widely applied manipulation to study brain immune interactions of relevance to depression has been the application of stressors of varying type, duration and intensity. Upon exposure to stress the brain sends signals to the periphery to stimulate the release of glucocorticoid hormone via the HPA axis and catecholamine transmitters via the sympathoadrenal medullary (SAM) axis in addition to a neuronal component that regulates peripheral activity by direct sympathetic innervations. These peripheral stress signals in turn have direct effects on various components of the immune system. Stressors of a psychological or psychological combined with physical nature are considered to more closely mimic psychological stress in humans and to be of greater relevance.

Stress has long been associated with changes in immune status and there are many reports of both immunoenhancement and immunosuppression where the impact of stressor exposure on the immune response depends on a variety of factors. The duration of stressor exposure seems to be important in determining the immune response. A meta-analysis of human studies carried out by Segerstrom and Miller (2004), reported that a range of altered immune profiles may be induced depending on the time frame of stressor exposure. Acute time limited stressors were associated with an up-regulation of some innate immune responses and suppression of specific immunity. Brief naturalistic stressors or real-life short-term challenges were associated with a shift in the profile of some cytokines whereas chronic stress was linked to global immunosuppression. The nature of the stressor is also influential, where there is a smaller behavioural and endocrine response upon repeated exposure to a stressor of a physical nature, the intensity of a psychological stressor increases with repeated exposures (Blanchard et al. 1998; Sato et al. 1996). For example, studies have shown that exposure to chronic social disruption stress results in altered splenocyte function including increased splenic monocytes and neutrophils and decreased lymphocytes, with implications for the healing of bite wounds that are often associated with social stress in rodents (Avitsur et al. 2002a, b). However in the studies of social stress, and in other studies that show an immunoenhancing effect of chronic stress exposure, the temporal proximity between the stressor termination and the immune challenge is delayed. Such studies have reported that an immunoenhanced state becomes evident within 24 h of stressor termination (Avitsur et al. 2005; Zalcman et al. 1988, 1989; Zalcman and Anisman 1993). Therefore an immune challenge administered immediately following stress and 24 h after stressor termination gives alternative immune responses suggestive of cellular alterations occurring during the recovery period following stress.

Experiments undertaken to address the mechanisms mediating stress-induced suppression of IL-12 and IFNγ in a murine model have shown that the glucocorticoid hormone corticosterone is a likely candidate to account for the suppressive effect of stress on IL-12 and IFNγ production (Curtin et al. 2009a). These and related studies (Glaser and Kiecolt-Glaser 2005; Konstantinos and Sheridan 2001) illustrate that the HPA axis is the main immunoregulatory pathway that influences the immune response to stressor exposure. In a further series of experiments Curtin et al. (2009b) determined the effect of the anxiolytic drug chlordiazepoxide, a drug commonly prescribed to relieve anxiety and stress perception. Prior treatment with chlordiazepoxide can prevent the stress-induced suppression of LPS-induced IFNγ and IL-12 and the stress related increase in LPS-induced IL-10 (Curtin et al. 2009a, b). These results support a mechanism where central perception of the stressor is critical to mediating the immunosuppressive effects of stress. It was noted that although LPS-induced IFNγ and IL-12p40 expression was suppressed in splenocytes immediately after exposure to chronic stress, an increase in splenic IFNγ and IL-12 concentrations was obtained 24 h later. In tandem with raised cytokine concentrations, a reduction in the expression of glucocorticoid and β_2-ARs in splenocytes 24 h after stressor termination was accompanied by a reduction in the sensitivity of LPS-induced IFNγ and IL-12 to the suppressive effects of corticosterone and the β_2-AR agonist salbutamol (Curtin 2008). Thus prolonged exposure to stress can precipitate a low grade inflammatory response associated with a down-regulation and reduced sensitivity of the glucocorticoid and β_2-AR on splenocytes.

It has long been demonstrated that stress induces adaptations in receptor expression and sensitivity during exposure to chronic stress (Quan et al. 2003). Alterations in immune response due to changes in receptor expression and sensitivity are commonly referred to as receptor desensitisation which involves alterations in receptor expression that remain for a period following stressor termination. As GR mediates tonic anti-inflammatory effects, down-regulation of these receptors in response to stress could promote the development of an inflammatory state. Exposure to chronic stress which chronically activates the HPA-axis has been shown to result in substantial increases in systematic glucocorticoid levels both by enhancing the release of glucocorticoid hormones and by decreasing the plasma binding capacity of corticosterone binding globulin (Stefanski 2000; Spencer et al. 1996). This has been demonstrated to lead to adaptive changes in immune cells with reduced nuclear translocation of GR and impaired immunosuppressive action of the receptor in monocytes. This stress-induced reduction of GR expression and function may be considered an adaptive response to chronic stress to alleviate immunosuppressive effects associated with sustained levels of glucocorticoids. When exposure to stress is terminated, the concentrations of glucocorticoids return to pre-stress basal levels. However the cellular adaptations persist for some time and during this interim an immunoenhanced/pro-inflammatory state may prevail as a result of GR desensitisation to normal physiological levels of glucocorticoids.

Pre-clinical studies have shown that exposure to a chronic stress can reduce the reactivity and amount of GR protein in immune cells (Curtin 2008; Quan et al. 2001), and human tests have shown that people who had long-term stressful exposure display glucocorticoid resistance which was accompanied by an increased pro-inflammatory profile following infection with a rhinovirus (Cohen et al. 2012). Reduced GR expression in mouse spleen 18 h following exposure to chronic social stress has been reported by Quan et al. (2001) who also reported that GR expression was reduced in the hippocampus, an important neural structure for the negative feedback action of glucocorticoids (De Kloet et al. 1998). Chronic stress-induced down-regulation of GR mRNA and protein in mouse hippocampus has also been demonstrated by other groups (Sapolsky et al. 1984; Sapolsky and Meaney 1986; Herman et al. 1995; Kitraki et al. 1999). The implications of stress-induced down-regulation of GR in the hippocampus for the activity, regulation and proliferation of resident microglia is a subject of considerable current interest in this field and will be discussed later.

Catecholamines, acting through α- and β-ARs, have been shown to have pro- and anti-inflammatory properties and play a role in the maintenance of an immunoregulated environment (Johnson et al. 2005; Nance and Sanders 2007; Feinstein et al. 2002). It has long been shown that following stimulation of β_2-ARs on immune cells subsequent production of cAMP promotes the production of cytokines that can suppress cell mediated immune responses (Suberville et al. 1996; Platzer et al. 2000). For instance, activated monocytes and dendritic cells exposed to NA have reduced IL-12 concentrations, in parallel with increased production of the anti-inflammatory cytokine IL-10 (Elenkov et al. 1996). Such an alteration in the cytokine network could potentially diminish the cell based immune response. In a similar fashion to persistent HPA-axis activation with chronic stress, chronic activation of the SAM-axis results in increases in circulating catecholamine concentrations and this can lead to adaptive changes in the expression of β_2-AR. For example, it has been demonstrated that continuous exposure to catecholamines leads to a reduced receptor response by a process of receptor phosphorylation, internalisation and down-regulation (Bawa-Khalfe et al. 2007). In addition, it has been reported that forced exercise which can be considered to provoke robust activation of the SAM-axis and to be a stressor for laboratory animals provokes a reduction in β_2-AR expression in peritoneal macrophages associated with an increase in IL-12 production 24 h following termination of the regime (Itoh et al. 2004). Therefore, in a similar fashion to GR, stress-induced reduction of β_2-AR could be involved in a diminished negative feedback response to catecholamines on immune cells. Conversely others have reported an increase in catecholamine reactivity in mice subjected to CMS (Edgar et al. 2002, 2003). Normally catecholamines exert a β_2-AR mediated inhibitory effect on mitogen induced T-cell proliferation and a stimulatory effect on B-cell proliferation in response to selective B-lymphocyte mitogens. Lymphocytes from mice subjected to CMS have an increased response to catecholamine mediated inhibition or enhancement of proliferation in T and B-cells coupled with an increase in β_2-AR density and responsivity. In this model, chronic stress is associated with an increased sympathetic influence on the immune response suggestive of an alternative mechanism through which stress can alter immunity.

In addition, the parasympathetic nervous system (PNS) may also play a role in the regulation of inflammation. Activation of efferent vagus nerve fibres by acetylcholine release and subsequent α_7-nicotinic receptor activation on immune cells can inhibit cytokine responses to endotoxin in laboratory animals (Pavlov and Tracey 2005) suggesting that the sympathetic and parasympathetic nervous systems may have opposing regulatory influences over innate immunity and inflammation during stress (Miller et al. 2009). Increased muscarinic receptor expression has been reported in T and B-cells isolated from mice exposed to chronic stress (Edgar et al. 2002). Thus long-term adaptive responses are also evident in the PNS in response to stress which may impact on immune function.

It is well established that reactivity to psychological stress is a contributing factor to depressive disorders and relapse and that severely depressed patients display hypercortisolaemia and glucocorticoid resistance related at least in part to impaired GR function (Pariante and Miller 2001; Pace et al. 2006; Maier and Watkins 1998). In this regard, it is known that inflammatory cytokines can promote glucocorticoid resistance (Pace et al. 2007). Pro-inflammatory cytokines have also been shown to affect GR signalling within the cell. Activation of p38 MAPK, JNK1/2/3, ERK1/2, STAT and NFκB signalling pathways have all been reported to reduce GR functioning within fibroblast cells. IL-1 treatment has been shown to reduce dexamethasone induced GR translocation to the nucleus and GR mediated transcription in vitro, via induction of p38 MAPK (Pariante et al. 1999; Wang et al. 2004). These reductions were reversed following co-treatment with IL-1ra or antisense oligonucleotides targeting p38. IL-2 has been reported to reduce translocation of GR (Goleva et al. 2002). Tliba et al. (2008) have also reported that IFN regulatory pathways can reduce GR gene transcription following treatment with TNFα and IFNγ. NFκB has been shown to interact with GR in the nucleus in a mutually inhibitory protein–protein interaction (McKay and Cidlowski 1999) as well as via competition for the co-activators CREB-binding protein and steroid receptor co-activator 1 (Sheppard et al. 1998). Thus activation of pro-inflammatory cytokine signalling pathways can interfere with GR signalling via disruption to binding in the nucleus that prevents transcription and by limiting the availability of common co-activators that would otherwise be available to activate GR signalling.

Thus the possibility exists that stress, which promotes glucocorticoid resistance, may precipitate a state of increased innate immune cytokines which in turn further drives glucocorticoid resistance in stress-related disorders like major depression. In that regard, chronic stress/depression has been consistently linked with the increased peripheral production of pro-inflammatory cytokines, most notably IL-6. Indeed high serum levels of IL-6 have been linked to risks for several conditions, such as cardiovascular disease, type 2 diabetes, mental health complications, and some cancers (Godbout and Glaser 2006).

Taken together, reports to date suggest that increased activity through the SAM-axis coupled with reduced sensitivity to the inhibitory effects of glucocorticoids may, during chronic stress, contribute to activation of the inflammatory response system. Given the role of glucocorticoids and catecholamines in regulating innate immune responses and the capacity of cytokines to influence behaviour, it is not unreasonable to suggest that the inter-relationship between SAM-axis activation, HPA-axis dysregulation and cytokines contributes to the pathophysiology of depression, at least in some patients where inflammation is evident (Miller et al. 2009; Pace and Miller 2009).

Evidence for Activation of Immune System in Depressed Patients

Patients with depression have been shown to have elevated levels of pro-inflammatory cytokines, chemokines, adhesion molecules and prostaglandins in the blood and cerebrospinal fluid (CSF) (Miller et al. 2009; Leonard 2007). Meta-analyses have reported that depressed patients have elevated levels of pro-inflammatory cytokines including IL-1β, IL-6 and TNFα circulating peripherally (Dowlati et al. 2010; Howren et al. 2009; Liu et al. 2012). IL-6 and IFNγ were found to be elevated in treatment resistant depressed patients when compared to controls, however no difference in TNFα concentrations were observed (Hughes et al. 2012). IL-6 has been the most consistently elevated cytokine in depressed patients and the presence of elevated IL-6 and C-reactive protein (CRP) can predict a subsequent development of depression over the course of a decade even in individuals with no background of depression when samples were taken (Gimeno et al. 2009; Pasco et al. 2010). Interestingly, while Simon et al. (2008) reported increased circulating concentrations of pro-inflammatory cytokines in depressed patients, they also reported increased concentrations of IL-10 and IL-4. Maes et al. (1997) also reported that the anti-inflammatory IL-1ra was elevated in conjunction with increased pro-inflammatory cytokines suggesting a complex profile of activation within the innate immune system.

People with autoimmune disorders have been shown to have a higher incidence of depressive disorders, including MS and RA, suggestive of an association between activation of the adaptive immune system and depression (Gibney and Drexhage 2013). Depressed patients have been reported to have fewer resting $CD3^+/CD25^-$ T_{reg}-cells, increased T-cells expressing activation markers, accompanied by increased expression of the soluble IL-2 receptor (sIL-2R), induction of IDO, higher levels of circulating tryptophan metabolites and lower levels of tryptophan (Maes 2011).

Investigations that have examined functional responses in peripheral blood mononuclear cells (PBMCs) have reported elevated levels of IFNγ and sIL-2R

(Seidel et al. 1995) and IL-1β and IL-6 (Maes et al. 1995), however other studies have reported the opposite effect, reporting reduced IL-1β, IL-2 and IL-3 in stimulated PBMCs isolated from depressed patients (Weizman et al. 1994). This inhibited response could be as a result of hypo-responsivity in PBMCs caused by raised circulating cytokines signalling through T-cells, suppressing the ability of cells to respond to stimulation (Maes et al. 2012; Haroon et al. 2012).

CSF of depressed patients has been found to contain elevated levels of pro-inflammatory cytokines and their soluble receptors (Müller et al. 1997; Raison et al. 2006). Increased concentrations of IL-1β have been reported in patients suffering from depression and a positive correlation found between concentration of IL-1β concentration and symptom severity (Levine et al. 1999). Elevated levels of IL-6 and TNFα have also been reported in patients who had attempted suicide (Lindqvist et al. 2009) while in a separate study IL-1, IL-6 and TNFα were also correlated with depression severity in patients (Martinez et al. 2012). In relation to cytokine therapy as mentioned previously, patients with hepatitis C treated with IFNα who subsequently developed depression related symptoms were reported to have elevated CSF levels of IFNα and IL-6 (Raison et al. 2009).

Post mortem, the expression of IFNα and its receptor IFNα/βR1 were reported to be elevated in dorsolateral prefrontal cortex of depressed patients when compared to non-depressed controls (Kang et al. 2007). In addition higher concentrations of the tryptophan/kynurenine metabolite, quinolinic acid (QUIN), an NMDA receptor agonist and potential excitotoxin, were observed in activated microglia in the subgenual anterior cingulated cortex and anterior midcingulate cortex of severely depressed patients (Steiner et al. 2011). These limited studies would suggest a role for microglial activation within the CNS in the progression of depressive disorders. Patients with depression have been shown to have a lower density of astrocytes when compared to healthy controls. A reduction in the astrocytic activation marker glial fibrillary acidic protein (GFAP) has been reported in post mortem tissue from depressed when compared to non-depressed controls (Muller and Schwarz 2007; Miguel-Hidalgo et al. 2010; Ongür et al. 1998). Astrocytes preferentially express kynurenine aminotransferase II (KAT II) which leads to the production of an NMDA receptor antagonist KYNA (Guillemin et al. 2001). Astrocytes also express the highest density of glutamate transporters and play an integral role in removal of glutamate from the synapse controlling the extent of excitation associated with glutamate receptors (Huang and Bergles 2004; Thomas et al. 2011). Reduced number and/or function of astrocytes could lead to reduced production of KYNA and decreased removal of glutamate from the synapse creating an excitotoxic extracellular environment.

Development of Animal Models of Depression Based on Immunological Activation

Animal models are most useful if they are based on the underlying cause and can elucidate mechanisms underlying the pathophysiology of depression and predict useful treatment strategies. A number of proposed mechanisms have emerged from animal studies bridging immune dysregulation to the pathophysiology of depression in the literature in recent times and these are summarised as follows:

1. Decompensation or dysregulation of the mechanisms that regulate sickness behaviours has been proposed to be of consequence for depression (Dantzer et al. 2008). Normally infectious episodes are reversible as the immune system combats the infectious pathogen and recovery mechanisms that oppose inflammation take effect. In depression however such mechanisms may be dysfunctional where the inflammatory response is more intense due to an imbalance between pro- and anti-inflammatory cytokines or glucocorticoid resistance. Depression may also be associated with a greater sensitivity to immune related changes due to pre-existing vulnerability such as changes in genotype, e.g. serotonergic receptors, transporters or metabolism or disturbances in HPA-axis or sleep quality. Overall Dantzer et al. (2008) suggest that depression may represent a maladaptive version of cytokine-induced sickness which could occur when activation of the innate immune response is exacerbated in intensity or duration.
2. Changes in brain plasticity associated with immune activation may also account for mechanisms underlying cytokine-induced depression. Cytokines are considered important for providing trophic support to neurons and enhancing neurogenesis while contributing to cognitive functions such as memory in laboratory animals. However with excessive or prolonged activation, cytokines such as IL-1β and TNFα can promote abnormalities that are thought to be relevant to the pathophysiology of depression, including cognitive impairment, diminished neurotrophic support, enhanced oxidative and apoptotic mechanisms and decreased neurogenesis in the hippocampus favouring impaired neuroplastic processes (Miller et al. 2009; Anisman 2009; Goshen et al. 2008). In support of a causative role for cytokines, blockade of cytokine activity through administration of IL-1ra, transplantation of IL-1ra secreting neural precursor cells into the hippocampus or through the use of IL-1 knockout mice, can prevent such changes (Barrientos et al. 2003; Koo and Duman 2008).
3. Stress and cytokines act synergistically and promote several common neurotransmitter, endocrine and behavioural measures and cross sensitisation of the effects of stressors on inflammatory immune system activation or vice versa may predispose to major depressive disorder. In this regard Anisman et al. (2008) propose that the CNS may perceive immune activation as a stressor and by virtue of stressor like effects may contribute to mood disorder vulnerability. For example, it has been reported that when an immune challenge (e.g. IL-1β, LPS or poly I:C) is administered against a backdrop of psychosocial stress, the central transmitter and endocrine changes ordinarily elicited by immune challenge are greatly enhanced coupled with a marked enhancement of pro-inflammatory cytokine expression in hypothalamic and extra-hypothalamic regions (Anisman et al. 2007; Gandhi et al. 2007; Gibb et al. 2008). Further work is required to clarify if sensitisation may also occur to behavioural changes upon exposure to a cytokine following a period of stress or vice versa.

4. Conditions of chronic inflammation are associated with priming of the CNS, e.g. age, diabetes or prion disease exacerbate signs of sickness and neurovegetative, somatic and psychological symptoms of depression that occur in response to acute peripheral inflammation. Such an exacerbation is associated with an exaggerated inflammatory response in the brain in response to peripheral immune activation and is proposed to reveal a primed state of the microglial compartment. Priming of macrophage in the periphery has been characterised, e.g. increased TLR4 signalling and NFκB activation in response to LPS following an initial exposure to IFNγ. Whilst the priming of microglia is also thought to occur, it is less well characterised and is considered to be associated with increased IL-1β, reduced IL-10 and IL-1 type II receptor and reduced IL-1ra, increased leucocytes and microglia in the brain coupled with mast cell migration to the perivascular space (Dantzer 2009).
5. Microglia have been reported to show an increased expression of IDO following treatment with TNFα and IFNγ (O'Connor et al. 2009a). Pre-treatment of mice with the microglial inhibitor minocycline was shown to attenuate LPS-induced increases of pro-inflammatory cytokines in the CNS, sickness behaviour and anhedonia (Henry et al. 2008). Microglia preferentially express kynurenine monooxygenase (KMO) which produces the NMDA receptor agonist QUIN, facilitating prolonged signalling of this receptor in the brain (Heyes et al. 1996). Recent studies involving positron emission topography (PET) imaging have been able to show microglial activation states in living brain tissue. Mice injected with LPS were shown to have activated microglia on day 2 and day 3 following LPS administration, which was accompanied by depressive-like behaviour on day 3 and an increased kynurenine/tryptophan ratio in the serum (Dobos et al. 2012).
6. Although a number of animal models support a role for inflammatory mediators in the induction of depression-like behaviours, it is also of interest to consider the possibility that cytokine-induced depression may occur in the absence of a direct immune stimulus. In this regard there is evidence from animal studies that stress promotes cytokine expression and microglial activation in the CNS through the induction of the pro-inflammatory cytokine IL-1β and the microglial activation marker CD11b, respectively, and inhibition of IL-1β receptors using IL-1ra blocks the ability of stress to cause cognitive dysfunction (Pugh et al. 1999). In a separate report, exposure of mice to CMS increases central IL-6 expression without the use of an immune stimulus to induce the cytokine response (Mormède et al. 2003). C57BL/6J mice subjected to a chronic prolonged restraint stress displayed a decrease in the expression of the anti-inflammatory cytokine IL-10 centrally accompanied by increased total immobility and decreased time to first period of immobility in the forced swim test paradigm. These depressive-like behavioural deficits were reversed via administration of recombinant IL-10 (Voorhees et al. 2013). The functional significance of such changes for the pathophysiology of stress or depression remains to be fully determined.

IL-1ra can inhibit stress-induced increases in hypothalamic NA, dopamine and 5-HT concentrations and stress-induced HPA-axis activation in rats (Shintani et al. 1995). These data suggest that IL-1β can mediate behavioural, neurochemical and endocrine effects of stress. Goshen et al. (2008) have shown that brain IL-1β mediates chronic stress-induced depression in mice via adrenocortical activation and the suppression of hippocampal neurogenesis. Mice subjected to CMS displayed increased depression related behaviours, increased circulating levels of corticosterone coupled with an increase in IL-1β concentrations in the hippocampus. Mice with a deletion of the IL-1 receptor type 1 (IL-1rKO) or mice with transgenic, brain restricted over-expression of IL-1ra did not display CMS-induced behavioural, neuroendocrine or neurogenesis changes. Removal of endogenous glucocorticoids by adrenalectomy abolished the depressive-like behaviours induced by CMS suggestive of a causal role of the reduced corticosterone response observed in IL-1rKO mice in their resistance to CMS-induced depression.

Given the ability of stress to activate IL-1β in the brain, it is conceivable that cytokines like IL-1β may be of relevance to the neurotrophic hypothesis of depression which proposes that stress-induced depression is mediated by a suppression of hippocampal BDNF that is counteracted following chronic antidepressant treatment (Duman et al. 1997). Such changes in BDNF have been proposed to underlie reductions in hippocampal volume and neuronal atrophy associated with depression. As pro-inflammatory cytokines increase glucocorticoid concentrations which in turn suppress the expression of growth factors in the brain including BDNF, it is not unreasonable to suggest that stress-induced activation of IL-1β may be related to reductions in hippocampal BDNF. In support of this mechanism it has previously been reported that systemic LPS administration to rats reduces BDNF expression in cortex and hippocampus (Guan and Fang 2006). Repeated administration of LPS into the hippocampus reduces mRNA expression of BDNF and its receptor TrkB (Tanaka et al. 2006). Moreover reductions in BDNF expression produced by social isolation in rats are blocked by intrahippocampal administration of IL-1ra (Barrientos et al. 2003). It is also of note that BDNF has also been reported to be reduced in the hippocampus of rats following central administration of IFNα which was mediated by IL-1β (Kaneko et al. 2006) indicating that other cytokines may converge on IL-1β to mediate changes in hippocampal BDNF.

The maternal immune activation (MIA) is a model in which a pregnant female is administered an immune stimulus, such as LPS or poly I:C, which subsequently leads to raised circulating pro-inflammatory cytokines, such as TNFα, IL-1β and IL-6, in the pregnant dam and offspring in utero (Boksa 2010; Patterson 2009; Wang et al. 2006). Changes in the levels of BDNF and other growth factors have been subsequently observed in the neonatal brain (Gilmore et al. 2003). The offspring have also been reported to have increased neuronal cell death, reduced neurogenesis and reduced brain size (Boksa 2010; Wang et al. 2006). Administration of pro-inflammatory cytokines individually is sufficient to induce MIA. IL-6 has been reported to cause deficits in offspring, including impaired spatial learning, neuronal loss, astrogliosis and changes in neurotransmitter expression (Samuelsson et al. 2006). Treatment with anti-IL-6 antibodies protected against these developmental and behavioural abnormalities and IL-6KO subjected to an immune stimulus were also protected against abnormal behavioural developments (Patterson 2009; Smith et al. 2007). Treatment with the anti-inflammatory cytokine IL-10 has been reported to protect against white matter damage observed in MIA (Pang et al. 2005). Similarly administration of anti-TNFα antibodies to the pregnant dam following LPS challenge reduced foetal death (Silver et al. 1994), and use of the drug pentoxifylline, to block TNFα synthesis, reduced foetal mortality and reversed previously observed growth and skeletal development restrictions of the foetus (Xu et al. 2006). The brain abnormalities and the associated behaviours, such as increased anxiety, impaired locomotor activity, altered social behaviour and deficits in learning and memory, associated with MIA appear to best mimic symptoms associated with schizophrenia (Meyer and Feldon 2009). In agreement with this, MIA leads to hyper-functionality of the dopaminergic system in the offspring, including increased tyrosine hydroxylase and dopamine transporter expression in neuronal cells (Meyer and Feldon 2009; Bakos et al. 2004). A recent study by Khan et al. (2014) reported an effect of poly I:C MIA on depressive-like behaviours in adult mouse offspring, accompanied by deficits in cognition and hippocampal long-term potentiation. Thus this provides a suitable model for assessing the long-term behavioural effects of an inflammatory stimulus during development with long-term phenotypic consequences.

An Emerging Role of Microglia?

The role of microglia in mediating inflammation in the brain following exposure to psychological stress has been substantiated by Blandino et al. (2006) who demonstrated that pre-treatment with the microglial activation inhibitor minocycline substantially attenuated the increase in IL-1β found following inescapable shock. Chronic stress has been reported to increase the expression of microglial activation markers and reduce the expression of glucocorticoid inducible genes in mice indicating glucocorticoid insensitivity. These changes, induced by repeated social defeat, were reversed by administration of propanolol suggesting a link between the SAM-axis, a GR signalling impairment and immune activation within the CNS (Wohleb et al. 2011). Increased microglial activation was also reported in a study by Couch et al. (2013), in stress sensitive C57BL/6J mice who displayed anhedonic and depressive-like behaviours following exposure to a chronic stress paradigm that included exposure to predation and restraint stress. When compared to stress resilient mice, increased TNFα, IDO and SERT expression were reported in the frontal cortex of stress susceptible mice (Couch et al. 2013). In rats, elevated TNFα concentrations have been reported in the CNS of rats subjected to a prolonged restraint stress (Madrigal et al. 2002), while an inescapable foot shock stress leads to an increase in the expression of the microglial activation markers CD11b and MHC II in the CNS and a down-regulation of CD200, which helps to keep microglia in a quiescent state (Frank et al. 2007). These changes were also accompanied by an increased response of hippocampal microglia to LPS. In addition immobilisation stress promotes changes to the number of microglia (Nair and Bonneau 2006) and the morphology of microglia from a ramified resting state to activated state that occurs in a number of stress sensitive regions of the brain (Sugama et al. 2007; Tynan et al. 2010). Stress-induced activation of cytokine responses in the CNS appears to be largely dependent on activation of microglia (Frank et al. 2007).

However despite these interesting leads from animal experiments indicating that stress can promote a pro-inflammatory phenotype and microglial activation in the CNS in the absence of an overt immune stimulus, to date there is no solid evidence of microglial activation or increased CNS cytokines in depressed humans. There is increasing recognition that psychosocial stress can activate the inflammatory response system in the periphery albeit with limited evidence for activation of central inflammatory pathways within clinical populations of relevance for depression. NFκB and IL-6 responses to psychosocial stress have been shown to be exaggerated in patients with depression (Pace et al. 2006; Glaser et al. 2003). In addition there are numerous reports that chronic stress in humans is associated with increases in the acute phase protein, CRP as well as IL-6 and other inflammatory markers (Kiecolt-Glaser et al. 2005; McDade et al. 2006; Miller 2008). There is evidence for the loss of glial elements including oligodendrocytes and astrocytes in multiple mood-relevant brain regions including the subgenual prefrontal cortex and amygdala which has emerged as a morphological abnormality in major depression (Ongür et al. 1998; Hamidi et al. 2004). Moreover recent post mortem data suggests that suicide is associated with increased microglial numbers in the prefrontal cortex (Steiner et al. 2008).

Microglia, when activated, are involved in the regulation of neuroinflammation but in the steady state they have important functions assisting in neurodevelopment, including synapse formation and pruning (Tremblay et al. 2011). There are now several lines of evidence showing that this physiological function in particular is perturbed in pre-psychiatric and psychiatric disease, resulting in an abnormal neurodevelopment and neuronal function due to an activated state of cells of the mononuclear phagocyte system (MPS), including the microglia in the brain. This concept is generally referred to as the "Macrophage-T cell or Immune Theory of Major Mental Disorders" first proposed in 1992 and supported by numerous reports of aberrant levels of inflammatory cytokines in the serum, plasma and CSF of patients with major mental illnesses (Beumer et al. 2012; Smith 1992). Monocytic activation has been reported to precede the onset of symptoms indicating that immune parameters may serve as biomarkers for an at-risk state for major mental illness (Drexhage et al. 2010). Histopathological studies, though limited and carried out on post mortem material, also indicate an abnormal activation of microglia in patients with major psychiatric disorders (Steiner et al. 2011). Moreover it has been proposed that microglial activation may underlie white matter integrity disturbances which accompany illness episodes over the patient's lifetime (Benedetti et al. 2011). Utilisation of diffusion tensor magnetic resonance imaging (DT-MRI) would allow for characterisation of white matter tracts in rodent models in a longitudinal fashion.

Implications for Treating Depression and Stress Related Disorders

The immunological activation state evident in some patients may be useful in predicting the response of patients to drug treatment. Patients with an activated immune state are less likely to respond to regular therapy and it is hypothesised that treatment resistant patients will respond to conventional treatment in combination with immunosuppressive therapy expected to dampen the inflammatory state of non-responders. Monocyte and microglial activation may be reversed by administration of immunosuppressive drugs including non-steroidal anti-inflammatories (NSAIDs) (Müller 2010; Sommer et al. 2012), *N*-acetylcysteine (NAC) (Dean et al. 2011) and minocycline (Levkovitz et al. 2010), although such treatments have not been systematically investigated in psychiatric disorders, with further potential for novel immunosuppressive interventions. The mechanisms responsible for the activation of microglia in psychiatric disease remain elusive. A direct microbe-driven activation of microglia is possible (Kristensson 2011), but apart from direct microbial/gDNA activation of microglia, inborn errors in the growth and differentiation of MPS cells can make the cells vulnerable for hyper-stimulation. Use of NSAIDs, NAC, minocycline, and novel immunomodulating drugs in the previously mentioned models could be examined for effectiveness in rectifying the behavioural abnormalities. Models will enable in-depth studies on the molecular mechanism of immune mediated behavioural abnormalities and their correction by drug treatment.

Understanding how stress and emotional distress modulates brain–immune–endocrine interactions has clear translational implications from laboratory data. Clearly there are numerous research opportunities in this area which would advance our current knowledge. The identification of genetic risk factors and the role that genetics might have in these complex relationships is largely unexplored territory to date. As the technology to manipulate the genome has come of age, such questions can be adopted from the clinic and readily explored in animal models. Strategies which attempt to address wide variation within animal studies by subdividing populations according to behavioural or physiological characteristics, e.g. coping styles, will also help to unravel factors underlying susceptibility or vulnerability. Additional research employing stressors with a greater degree of ecological validity which challenge the natural defence and/or adaptive capacity of animals, e.g. social stress, rather than pushing animals towards a physiological ceiling will serve to increase the face validity and relevance of existing models and practices (Koolhaas 2008). Moreover an investigation of the influence of early life stress, and its long-term impact on the developing endocrine and immune systems, is an important question that has not been well studied to date. The concept that stress may predispose to a premature ageing of the immune response has been proposed (Glaser and Kiecolt-Glaser 2005) and to explore this further, a greater emphasis on longitudinal investigations is required.

Leading on directly from developments with animal models, there is great potential for the development of biomarkers of inflammation to reflect pathophysiological mechanisms linking inflammation to sickness and depression. Such biomarkers would help to address the current inadequacy of psychological scales and over reliance on psychiatric diagnosis especially for the neurovegetative symptoms associated with depression. The availability of peripheral biomarkers that can both identify patients with specific pathophysiology and serve to objectively monitor therapeutic responses would represent a major advance in the diagnosis and treatment of depression. There are new possibilities for treatments that target pathways by which the immune system influences the brain such as cytokines or growth factors and their downstream mediators or the activation of relevant CNS immune cell types (e.g. microglia) to emerge. Results from trials to determine the efficacy of anti-inflammatory drugs such as the use of anti-TNFα in patients with psoriasis (Kannan et al. 2013) with antidepressant potential or the adjunctive use of COX-2 inhibitors for treating depression (Müller 2010) have been encouraging. Moreover treatments addressing the influence of stress and stress-induced activation of the SAM and HPA axes including behavioural interventions that address psychological and autonomic reactivity to stress such as psychotherapy, exercise and meditation may have efficacy regarding both treatment and prevention of depression (Fig. 3.1).

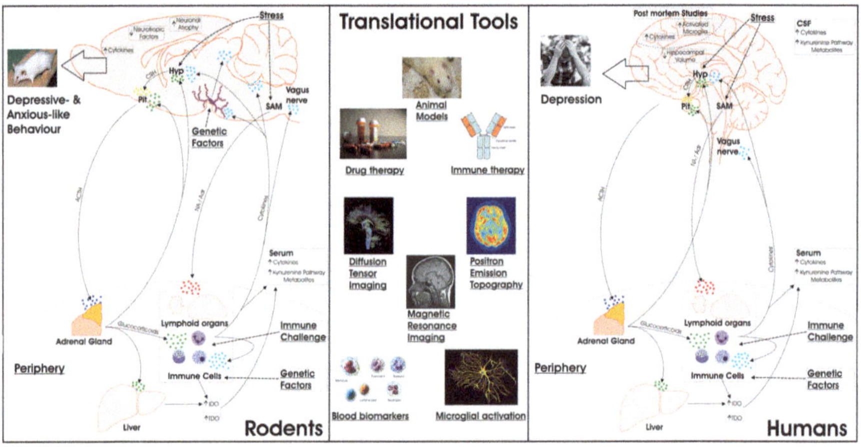

Fig. 3.1 Animal models play an integral role in the development of an understanding of the effect of immune activation and inflammation on behavioural traits. Investigation of mechanisms linking immune–brain communication can be approached in a more invasive way in animal studies to elucidate changes in the brain in response to immune challenge. Such mechanisms include changes to the sensitivity of glucocorticoid/catecholamine receptors and regulation of the hypothalamic pituitary adrenal (HPA) or sympathoadrenal medullary (SAM) axes. Furthermore animals may be selectively bred or genetically engineered to help assist in evaluating a role for immune related mechanisms underlying depression and anxiety-related behaviours with associated neurobiological changes including the expression of neurotrophic factors or neurotransmitters such as serotonin or tryptophan metabolites related to kynurenine. Investigations of the CNS in humans are limited to CSF and post mortem tissue although MR and PET neuroimaging are providing investigators with a window into the brain and parallels may be drawn between outcomes in animal and human investigations. The possibility to translate from animal to human experiments in this way will pave the way for the development of biomarkers related to a role for the immune system in the pathophysiology of mood and anxiety disorders and future development of therapeutics which may target the immune system or brain microglial cells directly for a desirable clinical outcome. *Abbreviations*: *ACTH* adrenocorticotropic hormone, *Adr* adrenaline, *CRH* corticotropin releasing hormone, *CSF* cerebrospinal fluid, *Hyp* hypothalamus, *IDO* indoleamine 2,3 dioxygenase, *NA* noradrenaline, *Pit* pituitary, *SAM* sympathoadrenal medullary axis, *TDO* tryptophan 2,3 dioxygenase

References

Anderson RJ, Freedland KE, Clouse RE, Lustman PJ. The prevalence of comorbid depression in adults with diabetes: a meta-analysis. Diabetes Care. 2001;24(6):1069–78.
Anisman H. Cascading effects of stressors and inflammatory immune system activation: implications for major depressive disorder. J Psychiatry Neurosci. 2009;34(1):4–20.
Anisman H, Kokkinidis L, Merali Z. Further evidence for the depressive effects of cytokines: anhedonia and neurochemical changes. Brain Behav Immun. 2002;16(5):544–56.
Anisman H, Poulter MO, Gandhi R, Merali Z, Hayley S. Interferon-alpha effects are exaggerated when administered on a psychosocial stressor backdrop: cytokine, corticosterone and brain monoamine variations. J Neuroimmunol. 2007;186(1–2):45–53.
Anisman H, Merali Z, Hayley S. Neurotransmitter, peptide and cytokine processes in relation to depressive disorder: comorbidity between depression and neurodegenerative disorders. Prog Neurobiol. 2008;85(1):1–74.
Avitsur R, Stark JL, Dhabhar FS, Padgett DA, Sheridan JF. Social disruption-induced glucocorticoid resistance: kinetics and site specificity. J Neuroimmunol. 2002a;124(1–2):54–61.
Avitsur R, Stark JL, Dhabhar FS, Sheridan JF. Social stress alters splenocyte phenotype and function. J Neuroimmunol. 2002b;132(1–2):66–71.
Avitsur R, Kavelaars A, Heijnen C, Sheridan JF. Social stress and the regulation of tumor necrosis factor-alpha secretion. Brain Behav Immun. 2005;19(4):311–7.
Bakos J, Duncko R, Makatsori A, Pirnik Z, Kiss A, Jezova D. Prenatal immune challenge affects growth, behavior, and brain dopamine in offspring. Ann N Y Acad Sci. 2004;1018:281–7.
Barrientos RM, Sprunger DB, Campeau S, Higgins EA, Watkins LR, Rudy JW, et al. Brain-derived neurotrophic factor mRNA downregulation produced by social isolation is blocked by intrahippocampal interleukin-1 receptor antagonist. Neuroscience. 2003;121(4):847–53.
Bawa-Khalfe T, Altememi GF, Mandyam CD, Schwarz LA, Eikenburg DC, Standifer KM. The presence of beta2-adrenoceptors sensitizes alpha2A-adrenoceptors to desensitization after chronic epinephrine treatment. BMC Pharmacol. 2007;7:16.
Benedetti F, Yeh PH, Bellani M, Radaelli D, Nicoletti MA, Poletti S, et al. Disruption of white matter integrity in bipolar depression as a possible structural marker of illness. Biol Psychiatry. 2011;69(4):309–17.
Beumer W, Gibney SM, Drexhage RC, Pont-Lezica L, Doorduin J, Klein HC, et al. The immune theory of psychiatric diseases: a key role for activated microglia and circulating monocytes. J Leukoc Biol. 2012;92(5):959–75.
Blanchard RJ, Nikulina JN, Sakai RR, McKittrick C, McEwen B, Blanchard DC. Behavioral and endocrine change following chronic predatory stress. Physiol Behav. 1998;63(4):561–9.
Blandino P, Barnum CJ, Deak T. The involvement of norepinephrine and microglia in hypothalamic and splenic IL-1beta responses to stress. J Neuroimmunol. 2006;173(1–2):87–95.
Bluthe RM, Laye S, Michaud B, Combe C, Dantzer R, Parnet P. Role of interleukin-1beta and tumour necrosis factor-alpha in lipopolysaccharide-induced sickness behaviour: a study with interleukin-1 type I receptor-deficient mice. Eur J Neurosci. 2000;12(12):4447–56.
Boksa P. Effects of prenatal infection on brain development and behavior: a review of findings from animal models. Brain Behav Immun. 2010;24(6):881–97.
Burton MD, Sparkman NL, Johnson RW. Inhibition of interleukin-6 trans-signaling in the brain facilitates recovery from lipopolysaccharide-induced sickness behavior. J Neuroinflammation. 2011;8:54.
Butterweck V, Prinz S, Schwaninger M. The role of interleukin-6 in stress-induced hyperthermia and emotional behaviour in mice. Behav Brain Res. 2003;144(1–2):49–56.
Cai W, Khaoustov VI, Xie Q, Pan T, Le W, Yoffe B. Interferon-alpha-induced modulation of glucocorticoid and serotonin receptors as a mechanism of depression. J Hepatol. 2005;42(6):880–7.
Capuron L, Ravaud A, Dantzer R. Early depressive symptoms in cancer patients receiving interleukin 2 and/or interferon alfa-2b therapy. J Clin Oncol. 2000;18(10):2143–51.
Capuron L, Raison CL, Musselman DL, Lawson DH, Nemeroff CB, Miller AH. Association of exaggerated HPA axis response to the initial injection of interferon-alpha with development of depression during interferon-alpha therapy. Am J Psychiatry. 2003;160(7):1342–5.

Capuron L, Ravaud A, Miller AH, Dantzer R. Baseline mood and psychosocial characteristics of patients developing depressive symptoms during interleukin-2 and/or interferon-alpha cancer therapy. Brain Behav Immun. 2004;18(3):205–13.

Carvalho LA, Bergink V, Sumaski L, Wijkhuijs J, Hoogendijk WJ, Birkenhager TK, et al. Inflammatory activation is associated with a reduced glucocorticoid receptor alpha/beta expression ratio in monocytes of inpatients with melancholic major depressive disorder. Transl Psychiatry. 2014;4:e344.

Chen B, Dowlatshahi D, MacQueen GM, Wang JF, Young LT. Increased hippocampal BDNF immunoreactivity in subjects treated with antidepressant medication. Biol Psychiatry. 2001;50(4):260–5.

Chung IY, Benveniste EN. Tumor necrosis factor-alpha production by astrocytes. Induction by lipopolysaccharide, IFN-gamma, and IL-1 beta. J Immunol. 1990;144(8):2999–3007.

Cohen S, Janicki-Deverts D, Doyle WJ, Miller GE, Frank E, Rabin BS, et al. Chronic stress, glucocorticoid receptor resistance, inflammation, and disease risk. Proc Natl Acad Sci U S A. 2012;109(16):5995–9.

Connor TJ, Brewer C, Kelly JP, Harkin A. Acute stress suppresses pro-inflammatory cytokines TNF-alpha and IL-1 beta independent of a catecholamine-driven increase in IL-10 production. J Neuroimmunol. 2005;159(1–2):119–28.

Couch Y, Anthony DC, Dolgov O, Revischin A, Festoff B, Santos AI, et al. Microglial activation, increased TNF and SERT expression in the prefrontal cortex define stress-altered behaviour in mice susceptible to anhedonia. Brain Behav Immun. 2013;29:136–46.

Curtin NM. Underlying stress-induced changes in the cytokine network: implications for cancer progression. Trinity College Dublin: University of Dublin; 2008.

Curtin NM, Boyle NT, Mills KH, Connor TJ. Psychological stress suppresses innate IFN-gamma production via glucocorticoid receptor activation: reversal by the anxiolytic chlordiazepoxide. Brain Behav Immun. 2009a;23(4):535–47.

Curtin NM, Mills KH, Connor TJ. Psychological stress increases expression of IL-10 and its homolog IL-19 via beta-adrenoceptor activation: reversal by the anxiolytic chlordiazepoxide. Brain Behav Immun. 2009b;23(3):371–9.

Dantzer R. Cytokine, sickness behavior, and depression. Immunol Allergy Clin North Am. 2009;29(2):247–64.

Dantzer R, O'Connor JC, Freund GG, Johnson RW, Kelley KW. From inflammation to sickness and depression: when the immune system subjugates the brain. Nat Rev Neurosci. 2008;9(1):46–56.

De Kloet ER, Vreugdenhil E, Oitzl MS, Joëls M. Brain corticosteroid receptor balance in health and disease. Endocr Rev. 1998;19(3):269–301.

Dean O, Giorlando F, Berk M. N-acetylcysteine in psychiatry: current therapeutic evidence and potential mechanisms of action. J Psychiatry Neurosci. 2011;36(2):78–86.

Dobos N, de Vries EF, Kema IP, Patas K, Prins M, Nijholt IM, et al. The role of indoleamine 2,3-dioxygenase in a mouse model of neuroinflammation-induced depression. J Alzheimers Dis. 2012;28(4):905–15.

Dowlati Y, Herrmann N, Swardfager W, Liu H, Sham L, Reim EK, et al. A meta-analysis of cytokines in major depression. Biol Psychiatry. 2010;67(5):446–57.

Drexhage RC, Knijff EM, Padmos RC, Heul-Nieuwenhuijzen L, Beumer W, Versnel MA, et al. The mononuclear phagocyte system and its cytokine inflammatory networks in schizophrenia and bipolar disorder. Expert Rev Neurother. 2010;10(1):59–76.

Duman RS, Heninger GR, Nestler EJ. A molecular and cellular theory of depression. Arch Gen Psychiatry. 1997;54(7):597–606.

Dunn AJ, Swiergiel AH. Effects of interleukin-1 and endotoxin in the forced swim and tail suspension tests in mice. Pharmacol Biochem Behav. 2005;81(3):688–93.

Dunn AJ, Swiergiel AH, de Beaurepaire R. Cytokines as mediators of depression: what can we learn from animal studies? Neurosci Biobehav Rev. 2005;29(4–5):891–909.

Edgar VA, Cremaschi GA, Sterin-Borda L, Genaro AM. Altered expression of autonomic neurotransmitter receptors and proliferative responses in lymphocytes from a chronic mild stress model of depression: effects of fluoxetine. Brain Behav Immun. 2002;16(4):333–50.

Edgar VA, Silberman DM, Cremaschi GA, Zieher LM, Genaro AM. Altered lymphocyte catecholamine reactivity in mice subjected to chronic mild stress. Biochem Pharmacol. 2003;65(1):15–23.
Eggert M, Zettl UK, Neeck G. Autoantibodies in autoimmune diseases. Curr Pharm Des. 2010;16(14):1634–43.
Elenkov IJ, Papanicolaou DA, Wilder RL, Chrousos GP. Modulatory effects of glucocorticoids and catecholamines on human interleukin-12 and interleukin-10 production: clinical implications. Proc Assoc Am Physicians. 1996;108(5):374–81.
Feinstein DL, Heneka MT, Gavrilyuk V, Dello Russo C, Weinberg G, Galea E. Noradrenergic regulation of inflammatory gene expression in brain. Neurochem Int. 2002;41(5):357–65.
Fortier ME, Kent S, Ashdown H, Poole S, Boksa P, Luheshi GN. The viral mimic, polyinosinic:polycytidylic acid, induces fever in rats via an interleukin-1-dependent mechanism. Am J Physiol Regul Integr Comp Physiol. 2004;287(4):R759–66.
Frank MG, Baratta MV, Sprunger DB, Watkins LR, Maier SF. Microglia serve as a neuroimmune substrate for stress-induced potentiation of CNS pro-inflammatory cytokine responses. Brain Behav Immun. 2007;21(1):47–59.
Frenois F, Moreau M, O'Connor J, Lawson M, Micon C, Lestage J, et al. Lipopolysaccharide induces delayed FosB/DeltaFosB immunostaining within the mouse extended amygdala, hippocampus and hypothalamus, that parallel the expression of depressive-like behavior. Psychoneuroendocrinology. 2007;32(5):516–31.
Fujioka H, Akema T. Lipopolysaccharide acutely inhibits proliferation of neural precursor cells in the dentate gyrus in adult rats. Brain Res. 2010;1352:35–42.
Gandhi R, Hayley S, Gibb J, Merali Z, Anisman H. Influence of poly I:C on sickness behaviors, plasma cytokines, corticosterone and central monoamine activity: moderation by social stressors. Brain Behav Immun. 2007;21(4):477–89.
Gibb J, Hayley S, Gandhi R, Poulter MO, Anisman H. Synergistic and additive actions of a psychosocial stressor and endotoxin challenge: circulating and brain cytokines, plasma corticosterone and behavioral changes in mice. Brain Behav Immun. 2008;22(4):573–89.
Gibb J, Hayley S, Poulter MO, Anisman H. Effects of stressors and immune activating agents on peripheral and central cytokines in mouse strains that differ in stressor responsivity. Brain Behav Immun. 2011;25(3):468–82.
Gibney SM, Drexhage HA. Evidence for a dysregulated immune system in the etiology of psychiatric disorders. J Neuroimmune Pharmacol. 2013;8(4):900–20.
Gibney SM, McGuinness B, Prendergast C, Harkin A, Connor TJ. Poly I:C-induced activation of the immune response is accompanied by depression and anxiety-like behaviours, kynurenine pathway activation and reduced BDNF expression. Brain Behav Immun. 2013;28:170–81.
Gilmore JH, Jarskog LF, Vadlamudi S. Maternal infection regulates BDNF and NGF expression in fetal and neonatal brain and maternal-fetal unit of the rat. J Neuroimmunol. 2003;138(1–2):49–55.
Gimeno D, Kivimäki M, Brunner EJ, Elovainio M, De Vogli R, Steptoe A, et al. Associations of C-reactive protein and interleukin-6 with cognitive symptoms of depression: 12-year follow-up of the Whitehall II study. Psychol Med. 2009;39(3):413–23.
Glaser R, Kiecolt-Glaser JK. Stress-induced immune dysfunction: implications for health. Nat Rev Immunol. 2005;5(3):243–51.
Glaser R, Robles TF, Sheridan J, Malarkey WB, Kiecolt-Glaser JK. Mild depressive symptoms are associated with amplified and prolonged inflammatory responses after influenza virus vaccination in older adults. Arch Gen Psychiatry. 2003;60(10):1009–14.
Godbout JP, Glaser R. Stress-induced immune dysregulation: implications for wound healing, infectious disease and cancer. J Neuroimmune Pharmacol. 2006;1(4):421–7.
Gold SM, Raji A, Huitinga I, Wiedemann K, Schulz KH, Heesen C. Hypothalamo-pituitary-adrenal axis activity predicts disease progression in multiple sclerosis. J Neuroimmunol. 2005;165(1–2):186–91.
Goleva E, Kisich KO, Leung DY. A role for STAT5 in the pathogenesis of IL-2-induced glucocorticoid resistance. J Immunol. 2002;169(10):5934–40.
Goshen I, Kreisel T, Ben-Menachem-Zidon O, Licht T, Weidenfeld J, Ben-Hur T, et al. Brain interleukin-1 mediates chronic stress-induced depression in mice via adrenocortical activation and hippocampal neurogenesis suppression. Mol Psychiatry. 2008;13(7):717–28.

Graff LA, Walker JR, Bernstein CN. Depression and anxiety in inflammatory bowel disease: a review of comorbidity and management. Inflamm Bowel Dis. 2009;15(7):1105–18.

Guan Z, Fang J. Peripheral immune activation by lipopolysaccharide decreases neurotrophins in the cortex and hippocampus in rats. Brain Behav Immun. 2006;20(1):64–71.

Guillemin GJ, Kerr SJ, Smythe GA, Smith DG, Kapoor V, Armati PJ, et al. Kynurenine pathway metabolism in human astrocytes: a paradox for neuronal protection. J Neurochem. 2001;78(4):842–53.

Hamidi M, Drevets WC, Price JL. Glial reduction in amygdala in major depressive disorder is due to oligodendrocytes. Biol Psychiatry. 2004;55(6):563–9.

Haroon E, Raison CL, Miller AH. Psychoneuroimmunology meets neuropsychopharmacology: translational implications of the impact of inflammation on behavior. Neuropsychopharmacology. 2012;37(1):137–62.

Henry CJ, Huang Y, Wynne A, Hanke M, Himler J, Bailey MT, et al. Minocycline attenuates lipopolysaccharide (LPS)-induced neuroinflammation, sickness behavior, and anhedonia. J Neuroinflammation. 2008;5:15.

Herman JP, Adams D, Prewitt C. Regulatory changes in neuroendocrine stress-integrative circuitry produced by a variable stress paradigm. Neuroendocrinology. 1995;61(2):180–90.

Heyes MP, Achim CL, Wiley CA, Major EO, Saito K, Markey SP. Human microglia convert l-tryptophan into the neurotoxin quinolinic acid. Biochem J. 1996;320(Pt 2):595–7.

Howren MB, Lamkin DM, Suls J. Associations of depression with C-reactive protein, IL-1, and IL-6: a meta-analysis. Psychosom Med. 2009;71(2):171–86.

Huang YH, Bergles DE. Glutamate transporters bring competition to the synapse. Curr Opin Neurobiol. 2004;14(3):346–52.

Hughes MM, Carballedo A, McLoughlin DM, Amico F, Harkin A, Frodl T, et al. Tryptophan depletion in depressed patients occurs independent of kynurenine pathway activation. Brain Behav Immun. 2012;26(6):979–87.

Hurwitz EL, Morgenstern H. Cross-sectional associations of asthma, hay fever, and other allergies with major depression and low-back pain among adults aged 20–39 years in the United States. Am J Epidemiol. 1999;150(10):1107–16.

Irwin MR. Human psychoneuroimmunology: 20 years of discovery. Brain Behav Immun. 2008;22(2):129–39.

Isik A, Koca SS, Ozturk A, Mermi O. Anxiety and depression in patients with rheumatoid arthritis. Clin Rheumatol. 2007;26(6):872–8.

Itoh CE, Kizaki T, Hitomi Y, Hanawa T, Kamiya S, Ookawara T, et al. Down-regulation of beta2-adrenergic receptor expression by exercise training increases IL-12 production by macrophages following LPS stimulation. Biochem Biophys Res Commun. 2004;322(3):979–84.

Jankord R, Zhang R, Flak JN, Solomon MB, Albertz J, Herman JP. Stress activation of IL-6 neurons in the hypothalamus. Am J Physiol Regul Integr Comp Physiol. 2010;299(1):R343–51.

Johnson JD, Campisi J, Sharkey CM, Kennedy SL, Nickerson M, Greenwood BN, et al. Catecholamines mediate stress-induced increases in peripheral and central inflammatory cytokines. Neuroscience. 2005;135(4):1295–307.

Kaneko N, Kudo K, Mabuchi T, Takemoto K, Fujimaki K, Wati H, et al. Suppression of cell proliferation by interferon-alpha through interleukin-1 production in adult rat dentate gyrus. Neuropsychopharmacology. 2006;31(12):2619–26.

Kang HJ, Adams DH, Simen A, Simen BB, Rajkowska G, Stockmeier CA, et al. Gene expression profiling in postmortem prefrontal cortex of major depressive disorder. J Neurosci. 2007;27(48):13329–40.

Kannan S, Heller MM, Lee ES, Koo JY. The role of tumor necrosis factor-alpha and other cytokines in depression: what dermatologists should know. J Dermatolog Treat. 2013;24(2):148–52.

Karson A, Demirtaş T, Bayramgürler D, Balci F, Utkan T. Chronic administration of infliximab (TNF-α inhibitor) decreases depression and anxiety-like behaviour in rat model of chronic mild stress. Basic Clin Pharmacol Toxicol. 2013;112(5):335–40.

Katon W, Sullivan MD. Depression and chronic medical illness. J Clin Psychiatry. 1990;51(Suppl):3–11. discussion 2–4.

Kentner AC, Miguelez M, James JS, Bielajew C. Behavioral and physiological effects of a single injection of rat interferon-alpha on male Sprague-Dawley rats: a long-term evaluation. Brain Res. 2006;1095(1):96–106.
Khan D, Fernando P, Cicvaric A, Berger A, Pollak A, Monje FJ, et al. Long-term effects of maternal immune activation on depression-like behavior in the mouse. Transl Psychiatry. 2014;4:e363.
Kiecolt-Glaser JK, Loving TJ, Stowell JR, Malarkey WB, Lemeshow S, Dickinson SL, et al. Hostile marital interactions, proinflammatory cytokine production, and wound healing. Arch Gen Psychiatry. 2005;62(12):1377–84.
Kitraki E, Karandrea D, Kittas C. Long-lasting effects of stress on glucocorticoid receptor gene expression in the rat brain. Neuroendocrinology. 1999;69(5):331–8.
Konat GW, Borysiewicz E, Fil D, James I. Peripheral challenge with double-stranded RNA elicits global up-regulation of cytokine gene expression in the brain. J Neurosci Res. 2009;87(6):1381–8.
Konstantinos AP, Sheridan JF. Stress and influenza viral infection: modulation of proinflammatory cytokine responses in the lung. Respir Physiol. 2001;128(1):71–7.
Koo JW, Duman RS. IL-1beta is an essential mediator of the antineurogenic and anhedonic effects of stress. Proc Natl Acad Sci U S A. 2008;105(2):751–6.
Koolhaas JM. Coping style and immunity in animals: making sense of individual variation. Brain Behav Immun. 2008;22(5):662–7.
Kristensson K. Microbes' roadmap to neurons. Nat Rev Neurosci. 2011;12(6):345–57.
Larson SJ, Romanoff RL, Dunn AJ, Glowa JR. Effects of interleukin-1beta on food-maintained behavior in the mouse. Brain Behav Immun. 2002;16(4):398–410.
Leonard BE. Inflammation, depression and dementia: are they connected? Neurochem Res. 2007;32(10):1749–56.
Lestage J, Verrier D, Palin K, Dantzer R. The enzyme indoleamine 2,3-dioxygenase is induced in the mouse brain in response to peripheral administration of lipopolysaccharide and superantigen. Brain Behav Immun. 2002;16(5):596–601.
Levine J, Barak Y, Chengappa KN, Rapoport A, Rebey M, Barak V. Cerebrospinal cytokine levels in patients with acute depression. Neuropsychobiology. 1999;40(4):171–6.
Levkovitz Y, Mendlovich S, Riwkes S, Braw Y, Levkovitch-Verbin H, Gal G, et al. A double-blind, randomized study of minocycline for the treatment of negative and cognitive symptoms in early-phase schizophrenia. J Clin Psychiatry. 2010;71(2):138–49.
Lindqvist D, Janelidze S, Hagell P, Erhardt S, Samuelsson M, Minthon L, et al. Interleukin-6 is elevated in the cerebrospinal fluid of suicide attempters and related to symptom severity. Biol Psychiatry. 2009;66(3):287–92.
Liu Y, Ho RC, Mak A. Interleukin (IL)-6, tumour necrosis factor alpha (TNF-α) and soluble interleukin-2 receptors (sIL-2R) are elevated in patients with major depressive disorder: a meta-analysis and meta-regression. J Affect Disord. 2012;139(3):230–9.
Madrigal JL, Hurtado O, Moro MA, Lizasoain I, Lorenzo P, Castrillo A, et al. The increase in TNF-alpha levels is implicated in NF-kappaB activation and inducible nitric oxide synthase expression in brain cortex after immobilization stress. Neuropsychopharmacology. 2002;26(2):155–63.
Maes M. Depression is an inflammatory disease, but cell-mediated immune activation is the key component of depression. Prog Neuropsychopharmacol Biol Psychiatry. 2011;35(3):664–75.
Maes M, Meltzer HY, Bosmans E, Bergmans R, Vandoolaeghe E, Ranjan R, et al. Increased plasma concentrations of interleukin-6, soluble interleukin-6, soluble interleukin-2 and transferrin receptor in major depression. J Affect Disord. 1995;34(4):301–9.
Maes M, Bosmans E, De Jongh R, Kenis G, Vandoolaeghe E, Neels H. Increased serum IL-6 and IL-1 receptor antagonist concentrations in major depression and treatment resistant depression. Cytokine. 1997;9(11):853–8.
Maes M, Lin AH, Delmeire L, Van Gastel A, Kenis G, De Jongh R, et al. Elevated serum interleukin-6 (IL-6) and IL-6 receptor concentrations in posttraumatic stress disorder following accidental man-made traumatic events. Biol Psychiatry. 1999;45(7):833–9.

Maes M, Ringel K, Kubera M, Berk M, Rybakowski J. Increased autoimmune activity against 5-HT: a key component of depression that is associated with inflammation and activation of cell-mediated immunity, and with severity and staging of depression. J Affect Disord. 2012;136(3):386–92.

Maier SF, Watkins LR. Cytokines for psychologists: implications of bidirectional immune-to-brain communication for understanding behavior, mood, and cognition. Psychol Rev. 1998;105(1):83–107.

Martinez JM, Garakani A, Yehuda R, Gorman JM. Proinflammatory and "resiliency" proteins in the CSF of patients with major depression. Depress Anxiety. 2012;29(1):32–8.

Mawdsley JE, Rampton DS. Psychological stress in IBD: new insights into pathogenic and therapeutic implications. Gut. 2005;54(10):1481–91.

McDade TW, Hawkley LC, Cacioppo JT. Psychosocial and behavioral predictors of inflammation in middle-aged and older adults: the Chicago health, aging, and social relations study. Psychosom Med. 2006;68(3):376–81.

McKay LI, Cidlowski JA. Molecular control of immune/inflammatory responses: interactions between nuclear factor-kappa B and steroid receptor-signaling pathways. Endocr Rev. 1999;20(4):435–59.

Merali Z, Brennan K, Brau P, Anisman H. Dissociating anorexia and anhedonia elicited by interleukin-1beta: antidepressant and gender effects on responding for "free chow" and "earned" sucrose intake. Psychopharmacology (Berl). 2003;165(4):413–8.

Meyer U, Feldon J. Neural basis of psychosis-related behaviour in the infection model of schizophrenia. Behav Brain Res. 2009;204(2):322–34.

Miguel-Hidalgo JJ, Waltzer R, Whittom AA, Austin MC, Rajkowska G, Stockmeier CA. Glial and glutamatergic markers in depression, alcoholism, and their comorbidity. J Affect Disord. 2010;127(1–3):230–40.

Miller AH. Elucidating the consequences of chronic stress on immune regulation and behavior in rheumatoid arthritis. Brain Behav Immun. 2008;22(1):22–3.

Miller AH, Maletic V, Raison CL. Inflammation and its discontents: the role of cytokines in the pathophysiology of major depression. Biol Psychiatry. 2009;65(9):732–41.

Minden SL, Schiffer RB. Affective disorders in multiple sclerosis. Review and recommendations for clinical research. Arch Neurol. 1990;47(1):98–104.

Morikawa O, Sakai N, Obara H, Saito N. Effects of interferon-alpha, interferon-gamma and cAMP on the transcriptional regulation of the serotonin transporter. Eur J Pharmacol. 1998;349(2–3):317–24.

Mormède C, Castanon N, Médina C, Dantzer R. Conditioned place aversion with interleukin-1beta in mice is not associated with activation of the cytokine network. Brain Behav Immun. 2003;17(2):110–20.

Müller N. COX-2 inhibitors as antidepressants and antipsychotics: clinical evidence. Curr Opin Investig Drugs. 2010;11(1):31–42.

Muller N, Schwarz MJ. The immune-mediated alteration of serotonin and glutamate: towards an integrated view of depression. Mol Psychiatry. 2007;12(11):988–1000.

Müller N, Dobmeier P, Empl M, Riedel M, Schwarz M, Ackenheil M. Soluble IL-6 receptors in the serum and cerebrospinal fluid of paranoid schizophrenic patients. Eur Psychiatry. 1997;12(6):294–9.

Nair A, Bonneau RH. Stress-induced elevation of glucocorticoids increases microglia proliferation through NMDA receptor activation. J Neuroimmunol. 2006;171(1–2):72–85.

Nance DM, Sanders VM. Autonomic innervation and regulation of the immune system (1987–2007). Brain Behav Immun. 2007;21(6):736–45.

O'Connor JC, Andre C, Wang Y, Lawson MA, Szegedi SS, Lestage J, et al. Interferon-gamma and tumor necrosis factor-alpha mediate the upregulation of indoleamine 2,3-dioxygenase and the induction of depressive-like behavior in mice in response to bacillus Calmette-Guerin. J Neurosci. 2009a;29(13):4200–9.

O'Connor JC, Lawson MA, Andre C, Briley EM, Szegedi SS, Lestage J, et al. Induction of IDO by Bacille Calmette–Guerin is responsible for development of murine depressive-like behavior. J Immunol. 2009b;182(5):3202–12.

O'Connor JC, Lawson MA, Andre C, Moreau M, Lestage J, Castanon N, et al. Lipopolysaccharide-induced depressive-like behavior is mediated by indoleamine 2,3-dioxygenase activation in mice. Mol Psychiatry. 2009c;14(5):511–22.
Olson JK, Miller SD. Microglia initiate central nervous system innate and adaptive immune responses through multiple TLRs. J Immunol. 2004;173(6):3916–24.
Ongür D, Drevets WC, Price JL. Glial reduction in the subgenual prefrontal cortex in mood disorders. Proc Natl Acad Sci U S A. 1998;95(22):13290–5.
Ormerod BK, Hanft SJ, Asokan A, Haditsch U, Lee SW, Palmer TD. PPARγ activation prevents impairments in spatial memory and neurogenesis following transient illness. Brain Behav Immun. 2013;29:28–38.
Pace TW, Miller AH. Cytokines and glucocorticoid receptor signaling. Relevance to major depression. Ann N Y Acad Sci. 2009;1179:86–105.
Pace TW, Mletzko TC, Alagbe O, Musselman DL, Nemeroff CB, Miller AH, et al. Increased stress-induced inflammatory responses in male patients with major depression and increased early life stress. Am J Psychiatry. 2006;163(9):1630–3.
Pace TW, Hu F, Miller AH. Cytokine-effects on glucocorticoid receptor function: relevance to glucocorticoid resistance and the pathophysiology and treatment of major depression. Brain Behav Immun. 2007;21(1):9–19.
Pang Y, Rodts-Palenik S, Cai Z, Bennett WA, Rhodes PG. Suppression of glial activation is involved in the protection of IL-10 on maternal *E. coli* induced neonatal white matter injury. Brain Res Dev Brain Res. 2005;157(2):141–9.
Pariante CM, Miller AH. Glucocorticoid receptors in major depression: relevance to pathophysiology and treatment. Biol Psychiatry. 2001;49(5):391–404.
Pariante CM, Pearce BD, Pisell TL, Sanchez CI, Po C, Su C, et al. The proinflammatory cytokine, interleukin-1alpha, reduces glucocorticoid receptor translocation and function. Endocrinology. 1999;140(9):4359–66.
Pasco JA, Nicholson GC, Williams LJ, Jacka FN, Henry MJ, Kotowicz MA, et al. Association of high-sensitivity C-reactive protein with de novo major depression. Br J Psychiatry. 2010;197(5):372–7.
Patterson PH. Immune involvement in schizophrenia and autism: etiology, pathology and animal models. Behav Brain Res. 2009;204(2):313–21.
Pavlov VA, Tracey KJ. The cholinergic anti-inflammatory pathway. Brain Behav Immun. 2005;19(6):493–9.
Platzer C, Döcke W, Volk H, Prösch S. Catecholamines trigger IL-10 release in acute systemic stress reaction by direct stimulation of its promoter/enhancer activity in monocytic cells. J Neuroimmunol. 2000;105(1):31–8.
Pugh CR, Nguyen KT, Gonyea JL, Fleshner M, Wakins LR, Maier SF, et al. Role of interleukin-1 beta in impairment of contextual fear conditioning caused by social isolation. Behav Brain Res. 1999;106(1–2):109–18.
Quan N, Banks WA. Brain–immune communication pathways. Brain Behav Immun. 2007;21(6):727–35.
Quan N, Avitsur R, Stark JL, He L, Shah M, Caligiuri M, et al. Social stress increases the susceptibility to endotoxic shock. J Neuroimmunol. 2001;115(1–2):36–45.
Quan N, Avitsur R, Stark JL, He L, Lai W, Dhabhar F, et al. Molecular mechanisms of glucocorticoid resistance in splenocytes of socially stressed male mice. J Neuroimmunol. 2003;137(1–2):51–8.
Raison CL, Capuron L, Miller AH. Cytokines sing the blues: inflammation and the pathogenesis of depression. Trends Immunol. 2006;27(1):24–31.
Raison CL, Borisov AS, Majer M, Drake DF, Pagnoni G, Woolwine BJ, et al. Activation of central nervous system inflammatory pathways by interferon-alpha: relationship to monoamines and depression. Biol Psychiatry. 2009;65(4):296–303.
Raison CL, Dantzer R, Kelley KW, Lawson MA, Woolwine BJ, Vogt G, et al. CSF concentrations of brain tryptophan and kynurenines during immune stimulation with IFN-alpha: relationship to CNS immune responses and depression. Mol Psychiatry. 2010;15(4):393–403.

Sadovnick AD, Remick RA, Allen J, Swartz E, Yee IM, Eisen K, et al. Depression and multiple sclerosis. Neurology. 1996;46(3):628–32.
Samuelsson AM, Jennische E, Hansson HA, Holmäng A. Prenatal exposure to interleukin-6 results in inflammatory neurodegeneration in hippocampus with NMDA/GABA(A) dysregulation and impaired spatial learning. Am J Physiol Regul Integr Comp Physiol. 2006;290(5):R1345–56.
Sapolsky RM, Meaney MJ. Maturation of the adrenocortical stress response: neuroendocrine control mechanisms and the stress hyporesponsive period. Brain Res. 1986;396(1):64–76.
Sapolsky RM, Krey LC, McEwen BS. Stress down-regulates corticosterone receptors in a site-specific manner in the brain. Endocrinology. 1984;114(1):287–92.
Sato Y, Suzuki N, Horita H, Wada H, Shibuya A, Adachi H, et al. Effects of long-term psychological stress on sexual behavior and brain catecholamine levels. J Androl. 1996;17(2):83–90.
Schott K, Schaefer JE, Richartz E, Batra A, Eusterschulte B, Klein R, et al. Autoantibodies to serotonin in serum of patients with psychiatric disorders. Psychiatry Res. 2003;121(1):51–7.
Segerstrom SC, Miller GE. Psychological stress and the human immune system: a meta-analytic study of 30 years of inquiry. Psychol Bull. 2004;130(4):601–30.
Seidel A, Arolt V, Hunstiger M, Rink L, Behnisch A, Kirchner H. Cytokine production and serum proteins in depression. Scand J Immunol. 1995;41(6):534–8.
Sheline YI, Gado MH, Kraemer HC. Untreated depression and hippocampal volume loss. Am J Psychiatry. 2003;160(8):1516–8.
Sheppard KA, Phelps KM, Williams AJ, Thanos D, Glass CK, Rosenfeld MG, et al. Nuclear integration of glucocorticoid receptor and nuclear factor-kappaB signaling by CREB-binding protein and steroid receptor coactivator-1. J Biol Chem. 1998;273(45):29291–4.
Shintani F, Nakaki T, Kanba S, Sato K, Yagi G, Shiozawa M, et al. Involvement of interleukin-1 in immobilization stress-induced increase in plasma adrenocorticotropic hormone and in release of hypothalamic monoamines in the rat. J Neurosci. 1995;15(3 Pt 1):1961–70.
Silver RM, Lohner WS, Daynes RA, Mitchell MD, Branch DW. Lipopolysaccharide-induced fetal death: the role of tumor-necrosis factor alpha. Biol Reprod. 1994;50(5):1108–12.
Simon NM, McNamara K, Chow CW, Maser RS, Papakostas GI, Pollack MH, et al. A detailed examination of cytokine abnormalities in Major Depressive Disorder. Eur Neuropsychopharmacol. 2008;18(3):230–3.
Skelly DT, Hennessy E, Dansereau MA, Cunningham C. A systematic analysis of the peripheral and CNS effects of systemic LPS, IL-1B, TNF-α and IL-6 challenges in C57BL/6 mice. PLoS One. 2013;8(7):e69123.
Smith RS. A comprehensive macrophage-T-lymphocyte theory of schizophrenia. Med Hypotheses. 1992;39(3):248–57.
Smith SE, Li J, Garbett K, Mirnics K, Patterson PH. Maternal immune activation alters fetal brain development through interleukin-6. J Neurosci. 2007;27(40):10695–702.
Sommer IE, de Witte L, Begemann M, Kahn RS. Nonsteroidal anti-inflammatory drugs in schizophrenia: ready for practice or a good start? A meta-analysis. J Clin Psychiatry. 2012;73(4):414–9.
Sparkman NL, Buchanan JB, Heyen JR, Chen J, Beverly JL, Johnson RW. Interleukin-6 facilitates lipopolysaccharide-induced disruption in working memory and expression of other proinflammatory cytokines in hippocampal neuronal cell layers. J Neurosci. 2006;26(42):10709–16.
Spencer RL, Miller AH, Moday H, McEwen BS, Blanchard RJ, Blanchard DC, et al. Chronic social stress produces reductions in available splenic type II corticosteroid receptor binding and plasma corticosteroid binding globulin levels. Psychoneuroendocrinology. 1996;21(1):95–109.
Stefanski V. Social stress in laboratory rats: hormonal responses and immune cell distribution. Psychoneuroendocrinology. 2000;25(4):389–406.
Steiner J, Bielau H, Brisch R, Danos P, Ullrich O, Mawrin C, et al. Immunological aspects in the neurobiology of suicide: elevated microglial density in schizophrenia and depression is associated with suicide. J Psychiatr Res. 2008;42(2):151–7.
Steiner J, Walter M, Gos T, Guillemin GJ, Bernstein HG, Sarnyai Z, et al. Severe depression is associated with increased microglial quinolinic acid in subregions of the anterior cingulate gyrus: evidence for an immune-modulated glutamatergic neurotransmission? J Neuroinflammation. 2011;8:94.

Suberville S, Bellocq A, Fouqueray B, Philippe C, Lantz O, Perez J, et al. Regulation of interleukin-10 production by beta-adrenergic agonists. Eur J Immunol. 1996;26(11):2601–5.
Sugama S, Fujita M, Hashimoto M, Conti B. Stress induced morphological microglial activation in the rodent brain: involvement of interleukin-18. Neuroscience. 2007;146(3):1388–99.
Sukoff Rizzo SJ, Neal SJ, Hughes ZA, Beyna M, Rosenzweig-Lipson S, Moss SJ, et al. Evidence for sustained elevation of IL-6 in the CNS as a key contributor of depressive-like phenotypes. Transl Psychiatry. 2012;2:e199.
Tanaka S, Matsunaga H, Kimura M, Tatsumi K, Hidaka Y, Takano T, et al. Autoantibodies against four kinds of neurotransmitter receptors in psychiatric disorders. J Neuroimmunol. 2003;141(1–2):155–64.
Tanaka S, Ide M, Shibutani T, Ohtaki H, Numazawa S, Shioda S, et al. Lipopolysaccharide-induced microglial activation induces learning and memory deficits without neuronal cell death in rats. J Neurosci Res. 2006;83(4):557–66.
Thomas CG, Tian H, Diamond JS. The relative roles of diffusion and uptake in clearing synaptically released glutamate change during early postnatal development. J Neurosci. 2011;31(12):4743–54.
Tliba O, Damera G, Banerjee A, Gu S, Baidouri H, Keslacy S, et al. Cytokines induce an early steroid resistance in airway smooth muscle cells: novel role of interferon regulatory factor-1. Am J Respir Cell Mol Biol. 2008;38(4):463–72.
Tremblay M, Stevens B, Sierra A, Wake H, Bessis A, Nimmerjahn A. The role of microglia in the healthy brain. J Neurosci. 2011;31(45):16064–9.
Tsao CW, Lin YS, Cheng JT, Lin CF, Wu HT, Wu SR, et al. Interferon-alpha-induced serotonin uptake in Jurkat T cells via mitogen-activated protein kinase and transcriptional regulation of the serotonin transporter. J Psychopharmacol. 2008;22(7):753–60.
Tynan RJ, Naicker S, Hinwood M, Nalivaiko E, Buller KM, Pow DV, et al. Chronic stress alters the density and morphology of microglia in a subset of stress-responsive brain regions. Brain Behav Immun. 2010;24(7):1058–68.
Valentine AD, Meyers CA. Neurobehavioral effects of interferon therapy. Curr Psychiatry Rep. 2005;7(5):391–5.
Voorhees JL, Tarr AJ, Wohleb ES, Godbout JP, Mo X, Sheridan JF, et al. Prolonged restraint stress increases IL-6, reduces IL-10, and causes persistent depressive-like behavior that is reversed by recombinant IL-10. PLoS One. 2013;8(3):e58488.
Walker JR, Graff LA, Dutz JP, Bernstein CN. Psychiatric disorders in patients with immune-mediated inflammatory diseases: prevalence, association with disease activity, and overall patient well-being. J Rheumatol Suppl. 2011;88:31–5.
Wang X, Wu H, Miller AH. Interleukin 1alpha (IL-1alpha) induced activation of p38 mitogen-activated protein kinase inhibits glucocorticoid receptor function. Mol Psychiatry. 2004;9(1):65–75.
Wang X, Rousset CI, Hagberg H, Mallard C. Lipopolysaccharide-induced inflammation and perinatal brain injury. Semin Fetal Neonatal Med. 2006;11(5):343–53.
Webster JC, Oakley RH, Jewell CM, Cidlowski JA. Proinflammatory cytokines regulate human glucocorticoid receptor gene expression and lead to the accumulation of the dominant negative beta isoform: a mechanism for the generation of glucocorticoid resistance. Proc Natl Acad Sci U S A. 2001;98(12):6865–70.
Weizman R, Laor N, Podliszewski E, Notti I, Djaldetti M, Bessler H. Cytokine production in major depressed patients before and after clomipramine treatment. Biol Psychiatry. 1994;35(1):42–7.
Wohleb ES, Hanke ML, Corona AW, Powell ND, Stiner LM, Bailey MT, et al. β-Adrenergic receptor antagonism prevents anxiety-like behavior and microglial reactivity induced by repeated social defeat. J Neurosci. 2011;31(17):6277–88.
Xu DX, Chen YH, Wang H, Zhao L, Wang JP, Wei W. Tumor necrosis factor alpha partially contributes to lipopolysaccharide-induced intra-uterine fetal growth restriction and skeletal development retardation in mice. Toxicol Lett. 2006;163(1):20–9.
Zalcman S, Anisman H. Acute and chronic stressor effects on the antibody response to sheep red blood cells. Pharmacol Biochem Behav. 1993;46(2):445–52.

Zalcman S, Minkiewicz-Janda A, Richter M, Anisman H. Critical periods associated with stressor effects on antibody titers and on the plaque-forming cell response to sheep red blood cells. Brain Behav Immun. 1988;2(3):254–66.

Zalcman S, Richter M, Anisman H. Alterations of immune functioning following exposure to stressor-related cues. Brain Behav Immun. 1989;3(2):99–109.

Zhu CB, Blakely RD, Hewlett WA. The proinflammatory cytokines interleukin-1beta and tumor necrosis factor-alpha activate serotonin transporters. Neuropsychopharmacology. 2006;31(10): 2121–31.

Zhu CB, Lindler KM, Owens AW, Daws LC, Blakely RD, Hewlett WA. Interleukin-1 receptor activation by systemic lipopolysaccharide induces behavioral despair linked to MAPK regulation of CNS serotonin transporters. Neuropsychopharmacology. 2010;35(13):2510–20.

Chapter 4
Experimental Human Endotoxemia, Sickness Behavior, and Neuropsychiatric Diseases

Jan-Sebastian Grigoleit, Harald Engler, and Manfred Schedlowski

Abstract Systemic inflammation is among the most prominent and most frequently observed responses of the immune system. From everyday challenges like mild infections to severe forms such as chronic inflammatory diseases or sepsis, inflammatory states can develop in many ways over the lifespan. Over the past few decades, it has become clear that inflammatory events can have a strong impact on brain functions. Inflammatory mediators released by activated immune cells induce not only adaptive behavioral responses such as sickness behavior but also can lead to cognitive impairment and may even promote the development of mood disorders. Since most of these aspects cannot be sufficiently modeled in laboratory animals, there is a need for experimental inflammatory models in humans. Intravenous or intramuscular injection of bacterial lipopolysaccharide (LPS, endotoxin) represents a model to induce transient systemic low-grade inflammation in healthy human subjects. During the past two decades, this model has been increasingly used for psychoneuroimmunology (PNI) research. After a brief introduction into sickness behavior and the history of immune-to-brain research, we will summarize work employing this promising model beginning with the first studies on sleep alterations and ending with studies on social behavior using cutting-edge neuroimaging techniques like functional magnetic resonance imaging (fMRI) or positron emission tomography (PET). We will discuss potential difficulties and limitations of the model and provide implications for future studies. In addition we give a short exposition on the immunologic properties of LPS, information pathways from the immune system to the brain, and on the cytokine hypothesis of depression.

Keywords Cytokines • Endotoxemia • fMRI • Human subjects • Inflammation • Lipopolysaccharide (LPS) • Sickness behavior

J.-S. Grigoleit
Institute of Medical Psychology & Behavioral Immunobiology, University Hospital Essen, University of Duisburg-Essen, Hufelandstr. 55, 45122 Essen, Germany

Laboratory of Neuronal Structure and Function, The Salk Institute for Biological Studies, La Jolla, CA 92037, USA

H. Engler • M. Schedlowski (✉)
Institute of Medical Psychology & Behavioral Immunobiology, University Hospital Essen, University of Duisburg-Essen, Hufelandstr. 55, 45122 Essen, Germany
e-mail: manfred.schedlowski@uk-essen.de

N. Müller et al. (eds.), *Immunology and Psychiatry*, Current Topics in Neurotoxicity 8, DOI 10.1007/978-3-319-13602-8_4

In 2013, a group of scientists from different medical research centers in Canada and the USA published a highly recognized study in which they demonstrated that, although acute inflammatory stresses from different etiologies result in highly consistent genomic responses in humans, the responses in corresponding mouse models poorly mimic human conditions (Seok et al. 2013). As this lack of correlation refers only to genomic responses to inflammation and inflammatory diseases, this suggests that the divide between mice and men multiplies exponentially when neural or behavioral responses are considered. Although we do not doubt the use or necessity of experimental studies on mice or rats, we think it is extremely important to verify data derived from animal studies by replicating it in humans wherever possible. Especially in PNI, which is at the interface of immunobiology, neurobiology, and behavior, a human model always should be favored if available. Here we present a human model of acute inflammation and its effects on neurobehavioral functions (Schedlowski et al. 2014). In addition, the results of the presented studies fit well into the concepts derived from animal studies in the literature.

The Beginnings of Sickness Behavior Research

While PNI is a relatively new branch of science, the notion that infections and states of illness affect our cognitive function is probably as old as the common cold itself. In the past, low mood, fatigue, increased sleep, and other phenomena that today we collectively call sickness behavior were believed to be caused by the infectious agents themselves or seen as collateral damage from the war between the pathogen and immune system.

In the 1940s, Eli Menkin and Paul Beeson induced fever in rabbits using supernatants from activated peritoneal exudate cells, showing for the first time that fever can be induced independently of the presence of pathogens by an endogenous protein, which they called pyrexin or granulocyte pyrogen and which is today known as interleukin (IL)-1 (Dinarello 2002). Later, it was demonstrated that the febrile response is mediated via direct action of IL-1 on hypothalamic brain regions and that IL-1 is also produced in the brain itself upon peripheral immunostimulation (Adler and Joy 1965; Bailey et al. 1976; Breder et al. 1988; Fontana et al. 1984). In the 1980s, evidence mounted that cytokines reaching the brain not only affect thermoregulation and hypothalamic–pituitary–adrenal (HPA) axis recruitment but also behavioral functions such as food intake or sleeping behavior (Krueger et al. 1984; McCarthy et al. 1985). In 1988, Benjamin Hart postulated that the febrile response and the observed behavioral changes in sick animals are evolutionarily conserved, are vital to their survival, and promote immune functioning during infection (Hart 1988). In subsequent years, parallel to the increasing knowledge on cytokines and neuroimmune interactions, research on sickness behavior started to grow and an increasing number of brain functions were demonstrated to be affected by acute and chronic inflammation (Besedovsky and del Rey 1996; Dantzer 2001). By the end of the 1990s, it became clear that immune activation not only affects brain functions directly related to sickness, but also learning and memory in infected or immunostimulated animals (Gibertini 1996; Gibertini et al. 1995; Pugh et al. 1998; Shaw et al. 2001). At the same time, cytokines became of interest to psychiatrists and clinical researchers for causing depression, mental confusion, and sleep alterations when applied for therapeutic purposes (Capuron et al. 2000; Smith et al. 1988; Spath-Schwalbe et al. 1998).

Endotoxin Administration as an Experimental Model to Induce Sickness Behavior and Cognitive Disturbances in Humans

The discovery that cytokine treatment may have psychological effects led to increased interest in how an immune response to an usual inflammatory challenge—like an acute infection—may affect human brain functioning. The use of infected subjects to address that question is limited, because individual differences and the pathogen itself bias the outcome measures, and methodical and ethical limitations constrict the establishment of a systematic and well-controlled study design. In contrast, the application of cytokines in subjects is less problematic but does not closely resemble a naturally occurring immune reaction, which is driven by a complex interplay of a number of various cytokines, all with distinct kinetics and local concentrations. Therefore, a model was needed which induces an inflammatory reaction similar to a systemic infection but without any pathogenic risk. Two models to mimic an acute bacterial infection in humans had previously been used for research or medical purposes: the intravenous injection of purified endotoxin from *Escherichia coli* or *Salmonella abortus equi* and the intramuscular administration of typhoid vaccine containing endotoxin from *Salmonella typhi* (Bahador and Cross 2007). Both of these elicit a transient response of the innate immune system via activation of toll-like receptor (TLR) 4 (see Box 4.1).

The first studies employing endotoxin to determine inflammatory effects on neuropsychological functions in humans focused on sleep. In one of the earliest endotoxin studies in humans, Pollmacher and colleagues reported that the intravenous administration of 0.4 ng LPS/kg of body weight reduced the amount of rapid eye movement (REM) sleep and promoted non-REM sleep during the night (Pollmacher et al. 1993). Subsequent studies reported similar results for daytime sleep as well as dose-dependent effects with significant sleep disruption at high doses and increases in non-REM sleep at low doses of endotoxin (Hermann et al. 1998; Korth et al. 1996; Mullington et al. 2000). In 2001, Reichenberg and colleagues published a highly recognized study in which they demonstrated for the first time that the mild immune response elicited by LPS at a dose of 0.8 ng/kg affects memory performance in humans independently from physical symptoms of sickness (Reichenberg et al. 2001). In addition to a decrease in verbal and non-verbal memory (with acquisition and retrieval within the course of the immune reaction), they also observed a decrease in mood and an increase in state anxiety. All of these changes correlated with those in cytokine secretion, predominantly that of tumor necrosis factor alpha (TNF-α).

Box 4.1 Lipopolysaccharide, Acute Inflammation, and Immune-to-Brain Communication

The acute inflammatory reaction of the innate immune system is initiated immediately after pathogen contact and represents the first line of defense against infections and possesses advantages over adaptive immunity. Cells of the innate immune system identify conserved pathogen-associated molecular patterns (PAMPs) employing soluble, surface-associated, or intracellular pattern recognition receptors (PRRs). PRRs encompass a variety of subfamilies among which the TLR family may be the most prominent. TLRs are integral membrane-bound glycoproteins which can be found in a variety of immune (e.g., monocytes, macrophages, mast cells, dendritic cells) and other cell types (Janeway and Medzhitov 2002). These receptors recognize bacterial and viral products like lipoteichoic acid (TLR-2), double-stranded RNA (TLR-3), or unmethylated CpG (TLR-9) (Akira and Takeda 2004). Among the 13 known TLRs, the one most abundant in the present literature is TLR-4, best known for its ability to detect bacterial lipopolysaccharide (LPS, endotoxin). LPS is an essential cell wall component of gram-negative bacteria like Escherichia or Salmonella and plays a major role in sepsis and a number of infectious diseases. LPS-induced activation of TLR-4 (together with associated CD14, which potentiates the response by orders of magnitude) triggers an MyD88-dependent signaling cascade which, via translocation of the transcription factor NFκB, results in increased expression and release of pro-inflammatory cytokines like IL-1β, IL-6, and TNF-α, initiating the acute inflammatory response (Akira and Takeda 2004). Typically such a response includes the recruitment of immune cells like neutrophils and macrophages, a rise in body temperature, activation of the hypothalamic–pituitary–adrenal (HPA) axis with the release of cortisol, an increase in heart rate, and, in response to overstimulation as observed during sepsis, critical decrease in blood pressure and hemostasis resulting in hemorrhagic shock (Cohen 2002).

The question of how cytokines and other inflammatory markers may access the brain is still the subject of ongoing research. Due to their relatively high molecular weights, cytokines normally cannot cross the blood–brain barrier (BBB). However, at certain sites like the circumventricular organs and the choroid plexus, the BBB is more permeable and allows cytokines to enter brain tissue via volume diffusion (Roth et al. 2004). An active transport via cytokine transporters within the brain endothelium is also possible (Banks 2006). Another important humoral pathway is the reception and transmission of inflammatory signals (cytokines and LPS itself) by brain endothelial cells and perivascular macrophages interacting with each other to release prostaglandin E2 (PGE2), which is mainly synthesized by the enzyme cyclooxygenase 2 (COX-2), a common drug target of non-steroidal anti-inflammatory drugs (NSAIDs). PGE2 activates neurons in different brain stem nuclei projecting to

(continued)

Box 4.1 (continued)

other brain regions and leading to altered temperature regulation and HPA axis activation (Ericsson et al. 1997; Hopkins 2007; Serrats et al. 2010). Under certain circumstances like sepsis or brain inflammation, the BBB may lose integrity and become permeable for cytokines and cells of the peripheral immune system (Banks 2006; Engelhardt 2006).

The second major pathway for the communication of inflammatory signals to the brain is mediated by the vagus nerve which receives signals through cytokine receptors on its peripheral nerve endings throughout the body and sends them to various regions of the brain stem (Bluthe et al. 1994; Hopkins 2007; Watkins et al. 1995). Interestingly, the efferent of the vagus nerve is the nucleus of the solitary tract, which also seems to play a major role in COX-2 mediated immune-to-brain communication, particularly as an essential relay for the inflammation-induced activation of the HPA axis (Ericsson et al. 1994, 1997; Serrats et al. 2010). A number of animal studies have shown that both the neural and the humoral pathways play an important role in communication from the immune system to the central nervous system (CNS) and that one system can only insufficiently compensate for each other (Dantzer et al. 2000).

To elicit an experimental inflammatory response it is possible to use isolated, sterile LPS and to either inject it directly into the circulation or to apply it intramuscularly. Depending on the employed dose, source, and method of administration, this kind of stimulation will result in a transient immune response as described above (Bahador and Cross 2007). Using intravenous administration of LPS, detectable signs of the response, like changes in temperature, numbers of circulating immune cells, or concentration of cytokines normally occur after approximately 1 h, peaking between 1.5 and 4 h and disappearing not later than 8 h post-injection (Bahador and Cross 2007; Grigoleit et al. 2011). The symptoms evoked by experimental endotoxemia at doses up to 2 ng/kg are usually mild (ranging from total absence of any affliction over mild discomfort to slight headache or transient nausea) with no reports on unforeseen side effects or critical incidents. Nevertheless, it needs to be mentioned that this model is not free of risk, since the abovementioned overreactions leading to shock and even multi-organ failure in theory also can be induced by LPS injection. Therefore it is of extreme importance to employ highest measures of security when applying LPS to human subjects. Since the potency of a certain source and batch of LPS cannot be foreseen, preliminary experiments with very low doses should precede every newly planned study employing LPS. For further detail, see methods sections of the papers listed in Tables 4.1 and 4.2.

Cytokine-associated, symptom-independent mood worsening was also observed after the more subtle typhoid vaccination (Strike et al. 2004; Wright et al. 2005) and is among the most consistent and most striking observations in studies employing TLR-4 activation in humans (Dellagioia and Hannestad 2010; Kullmann et al. 2011; Schedlowski et al. 2014; see also Box 4.2).

Box 4.2 Inflammation and Depressive Disorder

Based on the notion that sickness behavior resembles depressive symptomatology (e.g., lowered mood, fatigue, decrease in activity, social withdrawal, anhedonia, altered sleep patterns) it has been frequently discussed as to what extent both phenomena are connected to each other or whether they even may represent two sides of the same coin (Dantzer 2001; Dantzer et al. 2008; Miller et al. 2009). Prolonged pro-inflammatory cytokine release (like in chronic inflammation) is suspected of mediating negative effects on mental health, especially when released in high concentrations (as seen during sepsis or auto-immune disease), or in an inappropriate compartment of the body (e.g., after damage of the blood–brain barrier). Indeed, a number of studies suggest an increased incidence for depressive disorders in persons with chronically increased levels of pro-inflammatory cytokines (Dowlati et al. 2010). In addition, a number of patients receiving therapeutic administration of certain cytokines suffer from neuropsychological disturbances like cognitive impairments or emotional disorder (Capuron and Miller 2004; Spath-Schwalbe et al. 1998). The exact mechanism linking increased levels of cytokines with the etiology of depression is not well defined, however, it is believed that alterations in tryptophan metabolism and its effect on serotonin availability may play a crucial role (Dantzer et al. 2008) as well as up-regulation of the expression and activity of the serotonin transporter (SERT) (Zhu et al. 2006). This is supported by a study demonstrating that the selective serotonin reuptake inhibitor citalopram attenuated endotoxin-induced lassitude (Hannestad et al. 2011). Considering data from fMRI studies, the dopaminergic mesolimbic system is likely to be involved (see Table 4.2 and discussion). Another important system potentially connecting depression and inflammation bi-directionally is the HPA axis. The HPA axis responds to inflammation and the immune system responds to HPA axis activity. However, dysregulation of the HPA axis is known to accompany and promote depression (Pariante and Lightman 2008). Consequently, there is evidence that inflammation can contribute to depression via activation of the HPA axis or through cytokine-induced glucocorticoid-receptor resistance (Holsboer 2000; Pace et al. 2007). Finally, there is a number of depression risk genes also known to be involved in inflammatory processes (Raison and Miller 2013a)

(continued)

Box 4.2 (continued)
leading some authors to the assumption that "depression confers a critical compensatory advantage in immune protection that has offset notable disadvantages of depression for evolutionary fitness" (Anders et al. 2013).

Of note, depressive disorders are not the only psychiatric disorders linked with inflammation. The pathogeneses of schizophrenia, neurodegenerative disease, and age-related cognitive decline are all known to have an inflammatory component (Akiyama et al. 2000; Drexhage et al. 2010; Giovanoli et al. 2013; Weaver et al. 2002).

The Fine Print: What to Keep in Mind When Performing an Endotoxin Study

However, subsequent studies on the impact of inflammation on memory and cognitive functions turned out to be less consistent. In a study performed in our laboratory, we were not able to reproduce the results of Reichenberg et al., as there were no observable differences in memory performance or executive functions despite a profound increase in temperature, heart rate, and levels of pro-inflammatory cytokines (Grigoleit et al. 2010). In contrast, a study by Cohen et al., based on the same sample as the Reichenberg study, reported an enhancement of working memory (Cohen et al. 2003), whereas other studies found no such effect (Brydon et al. 2008; Grigoleit et al. 2010; Krabbe et al. 2005; van den Boogaard et al. 2010). However, our laboratory demonstrated that the dose of the employed immune stimulus may have a profound and quite selective impact on the appearance of cognitive changes. In a placebo-controlled crossover study, we compared cytokine, neuroendocrine, vital, mood, and cognitive responses to a dose of either 0.4 or 0.8 ng of *E. coli* endotoxin per kg of body weight (see figure Box 4.1). Whereas the physiological and neuroendocrine response was the same between both conditions, there were profound differences in mood worsening, anxiety, and cognitive test results. While accuracy and psychomotor speed in a working memory paradigm did not change in response to the lower dose, we observed a significantly improved reaction time after applying the higher dose (Grigoleit et al. 2011). In addition, the source and method of application of the immune stimulus might account for differences between studies; for example, Reichenberg et al. employed 0.8 ng LPS from *S. abortus equi* which resulted in a mean temperature increase of 0.5 °C without alterations in heart rate, whereas in our studies, 0.4 ng/kg *E. coli* LPS caused a rise in temperature of nearly 1 °C and profound increases in heart rate. In Table 4.1, most of the existing studies testing cognitive performance after endotoxin challenge in humans are listed together with the employed source and dosage of endotoxin. However, even the exact serotype and batch of an endotoxin might account for differences in the results, as well as the injection time point, the exact time between injection, type of employed

Table 4.1 Studies investigating changes in behavior and neuropsychological performance following LPS injection in humans

Study	Source and dose of LPS and way of application	N	Behavioral effects
Pollmacher et al. (1993)	*S. abortus equi*, 0.4 ng/kg i.v.	15M	Reduction in wakefulness and REM sleep, increase in non-REM sleep during the night
Korth et al. (1996)	*S. abortus equi* 0.8 ng/kg i.v.	10M	Slight reduction of non-REM sleep in daytime
Mullington et al. (2000)	*S. abortus equi* 0.2 or 0.4 or 0.8 ng/kg i.v.	19M	Non-linear, dose-dependent effects on wakefulness, REM sleep, and non-REM sleep
Reichenberg et al. (2001) and Cohen et al. (2003)	*S. abortus equi*, 0.8 ng/kg i.v.	20M 10M[a]	Decrease in mood, decrease in verbal and non-verbal memory correlating with increases in pro-inflammatory cytokines. Improvement of working memory associated with altered AChE cleavage
Strike et al. (2004)	*S. typhi* 25 μg i.m. (Typhim Vi®)	26X	Mood decrease
Wright et al. (2005)	*S. typhi* 25 μg i.m. (Typhim Vi®)	30M	Mood decrease, correlation with IL-6
Krabbe et al. (2005)	*E. coli* 0.2 ng/kg i.v.	12M	No effects on declarative memory, but negative correlation with IL-6
Brydon et al. (2008) and Harrison et al. (2009a, b)	*S. typhi* 25 μg i.m. (Typhim Vi®)	16M	Fatigue, mental confusion, increased reaction time, increased anxiety, decreased mood
Brydon et al. (2009a, b)	*S. typhi* 25 μg i.m. (Typhim Vi®)	59M	Decreased mood (more decreased when combined with stress, and less decreased in optimists), increased fatigue in combination with stress
Eisenberger et al. (2010b)	*E. coli* 0.8 ng/kg i.v.	39X	Feelings of social disconnection and depressed mood
Grigoleit et al. (2010)	*E. coli* 0.4 ng/kg i.v.	24M	No effect on memory
Van den Boogaard et al. (2010)	*E. coli* 2 ng/kg i.v.	15M	No effect on memory, slight EEG changes indicating increased attention
Hannestad et al. (2011)	*E. coli* 0.8 ng/kg i.v.	11X	Social anhedonia, lassitude, and depressed mood, reduced by pre-treatment with the antidepressant citalopram
Grigoleit et al. (2011)	*E. coli* 0.4 or 0.8 ng/kg i.v.	34M	Dose-dependent mood decrease, dose-dependently impaired emotional memory acquisition and improved reaction speed, no effect on non-emotional memory
Kullmann et al. (2014)	*E. coli* 0.4 ng/kg i.v.	18M	Decreased mood, no effect on social cognition performance

M male subjects, *F* female subjects, *X* mixed group of subjects, *i.v.* intravenously, *i.m.* intramuscular
[a]Cohen-study: sub-group of Reichenberg-study

tests, or the sex of the subjects. Most studies recruited only male volunteers, and to our knowledge there is only one study directly addressing sex differences in the neural response to endotoxin in humans (Eisenberger et al. 2009).

The dose of endotoxin is also of importance, as the effects are not linear. Mullington et al. were the first to report a non-linear response in the amount of non-REM sleep in response to different doses of endotoxin, which follow a bell-shaped curve with the greatest amount of non-REM sleep at an intermediate LPS dosage (Mullington et al. 2000; Pollmacher et al. 2002). In one of our studies, we observed an increase in reaction time after administration of a higher LPS dose, whereas in the same study, negative effects on long-term acquisition were solely visible after applying a lower dose (Grigoleit et al. 2011). Interestingly, in this study the physiological and neuroendocrine responses were not significantly different between doses, whereas the decrease in mood and self-reported alertness were far more pronounced in response to the higher dose. The non-linear effect of endotoxin may be related to its source, as we observed the same profound increase in heart rate both after applying 0.4 or 0.8 ng of *E. coli* LPS/kg of body weight, whereas Mullington et al. observed only a slight increase after administration of 0.4 but a marked rise after 0.8 ng of *S. abortus equi* LPS. Similar differences between the two studies are observable with IL-6 levels, cortisol, and body temperature. This observation highlights the importance of standardized protocols to receive comparable data and the difficulty of matching data from different studies.

Social Aspects of Sickness Behavior and the Illuminated Brain

Since the effects of sub-septic LPS doses on memory resulted in subtle effects, scientists began to ask more specific questions based on the following three observations:

1. Mood worsening and fatigue were the most constant and robust responses to an LPS injection.
2. The sickness response is believed to represent an adaptive and evolutionarily meaningful reaction, so most observations one would expect should be either beneficial or inevitable.
3. Humans are highly social beings. When inflammation induces changes in rodents' social behavior, it should, a fortiori, do so in humans but not necessarily in the same way and direction.

Two publications based on animal studies supported this change in orientation. First, a review by Cunningham and Sanderson raised the question whether prior experiments in rodents suggesting memory impairments after immune activation might in fact rather be representing changes in the animals' motivational state (Cunningham and Sanderson 2008). Second, a study by Willette et al. studied rhesus monkeys injected with low doses of LPS which showed a marked increase in social behavior (Willette et al. 2007), contrary to social withdrawal observed in sick rodents. Social withdrawal is a key feature of sickness behavior and is often assumed to be present in sick humans and other animals.

In 2009, two studies by Brydon et al. investigated the role of psychological stress and personality traits on the outcome of an acute immune challenge. They found that typhoid vaccination elicited a larger rise in participants' IL-6 serum levels when it was combined with a stressful task, and the stress response to the tasks was more pronounced in the vaccinated than in the stress-only group (Brydon et al. 2009a). Furthermore, in the combined vaccine/stress group, the authors found an inverse relationship between subjects' measures of optimism and IL-6 response and a positive correlation between optimism and antibody production after vaccination (Brydon et al. 2009b). This not only exemplifies the bi-directional brain-immune interactions in humans but also highlights a relationship that can only be investigated in a human experimental model, which underlines the importance and potential advantages of this model over animal models. Another example of a human-only study is a report in which participants showed an increase in feelings of social disconnection, which also was associated with the induction of depressed mood after receiving LPS (Eisenberger et al. 2010b). We recently investigated the effect of acute inflammation on social cognitive performance with two theory of mind (ToM) tests, which showed no effect (Kullmann et al. 2014; Grigoleit et al., unpublished).

One of the most intriguing examples for an adaptive change in social interaction in response to sickness and inflammation was recently published by Olsson et al. (2014). In this study, the authors found that the body odor of participants who received endotoxin was more aversive to blinded raters than that of control subjects, providing a potential non-verbal way of communicating sickness from one individual to another to avoid contagion.

The ability to create a well-controlled, reproducible immune response in a homogenous sample of subjects also allows for the use of imaging techniques to observe changes in brain activation and thereby identify to the neural origins of sickness and sickness-related behavior. Presently, there are data from one PET and three fMRI studies available, which are listed in Table 4.2. Using fMRI, it is possible to image and map differences in the level of blood oxygenation (BOLD signal) within the brain and thereby to draw conclusions about the degree of neural activity in respective brain regions. Brydon, Harrison, and colleagues were the first to publish an fMRI study in humans receiving endotoxin. They found that administration of the inflammatory stimulus (typhoid vaccine) had no effect on general neurovascular coupling upon visual stimulation, which was confirmed by findings from our group (Kullmann et al. 2012) and is important for the interpretation of specific changes in the BOLD response. In subjects who had received vaccination, higher activation in the left substantia nigra was observed which corresponded to the right hand which was used to press the response button during performance of the word-color Stroop task. This increase correlated with higher circulating IL-6 levels and prolonged response times in the task, which led the authors to conclude that IL-6-dependent modulations in the dopaminergic midbrain system may play a key role in

Table 4.2 Neuroimaging studies employing LPS injection in humans

Study	Dose of LPS and imaging method	N	LPS effects on neural activity
Brydon et al. (2008) and Harrison et al. (2009a, b)	*S. typhi* 25 μg i.m. (Typhim Vi®) fMRI (1.5T)	16M	Reduced substantia nigra activity. sACC activity and connectivity to other regions upon presentation of emotional faces predict inflammation-associated mood changes. Enhanced neural activity within brainstem, thalamus, amygdala, cingulate, and insula during Stroop color word task with additional recruitment of dlPFC and MCC in response to demanding incongruent trials
Eisenberger et al. (2009, 2010a) and Inagaki et al. (2012)	*E. coli* 0.8 ng/kg i.v. fMRI (3T)	39X	Significant reduction in ventral striatum activity to a monetary reward cue, increased activity in the amygdala in response to socially threatening images, no differences in neural response to social exclusion
Kullmann et al. (2012, 2014) *E. coli* 0.8 ng/kg i.v. PET (^{18}F-FDG)	*E. coli* 0.4 ng/kg i.v. fMRI (1.5T) 9X	18M Hannestad et al. (2012b)	Enhanced orbitofrontal cortex activity in response to emotional visual stimuli. Increased responses in the fusiform gyrus, temporo-parietal junction, superior temporal gyrus, and precuneus during a social cognition task (Reading the Mind in the Eyes) Higher NGM in the insula, trend towards higher NGM in cingulate, no effect on amygdaloid NGM

M male subjects, *X* mixed group of subjects, *i.v.* intravenously, *i.m.* intramuscular, *STS* sulcus temporalis superior, *sACC* subgenual anterior cingulate cortex, *dlPFC* dorsolateral prefrontal cortex, *MCC* midcingulate cortex, *fMRI* functional magnetic resonance imaging, *PET* positron emission tomography, *FDG* fludeoxyglucose, *NGM* normalized glucose metabolism

transmitting the motor deficits of inflammation (Brydon et al. 2008). In a further analysis of the same sample, the authors defined the Stroop task as a model for a set of high-demand cognitive processes typically compromised by sickness. During presentation of congruent as well as incongruent stimuli, they found that typhoid vaccination enhanced neural activity within the brainstem, thalamus, amygdala, and bilateral mid and anterior insula. Moreover, there was a significant correlation between insular activation and the amount of self-reported fatigue, which was only present in the vaccine group (Harrison et al. 2009b). This observation is supported by data from animal studies that also suggest a key role of the insula, amygdala, and various brain stem regions in afferent immune-to-brain communication (Doenlen et al. 2011; Engler et al. 2011; Serrats et al. 2010). In another study, Harrison et al. explored the neural mechanisms underlying inflammation-associated mood changes. Activity in the subgenual anterior cingulate cortex (sACC) in response to emotional face stimuli and its connectivity to several other brain regions correlated with changes in total mood 3 h after vaccine injection (Harrison et al. 2009a). Since the sACC and its functional connectivity to other regions are recognized for their role in the pathophysiology of major depressive disorder (MDD), this observation might contribute to the cytokine hypothesis of depression (see Box 4.2). Another fMRI study by Eisenberger, Inagaki, and colleagues dealt with the social aspects of sickness behavior using an i.v. injection of 0.8 ng *E. coli* endotoxin/kg of body weight as an inflammatory stimulus. Employing an online ball-tossing game as a model of social exclusion to explore differences in the neural processing of social pain in an inflammatory context, the authors did not find meaningful differences in neural responses between LPS and placebo conditions (Eisenberger et al. 2009). However, in the endotoxin group there was an association between increases in IL-6 and social pain-related neural activity in the dorsal anterior cingulate cortex and anterior insula, which was only present in female subjects. Based upon the observation that a decrease in reward-seeking behavior (anhedonia) is a key feature of sickness behavior in rodents, Eisenberger et al. examined the neural response to anticipation of monetary rewards and found a significant reduction in corresponding ventral striatum activity 2 h after endotoxin exposure (Eisenberger et al. 2010a). The ventral striatum, which contains the dopaminergic nucleus accumbens, is considered an important reward center and as such known to be involved in the etiology of depression (Nestler and Carlezon 2006). In their third paper on this study, the authors focused on the neural response to fearful stimuli (Inagaki et al. 2012). Previous studies have shown that experimental immune responses increase general state anxiety in humans and that the amygdala, a key structure for fear and anxiety processing, responds to systemic inflammation in rodents (Reichenberg et al. 2001; Grigoleit et al. 2011; Engler et al. 2011). Inagaki et al. presented a set of socially and non-socially threatening and non-threatening visual cues to their subjects and found that endotoxin led to greater amygdala activity in response to socially threatening (fear faces) versus all other types of images. This not only verifies an important immune-neural pathway but also highlights the specificity with which the brain reacts to inflammatory stimuli. A similar approach, however, using only non-socially threatening versus neutral images, was performed at our laboratory. Intravenous injection of 0.4 ng LPS/kg enhanced activation of the inferior orbitofrontal cortex, and (albeit at a lower statistical threshold) superior and medial prefrontal regions during aversive visual stimulation. These activations, which were not associated with changes in mood or anxiety, might reflect stronger engagement in emotion regulation, reappraisal, cognitive control of emotionally salient stimuli, and recognition of other individuals' emotions (Kullmann et al. 2012). In a subsequent report on this study, we investigated the effect of systemic inflammation on theory of mind (ToM) abilities and corresponding neural activity. Subjects were asked to detect facial expressions from pairs of eyes shown to them on a computer screen. Compared to a control condition (gender detection), all subjects showed enhanced activity in a set of regions typically engaged in ToM and face recognition with significantly higher activations within left fusiform gyrus, right temporo-parietal junction, right superior temporal gyrus, and right precuneus in the LPS compared to the placebo condition. In contrast, there was no effect on subjects' performance in this task, so that enhanced neural activity was interpreted as a compensatory mechanism to maintain social cognition abilities despite decreased overall brain performance or motivation (Kullmann et al. 2014). A different approach to imaging brain activity during inflammation was taken by Hannestad et al. who combined the human endotoxemia model with PET. Intravenous injection of the radioactive glucose analog fludeoxyglucose (18F-FDG) allows for the tracing and mapping of changes in local glucose metabolic rates, which indicate changes in energy consumption and neural activity in the respective region. In contrast to fMRI, PET records changes over a larger time frame of several minutes and independently of an experimental stimulus. Hence, it more closely represents a general engagement or tonus of a certain region than its recruitment in a specific task. Hannestad and colleagues found significantly increased normalized glucose metabolism (NGM) in the insula and a trend towards decreased NGM in the cingulate cortex without a general effect on global cerebral metabolic rate of glucose (CMRGlu), with secondary analyses suggesting the main effect occurred in the right anterior insula and right anterior cingulate (Hannestad et al. 2012b). Contrary to the authors' expectation, there was no effect on amygdala NGM. Unfortunately, there were no measures of anxiety employed in this study; however, in another study using a similar design of endotoxin application, the same authors did not find effects on anxiety (Hannestad et al. 2011). A great advantage of PET is that it is not restricted to parameters related to energy consumption like glucose metabolism or blood oxygenation to image changes in the brain. The density and engagement of neurotransmitter receptors or surface antigens might be explored as long as a sufficient tracer is available. Recently, the authors of the aforementioned studies demonstrated engagement of microglia in the baboon in response to peripheral endotoxin application using 11C-PBR28 as a ligand for translocator protein, a marker of microglial activation (Hannestad et al. 2012a).

Beyond Sickness: Other Uses of Human Endotoxemia in PNI Research

One of the strengths of the presented model is its versatility. Beyond exploring the effects of an inflammatory response on cognition, social behavior, and corresponding neural correlates, it was also used in other experiments addressing important questions in the field of PNI. In a study performed in our laboratory, we attempted to pair immunogenic effects of endotoxemia to a gustatory stimulus via classical conditioning, which had already been successfully utilized to condition immunosuppression in humans (Wirth et al. 2011). Although we failed to induce a conditioned inflammatory response, we observed a significant conditioned aversion against the conditioned stimulus solely in the endotoxin treated group, highlighting a potential evolutionary strategy to prevent the organism from pathogen-contaminated food (Grigoleit et al. 2012). In another study, we were able to demonstrate inflammation-induced release of salivary α-amylase, an enzyme which was recently suggested as a novel marker of sympathetic activity (Grigoleit et al. 2013). A few recent studies demonstrated that systemic inflammation dose-dependently increases visceral and somatic pain sensitivity (Benson et al. 2012; Hutchinson et al. 2013; Wegner et al. 2014). Exploring the interaction of sickness behavior and depression (see Box 4.2) Hannestad, DellaGioia, and colleagues investigated the effect of antidepressant treatment on endotoxin-induced fatigue and other depressant symptoms. Whereas pre-treatment with the selective serotonin-reuptake inhibitor citalopram significantly reduced endotoxin-induced lassitude and depressive symptomatology, pre-treatment with buprion, a norepinephrine and dopamine reuptake-inhibitor failed to display such an effect (Dellagioia et al. 2012; Hannestad et al. 2011). Although the latter study suffers from relatively low statistical power and is therefore considered preliminary, these reports demonstrate for the first time a direct connection between sickness behavior and depression with an involvement of the serotonergic system in humans. In addition, these findings may support the recently discussed possibility to adopt human endotoxemia as an experimental model of depression (DellaGioia and Hannestad 2010).

Conclusions

There are two fundamentally different concepts of how cytokines or other inflammatory mediators may alter brain functioning in mammals: adaptive changes and disturbance of normal cytokine communication within the brain. It is known that cytokines (e.g., IL-1β, IL-6) are expressed in the CNS not only in response to disease but also under physiological conditions and that they are involved in essential neural functions like long-term potentiation (LTP) (Balschun et al. 2004; Yirmiya and Goshen 2011). Recently, it was demonstrated that a network of at least five brain-borne pro- and anti-inflammatory cytokines operate during LTP and learning in freely moving rats (Del Rey et al. 2013). Hence, during inflammation the "cytokine storm" accessing the brain as well as brain-borne cytokines overexpressed in response to the inflammatory stimulus, may interfere with this network and thereby disturb CNS functions like memory. This effect might account for many of the results indicating profound memory impairments in laboratory animals, since the employed immune stimulus in those studies was rather high-dosed or applied directly into the intracerebrovascular space (for review see Yirmiya and Goshen 2011). Concerning the studies in humans presented here, we consider this kind of interference rather negligible. Since cytokine access to the brain under normal conditions is highly regulated (Banks 2006; Hopkins 2007; Yirmiya and Goshen 2011) and since the doses of LPS employed were relatively low and resembling an inflammatory stimulus one would encounter numerous times across the lifespan, it seems unlikely that evolution would not have developed strategies to circumvent or compensate for such an interference. More likely, such compensatory mechanisms themselves may account for alterations in cognitive performance as recently proposed by Yirmiya and Goshen: that the excitatory action of pro-inflammatory cytokines on neural activity in the CNS—as an adaptive reaction—extenuates neural excitability to prevent over-excitation and increased error rates (Yirmiya and Goshen 2011). Although this hypothesis still lacks experimental evidence, it might help explain some of the results presented in this review. Also involved may be the endotoxin effect on memory acquisition, which was seen with a lower but not with a higher dosage, which may be explained by a balance between preventive moderation and cytokine-induced over-excitation (Grigoleit et al. 2011). In a facial expression recognition task, we observed enhanced neural activity in the corresponding brain regions without an effect on task performance, which may reflect local compensation of a global down-regulation in neural excitability (Kullmann et al. 2014).

However, we assume that most results derived from endotoxemia studies in humans reflect adaptive responses helping the organism to encounter pathogen challenge as previously discussed (Dantzer 2001; Raison and Miller 2013b; Yirmiya and Goshen 2011). The most robust and reproducible endotoxin-induced changes in neuropsychological functioning were depressive-like mood worsening and fatigue, both reducing energy consumption and increased risk-taking. Notably, both of these occurred relatively independent from fever or other symptoms. As already exemplified by Dantzer more than 10 years ago, sickness can be considered a motivational state competing with other motivational states like initiative, anxiety, or parental care (Dantzer 2001). The assumption that sickness behavior does not per se impair an individual's abilities but rather his or her willingness to use them was supported by the observation that subjects could maintain or even slightly improve their performance in highly demanding cognitive tasks despite a profound decrease in self-reported attention (Cohen et al. 2003; Grigoleit et al. 2010, 2011; van den Boogaard et al. 2010). The observation that mechanisms to communicate illness and thereby reduce direct contact between infected and non-infected individuals seem to have evolved (Olsson et al. 2014) is underlining the general concept of adaptivity.

In concordance with psychomotor deficits, loss of motivation, increased anxiety, and decreased reward seeking behavior as key features of sickness behavior, results from the presented neuroimaging studies indicate an effect of inflammation on neural activity in dopaminergic nigro-striatal and mesolimbic structures including the amygdala (Brydon et al. 2008; Eisenberger et al. 2010a; Hannestad et al. 2012b; Harrison et al. 2009a; Inagaki et al. 2012). Whether these structures are directly affected by inflammatory signals or whether their activity is primarily altered through changes in other brain regions remains to be elucidated. However, the insular cortex, which is known to play a major role in interoceptive awareness, as well as a variety of brainstem nuclei might be crucially involved in the perception and transmission of information derived from our immune system, as indicated by a number of studies in humans and animals (Doenlen et al. 2011; Hannestad et al. 2012a; Harrison et al. 2009b; Kullmann et al. 2011; Serrats et al. 2010).

In addition to the dopaminergic system, the serotonergic system appears to be involved in the mediation of sickness behavior and the development of inflammation-related depressive disorder, which was indicated by the attenuation of endotoxin-induced depressive symptoms by pre-treatment with a selective serotonin-reuptake inhibitor (Hannestad et al. 2011). Inflammation also enhances activity of the enzyme indoleamine dioxygenase (IDO) and thereby causes decreased tryptophan levels and increased production of kynurenine (Dantzer et al. 2008). Additional evidence from humans comes from a study in elderly persons suffering from chronic low-grade inflammation (Capuron et al. 2011).

The next step in combined endotoxemia/brain imaging approaches may be to further investigate the exact neuronal structures and neurotransmitter systems involved in the mediation of sickness behavior. PET studies are especially needed as they can potentially measure distinct neurotransmitters.

As discussed in Box 4.2, inflammation is now widely believed to be involved in the pathogenesis and etiology of depressive and other neuropsychiatric disorders. However, it is still unknown what exact events or predispositions lead to inflammation-induced development or worsening of such disorders in one individual without affecting the other. Likewise, some subjects develop very strong depressive-like symptoms during experimental endotoxemia, whereas others are not affected or mildly affected (Grigoleit et al., unpublished). The opportunity to identify possible predictors for the severity of an inflammation-induced impact on our psyche is appealing. In this context, genetic approaches should be employed in future studies on experimental inflammation, since there are a number of genes known to be involved in immune processes as well as in the pathogenesis of psychiatric disorders like major depression (Raison and Miller 2013a).

In summary, administration of LPS to induce a transient experimental systemic inflammation is a useful model to study immune-to-brain communication in humans. It not only allows researchers to verify a number of hypotheses derived from animal studies but also to mimic aspects of human behavior, like feelings or social cognition, which cannot be sufficiently modeled in animals. The model is safe and provides—with some limitations—consistent and reproducible data, although it is important to mention that inflammatory and neural responses may vary between different dosages, sources, and methods of application.

References

Adler RD, Joy RJ. Febrile responses to the intracisternal injection of endogenous (leucocytic) pyrogen in the rabbit. Proc Soc Exp Biol Med. 1965;119:660–3.

Akira S, Takeda K. Toll-like receptor signalling. Nat Rev Immunol. 2004;4:499–511.

Akiyama H, Barger S, Barnum S, Bradt B, Bauer J, Cole GM, Cooper NR, Eikelenboom P, Emmerling M, Fiebich BL, Finch CE, Frautschy S, Griffin WS, Hampel H, Hull M, Landreth G, Lue L, Mrak R, Mackenzie IR, McGeer PL, O'Banion MK, Pachter J, Pasinetti G, Plata-Salaman C, Rogers J, Rydel R, Shen Y, Streit W, Strohmeyer R, Tooyoma I, Van Muiswinkel FL, Veerhuis R, Walker D, Webster S, Wegrzyniak B, Wenk G, Wyss-Coray T. Inflammation and Alzheimer's disease. Neurobiol Aging. 2000;21:383–421.

Anders S, Tanaka M, Kinney DK. Depression as an evolutionary strategy for defense against infection. Brain Behav Immun. 2013;31:9–22.

Bahador M, Cross AS. From therapy to experimental model: a hundred years of endotoxin administration to human subjects. J Endotoxin Res. 2007;13:251–79.

Bailey PT, Abeles FB, Hauer EC, Mapes CA. Intracerebroventricular administration of leukocytic endogenous mediators (LEM) in the rat. Proc Soc Exp Biol Med. 1976;153:419–23.

Balschun D, Wetzel W, Del Rey A, Pitossi F, Schneider H, Zuschratter W, Besedovsky HO. Interleukin-6: a cytokine to forget. FASEB J. 2004;18:1788–90.

Banks WA. The blood–brain barrier in psychoneuroimmunology. Neurol Clin. 2006;24:413–9.

Benson S, Kattoor J, Wegner A, Hammes F, Reidick D, Grigoleit JS, Engler H, Oberbeck R, Schedlowski M, Elsenbruch S. Acute experimental endotoxemia induces visceral hypersensitivity and altered pain evaluation in healthy humans. Pain. 2012;153:794–9.

Besedovsky HO, del Rey A. Immune-neuro-endocrine interactions: facts and hypotheses. Endocr Rev. 1996;17:64–102.

Bluthe RM, Walter V, Parnet P, Laye S, Lestage J, Verrier D, Poole S, Stenning BE, Kelley KW, Dantzer R. Lipopolysaccharide induces sickness behaviour in rats by a vagal mediated mechanism. C R Acad Sci III. 1994;317:499–503.

Breder CD, Dinarello CA, Saper CB. Interleukin-1 immunoreactive innervation of the human hypothalamus. Science. 1988;240:321–4.

Brydon L, Harrison NA, Walker C, Steptoe A, Critchley HD. Peripheral inflammation is associated with altered substantia nigra activity and psychomotor slowing in humans. Biol Psychiatry. 2008; 63:1022–9.

Brydon L, Walker C, Wawrzyniak A, Whitehead D, Okamura H, Yajima J, Tsuda A, Steptoe A. Synergistic effects of psychological and immune stressors on inflammatory cytokine and sickness responses in humans. Brain Behav Immun. 2009a;23:217–24.

Brydon L, Walker C, Wawrzyniak AJ, Chart H, Steptoe A. Dispositional optimism and stress-induced changes in immunity and negative mood. Brain Behav Immun. 2009b;23:810–6.

Capuron L, Miller AH. Cytokines and psychopathology: lessons from interferon-alpha. Biol Psychiatry. 2004;56:819–24.

Capuron L, Ravaud A, Dantzer R. Early depressive symptoms in cancer patients receiving interleukin 2 and/or interferon alfa-2b therapy. J Clin Oncol. 2000;18:2143–51.

Capuron L, Schroecksnadel S, Feart C, Aubert A, Higueret D, Barberger-Gateau P, Laye S, Fuchs D. Chronic low-grade inflammation in elderly persons is associated with altered tryptophan and tyrosine metabolism: role in neuropsychiatric symptoms. Biol Psychiatry. 2011;70:175–82.

Cohen J. The immunopathogenesis of sepsis. Nature. 2002;420:885–91.

Cohen O, Reichenberg A, Perry C, Ginzberg D, Pollmacher T, Soreq H, Yirmiya R. Endotoxin-induced changes in human working and declarative memory associate with cleavage of plasma "readthrough" acetylcholinesterase. J Mol Neurosci. 2003;21:199–212.

Cunningham C, Sanderson DJ. Malaise in the water maze: Untangling the effects of LPS and IL-1beta on learning and memory. Brain Behav Immun. 2008;22(8):1117–27.

Dantzer R. Cytokine-induced sickness behavior: where do we stand? Brain Behav Immun. 2001;15:7–24.

Dantzer R, Konsman JP, Bluthe RM, Kelley KW. Neural and humoral pathways of communication from the immune system to the brain: parallel or convergent? Auton Neurosci. 2000;85:60–5.

Dantzer R, O'Connor JC, Freund GG, Johnson RW, Kelley KW. From inflammation to sickness and depression: when the immune system subjugates the brain. Nat Rev Neurosci. 2008;9:46–56.

Del Rey A, Balschun D, Wetzel W, Randolf A, Besedovsky HO. A cytokine network involving brain-borne IL-1beta, IL-1ra, IL-18, IL-6, and TNFalpha operates during long-term potentiation and learning. Brain Behav Immun. 2013;33:15–23.

DellaGioia N, Hannestad J. A critical review of human endotoxin administration as an experimental paradigm of depression. Neurosci Biobehav Rev. 2010;34:130–43.

Dellagioia N, Devine L, Pittman B, Hannestad J. Bupropion pre-treatment of endotoxin-induced depressive symptoms. Brain Behav Immun. 2012;31:197–204.

Dinarello CA. The IL-1 family and inflammatory diseases. Clin Exp Rheumatol. 2002;20:S1–13.

Doenlen R, Krugel U, Wirth T, Riether C, Engler A, Prager G, Engler H, Schedlowski M, Pacheco-Lopez G. Electrical activity in rat cortico-limbic structures after single or repeated administration of lipopolysaccharide or staphylococcal enterotoxin B. Proc Biol Sci. 2011;278:1864–72.

Dowlati Y, Herrmann N, Swardfager W, Liu H, Sham L, Reim EK, Lanctot KL. A meta-analysis of cytokines in major depression. Biol Psychiatry. 2010;67:446–57.

Drexhage RC, Knijff EM, Padmos RC, Heul-Nieuwenhuijzen L, Beumer W, Versnel MA, Drexhage HA. The mononuclear phagocyte system and its cytokine inflammatory networks in schizophrenia and bipolar disorder. Expert Rev Neurother. 2010;10:59–76.

Eisenberger NI, Inagaki TK, Rameson LT, Mashal NM, Irwin MR. An fMRI study of cytokine-induced depressed mood and social pain: the role of sex differences. Neuroimage. 2009;47:881–90.
Eisenberger NI, Berkman ET, Inagaki TK, Rameson LT, Mashal NM, Irwin MR. Inflammation-induced anhedonia: endotoxin reduces ventral striatum responses to reward. Biol Psychiatry. 2010a;68:748–54.
Eisenberger NI, Inagaki TK, Mashal NM, Irwin MR. Inflammation and social experience: an inflammatory challenge induces feelings of social disconnection in addition to depressed mood. Brain Behav Immun. 2010b;24:558–63.
Engelhardt B. Molecular mechanisms involved in T cell migration across the blood–brain barrier. J Neural Transm. 2006;113:477–85.
Engler H, Doenlen R, Engler A, Riether C, Prager G, Niemi MB, Pacheco-Lopez G, Krugel U, Schedlowski M. Acute amygdaloid response to systemic inflammation. Brain Behav Immun. 2011;25:1384–92.
Ericsson A, Kovacs KJ, Sawchenko PE. A functional anatomical analysis of central pathways subserving the effects of interleukin-1 on stress-related neuroendocrine neurons. J Neurosci. 1994;14:897–913.
Ericsson A, Arias C, Sawchenko PE. Evidence for an intramedullary prostaglandin-dependent mechanism in the activation of stress-related neuroendocrine circuitry by intravenous interleukin-1. J Neurosci. 1997;17:7166–79.
Fontana A, Weber E, Dayer JM. Synthesis of interleukin 1/endogenous pyrogen in the brain of endotoxin-treated mice: a step in fever induction? J Immunol. 1984;133:1696–8.
Gibertini M. IL1 beta impairs relational but not procedural rodent learning in a water maze task. Adv Exp Med Biol. 1996;402:207–17.
Gibertini M, Newton C, Friedman H, Klein TW. Spatial learning impairment in mice infected with Legionella pneumophila or administered exogenous interleukin-1-beta. Brain Behav Immun. 1995;9:113–28.
Giovanoli S, Engler H, Engler A, Richetto J, Voget M, Willi R, Winter C, Riva MA, Mortensen PB, Schedlowski M, Meyer U. Stress in puberty unmasks latent neuropathological consequences of prenatal immune activation in mice. Science. 2013;339:1095–9.
Grigoleit JS, Oberbeck JR, Lichte P, Kobbe P, Wolf OT, Montag T, del Rey A, Gizewski ER, Engler H, Schedlowski M. Lipopolysaccharide-induced experimental immune activation does not impair memory functions in humans. Neurobiol Learn Mem. 2010;94:561–7.
Grigoleit JS, Kullmann JS, Wolf OT, Hammes F, Wegner A, Jablonowski S, Engler H, Gizewski E, Oberbeck R, Schedlowski M. Dose-dependent effects of endotoxin on neurobehavioral functions in humans. PLoS One. 2011;6:e28330.
Grigoleit JS, Kullmann JS, Winkelhaus A, Engler H, Wegner A, Hammes F, Oberbeck R, Schedlowski M. Single-trial conditioning in a human taste-endotoxin paradigm induces conditioned odor aversion but not cytokine responses. Brain Behav Immun. 2012;26:234–8.
Grigoleit JS, Kullmann JS, Oberbeck R, Schedlowski M, Engler H. Salivary alpha-amylase response to endotoxin administration in humans. Psychoneuroendocrinology. 2013;38(9):1819–23.
Hannestad J, DellaGioia N, Ortiz N, Pittman B, Bhagwagar Z. Citalopram reduces endotoxin-induced fatigue. Brain Behav Immun. 2011;25:256–9.
Hannestad J, Gallezot JD, Schafbauer T, Lim K, Kloczynski T, Morris ED, Carson RE, Ding YS, Cosgrove KP. Endotoxin-induced systemic inflammation activates microglia: [(1)(1)C]PBR28 positron emission tomography in nonhuman primates. Neuroimage. 2012a;63:232–9.
Hannestad J, Subramanyam K, Dellagioia N, Planeta-Wilson B, Weinzimmer D, Pittman B, Carson RE. Glucose metabolism in the insula and cingulate is affected by systemic inflammation in humans. J Nucl Med. 2012b;53:601–7.
Harrison NA, Brydon L, Walker C, Gray MA, Steptoe A, Critchley HD. Inflammation causes mood changes through alterations in subgenual cingulate activity and mesolimbic connectivity. Biol Psychiatry. 2009a;66:407–14.
Harrison NA, Brydon L, Walker C, Gray MA, Steptoe A, Dolan RJ, Critchley HD. Neural origins of human sickness in interoceptive responses to inflammation. Biol Psychiatry. 2009b;66:415–22.

Hart BL. Biological basis of the behavior of sick animals. Neurosci Biobehav Rev. 1988;12:123–37.
Hermann DM, Mullington J, Hinze-Selch D, Schreiber W, Galanos C, Pollmacher T. Endotoxin-induced changes in sleep and sleepiness during the day. Psychoneuroendocrinology. 1998;23:427–37.
Holsboer F. The corticosteroid receptor hypothesis of depression. Neuropsychopharmacology. 2000;23:477–501.
Hopkins SJ. Central nervous system recognition of peripheral inflammation: a neural, hormonal collaboration. Acta Biomed. 2007;78 Suppl 1:231–47.
Hutchinson MR, Buijs M, Tuke J, Kwok YH, Gentgall M, Williams D, Rolan P. Low-dose endotoxin potentiates capsaicin-induced pain in man: evidence for a pain neuroimmune connection. Brain Behav Immun. 2013;30:3–11.
Inagaki TK, Muscatell KA, Irwin MR, Cole SW, Eisenberger NI. Inflammation selectively enhances amygdala activity to socially threatening images. Neuroimage. 2012;59:3222–6.
Janeway Jr CA, Medzhitov R. Innate immune recognition. Annu Rev Immunol. 2002;20:197–216.
Korth C, Mullington J, Schreiber W, Pollmacher T. Influence of endotoxin on daytime sleep in humans. Infect Immun. 1996;64:1110–5.
Krabbe KS, Reichenberg A, Yirmiya R, Smed A, Pedersen BK, Bruunsgaard H. Low-dose endotoxemia and human neuropsychological functions. Brain Behav Immun. 2005;19:453–60.
Krueger JM, Walter J, Dinarello CA, Wolff SM, Chedid L. Sleep-promoting effects of endogenous pyrogen (interleukin-1). Am J Physiol. 1984;246:R994–9.
Kullmann JS, Grigoleit J-S, Schedlowski M. Effects of an acute inflammation on memory performance, mood and brain activity. Z Med Psychol. 2011;20:108–17.
Kullmann JS, Grigoleit JS, Lichte P, Kobbe P, Rosenberger C, Banner C, Wolf OT, Engler H, Oberbeck R, Elsenbruch S, Bingel U, Forsting M, Gizewski ER, Schedlowski M. Neural response to emotional stimuli during experimental human endotoxemia. Hum Brain Mapp. 2012;34(9):2217–27.
Kullmann JS, Grigoleit JS, Wolf OT, Engler H, Oberbeck R, Elsenbruch S, Forsting M, Schedlowski M, Gizewski ER. Experimental human endotoxemia enhances brain activity during social cognition. Soc Cogn Affect Neurosci. 2014;9(6):786–93.
McCarthy DO, Kluger MJ, Vander AJ. Suppression of food intake during infection: is interleukin-1 involved? Am J Clin Nutr. 1985;42:1179–82.
Miller AH, Maletic V, Raison CL. Inflammation and its discontents: the role of cytokines in the pathophysiology of major depression. Biol Psychiatry. 2009;65:732–41.
Mullington J, Korth C, Hermann DM, Orth A, Galanos C, Holsboer F, Pollmacher T. Dose-dependent effects of endotoxin on human sleep. Am J Physiol Regul Integr Comp Physiol. 2000;278:R947–55.
Nestler EJ, Carlezon Jr WA. The mesolimbic dopamine reward circuit in depression. Biol Psychiatry. 2006;59:1151–9.
Olsson MJ, Lundström JN, Kimball BA, Gordon AR, Karshikoff B, Hosseini N, Sorjonen K, Höglund CO, Solares C, Soop A, Axelsson J, Lekander M. The scent of disease: human body odor contains an early chemosensory cue of sickness. Psych Sci. 2014;25(3):817–23.
Pace TW, Hu F, Miller AH. Cytokine-effects on glucocorticoid receptor function: relevance to glucocorticoid resistance and the pathophysiology and treatment of major depression. Brain Behav Immun. 2007;21:9–19.
Pariante CM, Lightman SL. The HPA axis in major depression: classical theories and new developments. Trends Neurosci. 2008;31:464–8.
Pollmacher T, Schreiber W, Gudewill S, Vedder H, Fassbender K, Wiedemann K, Trachsel L, Galanos C, Holsboer F. Influence of endotoxin on nocturnal sleep in humans. Am J Physiol. 1993;264:R1077–83.
Pollmacher T, Haack M, Schuld A, Reichenberg A, Yirmiya R. Low levels of circulating inflammatory cytokines—do they affect human brain functions? Brain Behav Immun. 2002;16:525–32.

Pugh CR, Kumagawa K, Fleshner M, Watkins LR, Maier SF, Rudy JW. Selective effects of peripheral lipopolysaccharide administration on contextual and auditory-cue fear conditioning. Brain Behav Immun. 1998;12:212–29.

Raison CL, Miller AH. The evolutionary significance of depression in Pathogen Host Defense (PATHOS-D). Mol Psychiatry. 2013a;18:15–37.

Raison CL, Miller AH. Malaise, melancholia and madness: the evolutionary legacy of an inflammatory bias. Brain Behav Immun. 2013b;31:1–8.

Reichenberg A, Yirmiya R, Schuld A, Kraus T, Haack M, Morag A, Pollmacher T. Cytokine-associated emotional and cognitive disturbances in humans. Arch Gen Psychiatry. 2001;58:445–52.

Roth J, Harre EM, Rummel C, Gerstberger R, Hubschle T. Signaling the brain in systemic inflammation: role of sensory circumventricular organs. Front Biosci. 2004;9:290–300.

Schedlowski M, Engler H, Grigoleit J-S. Endotoxin-induced systemic inflammation in humans: a model to disentangle immune-to-brain communication. Brain Behav Immun. 2014;35:1–8.

Seok J, Warren HS, Cuenca AG, Mindrinos MN, Baker HV, Xu W, Richards DR, McDonald-Smith GP, Gao H, Hennessy L, Finnerty CC, Lopez CM, Honari S, Moore EE, Minei JP, Cuschieri J, Bankey PE, Johnson JL, Sperry J, Nathens AB, Billiar TR, West MA, Jeschke MG, Klein MB, Gamelli RL, Gibran NS, Brownstein BH, Miller-Graziano C, Calvano SE, Mason PH, Cobb JP, Rahme LG, Lowry SF, Maier RV, Moldawer LL, Herndon DN, Davis RW, Xiao W, Tompkins RG. Genomic responses in mouse models poorly mimic human inflammatory diseases. Proc Natl Acad Sci U S A. 2013;110:3507–12.

Serrats J, Schiltz JC, Garcia-Bueno B, van Rooijen N, Reyes TM, Sawchenko PE. Dual roles for perivascular macrophages in immune-to-brain signaling. Neuron. 2010;65:94–106.

Shaw KN, Commins S, O'Mara SM. Lipopolysaccharide causes deficits in spatial learning in the watermaze but not in BDNF expression in the rat dentate gyrus. Behav Brain Res. 2001;124:47–54.

Smith A, Tyrrell D, Coyle K, Higgins P. Effects of interferon alpha on performance in man: a preliminary report. Psychopharmacology (Berl). 1988;96:414–6.

Spath-Schwalbe E, Hansen K, Schmidt F, Schrezenmeier H, Marshall L, Burger K, Fehm HL, Born J. Acute effects of recombinant human interleukin-6 on endocrine and central nervous sleep functions in healthy men. J Clin Endocrinol Metab. 1998;83:1573–9.

Strike PC, Wardle J, Steptoe A. Mild acute inflammatory stimulation induces transient negative mood. J Psychosom Res. 2004;57:189–94.

van den Boogaard M, Ramakers BP, van Alfen N, van der Werf SP, Fick WF, Hoedemaekers CW, Verbeek MM, Schoonhoven L, van der Hoeven JG, Pickkers P. Endotoxemia-induced inflammation and the effect on the human brain. Crit Care. 2010;14:R81.

Watkins LR, Goehler LE, Relton JK, Tartaglia N, Silbert L, Martin D, Maier SF. Blockade of interleukin-1 induced hyperthermia by subdiaphragmatic vagotomy: evidence for vagal mediation of immune-brain communication. Neurosci Lett. 1995;183:27–31.

Weaver JD, Huang MH, Albert M, Harris T, Rowe JW, Seeman TE. Interleukin-6 and risk of cognitive decline: MacArthur studies of successful aging. Neurology. 2002;59:371–8.

Wegner A, Elsenbruch S, Maluck J, Grigoleit J-S, Engler H, Jäger M, Spreitzer I, Schedlowski M, Benson S. Inflammation-induced hyperalgesia: effects of timing, dosage, and negative affect on somatic pain sensitivity in human experimental endotoxemia. Brain Behav Immun. 2014;41:46–54.

Willette AA, Lubach GR, Coe CL. Environmental context differentially affects behavioral, leukocyte, cortisol, and interleukin-6 responses to low doses of endotoxin in the rhesus monkey. Brain Behav Immun. 2007;21:807–15.

Wirth T, Ober K, Prager G, Vogelsang M, Benson S, Witzke O, Kribben A, Engler H, Schedlowski M. Repeated recall of learned immunosuppression: evidence from rats and men. Brain Behav Immun. 2011;25(7):1444–51.

Wright CE, Strike PC, Brydon L, Steptoe A. Acute inflammation and negative mood: mediation by cytokine activation. Brain Behav Immun. 2005;19:345–50.

Yirmiya R, Goshen I. Immune modulation of learning, memory, neural plasticity and neurogenesis. Brain Behav Immun. 2011;25:181–213.

Zhu CB, Blakely RD, Hewlett WA. The proinflammatory cytokines interleukin-1beta and tumor necrosis factor-alpha activate serotonin transporters. Neuropsychopharmacology. 2006;31: 2121–31.

Part II
Translational Medicine

Chapter 5
Mild Encephalitis Theory of Psychiatric Disorders

Karl Bechter

Abstract *Introduction*: The mild encephalitis (ME) hypothesis addresses a separate category of clinically meaningful CNS inflammation of a lower degree compared to classical encephalitis (CE), patients presenting with predominant psychiatric symptoms. Inflammation is a dimensional entity over time and space within various bodily compartments or tissues, triggered by infections, trauma, toxicity, and immune challenges. Low level grade brain inflammation (LLNI) should be causally involved in ME. Main candidates of ME were severe psychiatric disorders of affective and schizophrenic type, including prodromal stages, acute exacerbations. Short-lived CE in small localized sites within the brain would match with ME. Clinical assessment of ME depends strongly upon the methods used to identify neuroinflammation. The international consensus that both categories, of encephalitis and of encephalopathy, can associate with neuroinflammation, demonstrates the present theoretical and practical dilemma.

Autoimmune Encephalitis (AE) and Limbic Encephalitis (LE): New insights into CE and LE pathophysiology emerged from the identification of neuronal autoantibodies, e.g. NMDAR antibodies. More and more neuronal autoantibodies become detected. AE cases present varying and various clinical syndromes including epilepsies. In some cases of AE defined autoantibodies maybe missing in the blood or CSF. Overall the threshold for the clinical diagnosis of LE, CE, and AE was lowered through improved laboratory methods. This recent development matches predictions of the ME hypothesis that with improved diagnostic methods a better clinical assessment of LLNI and ME will be possible and the definition of borders between CE and hypothetical ME may change.

K. Bechter (✉)
Clinic for Psychiatry and Psychotherapy II, Ulm University, Bezirkskrankenhaus Günzburg, Günzburg, Germany

Department of Psychosomatics/Psychotherapy, Ludwig-Heilmeyer-Str. 2, 89312 Günzburg, Germany
e-mail: karl.bechter@bkh-guenzburg.de

N. Müller et al. (eds.), *Immunology and Psychiatry*, Current Topics in Neurotoxicity 8, DOI 10.1007/978-3-319-13602-8_5

Utility and Limitations of Clinical Categorization: Behind the preferred strict categorization in the clinical field stands the important goal to obtain reliable guidelines for therapeutic management, which implies that dimensional aspects of a disease may be neglected. Present brain imaging is rather insensitive to detect LLNI, CSF investigation is the most sensitive approach, but protein concentrations in CSF are much lower than in the blood and the involvement of complicated CSF flow patterns makes interpretation difficult. Experimental models clearly support the view that grading of neuroinflammation into severity levels is needed. In addition topology is relevant for defining cause–response relationships. Definition of neuroinflammation differs between various research fields and clinical fields. Nevertheless with improved diagnostic approaches to ME, especially by CSF diagnostics, a clinical diagnosis of LLNI is becoming feasible. Idealized types of CE and ME in a dimensional scheme (severity by time) are presented.

Some Historical Remarks on Etiology Research in Psychiatric Disorders: The initial hypothesis around 1857 that syphilis might represent an important risk factor for general paresis (GP) was accepted only about 70 years later. Very important was the progress in laboratory diagnosis and introduction of lumbar puncture (with CSF analysis). Nevertheless up to now we cannot exactly explain why, after often long latency of years, GP may develop in a subgroup of the infected. Many infections are characterized by low overall pathogenicity (number of diseased per number of infected).

Multiple Interacting Systems and Factors: Severe psychiatric disorders are complex diseases with many contributing factors to be considered: certainly genes, age (of onset), timing, infection, autoimmunity, stress, and possibly chance. The overall scenario is similar to systemic autoimmune disorders a considerable subgroup of such patients suffering from comorbid psychiatric symptoms including severe psychosis. Respective pathomechanisms are only partially known yet. CSF analysis is the most sensitive diagnostic approach to neuroinflammatory disorders. With modern methods in about 70 % of cases with affective or schizophrenic spectrum (including bipolar) disorders various CSF abnormalities were found, demonstrating some LLNI process or immune activation in the intrathecal spaces, but also crosstalk between the periphery and the brain.

Brain Barriers and Systemic Crosstalk: Systemic inflammation can directly influence brain functions and elicit defined symptoms like sickness behavior. Neuroinflammation is primarily likely to change brain functions. Insights from experimental neuroimmunology show that chronic neuroinflammation develops when acute inflammation is not actively resolved. The choroid plexus (CP) represents an interface between blood and CSF spaces; immune cells are recruited and educated at the CP. The blood–brain barrier (BBB) and the blood–CSF barrier (BCSFB) are to be differentiated. CSF flow is pulsating forth and back with a flow distance up to 10 cm per pulse on certain sites. CSF cells are distributed by CSF flow and can follow the CSF outflow pathways (PCOP) through cribriform plate and along brain nerves and spinal nerves. Crosstalk between CNS immunity and systemic immunity is taking place at these barriers and the PCOP.

Outlook: Subgroups of severe psychiatric disorders present some neuroinflammatory process, which may at least contribute to, if not causally underlie, a variety of severe psychiatric disorders. LLNI pathogenesis remains incompletely understood. ME theory fits with epidemiological and clinical aspects, including the proven risk increase during lifetime by infections and autoimmune disorders, also head trauma. It remains a challenge to clinically assess and differentiate LLNI processes by neuroimaging, CSF investigation and new methods. Psychiatric research should focus on etiopathogenetic research including the ME theory.

Keywords Course of encephalitis • Severity grading of encephalitis • Mild encephalitis • Neuroinflammation • Affective disorders • Schizophrenia • Severe mental illness

Introduction

The mild encephalitis (ME) hypothesis (Bechter 2001, updated version Bechter 2013) has become more and more accepted as support from findings from various research areas grows. It can now be formulated as the mild encephalitis (ME) theory. The hypothesis was initially based on clinical aspects in psychoses, like course and symptoms, and on findings in experimental CNS infections, especially Borna disease virus (BDV) infection (Richt et al. 1997), and in spontaneous meningoencephalitis. The new term ME was proposed to address a separate category of clinically meaningful CNS inflammation, where the inflammation is of a lower degree compared to classical encephalitis (CE), and possibly present in patients with predominant psychiatric symptoms. However, it is difficult to define exact borders to separate CE and ME clinically. This is plausible a priori from the characteristics of any inflammatory process: inflammation is a dimensional entity over time and space, taking place possibly within various bodily compartments or tissues and representing a response to various well-known triggers including infections, trauma, toxicity, and immune challenges (compare Nathan 2004). This implies that, during the time periods before and after the defined classical inflammatory stage, some lower levels (or degrees) of inflammation have to occur in any type of classical inflammation, including neuroinflammation. It is also apparent from clinical experience that stages of low-level inflammation or low-level neuroinflammation (LLNI) may have clinical relevance (Johnson 1982). In addition, it is plausible that the clinical appearance of LLNI may be frequent and meaningful in cases without a transition to CE. Such a principle is seen in various clinical fields, for example in systemic autoimmune disorders, where there are many cases with incomplete signs of classical autoimmune disorder. These patients suffer from significant symptoms, although such cases remain difficult to diagnosis and research, because a clear categorization and treatment approaches are lacking (Mosca et al. 2002, Mosca 2014 oral presentation at the 9th International Congress on Autoimmunity, Nizza 2014 Autoimmunity Congress). In ME, some type of LLNI or autochthonous brain inflammation of low grade should be causally involved, not simply abnormal molecular signaling by cytokines and others from the blood to the brain, secondary to systemic inflammation. This means LLNI should be detectable within the brain area (parenchyma and/or CSF) using previously unavailable methods that have the required sensitivity. Main candidates of ME theory were severe forms of affective and schizophrenic disorders, including bipolar spectrum without neurological hard signs, though possibly with neurological soft signs. The LLNI process should be able to disturb brain function accordingly and be able to induce the symptoms observed. LLNI might exist over longer periods of time, not just hours or days, and cause meaningful clinical psychiatric syndromes. The prodromal stages frequent in the majority of these severe psychiatric disorders were assumed to match well with the stages of minor LLNI. However, acute exacerbations over short time periods with stronger types of LLNI, or even short-lived CE in localized sites within the brain, were assumed to be not infrequent. With advanced methods, one might detect not only such exacerbations, but also more chronic minor LLNI stages. Various known clinical conditions highlight the problem of missing sensitivity to detect the condition clinically, even though there is localized classical encephalitis. For example, in HIV infection, a number of histopathologically defined lesions that were only partially clinically meaningful during lifetime may be detected postmortem, thus in principle fulfilling the criteria of classical neuroinflammation (Cherner et al. 2007). The classical clinical signs of encephalitis are fever, disturbed consciousness, and others (Roos 1999). In histopathology the inflammatory lesion is usually defined as the local accumulation of (inflammatory) immune cells detected under the microscope. However, it is still rather difficult today to define in detail in an interdisciplinary perspective what neuroinflammation means. Apparently various scientific fields use different definitions, which cannot exactly fit together in a translational perspective. It is therefore not surprising that nonpathological neuroinflammation is claimed to exist (Wohleb and Godbout 2013). Here, despite no clear detection of clinical significance, it seems plausible that such cases are involved. However, no one is able to exclude the possibility that not all symptoms were carefully investigated, and therefore some neurological signs and symptoms were simply not observed or not correctly attributed, as can be concluded from experimental models (compare the discussion in Bechter 2001). Clinical assessment depends strongly upon the methods used, e.g., the available methods to identify inflammatory lesions within the brain. As well, it may not be easy or possible to attribute detectable symptoms to underlying defined lesions; this has to do with the redundant organization of the brain (Bechter et al. 2010a, b; Bechter 2012). In a very critical reconsideration of this problem, one may come to the view that brain lesions cannot occur without any consequences to brain function, although the consequences may not be easily clinically visible. Such a point of view can, for example, be underscored by the discussion over many years about the meaning of the so-called UBOs (unidentified bright objects). In the first years, these were considered to be nonpathologic, but now are usually counted as not irrelevant. They clearly indicate some brain pathology with varying, often minor, clinical relevance (Bechter et al. 1994; Osborn et al. 2010). In a reversed critical perspective, it may also be speculative or over-inclusive to attribute certain clinical symptoms simply to a certain underlying but undetected brain pathology; see the strict evaluation of the role of neuronal autoantibodies by Coutinho et al. (2014). An instructive example is the pathological relevance of lesions in cases of multiple sclerosis (see below). The compromise of defining HIV encephalitis in between clinical and theoretical aspects shows exactly this problem. From a theoretical point of view, it remains at least a weak compromise. Patients with postmortem-defined inflammatory brain lesions were not counted as fulfilling the criteria for HIV encephalitis because, during their lifetimes, they suffered only from severe psychiatric symptoms, but not from the symptoms of clinical acute encephalitis (Bechter 2013). Beyond the inflammatory brain lesions themselves, the duration of LLNI may contribute to clinical outcome. That longer-lasting LLNI status may have clinical relevance is strongly supported by various new insights, for example with multiple sclerosis (MS). Clinical symptoms of MS were previously understood mainly as a consequence of relapsing acute neuroinflammatory events, symptoms mainly or exclusively associated with brain damage from acute neuroinflammation. In the meantime, however, it turns out that the clinical outcome is determined more by a less well-defined chronic inflammatory process, which can be understood as LLNI. In MS the acute inflammatory exacerbations nevertheless play an important role, reflected for example in the updated McDonald criteria. But, even in MS, CSF investigation can lead to earlier and better diagnosis through improved quality of differential diagnosis (Stangel et al. 2013). It is apparent that CSF analysis is the most sensitive approach to neuroinflammatory disorders in general (Wildemann et al. 2010). Thus, CSF investigations represent the most interesting clinical approach to detecting possible LLNI in psychiatric patients.

Therefore, we have focused on CSF research for some years, and accumulated considerable evidence that large subgroups of psychiatric patients with affective and schizophrenic spectrum disorders show definite or probable signs of LLNI. About 70 % of cases with affective or schizophrenic spectrum (including bipolar) disorders with prolonged courses demonstrated some CSF abnormalities. These included rare cell number increases, signs of CSF cell activation by surface markers with normal cell numbers, definite LLNI by intrathecal oligoclonal bands or immunoglobulin production, blood CSF barrier dysfunction, and/or CSF neopterin increases (Maxeiner et al. 2009; Bechter et al. 2010a, b; Kühne et al. 2013). In addition, in a recent study we found in these patient groups clearly elevated lymphokines in both blood and CSF, but the focus of pathology may be within the CSF. Interestingly the cytokine patterns of the psychiatric patients showed similarities to those found in the study of patients with multiple sclerosis or idiopathic facial palsy, whereas the cytokine patterns differed clearly from those in bacterial or viral meningoencephalitis (Maxeiner et al. 2014). Together, these CSF findings clearly demonstrate some LLNI process or immune activation focused on the intrathecal spaces in a large subgroup of these psychiatric patients, and therefore LLNI is apparently involved in CNS functions. Overall, these results can be interpreted as signs of LLNI and not simply as an effect of peripheral inflammation, although these studies contained evidence of crosstalk between periphery and the brain.

Autoimmune Encephalitis (AE) and Limbic Encephalitis (LE)

Interestingly, CE was redefined after new insights into LE emerged from the use of newly discovered neuronal autoantibodies, especially NMDAR antibodies (Dalmau et al. 2008). In the meantime, more and more other neuronal autoantibodies were detected and still more are plausibly predicted (Prüss 2013). AE cases present varying and various clinical syndromes including epilepsies, which were previously not attributed to either LE or AE (Gable et al. 2012; Lancaster et al. 2010; Dalmau et al. 2011; Frisch et al. 2013; Prüss 2013). Some single cases were convincingly shown to be cases of AE, although defined autoantibodies were missing in the blood or CSF of these cases (Najjar et al. 2013). In effect, the threshold for the clinical diagnosis of LE, CE, and AE was lowered through improved laboratory methods. The present definition is broader and consequently more cases are included in the CE category than previously. This recent development matches the predictions of the ME hypothesis that with improved diagnostic methods the clinical assessment of LLNI and ME will be possible and the definitions and borders between CE and hypothetical ME may change (compare Bechter 2001). At the same time, other assumptions of the ME hypothesis gained more plausibility, for example that LLNI has to be considered a relevant clinical issue and not prematurely assigned as noninflammatory status.

Clinically undefined LE cases were collected by Gerd Huber during the 1950s: lethal cases of catatonic schizophrenia were probably not infrequent, largely because of limited intense medical care available at the time. In about 50 % of cases, postmortem histopathology detected slight signs of encephalitis (reviewed in Bechter 2001). These observations agree well with the recent insights into antiNMDAR encephalitis. The latter is graded now in clinical stages from prodromal to psychotic and/or seizure phase and unresponsive, hypokinetic phase, overall with major psychiatric symptoms (Peery et al. 2012). Also, the assumption that longer-lasting LLNI states may exist for at least weeks and be of clinical relevance is strongly supported by these new experiences with AE. Another supporting argument is also the new insights into the pathogenesis of multiple sclerosis (MS). MS is, however, a special case with relatively strong and frequently localized inflammatory lesions, for which neuroimaging techniques have the necessary sensitivity to identify. But neuroimaging is hardly able to detect LLNI (Bechter 2005).

In summary, it is apparent that CSF investigation is the most important method for the diagnosis of AE, CE, or LE and most cases of LLNI. One should note that some cases of AE may even present normal CSF and neuroimaging findings or inconsistent pathological findings (Venkatesan et al. 2013). The best diagnostic sensitivity is therefore achieved with a bundle of clinical methods in CE and AE (Dalmau et al. 2011; Najjar et al. 2013), and even more so in ME (Bechter 2013). It is also clear that the diagnostic sensitivity and its limitations directly impact the clinical categorization.

Utility and Limitations of Clinical Categorization

In clinical medicine, rather strict categorization of diseases is preferred. In neurology, this involves the differentiation between inflammatory and noninflammatory CNS disorders. CE was established in neurology by a combination of clinical and diagnostic methods; for example, fever, headache, and altered level of consciousness (Roos 1999) were important. However, classical clinical symptoms were losing their importance in comparison with criteria with refined methods, e.g., in AE (Dalmau et al. 2011; Peery et al. 2012; Lancaster et al. 2011; Prüss 2013). Behind the strict categorization in the clinical field stands the primary goal of obtaining reliable guidelines for a practical approach to therapeutic management and clinical studies. However, such an approach implies that other aspects may be neglected, especially the dimensional aspects of a disease over time and space that are a basic reality of inflammatory processes. In the case of neuroinflammation, an additional major disadvantage comes into play: it is hard to apply diagnostic methods to the brain, blood examinations are not very informative, and CSF is taken in cases only with serious clinical background, because lumbar puncture is a diagnostic procedure with more risks than simply taking blood. Furthermore, under some conditions brain imaging of CNS inflammation is sensitive but not highly sensitive, even in acute meningoencephalitis. For example, only in rare cases are lesions found directly in the brain parenchyma, whereas meningeal enhancement of contrast agent indicating increased meningeal blood flow is relatively frequent (Osborn et al. 2010). In LLNI, which is the subject of this section, brain imaging is rather insensitive, although some cases of suspected LLNI may present an increased prevalence of very small white matter lesions, as found, for example, in our study group of BDV antibody seropositive cases (Bechter et al. 1995).

It was stated above that the most important method for diagnosis of LLNI is CSF examination, the gold standard for the diagnosis of neuroinflammation in general (Glaser et al. 2006; Wildemann et al. 2010). In translational research there is, however, another problem: the typical laboratory animals are small and only minute volumes of CSF can be taken and only in the brain area, not at the lumbar site (with available methodology). This is different from humans, thus hampering considerably the translation of research results into the clinic and vice versa.

In the clinic, CSF is taken by lumbar puncture. Here several problems exist: (1) It is difficult to obtain groups of healthy controls and mostly patients with suspected or specific neurological disorders have to serve as controls. This causes a problem in interpreting the data. However, analysis of large amounts of data can in part compensate for the problem. Also, comparisons between different groups, for example patients with acute bacterial or viral meningoencephalitis or MS, are especially informative (Maxeiner et al. 2014; Bechter et al. 2010a, b; Kühne et al. 2013). (2) Another problem is the need for sophisticated CSF data analysis, because the protein concentrations in CSF are much lower than in the blood and CSF cells are selectively prevalent. There is clear evidence that the involvement of aspects of CSF flow patterns within the subarachnoid spaces can explain varying concentrations of blood-derived and also of brain-derived proteins (Reiber 1994; Kühne et al. 2013). Although well-established rules in neurology have been developed (Reiber and Peter 2001; Stangel et al. 2013; Wildemann et al. 2010), considerable problems remain. That means the limitations in the sensitivity of CSF examination and misinterpretation remain problems in detecting CNS inflammation. One important source of misinterpretation is the variability of CSF flow through the subarachnoid spaces. Driven primarily by the pulsating brain (Gupta et al. 2009, 2010; Kurtcuoglu 2012), the CSF flows, including the CSF outflow patterns, are only partially understood (Review in Bechter 2011; Carare et al. 2014, compare also the studies by the Seyfert et al. groups). (3) Experimental models, such as in BDV infection, clearly support the view that both grading of neuroinflammation into severity degrees and localization would be needed (Herden et al. 2000; Kramer et al. 2012) to better understand the clinical relevance and achieve more exact cause–response associations. Such a view has also been suggested for clinical staging in antiNMDAR encephalitis (Peery et al. 2012). (4) Presently available experimental studies almost never tell us what happens within CSF spaces in parallel to neuroinflammation, which has been carefully studied in the brain; translation to the clinic is thus difficult. Furthermore, there are considerable differences with regard to CSF flow between preferred laboratory animals and humans. The restricted conditions of studies, like missing motor activity due to anesthesia in most animal studies, further limit the extrapolation to humans (Bechter 2011).

An interesting spotlight on the problems involved in the clinical diagnostics of neuroinflammation is evidenced by recent findings in MS: MS was believed to represent mainly a white matter disease and to rarely involve the gray matter. However, in a careful study with experimental autoimmune encephalitis (EAE), it was shown that inflammation in EAE begins at the meninges (Brown and Sawchenko 2007; Ransohoff 2009). These findings represented something like a revolution (Bechter and Brown 2013), and were surprisingly rapidly confirmed in human MS from a worldwide collection of rare cases in which a brain biopsy had been taken (Lucchinetti et al. 2011). The importance of diagnostic aspects relating to clinical categorization is highlighted in AE (Najjar et al. 2013): In a case without any evidence of neuronal autoantibodies and with normal CSF, only a brain biopsy was sensitive enough to prove neuroinflammation. The chronic disease course over years showed predominant psychiatric symptoms. Onset of disease in this girl was at age 15; she was observed over 10 years with repeated careful broad-spectrum diagnostic approaches and experimental immune treatments. The evidence accumulated during a highly variable course is striking: Most diagnostic methods applied yielded normal results, but overall an intense diagnostic approach definitely supported the diagnosis of AE/CE by brain biopsy. One may assume that over a long time period some LLNI process or localized CE process at the cortex was clinically relevant.

In summary, it appears that the definition of neuroinflammation differs between research fields and clinical fields and even within sections of those fields, depending upon the methods in general and on the case characteristics, such as varying process and symptom severity and associated varying diagnostic sensitivity. Overall there is convincing evidence that, with improved and broadened diagnostic approaches, but especially with CSF diagnostics, a more reliable diagnosis of ME is becoming feasible. Methods need to be further improved and results to be confirmed in larger studies. Surprisingly, some have proposed that encephalitis should not be considered a clinical diagnosis in general. Instead, one should speak of encephalopathy (Deakin et al. 2014). This view is not shared here, as many specialists in neurology have left no doubt about the clinical diagnosis of encephalitis (see above) and of the evidence that LLNI is also meaningful. In addition, in cases of typical schizophrenia, a retrospective analysis demonstrated some cases that were rediagnosed as AE when the newly introduced neuronal antibody assays were applied (Steiner et al. 2013). The course and the outcome of AE clearly depends on the type of the antibodies, the pathogenetic mechanisms involved, and the cerebral areas involved (Frisch et al. 2013). A more selective approach to diagnostics seems to be behind the idea of encephalopathy. This approach is focused on neuronal autoimmune antibodies, which indeed may be toxic or may disturb function, and therefore may be considered encephalopathic. However, when a broader approach was used and more diagnostic procedures were involved, more pathological alterations were detected, thus contributing to the perspective of the ME theory. When we discuss the autoimmune encephalopathy model further, the critical evaluation of AE using modified Koch's postulates showed only one of four criteria being clearly fulfilled (Coutinho et al. 2014).

One may criticize the ME theory as being too weak a categorization. The difficulty of demonstrating the relevance of specific LLNI mechanisms, including what is preliminary considered as LLNI, is however nicely demonstrated in models of persisting CNS infections (Bentivoglio et al. 2011; Bechter 2013), or in studies on persistent infection of cerebral vasculature, as in cerebral malaria (Polimeni and Prato 2014). Despite a lot of research into malaria, much remains unknown about the mechanisms of various degrees of severity of increased blood–brain barrier permeability elicited by toxic molecules from the parasite or the host, although the parasite clearly is present in the brain vasculature.

Clinically mild encephalitis/encephalopathy with a reversible splenial lesion is another clinical entity (Okamoto et al. 2014) that is different from ME and fulfills largely the criteria for CE. But these examples demonstrate nevertheless the complexity and dimensional aspects of neuroinflammation as stressed with ME theory here.

Types of Classical or Mild Encephalitis in Idealized Models

Based on the above outline of CE (Fig. 5.1), AE (Fig. 5.2), LE (Fig. 5.1, 5.2), the focus now is on a model of ME (Fig. 5.3) with relevance for severe psychiatric disorders. The graphs are simplified or idealized models of hypothetical types of classical and mild encephalitis in a dimensional severity by time demonstration and with some idea about the underlying etiologies, which can be quite varied. This is speculative, but may nevertheless give some useful impression of the assumed clinical appearance and the suggested underlying pathogenetic processes, which surely include pro- and anti-inflammatory, neurodegenerative, and repair processes.

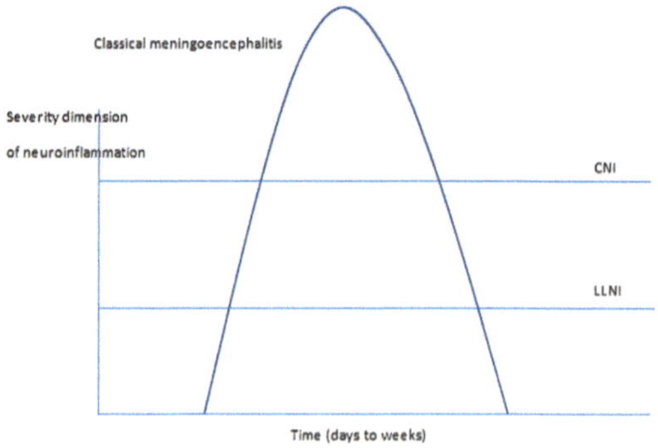

Fig. 5.1 CE usually develops within a short time period of hours to days and involves strong neuroinflammatory alterations within CSF spaces and the brain parenchyma. The latter has been demonstrated in neuroimaging in a subgroup of cases only (compare Osborn et al. 2010)

The models presented here focus on clinical course. Others have presented aspects of the pathogenesis involving local and systemic immune cell responses in acute and chronic CNS damage, including the resolution of neuroinflammation, which represents an active process (Schwartz and Baruch 2014).

Some Historical Remarks on Etiology Research in Psychiatric Disorders

After about 50 years of strong rejection and some additional 20 years of debate, the initial hypothesis that syphilis might represent an important risk factor for general paresis was accepted, and later turned into a strict view. The spirochete, although detected in the brain of only very few cases, represents the infectious cause of general paresis (GP) (compare Bechter 1995). A major factor contributing to the acceptance of the causality relationship in general was the similarity of findings in late syphilis with findings in the brains of dogs with experimental trypanosomiasis, as well as the successful treatment of the disease with malaria (and later penicillin). Also very important was considerable progress at the time in laboratory diagnosis with the introduction of the Wassermann reaction (a preliminary type of antibody testing) and of lumbar puncture and CSF analysis. To imagine the relevance of this breakthrough, one should recognize that, at the time, GP was very frequent and often fatal. For example, in Hamburg around 1900 about 20 % of psychiatric inpatients suffered from GP. Thus, the new principle of understanding GP as a late stage of an infectious disease was like a revolution. This also triggered the idea of searching for other infectious agents causing severe psychiatric disorders like schizophrenia, but this turned out to be difficult. Today, many evidently think that the pathogenesis of GP is very clear and causal. This is apparent in remarks from many colleagues who, when they are initially presented with the ME hypothesis,

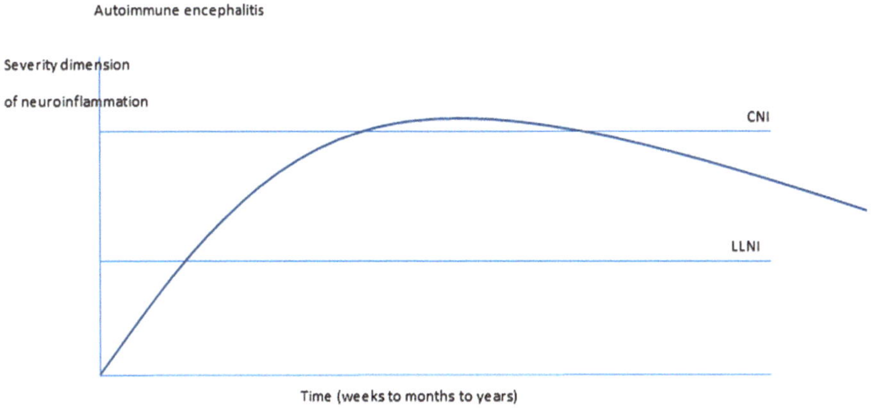

Fig. 5.2 AE includes prodromal symptoms over weeks, followed by more or less severe neurological symptoms after some weeks and variable, not infrequently long, recovery times (compare Dalmau et al. 2011; Najjar et al. 2013; Prüss 2013; Peery et al. 2012; Lancaster et al. 2011). In principle the graph resembles CE in Fig. 5.1, but the timeline, especially of the recovery phase, is more extended, and acute neuroinflammation is less severe

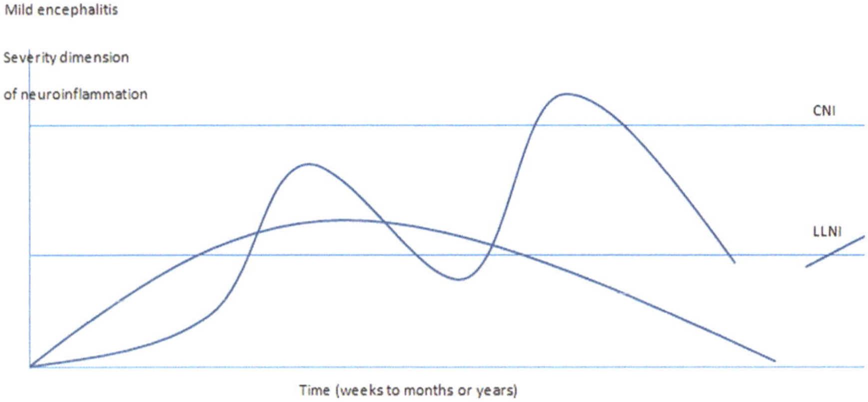

Fig. 5.3 The ME model is based on experimental findings that demonstrated slow onset and generally minor or localized classical encephalitis, the latter only poorly reflected in clinical symptoms. Human ME cases will often show normal brain images and normal CSF findings, if advanced methods are not used. Minor LLNI stages might persist for a long time before symptoms develop and coincide with long prodromal stages, as observed in schizophrenia (Häfner 1995; A. Riecher-Rössler). A similarity between basic symptoms in schizophrenia as described by Gerd Huber and initial symptoms in LE is worth noting (review in Bechter 2002). The major aspect of the ME definition is that some LLNI state should be detectable in a cross-sectional assessment during the active disease phase, especially by CSF examination. ME may not simply reflect secondary effects from systemic inflammation without any autochthonous neuroinflammation. Contributing factors may be many, not least neurodevelopmental alterations (Müller and Bechter 2013)

nevertheless view it critically or reject it. Yet, when we go into great detail about the present knowledge of the pathogenesis of GP, it appears that, in a strict scientific sense, our understanding of the pathogenesis remains quite limited. For example, we cannot explain why, after a long latency of years, GP develops only in a limited subgroup of the infected. This is a core problem of medical understanding that is involved in the further discussion of the ME theory here.

A similar problem was gastroduodenal ulcer disease in people infected with *Helicobacter pylori* (see Marshall 2006). Indeed, many infections are characterized by low overall pathogenicity (number of diseased per number of infected). Such was also the case with poliomyelitis, raising the debates over 20 years: only 2 % of patients get the terrible paresis but many show the serum antibodies. The same is true for EBV infection, CMV infections, or borreliosis (Bechter 2013). The recent studies on risk factors in schizophrenic and affective spectrum disorder represented by infections and autoimmune disorders and brain trauma (Benros et al. 2012, 2013, 2014; Sörensen et al. 2014) are therefore extremely important, because conclusive epidemiologic data from the Danish National Register provide a basis for reinforced clinical studies. These specific risk factors nicely match with the ME theory. But one should be aware of considerable problems with such research, when treating a patient. This is again highlighted by looking back into the history of GP: At the end of the twentieth century, leading newspapers unjustifiably criticized the awarding of the Nobel Prize to Julius Wagner-Jauregg for the introduction of malaria therapy. This appears to have been related to the often observed tendency to simplify questions of causality. The story of malaria therapy, then fever therapy, and later penicillin, raised a hype to search for other infectious causes of psychiatric disorders. The search remained unsuccessful for a long time. A similar revival followed after the description of the first slow virus infection by Gajdusek and others (Gajdusek 1965; Kurstak 1991; Bechter 1995).

Multiple Interacting Systems and Factors

Severe psychiatric disorders of the affective and schizophrenic (including bipolar) spectrum are complex diseases where many factors should be considered: certainly genes, age of onset, timing, stress, infections and autoimmunity, and possibly chance (Drexhage et al. 2010; Gibney and Drexhage 2013; Hertz et al. 2013; Rector et al. 2014). The situation is rather similar to systemic autoimmune disorders, where the idea is prevalent that "every autoimmune disorder has to be considered as of infectious origin unless otherwise proven" (repeated public remarks by Yehuda Shoenfeld, congress president at the ninth International Autoimmunity Congress, Nice, France, March 26–30, 2014), outlined in more detail in textbook "Infection and Autoimmunity," (Shoenfeld and Rose 2004).

Interestingly, in many systemic autoimmune disorders, there is a considerable subgroup of patients suffering from comorbid psychiatric symptoms. In some, there are even very severe psychoses over certain time periods, for example in lupus erythematosis (Diamond and Volpe 2012) and especially in cases with antiphospholipid autoantibodies (D'Ippolito et al. 2014). Time and duration of pathology is also definitely involved, as evidenced, for example, in multiple sclerosis (Rossi et al. 2014). Relevance of time was also directly investigated: the duration of systemic inflammation elicited different effects over the long term as compared to the short term (Maggio et al. 2013). But pathomechanisms are difficult to assess in individual cases. Risk genes of psychiatric disorders may be directly implicated in the life cycles of specific pathogens (Carter 2009). Some infections may be relevant in people with a specific genetic background (Kumarasinghe et al. 2014), or a specific genetic make-up like sphingomyelinase-ceramide system in depression (Gulbins et al. 2013), known to be involved in inflammatory responses. At-risk states showing an inflammatory molecular signature in both CSF and blood are associated with infectious agents in schizophrenia; however, abnormalities were even exacerbated in risk subjects (Hayes et al. 2014). Both systemic and CNS-specific tryptophan pathways appear to be involved. This has been demonstrated not only for IDO and cytokines, but also protective factors and cannabinoid receptor; all of these findings are linked to an activated inflammatory response (Myint and Kim 2014; Brietzke et al. 2009; Anderson et al. 2013; Sánchez-Blázquez et al. 2014). Stress represents another player, for example in psychosocial factors (Meyer-Lindenberg 2010). A relatively well-defined neurological autoimmune disorder, Guillain–Barre syndrome, associated with preceding *Campylobacter jejuni* infection, is frequently accompanied by psychiatric symptoms that include psychosis (Schielke et al. 2014). New insights into the role of the gut microbiome and the immune system of the host suggest a continuous battle taking place, with possible strong influences on CNS disorders (Wang and Kasper 2014). This also holds true for experimental autoimmune encephalomyelitis (Berer et al. 2011, 2014).

The role of prenatal infections has been reviewed repeatedly and the increase in risk from various specific infectious agents was considerable (Brown 2011), though not specific for schizophrenia but rather a spectrum of psychiatric syndromes (Bechter 2013). A set of experimental findings on infections during the intrauterine life showed subsequent disturbances of neurodevelopment or of immune development with schizophrenia-like symptoms in adulthood (Meyer 2013; Juckel et al. 2011). A recent model demonstrated ongoing inflammation within the brain, supporting the ME theory (Mattei et al. 2014). It should be recognized that neither models nor humans studies show an association between neurodevelopmental abnormalities and specific disorders. For example, minor neurodevelopmental abnormalities are also found in depression (Whittle et al. 2014). This points to the lack of specificity of brain seizures and strokes, with the consequence of overlapping scenarios (see also Buchsbaum and Rieder 1979; Bechter 2012). The recently discovered inflammasome is also involved in neuroinflammation, but is only beginning to be studied (Walsh et al. 2014). This will surely have relevant influences on microglia, which represents an important CNS cell system, relevant not only in pathology but also in normal physiology and in the developing brain (Kettenmann et al. 2013). Activation of microglia has been demonstrated in these psychiatric disorders (discussed in another chapter of this book).

Overall, there are many pathways and risk factors involved in the pathophysiology of LLNI, the basis of the psychiatric disorders in question here. Apparently, the scenario is extremely complex and difficult to dissect. To understand LLNI as either one common final pathway or alternatively several LLNI types or states appears plausible. Though it remains to be better defined, LLNI has the power to elicit and explain various symptoms or psychiatric disorders observed. The ME theory suggests that a subgroup of severe psychiatric disorders showing signs of LLNI are causally related to LLNI, although associated with varying symptom patterns, depending upon contributing and preexisting factors, e.g., genetic factors and many others (Bechter 2013; Müller and Bechter 2013).

Brain Barriers and Systemic Crosstalk

Systemic inflammation and inflammatory signaling can directly influence brain functions (Thomson et al. 2014) and elicit a pattern of symptoms described as sickness behavior (Dantzer 2012), in part fitting with symptoms observed in psychiatric spectrum disorders discussed here as the main candidates of ME theory. But sickness behavior cannot explain many aspects of these diseases nor their courses. Apparently, complex pathogenetic events are related plausibly to neuroinflammation. Of special interest here are recent insights from experimental neuroimmunology on the pathogenesis of chronic neuroinflammation (Schwartz and Baruch 2014): in general, acute inflammation needs to be actively resolved; otherwise, it becomes chronic, and this seems to be true also for neuroinflammation. The choroid plexus (CP) represents an interface between blood and CSF spaces; immune cells are recruited and educated at the CP and then released into the CSF. But there are several brain barriers. These barriers must be explained briefly here. One should differentiate between the blood–brain barrier (BBB) and the blood–CSF barrier (BCSFB). The BCSFB is defined as an anatomical structure, the CP, or the CP plus circumventricular organ (Schwartz and Baruch 2014; Wolburg and Paulus 2010).

The BBB is generally regarded as representing a barrier between brain vasculature and brain parenchyma (Engelhardt and Ransohoff 2012). However, the function of the BBB has to be further differentiated, one for cells and one for solutes, with anatomically differing sites (Bechmann et al. 2007). In clinical terminology, blood–CSF barrier (BCB) dysfunction is used differently. One paper (Reiber and Peter 2001) describes certain abnormal CSF findings, in effect a proportional albumin increase, which represents an increase of blood-derived proteins within the CSF. BCB dysfunction may mainly relate to blood CSF flow pattern changes or CSF production (Reiber 1994; Seyfert and Faulstich 2003; Seyfert et al. 2004, 2009), but not true dysfunction of the anatomically defined BCB or BBB. In addition, in such a complex scenario a new pathway has to be included, the so-called glymphatic pathway of CSF through the brain, characterized by periarterial inflow and perivenous outflow of CSF (Benveniste et al. 2014; Iliff et al. 2012). The CSF flow pulsates dramatically from the ventricles to the subarachnoid spaces and back;

the flow distance is about 10 cm per pulse back and forth at the aqueduct (Gupta et al. 2009, 2010; Kurtcuoglu 2012). It has so far been only partially investigated, and poorly down the neuraxis. In addition, the pulsating brain drives the flow of the extracellular CNS fluid, the flow has been studied in some detail and described as volume transmission mode balancing the wiring transmission (the latter mainly represented by synaptic transmission). This exchange includes not only molecular but also particulate fluid contents, e.g., exosomes (Fuxe et al. 2013). Understanding the CSF flow patterns and their respective influences on CSF content concentrations is of prime importance.

The long known absorption pathway of CSF to arachnoid granulation is not the only one. There is considerable outflow through the area called cribrosa to nasal submucosa and then to the cervical lymph nodes (Carare et al. 2014) and along all brain nerves and peripheral nerves into peripheral tissues (Bechter 2011). Interestingly, CSF cells can also follow these CSF outflow pathways through the area, cribrosa (Kaminski et al. 2012), and along lumbar nerves (Schmitt et al. 2011). Signal transmission by extracellular vesicles is now known to be relevant for CSF immune response (Robbins and Morelli 2014). The role of CSF cells as an independent player has been newly introduced. Immune cells apparently support normal brain functions, e.g., they support the cortex and cognitive functions directly (Kipnis et al. 2012; Baruch et al. 2013). CSF cells educated at the CP play a critical role in the interaction between neuroinflammation and neurodegeneration (Schwartz and Baruch 2014). This latter point is very interesting with regard to the historical controversies about the pathology of general paresis (see also above): the histopathologists in particular rejected the hypothesis that syphilis was a possible risk factor for general paresis (GP), not least because they preferred to distinguish between inflammatory and neurodegenerative GP (Review in Bechter 1995).

Outlook

From first ideas formulated as the mild encephalitis hypothesis, some experimental treatment approaches based on these ideas were developed. CSF filtration was adopted from neurological research, where it had been successful in treating therapy-resistant Guillain–Barre syndrome. Its use resulted in a considerable improvement over the long term in cases of therapy-resistant schizophrenia or depression, though only in a small number of cases (Bechter et al. 1999, 2000; Bechter 2007). Skepticism of many colleagues at that time about such experimental treatment approaches was evident. Now one can notice a dramatic change of scientific mainstream thinking. Many researchers accept the possibility that a considerable number of subgroups of severe psychiatric disorders include some neuroinflammatory process, which may at least contribute to, if not causally underlie, these disorders. However, the pathogenesis remains incompletely understood. Previous theories were based on specific or single neurotransmitter or receptor models. The ME theory also takes a simplified point of view by defining as a main criterion the ability to detect some LLNI process at a relevant stage of the disease in the individual patient. Such an approach seems nevertheless to be required in the clinic. Risk factors or contributing factors are an additional prerequisite. The ME theory fits with such aspects as increased risk through intrauterine injuries, through neurodevelopmental abnormalities or through immune developmental abnormalities, especially with the risk increase by infectious and autoimmune disturbances and head trauma during lifetime. These latter findings are in accordance with the association between infections and autoimmune disorders in general. The investigation of specific triggers of ME in an epidemiological scenario with respect to psychiatric disorders represents important progress. However, it remains a challenge to investigate the role of specific infections in the individual case, and to differentiate LLNI processes, perhaps by neuroimaging, CSF investigation and other new methods (Bechter 2013). Methods need to be improved. Newly combined methods like the TMS/EEG may improve direct insight into the specific pathology associated with certain symptoms. With combined long-term studies it may be possible to differentiate between different patterns of brain deficits, such as between normal age effects and prodromal and illness-related alterations (Kumarasinghe et al. 2014). Molecular profiling in blood and CSF may be helpful (Herberth et al. 2014). New pathways like microRNA dysregulation may need to be investigated.

In summary, psychiatric research should focus on etiopathogenetic research on LLNI and the ME theory.

References

Anderson G, Berk M, Dodd S, Bechter K, Altamura AC, Dell'osso B, et al. Immuno-inflammatory oxidative and nitrosative stress and neuroprogressive pathways in the etiology, course and treatment of schizophrenia. Prog Neuropsychopharmacol Biol Psychiatry. 2013;42:1–4.

Baruch K, Ron-Harel N, Gal H, Deczkowska A, Shifrut E, Ndifon W, et al. CNs-specific immunity at the choroid plexus shifts toward destructive Th2 inflammation in brain aging. Proc Natl Acad Sci U S A. 2013;110(6):2264–9.

Bechmann I, Galea I, Perry VH. What is the blood–brain barrier (not)? Trends Immunol. 2007;28(1):5–11.

Bechter K. Research strategies in "slow" infections in psychiatry. Hist Psychiatry. 1995;6: 503–11.

Bechter K. Mild encephalitis underlying psychiatric disorder – a reconsideration and hypothesis exemplified on Borna disease. Neurol Psychiatry Brain Res. 2001;9:55–70.

Bechter K. Basic symptoms in symptomatic schizophrenia. Neurol Psychiatry Brain Res. 2002;10: 35–40.

Bechter K. Neuroimaging in the diagnosis of encephalitis as a cause of psychoses. Neurol Croat. 2005;54 Suppl 2:59–63.

Bechter K. Cerebrospinal fluid may mediate pathogenic effects on nerves via efflux: a hypothesis from unexpected improved pain syndromes with cerebrospinal fluid filtration. Neurol Psychiatry Brain Res. 2007;14:37–42.

Bechter K. The peripheral cerebrospinal fluid outflow pathway - physiology and pathophysiology of CSF recirculation: a review and hypothesis. Neurol Psychiatry Brain Res. 2011;17(3):51–66.

Bechter K. Diagnosis of infectious or inflammatory psychosyndromes. Open Neurol J. 2012;6 (Suppl 1-M6):113–8.

Bechter K. Updating the mild encephalitis hypothesis of schizophrenia. Prog Neuropsychopharmacol Biol Psychiatry. 2013;42:71–91.
Bechter K, Brown D. Neuroinflammation in psychiatric disorders – evidence form research and clinic. Neurol Psychiatry Brain Res. 2013;19:139–40.
Bechter K, Bauer M, Estler HC, Herzog S, Schüttler R, Rott R. Erweiterte Kernspintomographische Untersuchungen bei Borna-disease-Virus seropositiven psychiatrischen Patienten und Kontrollen. Nervenarzt. 1994;65:169–74.
Bechter K, Herzog S, Behr W, Schüttler R. Investigations of cerebrospinal fluid in Borna disease virus seropositive psychiatric patients. Eur Psychiatry. 1995;10(5):250–8.
Bechter K, Herzog S, Schreiner V, Wollinsky KH, Schüttler R. Cerebrospinal fluid filtration in a case of schizophrenia related to 'subclinical' Borna disease virus encephalitis. In: Müller N, editor. Psychiatry, psychoneuroimmunology and viruses. Key topics in brain research. Wien: Springer; 1999. p. 19–35.
Bechter K, Herzog S, Schreiner V, Brinkmeier H, Aulkemeyer P, Weber F, Schüttler R. Borna disease virus-related therapy-resistant depression improved after cerebrospinal fluid filtration. J Psychiatr Res. 2000;34:393–6.
Bechter K, Wittek R, Seitz K, Antoniadis G. Personality disorders improved after arachnoid cyst neurosurgery, then re-diagnosed as 'minor' organic personality disorders. Psychiatry Res. 2010a;184:196–200.
Bechter K, Reiber H, Herzog S, Fuchs D, Tumani H, Maxeiner H-G. Cerebrospinal fluid analysis in affective and schizophrenic spectrum disorders: identification of subgroups with immune responses and blood–CSF barrier dysfunction. J Psychiatr Res. 2010b;44:321–30.
Benros ME, Mortensen PB, Eaton WW. Autoimmune diseases and infections as risk factors for schizophrenia. Ann N Y Acad Sci. 2012;1262:65–6.
Benros ME, Waltoft BL, Nordentoft M, Ostergaard SD, Eaton WW, Krogh J, Mortensen PB. Autoimmune diseases and severe infections as risk factors for mood disorders: a nationwide study. JAMA Psychiatry. 2013;70(8):812–20.
Benros ME, Eaton WW, Mortensen PB. The epidemiologic evidence linking autoimmune diseases and psychosis. Biol Psychiatry. 2014;75:300–6.
Bentivoglio M, Mariotti R, Bertini G. Neuroinflammation and brain infections: historical context and current perspectives. Brain Res Rev. 2011;66:152–73.
Benveniste H, Budassi M, Smith SD, Yu M, Lee H, Nedergaard M, Vaska P. Brain-wide glymphatic waste drainage characterized by PET-MRI. Neurol Psychiatry Brain Res. 2014;20:4.
Berer K, Mues M, Koutrolos M, Rasbi ZA, Boziki M, Johner C, et al. Commensal microbiota and myelin autoantigen cooperate to trigger autoimmune demyelination. Nature. 2011;479(7374):538–41.
Berer K, Imaoka A, Umesaki Y, Johner C, Wekerle H, Krishnamoorthy G. Commensal microbiota triggers spontaneous autoimmune encephalitis. Neurol Psychiatry Brain Res. 2014;20:5.
Brietzke E, Sterzt L, Fernandes BS, Kauer-Sant'Anna M, Mascarenhas M, Vargas AE. Comparison of cytokine levels in depressed, manic and euthymic patients with bipolar disorder. J Affect Disord. 2009;116:214–71.
Brown AS. Exposure to prenatal infection and risk of schizophrenia. Front Psychiatry. 2011;2:63.
Brown DA, Sawchenko PE. Time course and distribution of inflammatory and neurodegenerative events suggest structural bases for the pathogenesis of experimental autoimmune encephalomyelitis. J Comp Neurol. 2007;502(2):236–60.
Buchsbaum MS, Rieder R. Biologic heterogeneity and psychiatric research. Arch Gen Psychiatry. 1979;xxxvi:1163–9.
Carare RO, Hawkes CA, Well RO. Afferent and efferent immunological pathways of the brain. Anatomy, function and failure. Brain Behav Immun. 2014;36:9–14.
Carter CJ. Schizophrenia susceptibility genes directly implicated in the life cycles of pathogens: cytomegalovirus, influenza, herpes simplex, rubella, and Toxoplasma gondii. Schizophr Bull. 2009;35(6):1163–82.
Cherner M, Cysique L, Heaton RK, Marcotte TD, Ellis RJ, Masliah E, et al. Neuropathologic confirmation of definitional criteria for human immunodeficiency virus-associated neurocognitive disorders. J Neurovirol. 2007;13:23–8.

Coutinho E, Harrison P, Vincent A. Do neuronal autoantibodies cause psychosis? A neuroimmunological perspective. Biol Psychiatry. 2014;75:269–75.

D'Ippolito S, Meroni PL, Koike T, Veglia M, Scambia G, Di Simone N. Obstetric antiphospholipid syndrome: a recent classification for an old defined disorder. Autoimmun Rev. 2014;13(9):901–8.

Dalmau J, Gleichman AJ, Hughes EG, Rossi JE, Peng X, Lai M, et al. Anti-NMDA-receptor encephalitis: case series and analysis of the effects of antibodies. Lancet Neurol. 2008;7(12):1091–8.

Dalmau J, Lancaster E, Martinez-Hernandez E, Rosenfeld MR, Balice-Gordon R. Clinical experience and laboratory investigation in patients with anti-NMDAR encephalitis. Lancet Neurol. 2011;10:63–74.

Dantzer R. Depression and inflammation: an intricate relationship. Biol Psychiatry. 2012;71(1):4–5.

Deakin J, Lennox BR, Zandi MS. Antibodies to the *N*-methyl-D-aspartate receptor and other synaptic proteins in psychosis. Biol Psychiatry. 2014;75:284–91.

Diamond B, Volpe BT. A model for lupus brain disease. Immunol Rev. 2012;248(1):56–67.

Drexhage RC, van der Heul-Nieuwenhuijsen L, Padmos RC, van Beveren N, Cohen D, Versnel MA, et al. Inflammatory gene expression in monocytes of patients with schizophrenia: overlap and difference with bipolar disorder. A study in naturalistically treated patients. Int J Neuropsychopharmacol. 2010;13(10):1369–81.

Engelhardt B, Ransohoff RM. Capture, crawl, cross: the T cell code to breach the blood–brain barriers. Trends Immunol. 2012;33(12):579–89.

Frisch C, Malter MP, Elger CE, Helmstaedter C. Neuropsychological course of voltage-gated potassium channel and glutamic acid decarboxylase antibody related limbic encephalitis. Eur J Neurol. 2013;20(9):1297–304.

Fuxe K, Borroto-Escuela DO, Tarakanov A, Fernandez WR, Manger P, Rivera A, et al. Understanding the balance and integration of volume and synaptic transmission. Relevance for psychiatry. Neurol Psychiatry Brain Res. 2013;19:141–58.

Gable MS, Sheriff H, Dalmau J, Tilley DH, Glaser CA. The frequency of autoimmune *N-methyl-d*-aspartate receptor encephalitis surpasses that of individual viral etiologies in young individuals enrolled in the California encephalitis project. Clin Infect Dis. 2012;54(7):899–904.

Gajdusek DC. Kuru in New Guinea and the origin of the NINDB study of slow, latent and temperate virus infections of the nervous system of man. In: Gajdusek DC, Gibbs Jr CJ, Alpers M, editors. Slow, latent and temperate virus infections. NINDB2. Bethesda: NINDB; 1965. p. 3–12.

Gibney SM, Drexhage HA. Evidence for a dysregulated immune system in the etiology of psychiatric disorders. J Neuroimmune Pharmacol. 2013;8(4):900–20.

Glaser CA, Honarmand S, Anderson LJ, Schnurr DP, Forghani B, Cossen CK, et al. Beyond viruses: clinical profiles and etiologies associated with encephalitis. Clin Infect Dis. 2006;43(12):1565–77.

Gulbins E, Palmada M, Reichel M, Lüth A, Böhmer C, Amato D, et al. Acid sphingomyelinase-ceramide system mediates effects of antidepressant drugs. Nat Med. 2013;19:934–8.

Gupta S, Soellinger M, Boesiger P, Paulikakos D, Kurtcuoglu V. Three-dimensional computational modeling of subjects specific cerebrospinal fluid flow in the subarachnoid space. J Biomech Eng. 2009;131:021010.

Gupta S, Soellinger M, Grzybowski DM, Boesiger P, Biddiscombe J, Paulikakos D, et al. Cerebrospinal fluid dynamics in the human cranial subarachnoid space. An overlooked mediator of cerebral disease. I. Computational model. J R Soc Interface. 2010;7:1195–204.

Häfner H. Epidemiology of schizophrenia. The disease model of schizophrenia in the light of current epidemiological knowledge. Eur Psychiatry. 1995;10(5):217–27.

Hayes LN, Severance EG, Leek JT, Gressitt KL, Rohleder C, Coughlin JM, et al. Inflammatory molecular signature associated with infectious agents in psychosis. Schizophr Bull. 2014;40(5):963–72.

Herberth M, Rahmoune H, Schwarz E, Koethe D, Harris LW, Kranaster L, et al. Identification of a molecular profile associated with immune status in first-onset schizophrenia patients. Clin Schizophr Relat Psychoses. 2014;7(4):207–15.

Herden C, Herzog S, Richt JA, Nesseler A, Christ M, Failing K, Frese K. Distribution of Borna disease virus in the brain of rats infected with an obesity-inducing virus strain. Brain Pathol. 2000;10(1):39–48.

Hertz L, Song D, Li B, Yan E, Peng L. Importance of inflammatory molecules, but not necessarily of inflammation, in the pathophysiology of bipolar disorder and in the mechanisms of action of anti-bipolar drugs. Neurol Psychiatry Brain Res. 2013;19:174–9.

Iliff JJ, Wang M, Liao Y, Plogg BA, Peng W, Gundersen GA, et al. A paravascular pathway facilitates CSF flow through the brain parenchyma and the clearance of interstitial solutes, including amyloid β. Sci Transl Med. 2012;4(147):147ra111.

Johnson RT. Viral infections of the nervous system. New York: Raven; 1982.

Juckel G, Manitz MP, Brüne M, Friebe A, Heneka MT, Wolf RJ. Microglial activation in a neuroinflammational animal model of schizophrenia – a pilot study. Schizophr Res. 2011; 131:96–100.

Kaminski M, Bechmann I, Pohland M, Kiwit J, Nitsch R, Glumm J. Migration of monocytes after intracerebral injection at entorhinal cortex lesion site. J Leukoc Biol. 2012;92(1):31–9.

Kettenmann H, Kirchhoff F, Verkhratsky A. Microglia: new roles for the synaptic stripper. Neuron. 2013;77(1):10–18.

Kipnis J, Gadani S, Derecke NC. Pro-cognitive properties of T cells. Nat Rev Immunol. 2012;12(9):663–9.

Kramer K, Schaudien D, Eisel ULM, Herzog S, Richt JA, Baumgärtner W, et al. TNF-overexpression in Borna disease virus-infected mouse brains triggers inflammatory reaction and epileptic seizures. PLoS One. 2012;7(7):e41476.

Kuehne LK, Reiber H, Bechter K, Hagber L, Fuchs D. Cerebrospinal fluid neopterin is brain-derived and not associated with blood–CSF barrier dysfunction in non-inflammatory affective and schizophrenic spectrum disorders. J Psychiatr Res. 2013;47:1417–22.

Kumarasinghe N, Rasser PE, Mendis J, Bergman J, Knechtel L, Oxley S, Perera A, et al. Age effects on cerebral grey matter and their associations with psychopathology, cognition and treatment response in previously untreated schizophrenia patients. Neurol Psychiatry Brain Res. 2014;20(2):29–36.

Kurstak E. Introduction. In: Kurstag E, editor. Psychiatry and biological factors. New York: Plenum; 1991. p. 1–5.

Kurtcuoglu V. Pulsatile cerebrospinal fluid flow in the cranial subarachnoid space. Neurol Psychiatry Brain Res. 2012;18:66–7.

Lancaster E, Lai M, Peng X, Huges E, Constantinescu R, Raizer J, et al. Antibodies to the $GABA_B$ receptor in limbic encephalitis with seizures: case series and characterization of the antigen. Lancet Neurol. 2010;9:67–76.

Lancaster E, Martinez-Hernandez E, Dalmau J. Encephalitis and antibodies to synaptic and neuronal cell surface proteins. Neurology. 2011;77:179–89.

Lucchinetti CF, Popescu BF, Bunyan RF, Moll NM, Roemer SF, Lassmann H, et al. Inflammatory cortical demyelination in early multiple sclerosis. N Engl J Med. 2011;365(23):2188–97.

Maggio N, Shavit-Stein E, Dori A, Blatt I, Chapman J. Prolonged systemic inflammation persistently modifies synaptic plasticity in the hippocampus: modulation by the stress hormones. Front Mol Neurosci. 2013;6:46.

Marshall B. Helicobacter connections. ChemMedChem. 2006;1(8):783–802.

Mattei D, Djodari-Irani A, Hadar R, Pelz A, Fernandes de Cossio L, Goetz T, et al. Minocycline rescues decrease in neurogenesis, increase in microglia cytokines and deficits in sensorimotor gating in an animal model of schizophrenia. Brain Behav Immun. 2014;38:175–84.

Maxeiner H-G, Rojewski MT, Schmitt A, Tumani H, Bechter K, Schmitt M. Flow cytometric analysis of T cell subsets in paired samples of cerebrospinal fluid and peripheral blood from patients with neurological and psychiatric disorder. Brain Behav Immun. 2009;23:134–42.

Maxeiner HG, Schneider M, Kurfiss S-T, Brettschneider J, Tumani H, Bechter K. Cerebrospinal fluid and serum cytokine profiling to detect immune control of infectious and inflammatory neurological and psychiatric diseases. Cytokine 2014;69:62–67.

Meyer U. Developmental neuroinflammation and schizophrenia. Prog Neuropsychopharmacol Biol Psychiatry. 2013;42:20–34.

Meyer-Lindenberg A. From maps to mechanisms through neuroimaging of schizophrenia. Nature. 2010;468:194–202.

Mosca M, Neri R, Bencivelli W, Tavoni A, Bombardieri S. Undifferentiated connective tissue disease: analysis of 83 patients with a minimum followup of 5 years. J Rheumatol. 2002;29(11):2345–9.

Müller M, Bechter K. The mild encephalitis concept for psychiatric disorders revisited in the light of current psychoneuroimmunological findings. Neurol Psychiatry Brain Res. 2013;19:87–101.

Myint AM, Kim YK. Network beyond IDO in psychiatric disorders: revisiting neurodegeneration hypothesis. Prog Neuropsychopharmacol Biol Psychiatry. 2014;48:304–13.

Najjar S, Pearlman D, Devinsky O, Najjar A, Nadkarni S, Butler T, et al. Neuropsychiatric autoimmune encephalitis without VGKC-Complex, NMDAR, and GAD autoantibodies: Case report and Literature Review. Cogn Behav Neurol. 2013;26(1):36–49.

Nathan C. Antibiotics at the crossroads. Nature. 2004;431(7011):899–902.

Okamoto T, Sato Y, Yamazaki T, Hayashi A. Clinically mild encephalitis/encephalopathy with a reversible splenial lesion associated with febrile urinary tract infection. Eur J Pediatr. 2014;173:533–6.

Osborn AG, Salzman KL, Barkovich AJ. Diagnostic imaging: brain. Philadelphia: Lippincott Williams & Wilkins; 2010.

Peery HE, Day GS, Dunn S, Fritzler MJ, Prüss H, De Souza C, et al. Anti-NMDA receptor encephalitis. The disorder, the diagnosis and the immunobiology. Autoimmun Rev. 2012;11:863–72.

Polimeni M, Prato M. Host matrix metalloproteinases in cerebral malaria: new kids on the block against blood–brain barrier integrity? Fluids Barriers CNS. 2014;11:1.

Prüss H. Comment: infection antedating autoimmunity—shared mechanisms in the brain? Neurology. 2013;81(18):1639.

Ransohoff RM. Immunology: in the beginning. Nature. 2009;462(7269):41–2.

Rector JL, Dowd JB, Loerbroks A, Burns VE, Moss P, Jarczok MN, et al. Consistent associations between measures of psychological stress and CMV antibody levels in a large occupational sample. Brain Behav Immun. 2014;38:133–41.

Reiber H. Flow rate of cerebrospinal fluid (CSF)—a concept common to normal blood–CSF barrier function and to dysfunction in neurological diseases. J Neurol Sci. 1994;122(2):189–203.

Reiber H, Peter JB. Cerebrospinal fluid analysis: disease-related data patterns and evaluation programs. J Neurol Sci. 2001;184(2):101–22.

Richt JA, Pfeuffer I, Christ M, Frese K, Bechter K, Herzog S. Borna disease virus infection in animals and humans. Emerg Infect Dis. 1997;3(3):343–52 (Review).

Robbins PD, Morelli AE. Regulation of immune responses by extracellular vesicles. Nat Rev Immunol. 2014;14:195–208.

Roos KL. Encephalitis. Neurol Clin. 1999;17(4):813–33.

Rossi S, Studer V, Motta C, Germani G, Macchiarulo G, Buttari F, et al. Cerebrospinal fluid detection of interleukin-1 beta in phase of remission predicts disease progression in multiple sclerosis. J Neuroinflammation. 2014;11:32.

Sánchez-Blázquez P, Rodriguez-Munõz M, Herrero-Labrador R, Burgueñño J, Zamanillo D, Garzón J. The calcium-sensitive Sigma-1 receptor prevents cannabinoids from provoking glutamate NMDA receptor hypofunction: implications in antinociception and psychotic diseases. Int J Neuropsychopharmacol. 2014;31:1–13.

Schielke A, Rosner BM, Stark K. Epidemiology of campylobacteriosis in Germany – insights from 10 years of surveillance. BMC Infect Dis. 2014;14:30.

Schmitt M, Neubauer A, Greiner J, Xu X, Barth TF, Bechter K. Spreading of acute myeloid leukemia cells by trafficking along the peripheral outflow pathway of cerebrospinal fluid. Anticancer Res. 2011;31(6):2343–5.

Schwartz M, Baruch K. The resolution of neuroinflammation in neurodegeneration: leukocyte recruitment via the choroid plexus. EMBO J. 2014;33(7):7–22.

Seyfert S, Faulstich A. Is the blood–CSF barrier altered in disease? Acta Neurol Scand. 2003;108(4):252–6.

Seyfert S, Faulstich A, Marx P. What determines the CSF concentrations of albumin and plasma-derived IgG? J Neurol Sci. 2004;219(1–2):31–3.

Seyfert S, Quill S, Faulstich A. Variation of barrier permeability for albumin and immunoglobulin G influx into cerebrospinal fluid. Clin Chem Lab Med. 2009;47(8):955–8.

Shoenfeld Y, Rose NR. Infection and autoimmunity. Amsterdam: Elsevier; 2004.

Sörensen HJ, Nielsen PR, Pedersen CB, Benros ME, Nordentoft M, Mortensen PB. Population impact of familial and environmental risk factors for schizophrenia: a nationwide study. Schizophr Res. 2014;153(1–3):214–9.

Stangel M, Fredrikson S, Meinl E, Petzold A, Stüve O, Tumani H. The utility of cerebrospinal fluid analysis in patients with multiple sclerosis. Nat Rev Neurol. 2013;9:267–76.

Steiner J, Walter M, Glanz W, Sarnyai Z, Bernstein HG, Vielhaber S, et al. Increased prevalence of diverse *N*-methyl-D-aspartate glutamate receptor antibodies in patients with an initial diagnosis of schizophrenia: specific relevance of IgG NR1a antibodies for distinction from *N*-methyl-D-aspartate glutamate receptor encephalitis. JAMA Psychiatry. 2013;70(3):271–8.

Thomson CA, McColl A, Cavanagh J, Graham GJ. Peripheral inflammation is associated with remote global gene expression changes in the brain. J Neuroinflammation. 2014;11(1):73.

Venkatesan A, Tunkel AR, Bloch KC, Lauring AS, Sejvar J, Bitnun A, Stahl J-P, et al. Case definitions, diagnostic algorithms, and priorities in encephalitis: Consensus statement of the international encephalitis consortium. Clinical Infectious Diseases. 2013;57(8):1114–28.

Walsh JG, Muruve DA, Power C. Inflammasomes in the CNS. Nat Rev Neurosci. 2014;15: 84–97.

Wang Y, Kasper LH. The role of microbiome in central nervous system disorders. Brain Behav Immun. 2014;38C:1–12.

Whittle S, Lichter R, Dennison M, Yijayakumar N, Schwartz O, Byrne ML, et al. Structural brain development and depression onset during adolescence: a prospective longitudinal study. Am J Psychiatry. 2014;171(5):564–71.

Wildemann B, Oschmann P, Reiber H. Laboratory diagnosis in neurology. Stuttgart: Thieme; 2010.

Wohleb ES, Godbout JP. Basic aspects of the immunology of neuroinflammation. In: Halaris A, Leonard BE, editors. Inflammation in psychiatry. Mod trends pharmacopsychiatry, vol. 28. Basel: Karger; 2013. p. 1–19.

Wolburg H, Paulus W. Choroid plexus: biology and pathology. Acta Neuropathol. 2010;119(1): 75–88.

Chapter 6
The Role of Infections and Autoimmune Diseases for Schizophrenia and Depression: Findings from Large-Scale Epidemiological Studies

Michael Eriksen Benrós and Preben B. Mortensen

Abstract An immunologic component to schizophrenia and depression has been increasingly recognized, which has led to extensive research into the associations with infections and autoimmune diseases. Large-scale nationwide epidemiological studies have displayed an increased prevalence of both autoimmune diseases and infections among persons with schizophrenia and depression. Autoimmune diseases, and especially the number of infections requiring hospitalization, increase the risk of schizophrenia and depression in a dose–response relationship. Infections are a common exposure and a broad spectrum of infections are associated with schizophrenia and depression. Particularly the autoimmune diseases with a potential presence of brain-reactive antibodies were associated with psychiatric disorders. However, the associations seem to be bidirectional, since the risk of autoimmune diseases and infections is also increased after diagnosis with schizophrenia and depression. The risk of autoimmune diseases was particularly increased in individuals with prior hospital contacts for infections.

It has been suggested that inflammation and autoimmunity could be involved in the etiology and pathogenesis of some patients with symptoms of schizophrenia and depression. The psychiatric symptoms can be directly triggered by immune components, such as brain-reactive antibodies and cytokines, or infections reaching the central nervous system (CNS), or be secondary to systemic inflammation indirectly affecting the brain. However, the associations could also be caused by shared genetic factors, other environmental factors, or common etiological components. Nonetheless, autoimmune diseases and infections should be considered by clinicians in the treatment of individuals with psychiatric symptoms, since treatment would probably improve the psychiatric symptoms, quality of life, and the survival of the individuals.

M.E. Benrós (✉)
Faculty of Health Sciences, Mental Health Centre Copenhagen, Copenhagen University Hospital, Bispebjerg Hospital, Bispebjerg Bakke 23, 2400 Copenhagen NV, Denmark

National Centre for Register-Based Research, Aarhus University, Aarhus, Denmark
e-mail: benros@dadlnet.dk

P.B. Mortensen
National Centre for Register-Based Research, Aarhus University, Aarhus, Denmark

N. Müller et al. (eds.), *Immunology and Psychiatry*, Current Topics in Neurotoxicity 8, DOI 10.1007/978-3-319-13602-8_6

Keywords Schizophrenia • Depression • Bipolar disorder • Psychiatry • Autoimmune diseases • Infection • Inflammation • Immunology • Epidemiology • Register-based

Introduction

Immunological hypotheses have become increasingly prominent within psychiatric research (Muller and Schwarz 2010), suggesting that inflammation and autoimmunity could be involved in the etiology and pathogenesis of some patients with symptoms of schizophrenia and depression (Miller et al. 2009; Drexhage et al. 2011). Previously, the brain was regarded as an immune privileged site, protected by the blood–brain barrier, but particularly within the last decades, studies have demonstrated that the immune system may affect the brain and induce psychiatric symptoms. Many diverse immune alterations have been observed in persons with schizophrenia, such as elevated levels of cytokines and inflammation markers (Muller and Schwarz 2010; Nikkila et al. 2001; Potvin et al. 2008; Dowlati et al. 2010; Howren et al. 2009). Abnormalities of the blood–brain barrier have been indicated in studies of patients with schizophrenia and severe depression (Uranova et al. 2010), together with signs of central nervous system (CNS) inflammation (Bechter et al. 2010). Additionally, increased prevalence of autoimmune diseases has been observed in patients with both schizophrenia and mood disorders, and studies have indicated increased autoantibody reactivity and elevated autoantibody levels even in the patients with no known autoimmune diseases (Laske et al. 2008; Tanaka et al. 2003). Also, both schizophrenia and depression have been associated with genetic markers related to the immune system (Marballi et al. 2010; Stefansson et al. 2009; Shelton et al. 2011).

Inflammatory mechanisms can affect the brain through many different pathways that are not necessarily mutually exclusive (Jones et al. 2005; Dalman et al. 2008; Yolken and Torrey 2008; Niebuhr et al. 2008a; Eaton et al. 2006). During inflammation and infections the permeability of the blood–brain barrier might be increased, making the brain particularly vulnerable to inflammation and immune components, such as antibodies and cytokines. Infections and immune components can particularly affect the brain during periods with increased permeability of the blood–brain barrier, which occurs during trauma and inflammation (Margutti et al. 2006; Irani and Lang 2008). Peripheral inflammation or infections can additionally interact with the neuroendocrine system regarded to be involved in psychiatric disorders through multiple pathways (Dantzer et al. 2008; Rivest 2010). Furthermore, experimental animal studies have found that symptoms of depression and psychosis can be induced by inflammation or brain-reactive antibodies (Katzav et al. 2007; Diamond et al. 2009; Chen et al. 2009). In summary, genetically vulnerable individuals might be at risk of developing neuropsychiatric symptoms like schizophrenia and depression as a consequence of inflammation and immune components affecting the brain. If the immunological alterations are confirmed to play a role in the pathogenesis of schizophrenia and depression, it could provide an interesting and promising target of future prevention and treatment.

A Historical Overview of the Associations

The possibility that bacterial infections have a causal relationship to psychoses was reported as early as 1896 (Noll 2007), and virus infections have been suspected since the 1918 influenza pandemic was followed by multiple reports of post-influenza psychoses and schizophrenia-like symptoms (Torrey et al. 2006). Later, when antibiotics were introduced, anecdotal data suggests that significant proportions of patients with psychiatric symptoms caused by neurosyphilis were cured also for their psychiatric symptoms (Sullivan et al. 2012). However, interest in the possible connection between infections and the occurrence of psychiatric disorders waned due to lack of further relevant treatment methods, but during the last decades, research on the relationship between infections and psychiatric disorders has re-emerged (Yolken and Torrey 2008).

Research on the associations with autoimmune diseases began in the 1950s, where investigators were puzzled by the apparent protective effect of schizophrenia on rheumatoid arthritis (Trevathan and Tatum 1953; Pilkington 1955). Furthermore, in the 1950s and 1960s, clinicians noticed what seemed to be an unusually high occurrence of celiac disease in persons with schizophrenia (Bender 1953; Graff and Handford 1961). Also, as early as the 1960s, a variety of autoantibodies with cross-reactivity against brain antigens were described in the sera and CSF of patients with schizophrenia (Fessel 1962; Heath and Krupp 1967a, b). In the last decade, a wider range of autoimmune diseases, infections, and autoantibodies have been implicated in population-based prospective studies as increasing the risk for schizophrenia and depression.

How Infections Can Induce Psychiatric Symptoms?

Various infectious agents have the potential to penetrate the blood–brain barrier and invade the CNS directly, possibly after reaching a threshold level of bacteremia (Kim 2008). Furthermore, inflammation in response to infections can affect the brain through many different pathways, hereunder increased permeability of the blood–brain barrier. Peripheral infections and inflammation can also affect the brain without passing the blood–brain barrier through proinflammatory cytokines activating the tryptophan–kynurenine pathway that regulates NMDA glutamate receptor activity together with serotonin production (Dantzer et al. 2008), and this may indirectly also affect dopamine regulation (Muller and Schwarz 2010).

Studies have additionally indicated that peripheral inflammation can affect the brain through stimulation of peripheral nerves such as the vagal nerve (Dantzer et al. 2008) or through alterations of the microbioma (Cryan and Dinan 2012). Several infectious agents can persist in the CNS and may not present with neurological symptoms, such as *toxoplasma* and certain viruses like Borna disease virus, HIV and hepatitis C virus, and even if not directly involved in destruction of CNS tissue, it might trigger CNS immune responses and thereby indirectly cause damage (Wilkinson et al. 2010; Shankar et al. 1992; Fishman et al. 2008). Furthermore, inflammation might act as a priming event on microglia, inducing a long-term development of abnormal signal patterns possibly involved in schizophrenia and depression (Hickie et al. 2009). Infections can additionally induce the development of autoimmune diseases and autoantibodies, possibly affecting the brain through a mechanism called molecular mimicry (Diamond et al. 2009; Rose 1998).

How Autoimmunity Can Induce Psychiatric Symptoms?

Autoimmune diseases are characterized by "self-reactivity" induced by autoantibodies and T-cells that can react against the body's own tissues and induce diverse symptoms, depending on the affected part of the body (Davidson and Diamond 2001). Many autoimmune diseases involve multiple organs and general dysfunction of the immune system which could affect the brain and induce psychiatric symptoms. CNS symptoms associated with autoantibodies have mostly been recognized in cancer patients with paraneoplastic symptoms that may in part be caused by an immunologic reaction where antibodies against tumor antigens cross-react with elements of the nervous system (Kayser et al. 2010; Darnell and Posner 2003). Brain-reactive antibodies have also been associated with some autoimmune diseases and are suspected to induce the high prevalence of neuropsychiatric symptoms observed in some autoimmune diseases, such as systemic lupus erythematosus (Margutti et al. 2006; Ballok 2007; Rice et al. 2005; Sundquist et al. 2008). Additionally, experimental studies have demonstrated that neuropsychiatric syndromes can be induced after an influx of brain-reactive antibodies into the brain (Kowal et al. 2004).

Associations Between Autoimmune Diseases and Schizophrenia

Large-scale Danish population-based studies with up to 20,317 patients with schizophrenia and 39,076 patients with non-affective psychosis have shown that individuals with schizophrenia are associated with a nearly 50 % higher lifetime prevalence of autoimmune disorders (Eaton et al. 2006; Benros et al. 2011). A cross-sectional analysis of a national sample from Taiwan on 10,811 individuals with schizophrenia replicated the association of a range of autoimmune diseases with schizophrenia, including specific positive associations with celiac disease, Graves' disease, psoriasis, pernicious anemia, hypersensitivity vasculitis, and the negative association with rheumatoid arthritis (Chen et al. 2012). Screening studies of persons with schizophrenia have found antibodies to the self-antigen tissue transglutaminase, indicative of celiac disease, in about five times as many persons as expected (5.4 vs 0.8 % in the CATIE study, $n = 1{,}401$) (Reichelt and Landmark 1995; Cascella et al. 2011; Jin et al. 2012; Samaroo et al. 2010). Furthermore, antibodies to gliadin, indicating sensitivity to wheat not necessarily associated with autoimmune disease, are also found in much higher proportion in patients with schizophrenia (23.1 vs 3.1 % in the CATIE study) (Cascella et al. 2011; Samaroo et al. 2010; Dickerson et al. 2010), which is interesting since a wide range of neurological complications are associated with antibodies to gliadin, even in the absence of autoimmune disease (Irani and Lang 2008; Hadjivassiliou et al. 2010; Jackson et al. 2012). Clinical studies have estimated the prevalence of celiac disease to be 2.1–2.6 % in patients with schizophrenia compared to 0.3–1 % in the general population (Cascella et al. 2011; Kalaydjian et al. 2006).

Based on the Danish register data, hospital contacts because of autoimmune diseases had occurred in 2.4 % of the patients before a schizophrenia diagnosis and autoimmune diseases occurred in 3.6 % of patients with schizophrenia after the diagnosis, resulting in 6 % of people with schizophrenia who had a hospital contact with autoimmune diseases during follow-up (Benros et al. 2011, 2014). Based on data from the study in Taiwan (Chen et al. 2012), 3.4 % of persons with a hospital contact for autoimmune diseases also had a hospital contact with schizophrenia during the follow-up period, which was shorter than the Danish studies. These prevalence estimates are based on hospital contacts only, and the actual prevalence of autoimmune diseases in people with schizophrenia is likely much higher if one were to screen the individuals, as exemplified by the clinical studies investigating the prevalence of celiac disease, which was much more common than in the register-based studies.

The Risk of Schizophrenia After an Autoimmune Disease Diagnosis

A Danish population-based study on 7,704 patients with schizophrenia showed that the relative risk of schizophrenia for an individual with a history of autoimmune disease, in themselves or in their family, was elevated by about 45 % (Eaton et al. 2006). Subsequent larger Danish population-based studies on 20,317 patients with schizophrenia and a total of 39,076 patients with non-affective psychosis confirmed that the risk of autoimmune diseases was increased by 45 % after an autoimmune disease diagnosis (Benros et al. 2011, 2014; Eaton et al. 2010). However, when restricting to persons without a history of infection, the increased risk of schizophrenia diminished from 45 to 29 % (Table 6.1) (Benros et al. 2011). Additionally, the study found that when autoimmune diseases and severe infections occurred together they interacted in synergy and increased the risk of schizophrenia by 2.25 times, which did not seem to be confined to one particular pathogen (Benros et al. 2011).

Table 6.1 Relative risk of schizophrenia spectrum diagnosis in persons with a hospital contact for autoimmune diseases and infections, Denmark, 1977–2006

	Schizophrenia spectrum disorders in persons without infections			Schizophrenia spectrum disorder in persons with hospital contact with infections		
Autoimmune disease	RR[a]	95 % CI	Cases	RR[a]	95 % CI	Cases
Reference without autoimmune disease or infection	1.00	(Reference)	29,372	1.00	(Reference)	29,372
Any autoimmune disease	**1.29**	1.18–1.41	483[b]	**2.25**	2.04–2.46	444[b]
Autoimmune diseases with suspected presence of brain-reactive antibodies:	**1.48**	1.31–1.68	244	**2.56**	2.25–2.89	243
Autoimmune hepatitis	**2.75**	1.38–4.83	10	**8.91**	6.50–11.84	43
Autoimmune thyroiditis	–	–	3	**4.57**	2.09–8.51	8
Celiac disease	**2.11**	1.09–3.61	11	**2.47**	1.13–4.61	8
Guillain Barre syndrome	1.22	0.58–2.19	9	**2.84**	1.52–4.76	12
Multiple sclerosis	**1.44**	1.03–1.94	39	**2.10**	1.37–3.06	24
Sjögren's syndrome	2.07	0.82–4.20	6	–	–	4
Systemic lupus erythematosis	1.84	0.92–3.23	10	**2.11**	1.06–3.70	10
Thyrotoxicosis (Graves disease)	**1.94**	1.47–2.49	56	**2.47**	1.68–3.49	29
Type 1 diabetes	**1.27**	1.04–1.53	104	**2.04**	1.68–2.44	109
Other autoimmune diseases listed below:	**1.19**	1.05–1.34	256	**1.95**	1.70–2.23	212
Ankylosing spondylitis	1.38	0.79–2.20	15	1.68	0.72–3.25	7
Crohn's disease	1.22	0.88–1.65	39	**1.67**	1.18–2.27	36
Iridocyclitis	1.32	0.87–1.91	25	**1.99**	1.21–3.06	18
Juvenile arthritis	1.00	0.52–1.71	11	1.77	0.95–2.97	12
Psoriasis vulgaris	**1.37**	1.01–1.80	47	**2.77**	2.07–3.63	49
Seropositive rheumatoid arthritis	0.75	0.51–1.06	28	**2.15**	1.52–2.95	35
Ulcerative colitis	1.22	0.97–1.51	80	**1.65**	1.24–2.14	52
Autoimmune diseases with too few cases to calculate the individual risk[c]	**1.59**[d]	1.13–2.17	36	**2.21**[d]	1.53–3.07	32

Source: Benros et al. Am J Psychiatry. 2011

[a]Boldface indicates that the 95 % confidence interval did not include 1.0. Relative risks were not estimated when there were less than five exposed cases

[b]Cases do not add up as one can have multiple autoimmune diseases

[c]Alopecia areata, autoimmune hemolytic anemia, dermatopolymyositis, idiopathic thrombocytopenic purpura, myasthenia gravis, pernicious anemia, primary adrenocortical insufficiency, primary biliary cirrhosis, pemphigus, pemphigoid, polymyalgia rheumatica, scleroderma, vitiligo, Wegener's granulomatosis

[d]Estimates should be interpreted with caution, but these autoimmune diseases were estimated together for completeness

The Risk of Autoimmune Diseases After a Diagnosis with Schizophrenia

Individuals diagnosed with schizophrenia had a 53 % increased risk of subsequent diagnoses with autoimmune diseases, particularly the group with suspected presence of brain-reactive antibodies had a 91 % increase of risk (Table 6.2). There was a significant multiplicative interaction between having both a schizophrenia

Table 6.2 Relative risk of autoimmune diseases after the diagnosis with schizophrenia spectrum disorder in Denmark (1987–2010)[a]

Autoimmune diseases		Cases	RR	95 % CI
Persons without schizophrenia (reference)			1.00	(Reference)
Any autoimmune disease	142,328	1,401	**1.53**	1.46–1.62
Autoimmune disease with suspected presence of brain-reactive antibodies	75,087	849	**1.91**	1.78–2.04
Autoimmune hepatitis	1,878	40	**3.51**	2.51–4.73
Autoimmune thyroiditis	3,386	23	0.90	0.58–1.33
Celiac disease	2,350	20	1.33	0.82–2.03
Guillain Barre syndrome	1,648	24	**2.73**	1.77–3.99
Multiple sclerosis	9,759	83	**1.57**	1.29–1.90
Sjögren's syndrome	1,994	19	1.31	0.80–2.00
Systemic lupus erythematosis	2,101	19	1.57	0.96–2.39
Thyrotoxicosis	17,308	136	1.10	0.93–1.30
Type 1 diabetes	28,272	478	**2.83**	2.58–3.10
Other autoimmune diseases	80,979	642	**1.21**	1.11–1.30
Alopecia areata	1,377	9	1.05	0.50–1.90
Ankylosing spondylitis	3,661	21	0.78	0.49–1.16
Crohn's disease	12,117	98	**1.33**	1.08–1.61
Idiopathic thrombocytopenic purpura	1,660	14	1.36	0.76–2.21
Iridocyclitis	9,220	80	1.18	0.94–1.46
Pernicious anemia	1,082	19	**2.59**	1.58–3.95
Polymyalgia rheumatica	2,396	11	0.61	0.31–1.04
Primary adrenocortical insufficiency	785	20	**3.81**	2.36–5.79
Primary biliary cirrhosis	478	9	**2.65**	1.26–4.81
Psoriasis vulgaris	13,269	190	**2.13**	1.84–2.45
Seropositive rheumatoid arthritis	15,768	82	**0.75**	0.60–0.93
Ulcerative colitis	22,289	146	0.99	0.84–1.16
Autoimmune diseases with too few cases to calculate the individual risk[b, c]	6,701	25	0.72	0.47–1.04

Source: Benros et al. Am J Psychiatry 2014

[a]Incidence rate ratios were adjusted for age and its interaction with sex, and calendar year. Persons without a history of schizophrenia spectrum diagnoses were chosen as reference category. Boldface indicates a significant result

[b]Only estimates building on five or more exposed cases are shown

[c]Estimates should be interpreted with caution, but the following autoimmune diseases were estimated together for completeness: autoimmune hemolytic anemia, pemphigus, pemphigoid, vitiligo, juvenile arthritis, Wegener's granulomatosis, dermatopolymyositis, myasthenia gravis, scleroderma

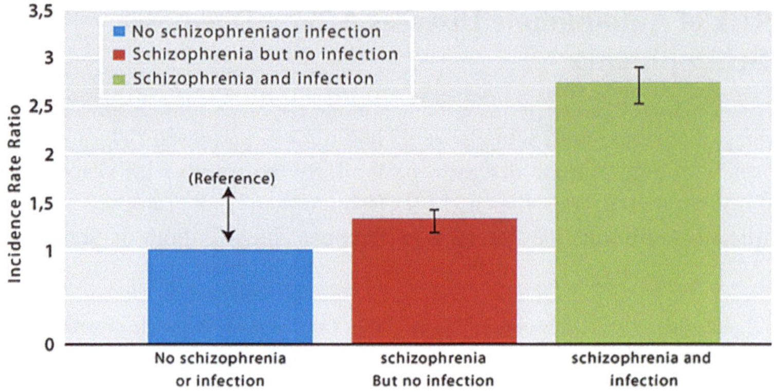

Fig. 6.1 Relative risk of subsequent autoimmune diseases in people with schizophrenia (significant multiplicative interaction between schizophrenia and infection on the risk autoimmune diseases ($p=0.004$). *Source*: *Benros et al.* Am J Psychiatry 2014

diagnosis and hospital contacts due to infections which increased the risk of subsequent autoimmune diseases by 2.7 times (Fig. 6.1) (Benros et al. 2014). In persons with schizophrenia, but no hospital contacts due to infections, the risk of autoimmune diseases was elevated with 32 % and diminished with time to a non-significant level in the time period 15 or more years after the schizophrenia diagnosis, whereas for persons with schizophrenia and infections the risk remained elevated. The increased incidence of autoimmune diseases following a diagnosis of schizophrenia might in some cases reflect symptoms of schizophrenia resulting from neuropsychiatric manifestations from the not-yet diagnosed autoimmune disease, particularly in the group with suspected presence of brain-reactive antibodies.

Associations with a Family History of Either Autoimmune Diseases or Schizophrenia

A family history with autoimmune diseases has been shown to increase the risk of schizophrenia by 10 % and a family history with schizophrenia increases the risk of autoimmune diseases by 6 % (Benros et al. 2014; Eaton et al. 2010). However, a family history with bipolar disorder was not significantly associated with autoimmune diseases, and there was no association in the reverse direction either (Benros et al. 2014; Eaton et al. 2010). A family history with the following specific autoimmune diseases has been associated with an increased incidence of schizophrenia in a nationwide Danish study (Eaton et al. 2010): autoimmune hepatitis, type 1 diabetes, Sjögrens syndrome, iridocyclitis, multiple sclerosis, psoriasis vulgaris, and dermatopolymyositis, whereas only a family history with pernicious anemia was associated with bipolar disorder out of the 30 autoimmune diseases studied. The association with a family history of diabetes type 1 and autoimmune thyrotoxicosis with schizophrenia has been confirmed in other populations as well (Wright et al. 1996; Gilvarry et al. 1996). A family history of schizophrenia was associated with pernicious anemia, diabetes type 1, iridocyclitis, autoimmune hepatitis, systemic lupus erythematosus, Sjögren's syndrome, and primary biliary cirrhosis.

Associations Between Autoimmune Diseases and Depression

Several autoimmune diseases, such as diabetes type 1, multiple sclerosis, systemic lupus erythematosus, and autoimmune thyroid disease, have been associated with depression in smaller studies (Korczak et al. 2011; Strous and Shoenfeld 2007; Gold and Irwin 2009; Padmos et al. 2004; Pop et al. 1998; Vonk et al. 2007). Rheumatoid arthritis has been associated with depression in several studies and in a meta-analysis (Dickens et al. 2002). A Danish nationwide study on 91,637 cases with a first-time hospital contact due to mood disorders showed that a prior hospital contact because of autoimmune diseases increased the risk of a subsequent mood disorder diagnosis by 57 %, but when separating the effect of infections, autoimmune diseases were associated with an increased risk by 45 % compared to the general population (Table 6.3) (Benros et al. 2013). The risk of developing mood disorders was elevated to the greatest degree in the group of autoimmune diseases with suspected presence of brain-reactive antibodies (58 %), particularly when combined with an infection (2.49 times increase of risk). In another population-based study (Eaton et al. 2010), a 70 % increased risk was found of developing bipolar disorder within 4 years of an autoimmune disease diagnosis and a 20 % increased risk in the time span from 5 years and onwards after the diagnosis, compared to the background population.

Table 6.3 Relative risk of mood disorders with a hospital contact in persons with hospital contact for autoimmune diseases and infections in Denmark (1977–2010)[a]

	Mood disorders in persons without infections			Mood disorder in persons with infections		
Autoimmune disease	Relative risk[b]	95 % CI	Case patients	Relative risk[b]	95 % CI	Case patients
Persons without autoimmune disease (reference)	1.00	(Reference)	60,361	**1.62**	**1.60–1.64**	27,081
Any autoimmune disease	**1.45**	1.39–1.52	2,082[c]	**2.35**	2.25–2.46	2,113[c]
Autoimmune diseases with suspected presence of brain-reactive antibodies:	**1.58**	1.49–1.68	1,057	**2.49**	2.35–2.65	1,123
Autoimmune hepatitis	**2.28**	1.53–3.41	24	**3.13**	2.39–4.11	52
Autoimmune thyroiditis	1.05	0.72–1.52	28	**1.63**	1.09–2.43	24
Celiac disease	**1.91**	1.41–2.60	41	**1.90**	1.32–2.73	29
Guillain Barre syndrome	**1.61**	1.14–2.26	33	**2.24**	1.58–3.17	32
Multiple sclerosis	**1.52**	1.30–1.77	162	**2.42**	2.06–2.86	142

(continued)

Table 6.3 (continued)

Autoimmune disease	Mood disorders in persons without infections			Mood disorder in persons with infections		
	Relative risk[b]	95 % CI	Case patients	Relative risk[b]	95 % CI	Case patients
Sjögren's syndrome	**1.79**	1.23–2.61	27	**2.58**	1.79–3.71	29
Systemic lupus erythematosis	**2.16**	1.61–2.89	45	**2.19**	1.65–2.92	47
Thyrotoxicosis (Graves disease)	**1.28**	1.12–1.45	228	**1.90**	1.63–2.21	165
Type 1 diabetes	**1.77**	1.61–1.94	469	**2.84**	2.62–3.07	603
Other autoimmune diseases listed below:	**1.35**	1.27–1.43	1,206	**2.22**	2.08–2.36	1,282
Ankylosing spondylitis	1.23	0.91–1.65	44	**2.02**	1.46–2.81	36
Crohn's disease	**1.75**	1.52–2.01	191	**2.32**	2.04–2.65	224
Iridocyclitis	1.22	1.00–1.48	101	**2.08**	1.70–2.54	94
Juvenile arthritis	1.20	0.88–1.64	39	**2.48**	1.93–3.19	61
Psoriasis vulgaris	**1.58**	1.38–1.81	203	**2.60**	2.27–2.97	212
Seropositive rheumatoid arthritis	1.08	0.93–1.25	167	**2.18**	1.89–2.52	183
Ulcerative colitis	**1.41**	1.27–1.57	331	**2.27**	2.04–2.54	315
Alopecia areata	1.08	0.66–1.77	16	**2.43**	1.60–3.69	22
Autoimmune hemolytic anemia	**2.53**	1.27–5.06	8	**2.28**	1.02–5.07	6
Dermatopolymyositis	1.25	0.60–2.62	7	**3.40**	2.01–5.74	14
Idiopathic thrombocytopenic purpura	1.18	0.71–1.96	15	**2.13**	1.37–3.30	20
Myasthenia gravis	1.19	0.62–2.30	9	–	–	4
Pernicious anemia	1.37	0.81–2.30	14	**2.14**	1.24–3.68	13
Primary adrenocortical insufficiency	**2.58**	1.53–4.35	14	1.64	0.95–2.82	13
Primary biliary cirrhosis	1.74	0.78–3.88	6	–	–	3
Pemphigus	–	–	2	**4.31**	1.94–9.60	6
Pemphigoid	–	–	1	–	–	1
Polymyalgia rheumatica	1.30	0.77–2.19	14	**3.81**	2.53–5.73	23
Scleroderma	1.03	0.56–1.92	10	**3.18**	2.05–4.93	20
Vitiligo	1.24	0.70–2.17	12	**2.01**	1.14–3.54	12
Wegener's granulomatosis	–	–	2	**2.37**	1.18–4.75	8

Source: Benros et al. JAMA Psychiatry. 2013

[a]Analyses were adjusted for sex, age, and calendar period

[b]Boldface indicates that the 95 % confidence interval did not include 1.0. Relative risks were not estimated when there were less than five exposed cases. Each separate autoimmune disease gives rise to one analysis adjusted for all other autoimmune diagnoses

[c]The data reflect that an individual can have multiple autoimmune diseases

Associations Between Infections and Schizophrenia

Epidemiological studies have indicated a dose–response relationship between urbanicity during upbringing and the risk of schizophrenia (Pedersen and Mortensen 2001), which could be related to, for instance, an increased probability of acquiring infections in urban environments (Yolken and Torrey 2008). An increased risk of schizophrenia has been associated with many different infectious agents, and a recent meta-analysis displayed significant associations between schizophrenia with *Toxoplasma gondii*, human herpesvirus 2, Borna disease virus, human endogenous retrovirus W, *Chlamydophila psittaci*, and *Chlamydophila* pneumonia (Arias et al. 2012). *T. gondii* infection has in many studies been associated with schizophrenia (Yolken and Torrey 2008; Niebuhr et al. 2008a; Torrey et al. 2007), and a recent population-based study indicated a dose–response relationship correlating with the serum titer of *toxoplasma* and the subsequent risk of schizophrenia (Pedersen et al. 2011). An increased risk of schizophrenia has also been associated with herpes simplex virus infection, detected by both serum antibodies (Dickerson et al. 2003; Niebuhr et al. 2008b) and CSF antibodies (Bartova et al. 1987). Cytomegalovirus (CMV) antibody titers have been found to be higher in the serum of patients with schizophrenia (Torrey et al. 2006; Leweke et al. 2004) and in the CSF (Torrey et al. 1982). However, studies have shown contradictory associations between CMV and schizophrenia, with stronger associations in newly diagnosed and untreated patients (Leweke et al. 2004), and to date, no neuropathologic evidence of CMV in the brains of patients with schizophrenia has been found (Torrey et al. 2006). Retroviral antigens and products have been identified in patients with schizophrenia (Karlsson et al. 2001; Hart et al. 1999), and other viruses associated with schizophrenia include Borna virus, where an increased serum prevalence having been observed (Chen et al. 1999). An increased prevalence of *Chlamydophila* infection has also been observed in patients with schizophrenia, especially when linked to genetic markers of the immune system (Fellerhoff et al. 2007). Additionally, post-mortem studies have found increased prevalence of *Chlamydophila* DNA in brains from patients with schizophrenia (Fellerhoff and Wank 2011). Psychotic disorders have also in population-based studies been associated with higher rates of several infections, such as pneumonia and pneumococcal disease (Crump et al. 2013a, b; Smith et al. 2013; Seminog and Goldacre 2013).

The Risk of Schizophrenia After Infections

A large-scale Danish nationwide study on 39,076 persons with a diagnosis of schizophrenia spectrum disorders showed that any history of hospitalization with infection increased the risk of schizophrenia by 60 % (Benros et al. 2011).

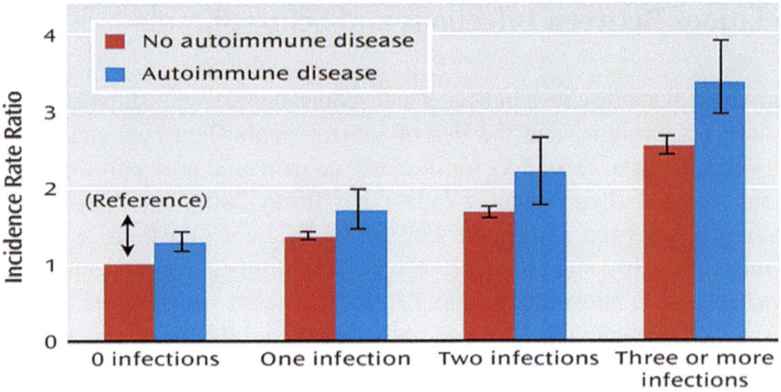

Fig. 6.2 Relative risk of schizophrenia spectrum disorders in persons with autoimmune disease and infections. *Source*: Benros et al. *Am J Psychiatry*. 2011

The risk of schizophrenia increased in a significant dose–response relationship with the number of infections (Fig. 6.2) and were increased the most with the temporal proximity of the last infection. The results remained significant after excluding persons diagnosed with substance use disorders, and there were no important differences in the relative risk added in persons with or without a psychiatric family history. Hospital contact due to infection had previously occurred in 23.6 % of persons diagnosed with schizophrenia spectrum disorders, yielded a population-attributable risk of 9 % associated with hospital contacts due to infections. A subsequent nationwide studies from Sweden confirmed the associations with previous hospital contacts with infections and schizophrenia spectrum disorders (Blomstrom et al. 2014). A subsequent Danish nationwide study on a narrower cohort during a period with complete follow-up of all hospital contacts showed that 45 % of persons with schizophrenia had a previous hospital contact with infection, which increased the risk of schizophrenia by 41 %, with bacterial infections increasing the risk by 63 % (Nielsen et al. 2014).

A recent meta-analysis has reported significant associations between childhood CNS infections and schizophrenia (Khandaker et al. 2012). Most studies on individuals with hospitalizations for CNS infections have found an increased risk of schizophrenia, including population-based studies from Denmark, Sweden, Finland, and Australia (Dalman et al. 2008; Benros et al. 2011; Liang and Chikritzhs 2012; Abrahao et al. 2005; Koponen et al. 2004). However, some studies did not find a significant association with CNS infections (Weiser et al. 2010; Suvisaari et al. 2003). In nationwide studies a broad spectrum of infections have been associated with schizophrenia, where sepsis and hepatitis infections were associated with the most elevated risk (2 and 4.9 times, respectively Table 6.4).

Table 6.4 Relative risk of schizophrenia spectrum diagnosis in persons with autoimmune diseases depending on the site of infection, Denmark, 1977–2006

	Only infection (no autoimmune disease)			Autoimmune disease		
Site of infection	IRR	95 % CI	Cases	IRR	95 % CI	Cases
Sepsis infections	**1.95**	1.47–2.51	55	**4.98**	2.49–8.73	10
Hepatitis infections	**4.89**	4.26–5.58	212	**8.89**	6.03–12.53	29
Gastrointestinal infections	**1.32**	1.26–1.39	1,847	**1.82**	1.46–2.24	83
Skin infection	**1.71**	1.62–1.80	1,427	**2.14**	1.69–2.66	74
Pregnancy related infection	1.14	0.98–1.31	185	1.22	0.48–2.47	6
Respiratory infections	**1.53**	1.46–1.61	1,885	**2.25**	1.79–2.79	77
Urogenital infections	**1.90**	1.79–2.01	1,200	**2.70**	2.10–3.41	66
CNS infections	**1.28**	1.09–1.50	148	**2.62**	1.31–4.60	10
Other types of infections	**1.70**	1.62–1.78	1,818	**1.99**	1.60–2.43	89
Persons without a hospital contact with infection (reference)	1.00 (reference)		29,372	**1.30**	1.18–1.42	483

Source: Benros et al. Am J Psychiatry. 2011

Associations Between Infections and Depression

Depression has also been associated with infections in several studies (Okusaga et al. 2011; Goodwin 2011; Steiner et al. 2012). In a study in which patients with depression were asked to describe in their own words their current state, the next most common descriptor was of physical changes that were described in terms of "feeling that the subject was coming down with a viral illness, either influenza or glandular fever, along with descriptions of aches and pains and, in particular, headaches or numbness of the head" (Maj 2011). These symptoms appear to be remarkably similar to symptoms of inflammation and infections. A study has demonstrated the associations of seropositivity for influenza and coronaviruses with both bipolar disorder and unipolar depression (Okusaga et al. 2011). Increased prevalence of anti-Borna disease virus antibodies has been detected in studies of patients with mood disorders (Terayama et al. 2003; Bode et al. 2001); however, a larger, blinded case–control study did not find any associations (Hornig et al. 2012). Chronic hepatitis C virus infections have consistently been associated with an increased incidence of depression, but other factors may be involved, for instance, the association may be due to alcohol and substance abuse or to treatment of hepatitis C with interferon, which is known to induce symptoms of depression. However, a recent study took all these factors into account and still found an increased risk of recurrent brief depression in patients with chronic hepatitis C (Carta et al. 2012). HIV infection and AIDS have also been associated with increased incidence of depressive symptoms, and neuro-HIV infection can replicate in the CNS, which might induce neuropsychiatric symptoms. However, the possible social stigmatization or psychological effect of a HIV/AIDS diagnosis may also increase the risk of mood disorders, which have led to exclusion of HIV/AIDS in larger prospective studies.

The Risk of Depression After an Infection

A Danish nationwide study on 91,637 cases with a first-time hospital contact due to mood disorders showed that any history of hospitalization with infection increased the IRR of later mood disorders by 62 %. The IRR of mood disorders was increased irrespective of the site of infection, with hepatitis resulting in the most elevated risk for mood disorders (2.8 times), followed by sepsis (2.1 times), and urogenital infections (2.1 times) (Table 6.5). The number of infections increased the risk of mood disorders in a dose–response relationship and the risk were increased the most with the temporal proximity of the last infection. Eight or more hospital contacts due to infections were associated with an increased risk for mood disorders by 3.4 times

Table 6.5 Relative risk of mood disorders with a hospital contact among persons with infections according to the infection site[a]

	Infection but no autoimmune disease			Infection and autoimmune disease		
Site of infection	Relative risk[b]	95 % CI	Case patients[c]	Relative risk[b]	95 % CI	Case patients[c]
Sepsis infections	**2.06**	1.85–2.29	332	**3.24**	2.61–4.03	82
Hepatitis infections	**2.82**	2.58–3.08	494	**3.01**	2.33–3.89	58
Gastrointestinal infections	**1.62**	1.58–1.66	7,598	**2.52**	2.34–2.72	676
Skin infection	**1.70**	1.65–1.74	5,865	**2.63**	2.43–2.84	621
Pregnancy related infection	**1.68**	1.60–1.76	1,708	**2.26**	1.90–2.69	128
Respiratory infections	**1.69**	1.65–1.74	7,035	**2.72**	2.50–2.97	527
Urogenital infections	**2.05**	2.00–2.10	7,037	**2.83**	2.62–3.05	660
CNS infections	**1.65**	1.54–1.78	716	**2.72**	2.17–3.42	74
Other types of infections	**1.81**	1.76–1.86	5,729	**2.53**	2.32–2.75	531
Persons without a hospital contact with infection (reference)	1.00	(reference)	60,361	**1.45**	1.39–1.52	2,082

[a]Analyses were adjusted for sex, age, and calendar period
[b]Calculated in nine separate analyses adjusted for sex, age, calendar period, and other infections. Boldface indicates that the 95 % confidence interval did not include 1.0
[c]Notice that one individual may have more than one diagnosis

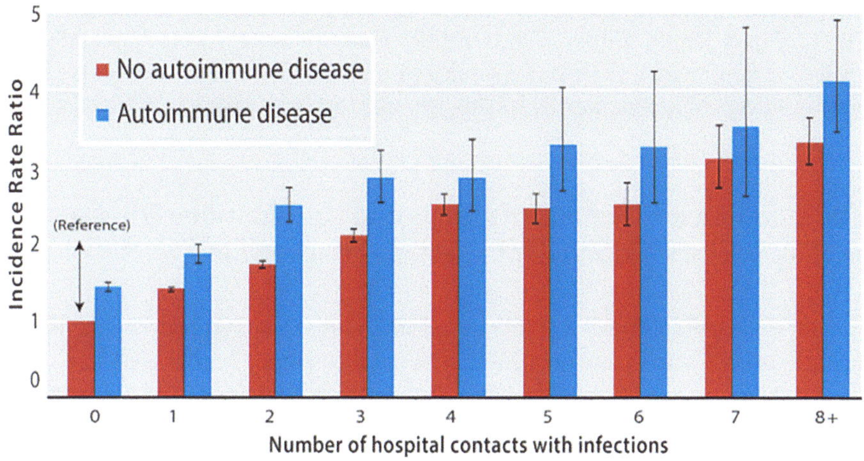

Fig. 6.3 Relative risk of mood disorders in persons with autoimmune disease and infections. *Source*: *Benros et al. JAMA Psychiatry.* 2013

(Fig. 6.3), and five or more different types of infections raised the risk of mood disorders by 4.8 times. Another recent study found an association between infections in early life and the occurrence of a broad range of mental disorders, including major depression, during youth (Goodwin 2011).

Associations Between a Family History of Infections and Schizophrenia

A number of epidemiological studies suggest that maternal infection during pregnancy is a risk factor for schizophrenia (Yolken and Torrey 2008; Brown et al. 2005; Brown and Derkits 2010; Mednick et al. 1988; Mortensen et al. 2007). It has been suggested that it is not a specific infectious agent causing the disorder, but instead a maternal immune response common to various infectious agents influencing fetal brain development and, in consequence, leading to schizophrenia in later life, most likely in genetically susceptible individuals (Ozawa et al. 2006). As part of the immune response, elevated levels of antibodies are found in mothers of schizophrenic offspring at the end of pregnancy (Buka et al. 2001).

A Danish nationwide study on 3,722 individuals with schizophrenia showed that maternal infections were associated with a 39 % increased risk of schizophrenia in the offspring, and the risk was increased with 23 % after adjustments for parental mental illness (Nielsen et al. 2013). However, there was no significant difference in the increased risk of schizophrenia if the infections occurred during or outside the pregnancy period, and the risk of schizophrenia was similarly increased when comparing infections in the mother or father (39 and 46 %, respectively). The data on a family history with infections could reflect a genetic susceptibility towards acquiring infections in individuals with schizophrenia, which is in line with genetic studies of individuals with schizophrenia (Stefansson et al. 2009).

The Combined Effect of Infections and Autoimmune Diseases as Risk Factors for Schizophrenia and Depression

Animal studies have shown that if brain-reactive antibodies are present in the blood and agents that increase the permeability of the blood–brain barrier are given, they cause a temporary opening of the blood–brain barrier, with influx of brain-reactive antibodies into the brain and a subsequent development of a neuropsychiatric syndrome (Kowal et al. 2004). This indicates that brain-reactive antibodies in the circulation might not have pathological consequences until there is a breach of blood–brain barrier integrity (Diamond et al. 2009).

Therefore, studies have investigated the combined effect of hospital contacts due to autoimmune diseases and infections on the risk of developing schizophrenia and mood disorders. The combined effect was larger than predicted by a combination of the single effects of the two disease groups, indicating the presence of a synergistic effect of the two exposures, which increased the risk of schizophrenia by 2.3 times and the risk of mood disorders by 2.4 times (Tables 6.1 and 6.3) (Benros et al. 2011, 2013). The risk of developing schizophrenia or mood disorders was elevated to the greatest degree in the group of autoimmune diseases with a suspected presence of brain-reactive antibodies and infections. The findings could support that CNS autoimmune disorders may require a "double-insult" of circulating pathogenic serum antibodies present at the time when the blood–brain barrier is compromised by, for instance, infection, fever, or stress (Irani and Lang 2008). In persons with an autoimmune disease, three or more hospital contacts due to infections increased the risk of schizophrenia by 3.4 times and eight or more hospital contacts due to infections increased the risk of mood disorders by 4.1 times (Figs. 6.2 and 6.3). Sepsis is the type of infection that would probably increase the blood–brain barrier permeability the most, and it is interesting that persons with a sepsis infection and an autoimmune disease had a fivefold increased risk of schizophrenia and a 3.3 times increased risk of mood disorders. Of other notice is that having both a hepatitis infection and an autoimmune disease, the risk of schizophrenia is increased by 8.9 times, whereas the risk of mood disorders is increased by 3.0 times. If a person had an autoimmune hepatitis diagnosis and an infection, the risk of schizophrenia was increased by 8.9 times and the risk of mood disorders by 3.1 times. The association between inflammatory diseases in the liver and particularly schizophrenia (Benros et al. 2011), but also mood disorders, is interesting, since autoimmune hepatitis is also associated with brain-reactive antibodies (Kimura et al. 2010), and in patients with severe affection of the liver, as seen in coma hepaticum in the initial phases, psychiatric symptoms are dominating (Butterworth 2011).

Somatic Co-morbidity in Patients with Schizophrenia and Depression

Both schizophrenia and depression are associated with excess somatic co-morbidity and mortality (Goldman 1999; Laursen et al. 2007, 2011; Moussavi et al. 2007). However, recent research indicates that the somatic co-morbidity, as exemplified in this chapter by autoimmune diseases and infections, is not only increased after the diagnoses of psychiatric disorders but also before the diagnosis and therefore probably not only due to the effects of psychiatric medication or lifestyle (Benros et al. 2009, 2011, 2013; Pedersen et al. 2012). Both schizophrenia and mood disorders have an excess co-occurrence of medical diseases involving inflammatory pathophysiological mechanisms, such as atopic disorders, autoimmune diseases, type 2 diabetes, and cardiovascular diseases (Chen et al. 2012; Moussavi et al. 2007; Pedersen et al. 2012). Depression is also a frequent co-morbidity to many chronic physical illnesses with inflammation, such as cancer (Moussavi et al. 2007). Additionally, Danish large-scale prospective population studies have shown that elevated C-reactive protein levels in the general population are associated with an increased risk of depression but also late-onset schizophrenia (Wium-Andersen et al. 2013, 2014).

Genetics

Inflammation-related genes have been associated with susceptibility to both schizophrenia and mood disorders (Shelton et al. 2011), in which environmental influences, such as infections and autoimmune diseases, may interact with genetic factors. Genetic markers within the human leukocyte antigen (HLA) region, which contains genes related to immune function, have been associated with occurrence of both autoimmune diseases and schizophrenia (Stefansson et al. 2009; Jones et al. 2005; Ripke et al. 2011). GWAS studies on schizophrenia patients have consistently implicated chromosome 6 in the HLA region as a common susceptibility factor, and this has been shown in three large-scale GWAS studies (Stefansson et al. 2009; Ripke et al. 2011). Nonetheless, studies of autoimmune diseases and infections as risk factors for either schizophrenia or mood disorders stratified by a psychiatric family history showed that no additional increase in risk was added in persons with a psychiatric family history (Benros et al. 2011, 2013). However, several of the autoimmune diseases have been associated with, for instance, different markers in the HLA region, and these markers might be differently associated with psychiatric disorders. This could, for instance, explain the negative association between schizophrenia and rheumatoid arthritis, which has been shown in more than a dozen studies; a fact that could be due to the interplay of genetic influences (Eaton et al. 1992; Benros et al. 2012; Vinogradov et al. 1991). However, ascertainment bias or anti-inflammatory and analgesic effects of antipsychotics might also be involved in this negative association (Mors et al. 1999; Sellgren et al. 2014; Torrey and Yolken 2001).

Influence of Maternal Factors

Infections during pregnancy might permanently alter the peripheral immune system of the fetus, and thereby change the response and vulnerability to future infections (Patterson 2009). Early-life exposure to infection and/or immune activation may induce sensitizing or preconditioning effects that can cause the organism to exacerbate reactions to subsequent immunological challenges in later life (Meyer et al. 2011a). This could be an alternative explanation for the increased incidence of infections prior to a diagnosis of schizophrenia and depression. Furthermore, epigenetic modifications after exposures to infections or other environmental factors could also induce dysregulation of the immune system.

The Influence of Psychological Stress

Psychological stress can affect the immune system functioning and might increase the risk of acquiring infections and enhancing immunological responses in the individual (Pedersen et al. 2010; Ader et al. 1995). Both schizophrenia and depression are associated with psychological stress particularly preceding the onset of the diagnosis. Thus, the inflammatory response might simply be a parallel finding and not a causal relationship. The relationship might also be bidirectional with immune responses leading to psychological stress preceding the more severe psychiatric symptoms. The psychological effects of having an infection or autoimmune disease may also affect the associations with schizophrenia and depression. However, several of the chronic autoimmune diseases did not by themselves increase the risk, which indicates that the associations are not only due to the psychological stress of living with a chronic disease or being hospitalized.

The Influence of Medical Treatment

An iatrogenic effect of medical treatment seems unlikely to explain the major associations, since only some of the autoimmune diseases would be treated with, for instance, steroids or interferon, which may increase the risk of psychosis and depression. Additionally, a decreased risk of psychosis associated with the use of steroids has also been reported (Laan et al. 2009), and newer biological treatment of autoimmune diseases has been shown to decrease the risk of mood disorders (Tyring et al. 2006). Furthermore, anti-inflammatory agents have been suggested to improve depression symptoms in patients with inflammatory disorders and enhance responsiveness to antidepressants (Tyring et al. 2006; Miller 2010; Meyer et al. 2011b; Henry et al. 2008). Antibiotics have not been associated with an increased risk of schizophrenia or depression. However, one could speculate that certain types of antibiotics might alter, for instance, the gut microflora, and thereby have positive or negative effects on the immune system, and possibly also on psychiatric disorders (Cryan and Dinan 2012). Additionally, some antipsychotics can also affect immune responses possibly affecting the risk of subsequent autoimmune diseases and infections (Goldsmith and Rogers 2008).

The Influence of Social and Lifestyle Factors

Social and lifestyle factors of persons not yet diagnosed with psychiatric disorders may increase the probability of smoking and alcohol and drug abuse, which can suppress the immune system and thereby increase the vulnerability to infection or have immune-activating effects resulting in autoimmunity (Margutti et al. 2006; Sperner-Unterweger 2005; Sopori 2002). Furthermore, social and lifestyle factors may affect the way individuals seek help in the health care system, and decreased compliance of treatment initiated by the general practitioner might lead to hospital treatment for autoimmune diseases and infections. Nevertheless, the added risk from infections or autoimmune diseases was not more pronounced in persons with a psychiatric family history or a personal history with substance abuse, which could be used as proxy variables for social and lifestyle factors (Benros et al. 2011, 2013).

Views on the Role of Infections and Autoimmunity in Association with Schizophrenia and Depression

The underlying etiological mechanisms of the associations between infections and autoimmune diseases with schizophrenia and depression may be numerous and speculative, but not necessarily mutually exclusive and may in fact be interconnected. Both innate and adaptive immune responses might be involved, and there are many different routes of communication between the periphery and the brain (Dantzer et al. 2008). The psychiatric symptoms can be directly triggered by immune components, such as brain-reactive antibodies and cytokines, or infections reaching the CNS, possibly through increased permeability of the blood–brain barrier, or be secondary to systemic inflammation indirectly affecting the brain (Muller and Schwarz 2010; Dantzer et al. 2008; Diamond et al. 2009; Kayser et al. 2010; Bechter 2013). The associations with a range of autoimmune diseases and infections may reflect inflammation as a common pathway to schizophrenia and depression. Increased inflammation may increase the blood–brain barrier permeability, making the brain vulnerable to immune components such as autoantibodies and cytokines, or possibly the effect of specific T-cell subsets that are involved in immune surveillance of the brain (Goverman 2009). Furthermore, an imbalance between the Th1 and Th2 systems has also been proposed as an etiological component (Muller and Schwarz 2010), which would fit with the increased prevalence of autoimmune diseases and atopic disorders in people with schizophrenia

(Eaton et al. 2006; Pedersen et al. 2012). Inflammation can additionally affect the brain without passing the blood–brain barrier through stimulation of peripheral nerves (Dantzer et al. 2008) or proinflammatory cytokines activating the tryptophan–kynurenine pathway involved in regulation of the glutamate and serotonin system (Dantzer et al. 2008), and probably also indirectly the dopamine system (Muller and Schwarz 2010). Low-grade brain inflammation has also been proposed as an underlying causal mechanism of subgroups of patients with schizophrenia and severe depression, which could be triggered by infections and autoimmune diseases (Bechter et al. 2010; Bechter 2013). Additionally, inflammation might act as a priming event on microglia, inducing a long-term development of abnormal signal patterns (Hickie et al. 2009). The inflammation associated with severe infections and autoimmune diseases may also reactivate less severe infections that have been associated with schizophrenia and depression, such as herpes virus and *toxoplasma* infections. Some infectious agents escape surveillance by the immune system, but after an acute infection or inflammation, symptoms may flare up from the latent infection (Fellerhoff and Wank 2011). Additionally, alterations of the gut microbiota can also influence brain function and behavior possibly through neural, endocrine, and immune pathways (Cryan and Dinan 2012).

Implications from Large-Scale Epidemiological Studies

The observed associations in the population-based register studies support a possible immunological contribution in subgroups of patients with schizophrenia and depression. However, whether it is a causal relationship or an epiphenomenon due to, for instance, other environmental factors or common genetic vulnerability remains to be proven. Register-based studies cannot identify specific etiological pathways; nonetheless, the observed associations seem robust. Additionally, there is a temporal and a dose–response relationship particularly for infections, which could indicate causal associations. Furthermore, the associations seem biologically plausible based on experimental animal and human studies. However, the associations between several somatic diseases seem to be of a bidirectional nature with increased incidence both before and after the psychiatric diagnosis. Additionally, new studies indicate that an even broader range of hospital contacts for somatic diseases than previously described seem to be associated with an elevated risk of psychiatric disorders. Even though most people with infections or autoimmune diseases do not develop psychiatric symptoms, immune exposures are rather common, and, in particular, infections might prove to be an important risk factor that is very common.

Schizophrenia and depression have complex, multifactorial, and to a large extent unknown etiologies (Yirmiya and Goshen 2011). A main contributing risk factor in population-based studies is a psychiatric family history that is, however, unlikely to be a sufficient cause and probably increases the risk of psychiatric disorders through interaction with environmental risk factors (Agerbo et al. 2012; Mortensen et al. 2010). Genetically vulnerable individuals might be at a particular risk of developing psychiatric disorders like schizophrenia and depression as a consequence of inflammation and immune components affecting the brain. Subgroups of people with schizophrenia and depression may demonstrate features of an autoimmune process, and the hypothesis is strengthened by the findings of an increased familial association between autoimmune diseases and schizophrenia (Eaton et al. 2006; Wright et al. 1996; Benros et al. 2014). Additionally, it is interesting that etiological mechanisms similar to those of schizophrenia and depression are hypothesized to be involved in the initiation of autoimmunity where genetic susceptibility is required along with triggering events such as infections (Goldsmith and Rogers 2008).

Perspectives

Diagnostic delay and under-treatment of somatic co-morbidity is a general problem for psychiatric patients (Laursen et al. 2007; Goldman 1999; Harris and Barraclough 1998), and possibly explains the increased mortality (Laursen et al. 2007, 2011). The increased mortality in psychiatric patients is actually mainly due to somatic diseases, even though suicide has received the most attention (Laursen et al. 2011; Laursen 2011). Hence, a thorough clinical examination and frequent somatic check-ups also seem of the utmost importance in patients with schizophrenia and depression. Screening for somatic diseases affecting the brain, such as autoimmune diseases and infections, preferentially with material closer to the brain, such as CSF in addition to sera, could in the near future prove to be helpful in diagnosing and treatment planning for persons with first-onset symptoms of schizophrenia and depression. However, the CNS is well protected and still difficult to access in vivo with regard to investigating signs of inflammation or relevant immune components.

Interestingly, recent CSF screening studies of patients with schizophrenia, and no known autoimmune diseases or infection, have detected autoantibodies or antibodies against infectious agents in the CSF of 3.2–6 % of patients with schizophrenia (Bechter et al. 2010; Kranaster et al. 2011). Furthermore, an increasing number of previously unknown antibodies with reactivity against the CNS are being reported in recent years (Steiner et al. 2012; Graus et al. 2010). Some of the strongest evidence for the potential for autoimmunity and immune components to cause psychiatric symptoms comes from the NMDA antibody-induced limbic encephalitis where psychiatric symptoms are often dominant in the initial and the remission phase of the disorder in up to 70 % of the cases (Kayser and Dalmau 2011), and which has been demonstrated to be treatable with immune therapies (Graus et al. 2010; Kayser and Dalmau 2011).

Potentially, immune and autoimmune processes could be involved in the prodrome and perhaps etiology of a non-negligible proportion of individuals with schizophrenia and depression. Symptom manifestations of autoimmune conditions might particularly resemble the subtype of psychosis and depression with chronic relapsing remitting illness, for instance (Yum et al. 2009). Subgroups of schizophrenia and depression could prove to be of a more systemic character, with inflammation as a common etiological mechanism that, besides the neuropsychiatric symptoms, includes physical diseases such as autoimmune disease, cardiovascular disease, diabetes, and cancer (Dantzer et al. 2008; Leboyer et al. 2012). In addition, some psychiatric disorders and somatic diseases might have shared causes and be manifestations of the same underlying disease process or be exacerbated by each other. Whether or not the co-occurrence of somatic diseases in people with psychiatric symptoms is causally related to the psychiatric symptoms, the individuals would benefit from treatment for their somatic co-morbidity to reduce mortality and improve quality of life.

Conclusion

In summary, experimental animal studies as well as human studies indicate that many diverse immune challenges can induce symptoms of schizophrenia and depression. There seems to be a solid association between autoimmune diseases and infections with both schizophrenia and depression based on large-scale nationwide studies. A prior autoimmune disease is associated with an increased risk of schizophrenia by 29 % and depression by 45 %. Any history of hospitalization with infection is associated with an increased risk of schizophrenia with 60 % and depression by 62 %. The combined effect of having had hospital contacts with both autoimmune diseases and infections increased the risk of schizophrenia by 2.3 times and the risk of depression by 2.4 times. Particularly the number of infections requiring hospitalization increases the risk of developing schizophrenia and depression in a dose–response relationship. Hospital contact because of infections was the most common risk factor, occurring in nearly 24 % of all patients before a schizophrenia diagnosis and in nearly 32 % of all patients before a depression diagnosis. Hospital contacts because of autoimmune diseases had occurred in 2.4 % of the patients before a schizophrenia diagnosis and in 5 % before a mood disorder diagnosis. After the diagnosis with schizophrenia the subsequent risk of autoimmune diseases is increased by 53 %, and in individuals with schizophrenia and hospital contacts for infections the risk is increased by 2.7 times. Autoimmune diseases occurred in 3.6 % after the schizophrenia diagnosis.

References

Abrahao AL, Focaccia R, Gattaz WF. Childhood meningitis increases the risk for adult schizophrenia. World J Biol Psychiatry. 2005;6 Suppl 2:44–8.

Ader R, Cohen N, Felten D. Psychoneuroimmunology: interactions between the nervous system and the immune system. Lancet. 1995;345(8942):99–103.

Agerbo E, Mortensen PB, Wiuf C, et al. Modelling the contribution of family history and variation in single nucleotide polymorphisms to risk of schizophrenia: a Danish national birth cohort-based study. Schizophr Res. 2012;134(2–3):246–52.

Arias I, Sorlozano A, Villegas E, et al. Infectious agents associated with schizophrenia: a meta-analysis. Schizophr Res. 2012;136(1–3):128–36.

Ballok DA. Neuroimmunopathology in a murine model of neuropsychiatric lupus. Brain Res Rev. 2007;54(1):67–79.

Bartova L, Rajcani J, Pogady J. Herpes simplex virus antibodies in the cerebrospinal fluid of schizophrenic patients. Acta Virol. 1987;31(5):443–6.

Bechter K. Updating the mild encephalitis hypothesis of schizophrenia. Prog Neuropsychopharmacol Biol Psychiatry. 2013;42:71–91.

Bechter K, Reiber H, Herzog S, Fuchs D, Tumani H, Maxeiner HG. Cerebrospinal fluid analysis in affective and schizophrenic spectrum disorders: identification of subgroups with immune responses and blood–CSF barrier dysfunction. J Psychiatr Res. 2010;44(5):321–30.

Bender L. Childhood schizophrenia. Psychiatr Q. 1953;27:663–81.

Benros ME, Laursen TM, Dalton SO, Mortensen PB. Psychiatric disorder as a first manifestation of cancer: a 10-year population-based study. Int J Cancer. 2009;124(12):2917–22.

Benros ME, Nielsen PR, Nordentoft M, Eaton WW, Dalton SO, Mortensen PB. Autoimmune diseases and severe infections as risk factors for schizophrenia: a 30-year population-based register study. Am J Psychiatry. 2011;168(12):1303–10.

Benros ME, Mortensen PB, Eaton WW. Autoimmune diseases and infections as risk factors for schizophrenia. Ann N Y Acad Sci. 2012;1262(1):56–66.

Benros ME, Waltoft BL, Nordentoft M, et al. Autoimmune diseases and severe infections as risk factors for mood disorders: a nationwide study. JAMA Psychiatry. 2013;70(8):812–20.

Benros ME, Pedersen MG, Rasmussen H, Eaton WW, Nordentoft M, Mortensen PB. A nationwide study on the risk of autoimmune diseases in individuals with a personal or a family history of schizophrenia and related psychosis. Am J Psychiatry. 2014;171(2):218–26.

Blomstrom A, Karlsson H, Svensson A, et al. Hospital admission with infection during childhood and risk for psychotic illness—a population-based cohort study. Schizophr Bull. 2014;40(6): 1518–25.

Bode L, Reckwald P, Severus WE, et al. Borna disease virus-specific circulating immune complexes, antigenemia, and free antibodies—the key marker triplet determining infection and prevailing in severe mood disorders. Mol Psychiatry. 2001;6(4):481–91.

Brown AS, Derkits EJ. Prenatal infection and schizophrenia: a review of epidemiologic and translational studies. Am J Psychiatry. 2010;167(3):261–80.

Brown AS, Schaefer CA, Quesenberry Jr CP, Liu L, Babulas VP, Susser ES. Maternal exposure to toxoplasmosis and risk of schizophrenia in adult offspring. Am J Psychiatry. 2005;162(4): 767–73.

Buka SL, Tsuang MT, Torrey EF, Klebanoff MA, Bernstein D, Yolken RH. Maternal infections and subsequent psychosis among offspring. Arch Gen Psychiatry. 2001;58(11):1032–7.

Butterworth RF. Hepatic encephalopathy: a central neuroinflammatory disorder? Hepatology. 2011;53(4):1372–6.

Carta MG, Angst J, Moro MF, et al. Association of chronic hepatitis C with recurrent brief depression. J Affect Disord. 2012;141(2–3):361–6.

Cascella NG, Kryszak D, Bhatti B, et al. Prevalence of celiac disease and gluten sensitivity in the United States clinical antipsychotic trials of intervention effectiveness study population. Schizophr Bull. 2011;37(1):94–100.

Chen CH, Chiu YL, Wei FC, et al. High seroprevalence of Borna virus infection in schizophrenic patients, family members and mental health workers in Taiwan. Mol Psychiatry. 1999;4(1):33–8.

Chen P, Jiang T, Ouyang J, Chen Y. Depression, another autoimmune disease from the view of autoantibodies. Med Hypotheses. 2009;73(4):508–9.

Chen SJ, Chao YL, Chen CY, et al. Prevalence of autoimmune diseases in in-patients with schizophrenia: nationwide population-based study. Br J Psychiatry. 2012;200(5):374–80.

Crump C, Sundquist K, Winkleby MA, Sundquist J. Comorbidities and mortality in bipolar disorder: a Swedish national cohort study. JAMA Psychiatry. 2013a;70(9):931–9.

Crump C, Winkleby MA, Sundquist K, Sundquist J. Comorbidities and mortality in persons with schizophrenia: a Swedish national cohort study. Am J Psychiatry. 2013b;170(3):324–33.

Cryan JF, Dinan TG. Mind-altering microorganisms: the impact of the gut microbiota on brain and behaviour. Nat Rev Neurosci. 2012;13(10):701–12.

Dalman C, Allebeck P, Gunnell D, et al. Infections in the CNS during childhood and the risk of subsequent psychotic illness: a cohort study of more than one million Swedish subjects. Am J Psychiatry. 2008;165(1):59–65.

Dantzer R, O'Connor JC, Freund GG, Johnson RW, Kelley KW. From inflammation to sickness and depression: when the immune system subjugates the brain. Nat Rev Neurosci. 2008;9(1):46–56.

Darnell RB, Posner JB. Paraneoplastic syndromes involving the nervous system. N Engl J Med. 2003;349(16):1543–54.

Davidson A, Diamond B. Autoimmune diseases. N Engl J Med. 2001;345(5):340–50.

Diamond B, Huerta PT, Mina-Osorio P, Kowal C, Volpe BT. Losing your nerves? Maybe it's the antibodies. Nat Rev Immunol. 2009;9(6):449–56.

Dickens C, McGowan L, Clark-Carter D, Creed F. Depression in rheumatoid arthritis: a systematic review of the literature with meta-analysis. Psychosom Med. 2002;64(1):52–60.

Dickerson FB, Boronow JJ, Stallings C, Origoni AE, Ruslanova I, Yolken RH. Association of serum antibodies to herpes simplex virus 1 with cognitive deficits in individuals with schizophrenia. Arch Gen Psychiatry. 2003;60(5):466–72.

Dickerson F, Stallings C, Origoni A, et al. Markers of gluten sensitivity and celiac disease in recent-onset psychosis and multi-episode schizophrenia. Biol Psychiatry. 2010;68(1):100–4.

Dowlati Y, Herrmann N, Swardfager W, et al. A meta-analysis of cytokines in major depression. Biol Psychiatry. 2010;67(5):446–57.

Drexhage RC, Weigelt K, van Beveren N, et al. Immune and neuroimmune alterations in mood disorders and schizophrenia. Int Rev Neurobiol. 2011;101:169–201.

Eaton WW, Hayward C, Ram R. Schizophrenia and rheumatoid arthritis: a review. Schizophr Res. 1992;6(3):181–92.

Eaton WW, Byrne M, Ewald H, et al. Association of schizophrenia and autoimmune diseases: linkage of Danish national registers. Am J Psychiatry. 2006;163(3):521–8.

Eaton WW, Pedersen MG, Nielsen PR, Mortensen PB. Autoimmune diseases, bipolar disorder, and non-affective psychosis. Bipolar Disord. 2010;12(6):638–46.

Fellerhoff B, Wank R. Increased prevalence of *Chlamydophila* DNA in post-mortem brain frontal cortex from patients with schizophrenia. Schizophr Res. 2011;129(2–3):191–5.

Fellerhoff B, Laumbacher B, Mueller N, Gu S, Wank R. Associations between *Chlamydophila* infections, schizophrenia and risk of HLA-A10. Mol Psychiatry. 2007;12(3):264–72.

Fessel WJ. Autoimmunity and mental illness. A preliminary report. Arch Gen Psychiatry. 1962;6:320–3.

Fishman SL, Murray JM, Eng FJ, Walewski JL, Morgello S, Branch AD. Molecular and bioinformatic evidence of hepatitis C virus evolution in brain. J Infect Dis. 2008;197(4):597–607.

Gilvarry CM, Sham PC, Jones PB, et al. Family history of autoimmune diseases in psychosis. Schizophr Res. 1996;19(1):33–40.

Gold SM, Irwin MR. Depression and immunity: inflammation and depressive symptoms in multiple sclerosis. Immunol Allergy Clin North Am. 2009;29(2):309–20.

Goldman LS. Medical illness in patients with schizophrenia. J Clin Psychiatry. 1999;60 Suppl 21:10–5.

Goldsmith CA, Rogers DP. The case for autoimmunity in the etiology of schizophrenia. Pharmacotherapy. 2008;28(6):730–41.

Goodwin RD. Association between infection early in life and mental disorders among youth in the community: a cross-sectional study. BMC Public Health. 2011;11:878.

Goverman J. Autoimmune T, cell responses in the central nervous system. Nat Rev Immunol. 2009;9(6):393–407.

Graff H, Handford A. Celiac syndrome in the case histories of five schizophrenics. Psychiatr Q. 1961;35:306–13.

Graus F, Saiz A, Dalmau J. Antibodies and neuronal autoimmune disorders of the CNS. J Neurol. 2010;257(4):509–17.

Hadjivassiliou M, Sanders DSGRA, Woodroofe N, Boscolo S, Aeschlimann D. Gluten sensitivity: from gut to brain. Lancet Neurol. 2010;9(3):318–30.

Harris EC, Barraclough B. Excess mortality of mental disorder. Br J Psychiatry. 1998;173:11–53.
Hart DJ, Heath RG, Sautter Jr FJ, et al. Antiretroviral antibodies: implications for schizophrenia, schizophrenia spectrum disorders, and bipolar disorder. Biol Psychiatry. 1999;45(6):704–14.
Heath RG, Krupp IM. Schizophrenia as an immunologic disorder. I. Demonstration of antibrain globulins by fluorescent antibody techniques. Arch Gen Psychiatry. 1967a;16(1):1–9.
Heath RG, Krupp IM. Catatonia induced in monkeys by antibrain antibody. Am J Psychiatry. 1967b;123(12):1499–504.
Henry CJ, Huang Y, Wynne A, et al. Minocycline attenuates lipopolysaccharide (LPS)-induced neuroinflammation, sickness behavior, and anhedonia. J Neuroinflammation. 2008;5:15.
Hickie IB, Banati R, Stewart CH, Lloyd AR. Are common childhood or adolescent infections risk factors for schizophrenia and other psychotic disorders? Med J Aust. 2009;190(4 Suppl):S17–21.
Hornig M, Briese T, Licinio J, et al. Absence of evidence for bornavirus infection in schizophrenia, bipolar disorder and major depressive disorder. Mol Psychiatry. 2012;17(5):486–93.
Howren MB, Lamkin DM, Suls J. Associations of depression with C-reactive protein, IL-1, and IL-6: a meta-analysis. Psychosom Med. 2009;71(2):171–86.
Irani S, Lang B. Autoantibody-mediated disorders of the central nervous system. Autoimmunity. 2008;41(1):55–65.
Jackson JR, Eaton WW, Cascella NG, Fasano A, Kelly DL. Neurologic and psychiatric manifestations of celiac disease and gluten sensitivity. Psychiatr Q. 2012;83(1):91–102.
Jin SZ, Wu N, Xu Q, et al. A study of circulating gliadin antibodies in schizophrenia among a Chinese population. Schizophr Bull. 2012;38(3):514–8.
Jones AL, Mowry BJ, Pender MP, Greer JM. Immune dysregulation and self-reactivity in schizophrenia: do some cases of schizophrenia have an autoimmune basis? Immunol Cell Biol. 2005;83(1):9–17.
Kalaydjian AE, Eaton W, Cascella N, Fasano A. The gluten connection: the association between schizophrenia and celiac disease. Acta Psychiatr Scand. 2006;113(2):82–90.
Karlsson H, Bachmann S, Schroder J, McArthur J, Torrey EF, Yolken RH. Retroviral RNA identified in the cerebrospinal fluids and brains of individuals with schizophrenia. Proc Natl Acad Sci U S A. 2001;98(8):4634–9.
Katzav A, Solodeev I, Brodsky O, et al. Induction of autoimmune depression in mice by anti-ribosomal P antibodies via the limbic system. Arthritis Rheum. 2007;56(3):938–48.
Kayser MS, Dalmau J. Anti-NMDA receptor encephalitis in psychiatry. Curr Psychiatry Rev. 2011;7:189–93.
Kayser MS, Kohler CG, Dalmau J. Psychiatric manifestations of paraneoplastic disorders. Am J Psychiatry. 2010;167(9):1039–50.
Khandaker GM, Zimbron J, Dalman C, Lewis G, Jones PB. Childhood infection and adult schizophrenia: a meta-analysis of population-based studies. Schizophr Res. 2012;139(1–3):161–8.
Kim KS. Mechanisms of microbial traversal of the blood–brain barrier. Nat Rev Microbiol. 2008;6(8):625–34.
Kimura A, Sakurai T, Koumura A, et al. High prevalence of autoantibodies against phosphoglycerate mutase 1 in patients with autoimmune central nervous system diseases. J Neuroimmunol. 2010;219(1–2):105–8.
Koponen H, Rantakallio P, Veijola J, Jones P, Jokelainen J, Isohanni M. Childhood central nervous system infections and risk for schizophrenia. Eur Arch Psychiatry Clin Neurosci. 2004;254(1):9–13.
Korczak DJ, Pereira S, Koulajian K, Matejcek A, Giacca A. Type 1 diabetes mellitus and major depressive disorder: evidence for a biological link. Diabetologia. 2011;54(10):2483–93.
Kowal C, Degiorgio LA, Nakaoka T, et al. Cognition and immunity; antibody impairs memory. Immunity. 2004;21(2):179–88.
Kranaster L, Koethe D, Hoyer C, Meyer-Lindenberg A, Leweke FM. Cerebrospinal fluid diagnostics in first-episode schizophrenia. Eur Arch Psychiatry Clin Neurosci. 2011;261(7):529–30.
Laan W, Smeets H, de Wit NJ, Kahn RS, Grobbee DE, Burger H. Glucocorticosteroids associated with a decreased risk of psychosis. J Clin Psychopharmacol. 2009;29(3):288–90.
Laske C, Zank M, Klein R, et al. Autoantibody reactivity in serum of patients with major depression, schizophrenia and healthy controls. Psychiatry Res. 2008;158(1):83–6.

Laursen TM. Life expectancy among persons with schizophrenia or bipolar affective disorder. Schizophr Res. 2011;131(1–3):101–4.

Laursen TM, Munk-Olsen T, Nordentoft M, Mortensen PB. Increased mortality among patients admitted with major psychiatric disorders: a register-based study comparing mortality in unipolar depressive disorder, bipolar affective disorder, schizoaffective disorder, and schizophrenia. J Clin Psychiatry. 2007;68(6):899–907.

Laursen TM, Munk-Olsen T, Gasse C. Chronic somatic comorbidity and excess mortality due to natural causes in persons with schizophrenia or bipolar affective disorder. PLoS One. 2011;6(9):e24597.

Leboyer M, Soreca I, Scott J, et al. Can bipolar disorder be viewed as a multi-system inflammatory disease? J Affect Disord. 2012;141(1):1–10.

Leweke FM, Gerth CW, Koethe D, et al. Antibodies to infectious agents in individuals with recent onset schizophrenia. Eur Arch Psychiatry Clin Neurosci. 2004;254(1):4–8.

Liang W, Chikritzhs T. Early childhood infections and risk of schizophrenia. Psychiatry Res. 2012;200(2–3):214–7.

Maj M. When does depression become a mental disorder? Br J Psychiatry. 2011;199(2):85–6.

Marballi K, Quinones MP, Jimenez F, et al. In vivo and in vitro genetic evidence of involvement of neuregulin 1 in immune system dysregulation. J Mol Med (Berl). 2010;88(11):1133–41.

Margutti P, Delunardo F, Ortona E. Autoantibodies associated with psychiatric disorders. Curr Neurovasc Res. 2006;3(2):149–57.

Mednick SA, Machon RA, Huttunen MO, Bonett D. Adult schizophrenia following prenatal exposure to an influenza epidemic. Arch Gen Psychiatry. 1988;45(2):189–92.

Meyer U, Feldon J, Dammann O. Schizophrenia and autism: both shared and disorder-specific pathogenesis via perinatal inflammation? Pediatr Res. 2011a;69(5 Pt 2):26R–33.

Meyer U, Schwarz MJ, Muller N. Inflammatory processes in schizophrenia: a promising neuroimmunological target for the treatment of negative/cognitive symptoms and beyond. Pharmacol Ther. 2011b;132(1):96–110.

Miller AH. Depression and immunity: a role for T cells? Brain Behav Immun. 2010;24(1):1–8.

Miller AH, Maletic V, Raison CL. Inflammation and its discontents: the role of cytokines in the pathophysiology of major depression. Biol Psychiatry. 2009;65(9):732–41.

Mors O, Mortensen PB, Ewald H. A population-based register study of the association between schizophrenia and rheumatoid arthritis. Schizophr Res. 1999;40(1):67–74.

Mortensen PB, Norgaard-Pedersen B, Waltoft BL, et al. *Toxoplasma gondii* as a risk factor for early-onset schizophrenia: analysis of filter paper blood samples obtained at birth. Biol Psychiatry. 2007;61(5):688–93.

Mortensen PB, Pedersen MG, Pedersen CB. Psychiatric family history and schizophrenia risk in Denmark: which mental disorders are relevant? Psychol Med. 2010;40(2):201–10.

Moussavi S, Chatterji S, Verdes E, Tandon A, Patel V, Ustun B. Depression, chronic diseases, and decrements in health: results from the World Health Surveys. Lancet. 2007;370(9590):851–8.

Muller N, Schwarz MJ. Immune system and schizophrenia. Curr Immunol Rev. 2010;6(3): 213–20.

Niebuhr DW, Millikan AM, Cowan DN, Yolken R, Li Y, Weber NS. Selected infectious agents and risk of schizophrenia among U.S. military personnel. Am J Psychiatry. 2008a;165(1):99–106.

Niebuhr DW, Millikan AM, Yolken R, Li Y, Weber NS. Results from a hypothesis generating case–control study: herpes family viruses and schizophrenia among military personnel. Schizophr Bull. 2008b;34(6):1182–8.

Nielsen PR, Laursen TM, Mortensen PB. Association between parental hospital-treated infection and the risk of schizophrenia in adolescence and early adulthood. Schizophr Bull. 2013;39(1): 230–7.

Nielsen PR, Benros ME, Mortensen PB. Hospital contacts with infection and risk of schizophrenia: a population-based cohort study with linkage of Danish national registers. Schizophr Bull. 2014; 40(6):1526–32.

Nikkila HV, Muller K, Ahokas A, Rimon R, Andersson LC. Increased frequency of activated lymphocytes in the cerebrospinal fluid of patients with acute schizophrenia. Schizophr Res. 2001;49(1–2):99–105.

Noll R. Kraepelin's 'lost biological psychiatry'? Autointoxication, organotherapy and surgery for dementia praecox. Hist Psychiatry. 2007;18(71 Pt 3):301–20.

Okusaga O, Yolken RH, Langenberg P, et al. Association of seropositivity for influenza and coronaviruses with history of mood disorders and suicide attempts. J Affect Disord. 2011;130(1–2):220–5.

Ozawa K, Hashimoto K, Kishimoto T, Shimizu E, Ishikura H, Iyo M. Immune activation during pregnancy in mice leads to dopaminergic hyperfunction and cognitive impairment in the offspring: a neurodevelopmental animal model of schizophrenia. Biol Psychiatry. 2006;59(6): 546–54.

Padmos RC, Bekris L, Knijff EM, et al. A high prevalence of organ-specific autoimmunity in patients with bipolar disorder. Biol Psychiatry. 2004;56(7):476–82.

Patterson PH. Immune involvement in schizophrenia and autism: etiology, pathology and animal models. Behav Brain Res. 2009;204(2):313–21.

Pedersen CB, Mortensen PB. Evidence of a dose–response relationship between urbanicity during upbringing and schizophrenia risk. Arch Gen Psychiatry. 2001;58(11):1039–46.

Pedersen A, Zachariae R, Bovbjerg DH. Influence of psychological stress on upper respiratory infection—a meta-analysis of prospective studies. Psychosom Med. 2010;72(8):823–32.

Pedersen MG, Stevens H, Pedersen CB, Norgaard-Pedersen B, Mortensen PB. Toxoplasma infection and later development of schizophrenia in mothers. Am J Psychiatry. 2011;168(8):814–21.

Pedersen MS, Benros ME, Agerbo E, Borglum AD, Mortensen PB. Schizophrenia in patients with atopic disorders with particular emphasis on asthma: a Danish population-based study. Schizophr Res. 2012;138(1):58–62.

Pilkington T. The coincidence of rheumatoid arthritis and schizophrenia. J Nerv Ment Dis. 1955;124:604–6.

Pop VJ, Maartens LH, Leusink G, et al. Are autoimmune thyroid dysfunction and depression related? J Clin Endocrinol Metab. 1998;83(9):3194–7.

Potvin S, Stip E, Sepehry AA, Gendron A, Bah R, Kouassi E. Inflammatory cytokine alterations in schizophrenia: a systematic quantitative review. Biol Psychiatry. 2008;63(8):801–8.

Reichelt KL, Landmark J. Specific IgA antibody increases in schizophrenia. Biol Psychiatry. 1995;37(6):410–3.

Rice JS, Kowal C, Volpe BT, Degiorgio LA, Diamond B. Molecular mimicry: anti-DNA antibodies bind microbial and nonnucleic acid self-antigens. Curr Top Microbiol Immunol. 2005;296:137–51.

Ripke S, Sanders AR, Kendler KS, et al. Genome-wide association study identifies five new schizophrenia loci. Nat Genet. 2011;43(10):969–76.

Rivest S. Interactions between the immune and neuroendocrine systems. Prog Brain Res. 2010;181:43–53.

Rose NR. The role of infection in the pathogenesis of autoimmune disease. Semin Immunol. 1998;10(1):5–13.

Samaroo D, Dickerson F, Kasarda DD, et al. Novel immune response to gluten in individuals with schizophrenia. Schizophr Res. 2010;118(1–3):248–55.

Sellgren C, Frisell T, Lichtenstein P, Landen M, Askling J. The association between schizophrenia and rheumatoid arthritis: a nationwide population-based Swedish study on intraindividual and familial risks. Schizophr Bull. 2014;40(6):1552–9.

Seminog OO, Goldacre MJ. Risk of pneumonia and pneumococcal disease in people with severe mental illness: English record linkage studies. Thorax. 2013;68(2):171–6.

Shankar V, Kao M, Hamir AN, Sheng H, Koprowski H, Dietzschold B. Kinetics of virus spread and changes in levels of several cytokine mRNAs in the brain after intranasal infection of rats with Borna disease virus. J Virol. 1992;66(2):992–8.

Shelton RC, Claiborne J, Sidoryk-Wegrzynowicz M, et al. Altered expression of genes involved in inflammation and apoptosis in frontal cortex in major depression. Mol Psychiatry. 2011;16(7):751–62.

Smith DJ, Langan J, McLean G, Guthrie B, Mercer SW. Schizophrenia is associated with excess multiple physical-health comorbidities but low levels of recorded cardiovascular disease in primary care: cross-sectional study. BMJ Open. 2013;3(4).

Sopori M. Effects of cigarette smoke on the immune system. Nat Rev Immunol. 2002;2(5): 372–7.

Sperner-Unterweger B. Immunological aetiology of major psychiatric disorders: evidence and therapeutic implications. Drugs. 2005;65(11):1493–520.

Stefansson H, Ophoff RA, Steinberg S, et al. Common variants conferring risk of schizophrenia. Nature. 2009;460(7256):744–7.

Steiner J, Bogerts B, Sarnyai Z, et al. Bridging the gap between the immune and glutamate hypotheses of schizophrenia and major depression: potential role of glial NMDA receptor modulators and impaired blood–brain barrier integrity. World J Biol Psychiatry. 2012;13(7):482–92.

Strous RD, Shoenfeld Y. Behavioral changes in systemic lupus erythematosus are of an autoimmune nature. Nat Clin Pract Rheumatol. 2007;3(11):592–3.

Sullivan PF, Daly MJ, O'Donovan M. Genetic architectures of psychiatric disorders: the emerging picture and its implications. Nat Rev Genet. 2012;13(8):537–51.

Sundquist K, Li X, Hemminki K, Sundquist J. Subsequent risk of hospitalization for neuropsychiatric disorders in patients with rheumatic diseases: a nationwide study from Sweden. Arch Gen Psychiatry. 2008;65(5):501–7.

Suvisaari J, Mautemps N, Haukka J, Hovi T, Lonnqvist J. Childhood central nervous system viral infections and adult schizophrenia. Am J Psychiatry. 2003;160(6):1183–5.

Tanaka S, Matsunaga H, Kimura M, et al. Autoantibodies against four kinds of neurotransmitter receptors in psychiatric disorders. J Neuroimmunol. 2003;141(1–2):155–64.

Terayama H, Nishino Y, Kishi M, Ikuta K, Itoh M, Iwahashi K. Detection of anti-Borna disease virus (BDV) antibodies from patients with schizophrenia and mood disorders in Japan. Psychiatry Res. 2003;120(2):201–6.

Torrey EF, Yolken RH. The schizophrenia-rheumatoid arthritis connection: infectious, immune, or both? Brain Behav Immun. 2001;15(4):401–10.

Torrey EF, Yolken RH, Winfrey CJ. Cytomegalovirus antibody in cerebrospinal fluid of schizophrenic patients detected by enzyme immunoassay. Science. 1982;216(4548):892–4.

Torrey EF, Leweke MF, Schwarz MJ, et al. Cytomegalovirus and schizophrenia. CNS Drugs. 2006;20(11):879–85.

Torrey EF, Bartko JJ, Lun ZR, Yolken RH. Antibodies to *Toxoplasma gondii* in patients with schizophrenia: a meta-analysis. Schizophr Bull. 2007;33(3):729–36.

Trevathan R, Tatum JC. Rarity of concurrence of psychosis and rheumatoid arthritis in individual patients. J Nerv Ment Dis. 1953;120:83–4.

Tyring S, Gottlieb A, Papp K, et al. Etanercept and clinical outcomes, fatigue, and depression in psoriasis: double-blind placebo-controlled randomised phase III trial. Lancet. 2006;367(9504): 29–35.

Uranova NA, Zimina IS, Vikhreva OV, Krukov NO, Rachmanova VI, Orlovskaya DD. Ultrastructural damage of capillaries in the neocortex in schizophrenia. World J Biol Psychiatry. 2010;11(3):567–78.

Vinogradov S, Gottesman II, Moises HW, Nicol S. Negative association between schizophrenia and rheumatoid arthritis. Schizophr Bull. 1991;17(4):669–78.

Vonk R, van der Schot AC, Kahn RS, Nolen WA, Drexhage HA. Is autoimmune thyroiditis part of the genetic vulnerability (or an endophenotype) for bipolar disorder? Biol Psychiatry. 2007;62(2):135–40.

Weiser M, Werbeloff N, Levine A, et al. CNS infection in childhood does not confer risk for later schizophrenia: a case–control study. Schizophr Res. 2010;124(1–3):231–5.

Wilkinson J, Radkowski M, Eschbacher JM, Laskus T. Activation of brain macrophages/microglia cells in hepatitis C infection. Gut. 2010;59(10):1394–400.

Wium-Andersen MK, Orsted DD, Nielsen SF, Nordestgaard BG. Elevated C-reactive protein levels, psychological distress, and depression in 73 131 individuals. JAMA Psychiatry. 2013; 70(2):176–84.

Wium-Andersen MK, Orsted DD, Nordestgaard BG. Elevated C-reactive protein associated with late- and very-late-onset schizophrenia in the general population: a prospective study. Schizophr Bull. 2014;40(5):1117–27.

Wright P, Sham PC, Gilvarry CM, et al. Autoimmune diseases in the pedigrees of schizophrenic and control subjects. Schizophr Res. 1996;20(3):261–7.

Yirmiya R, Goshen I. Immune modulation of learning, memory, neural plasticity and neurogenesis. Brain Behav Immun. 2011;25(2):181–213.

Yolken RH, Torrey EF. Are some cases of psychosis caused by microbial agents? A review of the evidence. Mol Psychiatry. 2008;13(5):470–9.

Yum SY, Yum SK, Kim T, Hwang MY. Clinical perspectives on autoimmune processes in schizophrenia. Psychiatr Clin North Am. 2009;32(4):795–808.

Chapter 7
Exposure to Microorganisms and Adult Psychiatric Disorders: The Case for a Causal Role of *Toxoplasma gondii*

Robert H. Yolken and E. Fuller Torrey

Abstract Since the 1970s there has been an increasing interest in infectious agents as possible causes of serious mental illness. The microorganism which has been most intensively studied is *Toxoplasma gondii*, a parasite carried by felines. Several lines of research have suggested that *T. gondii* is associated with schizophrenia: increased prevalence of *T. gondii* antibodies in individuals with this disease, and in mothers who give birth to individuals who later are so diagnosed; increased childhood contact with cats by individuals who later are so diagnosed; the fact that *T. gondii* affects several neurotransmitters and makes dopamine; and the fact that some antipsychotics suppress *T. gondii*. But is there any evidence that this association is etiological? Austin Bradford Hill published criteria for assessing possible etiological associations between environmental factors and complex diseases. His criteria include strength of association, consistency of findings, specificity of findings, temporality of relationship, biological gradient, plausibility of relationship, coherence of relationship, analogy with other data, and experimental data. Using these criteria, it is concluded that an etiological relationship between *T. gondii* and some cases of schizophrenia is reasonable.

Keywords *Toxoplasma gondii* • Hill criteria • Causation • Infectious disease • Schizophrenia

R.H. Yolken
Stanley Division of Developmental Neurovirology, Department of Pediatrics, Johns Hopkins University School of Medicine, 600 N. Wolfe Street, Blalock 1105, Baltimore, MD 21287-4933, USA

E.F. Torrey (✉)
The Stanley Medical Research Institute, 8401 Connecticut Avenue, Suite 200, Chevy Chase, MD 20815, USA
e-mail: torreyf@stanleyresearch.org

N. Müller et al. (eds.), *Immunology and Psychiatry*, Current Topics in Neurotoxicity 8, DOI 10.1007/978-3-319-13602-8_7

Introduction

The idea that infectious agents may cause serious psychiatric disorders has a long history. As early as the fifteenth century Bartholomew, a Franciscan friar in England, wrote that "insanity may come as the result of an infection, from the bite of a mad dog or some other venomous animal" (Graham 1967). In 1874, as bacteria were being discovered, the *American Journal of Insanity* published a long article "On the Germ-Theory of Disease" (Drecke 1874). In 1896, the *Scientific American* published a paper with a title asking "Is Insanity Due to a Microbe?" It described the work of American researchers who believed that bacteria caused insanity and were injecting sera from individuals with insanity into rabbits to ascertain the effect. It reported that the rabbits became sick, "although it is not alleged that they were insane" (*Scientific American* 1896).

The subsequent discovery of spirochetes as the cause of syphilis, which in its late stages often includes psychotic symptoms, strengthened the infectious theory. This was followed by the case of psychoses associated with the 1918–1919 pandemic of influenza. Karl Menninger summarized 39 such cases and said that he was "persuaded that dementia praecox is at least in most instances a somatopsychosis, the psychic manifestations of an encephalitis" (Menninger 1922, 1926). A Danish psychiatrist agreed that the "toxi-infection" hypothesis of dementia praecox "might… explain the difficulties which have arisen owing to heredo-biological and endocrinological investigations and their divergent results (Reiter 1926).

For the next half century there was comparatively little interest in infectious theories of serious psychiatric disorders. However, beginning in the early 1970s there was a revival of interest in this hypothesis (Torrey and Peterson 1973) and in 1983 the World Health Organization sponsored a symposium on "Research on the Viral Hypothesis of Mental Disorders" (Morozov 1983). Since that time, interest in infectious agents, interacting with predisposing genes, has steadily increased as possible etiological agents for serious mental illness, especially in recent years as inflammatory aspects of these diseases have become prominent.

In a single chapter it is impossible to review the large number of infectious agents that have been investigated during the last three decades. We will therefore focus on *Toxoplasma gondii* the infectious agent which has been most extensively studied in terms of association with psychiatric disorders. We will then discuss the nature of causal relationships for infectious agents and chronic diseases, including on the use of the Hill criteria of causation, first proposed by Austin Bradford Hill in 1965 and widely used to define associations between environmental exposures and chronic disorders (Hill 1965).

The Biology of *T. gondii*

T. gondii is a coccidian protozoan of the apicomplexa family, first described in 1908. Felines, including domestic cats, are its definitive host, and the organism can only complete the sexual part of its life cycle within feline hosts. *T. gondii* oocysts are excreted in the feces of cats at the time they are initially infected. The oocysts may then become aerosolized and infect humans who are changing cat litter boxes, gardening, or playing in sandboxes. The cat may also deposit feces on the ground or in animal feed in barns; domestic animals may eat it, producing *T. gondii* tissue cysts in their muscles, which then infect humans who eat undercooked meat. Oocysts shed by infected felines can also contaminate water, with subsequent infection of humans through drinking water, eating vegetables, washed with contaminated water, or eating undercooked fish that had lived in the contaminated water. The relative importance of different modes of transmission depends upon local geographic, cultural, and socioeconomic facts. *T. gondii* is distributed worldwide and infects between 10 and 80 % of the adult population; in the USA, the rate is between 10 and 20 % in young adults.

Infection of pregnant women with *T. gondii* can produce stillbirths as well as deafness, seizures, cerebral palsy, damage to the retina, or mental retardation in the offspring. In individuals who are immunosuppressed, such as those with AIDS or those undergoing organ transplantation or cancer chemotherapy, infection with or reactivation of *T. gondii* can produce psychiatric symptoms, including delusions and hallucinations (Kramer 1966). Until recently it was assumed that the infection of immunocompetent, non-pregnant individuals did not pose a health risk but that assumption is now in question. For example, individuals exposed to *Toxoplasma* through ingestion of contaminated water have shown clear signs of systemic infection at defined times after their exposure (Jones and Dubey 2010).

Association Between *Toxoplasma* Exposure and Risk of Psychiatric Disorders

Studies indicating that exposure to *T. gondii* might be associated with increased risk of schizophrenia are summarized as follows:

Antibodies against *T. gondii* are increased in individuals with serious mental illness, especially schizophrenia.

A meta-analysis of 38 studies of *T. gondii* antibodies in individuals with schizophrenia, compared to non-affected controls, reported that the prevalence of *T. gondii* antibodies was higher in 36 of the studies (OR 2.71; 95 % CI 1.93–3.80) (Torrey et al. 2012). Another study collected sera from 45,609 women at the time they gave birth. During the following 16 years, 80 of the women developed schizophrenia; those with the highest *T. gondii* IgG antibody levels at the time they gave birth had a significantly increased chance of developing schizophrenia (Pedersen et al. 2011). Four studies have reported an increased prevalence of *T. gondii* antibodies in individuals with bipolar disorder (Fekadu et al. 2010; Pearce et al. 2012; Hamdani et al. 2013; Nascimento et al. 2012) and one study reported that such antibodies are increased in individuals with obsessive-compulsive disorder (Miman et al. 2010)

T. gondii is known to affect some neurotransmitters known to be abnormal in serious mental illnesses.

It is known that *T. gondii* is highly neurotropic and infects both neurons and glia. One study in cell culture reported that *T. gondii* affected the expression of RNA sequences associated with dopamine, glutamate, and serotonin and that these effects varied significantly depending on the strain of *T. gondii* (Xiao et al. 2013). Another study in mice reported that *T. gondii* affects the GABA neurotransmitter system (Fuks et al. 2012). The effect of *T. gondii* on dopamine is especially interesting since dopamine has been assumed to be increased in individuals with schizophrenia and many of the antipsychotic drugs used to successfully treat schizophrenia block dopamine. It is now known that *T. gondii* makes dopamine, raising the possibility that the excess dopamine which occurs in schizophrenia may be a product of the infecting agent and not being made by the individual affected (Prandovszky et al. 2011).

Exposure to cats in childhood increases the chances of being later diagnosed with schizophrenia or bipolar disorder.

Although felines are the definitive host for *T. gondii*, it is not necessary to have had contact with cats to become infected with *T. gondii*. The reason, as noted above, is that *T. gondii* is spread widely in the environment and infective oocysts may remain viable for several years under certain environmental conditions. Studies have quantified the *T. gondii* oocyst burden in the environment. Studies of community soil samples in California and France reported between 4 and 434 oocysts per square foot. A study of public sandboxes, where cats selectively defecate, reported that the *T. gondii* oocyst burden may exceed one million oocysts per square foot (Torrey and Yolken 2013).

Given the many ways in which individuals may become infected with *T. gondii* without having had direct cat contact, it is surprising that two studies reported that close cat contact in childhood increases the chances of later being diagnosed with schizophrenia or bipolar disorder. In one study of 165 individuals with schizophrenia or bipolar disorder, 51 % of the families owned a cat during the child's first 10 years compared to 38 % of controls ($p=0.02$ but not corrected for multiple questions asked) (Torrey and Yolken 1995). In the other study of 264 individuals with schizophrenia or bipolar disorder and 528 matched controls, 52 % of the families owned a cat during the child's first 13 years compared to 42 % of the controls ($p=0.007$) (Torrey et al. 2000). For dog ownership, there was no significant difference, and in fact more of the control families owned dogs. These studies suggest that *T. gondii* does in fact play a role in the etiology of schizophrenia or bipolar disorder, transmission of the parasite may occur during childhood.

The historical periods encompassing the increased prevalence of schizophrenia and the rise of cat-keeping as pets are similar

Prior to about 1800, schizophrenia was a rare disease. It then increased sharply in prevalence in the nineteenth and first half of the twentieth centuries; since that time, its prevalence appears to have remained steady (Torrey and Miller 2002). Similarly, prior to about 1800, most cats were confined to barns since they were regarded not as pets but rather as guardians of the grain from rodents. With the exception of the earlier Egyptian empire, the first widespread keeping of cats as pets took place in Europe and America in the nineteenth century, during which time it increased sharply and has continued to increase until the present.

Some medications effective in treating serious mental illness also suppress *T. gondii*.

As early as 1891 it was reported that a phenothiazine dye effectively killed the parasite *Plasmodium vivax*. In more recent years, phenothiazines have been shown to be effective in inhibiting the growth of *Leishmania donovani*, *Trypanosoma brucei*, and other parasites. Four in vitro studies have assessed the effectiveness of various antipsychotics, used to treat schizophrenia, and mood stabilizers, used to treat bipolar disorder, in suppressing the growth of *T. gondii*; many but not all of the antipsychotics and mood stabilizers demonstrated some effectiveness (Jones-Brando et al. 2003; Goodwin et al. 2008, 2011; Fond et al. 2014). This raises the possibility that at least part of the effectiveness of these drugs, which has been assumed to be related to their effect on neurotransmitters, may be partly related to their direct effect in suppressing the infectious agents, possibly through the interference of protozoan related dopamine interactions.

One way of assessing the relationship between is by the application of the "Hill criteria" which were developed by Austin Bradford Hill to quantify the association between environmental factors and complex disorders. The criteria are enumerated below along with discussions of how the *Toxoplasma*-schizophrenia link might apply.

Strength: The odds ratios describing the association between *Toxoplasma* and schizophrenia are in the range of 2–2.5. For most complex disorders this would be considered a moderate association. In the case of schizophrenia, this level of association is greater than that of common single nucleotide polymorphisms and in the same range as other proposed environmental factors such as exposure to marijuana and immigration. From a practical point of view, the moderate level of association means that many individuals exposed to *Toxoplasma* will not develop psychiatric symptoms. Some possible explanations for this moderate level of association are strain differences among *Toxoplasma* organisms, i.e., it may be that only some strains cause psychiatric disorders. There are also suggestions that the timing and route of the initial infection may be crucial. For example, initial infection at age 5–10 with a *T. gondii* oocyst may lead to a psychiatric illness but an initial infection at ages 20–25 with a tissue cyst does not. Finally, genetics certainly play a role, as it does for more infectious agents, with some people having predisposing genes and others not.

Consistency: The association between *Toxoplasma* and schizophrenia has been found in a large number of different populations, including those in North America, South America, Europe, and Asia and in studies reported as early as 1953 and as late as 2013. Thus the *Toxoplasma*-schizophrenia association rates highly in terms of both temporal and geographic consistency

Specificity: In addition to an association with schizophrenia, exposure to *Toxoplasma* has been associated with a number of additional psychiatric disorders including bipolar disorder, post-partum psychosis suicidal intent, and cognitive impairment. These findings, while not consistent with the usual classification of psychiatric disorders, are consistent with genetic studies indicating shared genetic risk among different psychiatric disorders. As noted above, the reasons for the variable expression of *Toxoplasma* infection in different individuals are not known with certainty but is likely to be related to genetic variability in both the host and the parasite as well as the timing of *Toxoplasma* exposure.

Temporality: The temporal association between an increase in cat ownership and rates of schizophrenia is discussed above. In terms of the temporality of an individual's exposure to *Toxoplasma*, most studies of *Toxoplasma* exposure have examined individuals with established schizophrenia and thus cannot be used to document temporality. However several studies have indicated a particular strong association between *Toxoplasma* and schizophrenia close to illness onset. Furthermore, several studies have documented association between exposures in pregnancy or early life (Mortensen et al. 2007; Blomström et al. 2012; Brown et al. 2005) and disease risk. While these exposures occurred many years prior to illness onset in adulthood, they are consistent with the neurodevelopmental hypothesis of schizophrenia which posits that exposures in early life do not become manifest until later life due to postnatal changes in brain development. Temporality is also suggested by studies documenting increased rates of psychiatric symptoms in individuals with acute *Toxoplasma* lymphadenitis (Wong et al. 2013) as well as a study documenting an increased rate of psychiatric disorders in a cohort of otherwise healthy adults who were *Toxoplasma* seropositive (Niebuhr et al. 2008), although other cohort studies have not found this association (Li et al. 2013).

Biological gradient: Both *Toxoplasma* exposure and schizophrenia are found in virtually every area of the world. However, geographic differences in *Toxoplasma* exposure I areas of the world such as Southern Europe, Eastern Europe, and Latin America have generally not been associated with increased risks of schizophrenia. The reasons for this lack of association are not known with certainty but may be related to differences among *Toxoplasma* strains, life stage of the infecting *Toxoplasma* (oocyst vs tissue cyst), genetic differences among the host populations, or incomplete disease ascertainment. Interestingly, another manifestation of *Toxoplasma* infection, suicide, has been geographically linked to differences in *Toxoplasma* exposure (Lester 2010).

Plausibility: The plausibility of the link between *Toxoplasma* exposure and schizophrenia is supported by studies in animal models indicating that *Toxoplasma*

changes brain dopamine levels following infection. The mechanisms by which this occurs are not known with certainty but may be related to dopamine generating enzymes in the *Toxoplasma* organisms or to the immune response to infection. *Toxoplasma* infection in experimental animals is also associated with a range of behavioral effects, some of which are specific to interaction with cats and some of which have more general effects on cognition and behavior.

Coherence: The association between *Toxoplasma* infections and schizophrenia is supported by numerous studies in animal model systems. In addition to the association with dopamine cited above, *Toxoplasma* infection has been shown to alter cognitive functioning and other behavioral traits.

Analogy: As noted above, other infectious agents such as herpes viruses have been associated with increased risk of schizophrenia. All share the ability to infect the brain, particularly during the pre-natal or early post-natal period and to establish persistent infection within the central nervous system.

Experiment: The most direct way to examine an association between *Toxoplasma* infection and schizophrenia would be to treat *Toxoplasma* infection and document an improvement in symptoms, in a manner similar to how the association between Helicobacter pylori and peptic ulcers was proven. However, most immune competent individuals are largely infected with the tissue cyst form of *Toxoplasma* and this form is resistant to currently available anti-*Toxoplasma* medications. One recent study using artemether, an anti-malarial with anti-*Toxoplasma* properties, did improve the symptoms of schizophrenia (Wang et al. 2014). The recent discovery of quinolone compounds which can remove the cyst form of *Toxoplasma* from the brains of experimentally infected mice offers the promise of future effective therapies which might be employed to eradicate tissue cysts from the brains of individuals with schizophrenia (Doggett et al. 2012). A resulting improvement in clinical symptoms following treatment would effectively establish a link between *Toxoplasma* infection and disease pathogenesis. This outcome, combined with the epidemiological and experimental data cited above would conclusively prove a link between *Toxoplasma* infection and serious psychiatric disorders and would provide novel methods for the prevention and treatment of these disorders.

References

Blomström A, Karlsson H, Wicks S, Yang S, Yolken RH, Dalman C. Maternal antibodies to infectious agents and risk for non-affective psychoses in the offspring—a matched case–control study. Schizophr Res. 2012;140:25–30.

Brown AS, Schaefer CA, Quesenberry Jr CP, Liu L, Babulas VP, Susser ES. Maternal exposure to toxoplasmosis and risk of schizophrenia in adult offspring. Am J Psychiatry. 2005; 162:767–73.

Doggett JS, Nilsen A, Forquer I, et al. Endochin-like quinolones are highly efficacious against acute and latent experimental toxoplasmosis. Proc Natl Acad Sci U S A. 2012;109:15936–41.

Drecke T. On the germ-theory of disease. Am J Insanity. 1874;30:443–6.

Fekadu A, Shibre T, Cleare AJ. Toxoplasmosis as a cause for behaviour disorders—overview of evidence and mechanisms. Folia Parasitol (Praha). 2010;57:105–13.

Fond G, Macgregor A, Tamouza R, et al. Comparative analysis of anti-toxoplasmic activity of antipsychotic drugs and valproate. Eur Arch Psychiatry Clin Neurosci. 2014;264:179–83.

Fuks JM, Arrighi RB, Weidner JM, et al. GABAergic signaling is linked to a hypermigratory phenotype in dendritic cells infected by *Toxoplasma gondii*. PLoS Pathog. 2012;8:e1003051.

Goodwin DG, Strobl J, Mitchell SM, Zajac AM, Lindsay DS. Evaluation of the mood-stabilizing agent valproic acid as a preventative for toxoplasmosis in mice and activity against tissue cysts in mice. J Parasitol. 2008;94:555–7.

Goodwin DG, Strobl JS, Lindsay DS. Evaluation of five antischizophrenic agents against *Toxoplasma gondii* in human cell cultures. J Parasitol. 2011;97:148–51.

Graham TF. Medieval minds: mental health in the middle ages. London: George Allen & Unwin; 1967. p. 64.

Hamdani N, Daban-Huard C, Lajnef M, et al. Relationship between *Toxoplasma gondii* infection and bipolar disorder in a French sample. J Affect Disord. 2013;148(2–3):444–8.

Hill AB. The environment and disease: association or causation? Proc R Soc Med. 1965;58:295–300.

Jones JL, Dubey JP. Waterborne toxoplasmosis—recent developments. Exp Parasitol. 2010;124:10–25.

Jones-Brando L, Torrey EF, Yolken R. Drugs used in the treatment of schizophrenia and bipolar disorder inhibit the replication of *Toxoplasma gondii*. Schizophr Res. 2003;62:237–44.

Kramer W. Frontiers of neurological diagnosis in acquired toxoplasmosis. Psychiatr Neurol Neurochir. 1966;69:43–64.

Lester D. Brain parasites and suicide. Psychol Rep. 2010;107:424.

Li Y, Weber NS, Fisher JA, et al. Association between antibodies to multiple infectious and food antigens and new onset schizophrenia among US military personnel. Schizophr Res. 2013;151:36–42.

Menninger KA. Reversible schizophrenia. Am J Psychiatry. 1922;1:573–88.

Menninger KA. Influenza and schizophrenia. Am J Psychiatry. 1926;5:469–529.

Miman O, Mutlu EA, Ozcan O, Atambay M, Karlidaq R, Unal S. Is there any role of *Toxoplasma gondii* in the etiology of obsessive-compulsive disorder? Psychiatry Res. 2010;177:263–5.

Morozov PV, editor. Research on the viral hypothesis of mental disorders. Basel: Karger; 1983.

Mortensen PB, Nørgaard-Pedersen B, Waltoft BL, et al. *Toxoplasma gondii* as a risk factor for early-onset schizophrenia: analysis of filter paper blood samples obtained at birth. Biol Psychiatry. 2007;61:688–93.

Nascimento FS, de Rosalmeida Dantas C, Netto MP, Mella LF, Suzuki LA, Banzato CE, Rossi CL. Prevalence of antibodies to *Toxoplasma gondii* in patients with schizophrenia and mood disorders. Schizophr Res. 2012;142(1–3):244–5.

Niebuhr DW, Millikan AM, Cowan DN, Yolken R, Li Y, Weber NS. Selected infectious agents and risk of schizophrenia among U.S. military personnel. Am J Psychiatry. 2008;165:99–106.

Pearce BD, Kruszon-Moran D, Jones JL. The relationship between *Toxoplasma gondii* infection and mood disorders in the third National Health and Nutrition survey. Biol Psychiatry. 2012;72:290–5.

Pedersen MG, Stevens H, Pedersen CB, Nørgaard-Pedersen B, Mortensen PB. *Toxoplasma* infection and later development of schizophrenia in mothers. Am J Psychiatry. 2011;168:814–21.

Prandovszky E, Gaskell E, Martin H, Dubey JP, Webster JP, McConkey GA. The neurotropic parasite *Toxoplasma gondii* increases dopamine metabolism. PLoS One. 2011;6:e23866.

Reiter PJ. Extrapyramidal-motor-disturbances in dementia praecox. Acta Psychiatr Neurol. 1926;1:287–309.

Scientific American. Is insanity due to a microbe? Sci Am. 1896;75:303.

Torrey EF, Miller J. The invisible plague: the rise of mental illness from 1750 to the present. New Brunswick: Rutgers University Press; 2002.

Torrey EF, Peterson MR. Slow and latent viruses in schizophrenia. Lancet. 1973;2:22–4.

Torrey EF, Yolken RH. Could schizophrenia be a viral zoonosis transmitted from house cats? Schizophr Bull. 1995;21:167–71.

Torrey EF, Yolken RH. *Toxoplasma* oocysts as a public health problem. Trends Parasitol. 2013;29:380–4.

Torrey EF, Rawlings R, Yolken RH. The antecedents of psychoses: a case–control study of selected risk factors. Schizophr Res. 2000;46:17–23.

Torrey EF, Bartko JJ, Yolken RH. *Toxoplasma gondii* and other risk factors for schizophrenia: an update. Schizophr Bull. 2012;38:642–7.

Wang HL, Xiang YT, Li QY, et al. The effect of artemether on psychotic symptoms and cognitive impairment in first-episode, antipsychotic drug-naive persons with schizophrenia seropositive to *Toxoplasma gondii*. J Psychiatr Res. 2014; pii: S0022-3956(14)00056-9.

Wong WK, Upton A, Thomas MG. Neuropsychiatric symptoms are common in immunocompetent adult patients with *Toxoplasma gondii* acute lymphadenitis. Scand J Infect Dis. 2013;45: 357–61.

Xiao J, Li Y, Jones-Brando L, Yolken RH. Abnormalities of neurotransmitter and neuropeptide systems in human neuroepithelioma cells infected by three *Toxoplasma* strains. J Neural Transm. 2013;120:1631–9.

Chapter 8
Major Depression: An Immune-Inflammatory Disorder

Cristiano Noto, Lucas B. Rizzo, Rodrigo Mansur, Elisa Brietzke, and Michael Maes

Abstract Major depressive disorder (MDD) is a high prevalent mental disorder. Several lines of evidence have demonstrated that MDD is accompanied by cell-mediated immune (CMI) activation and a chronic mild inflammatory response, characterized by an increased production of pro-inflammatory cytokines, such as interleukin (IL)-1β, IL-6, and tumor necrosis factor (TNF)-α. Immune-inflammatory pathways are involved in the onset of depressive symptoms and partly explain the high comorbidity of MDD and general medical diseases. Moreover, antidepressants have anti-inflammatory and negative immunoregulatory effects, while anti-inflammatory agents may improve depressive symptoms. Despite the large evidence describing activated immune-inflammatory responses in MDD, the pathophysiological mechanisms are unclear yet. Future studies should examine the clinical implications of those findings by targeting immune-inflammatory pathways as a therapeutic tool for individuals with MDD.

Keywords Major depressive disorder • Inflammation • Cytokines • Immune • Oxidative stress

C. Noto • L.B. Rizzo • E. Brietzke
Interdisciplinary Laboratory of Clinical Neuroscience (LINC), Department of Psychiatry, Universidade Federal de São Paulo, São Paulo, Brazil

R. Mansur
Interdisciplinary Laboratory of Clinical Neuroscience (LINC), Department of Psychiatry, Universidade Federal de São Paulo, São Paulo, Brazil

Mood Disorders Psychopharmacology Unit (MDPU), University Health Network, University of Toronto, Toronto, ON, Canada

M. Maes (✉)
IMPACT Strategic Research Center, Deakin University, Geelong, VIC, Australia

Faculty of Medicine, Department of Psychiatry, Chulalongkorn University, Bangkok, Thailand

Health Sciences Graduate Program, Health Sciences Center, State University of Londrina, Londrina, Paraná, Brazil
e-mail: dr.michaelmaes@hotmail.com

N. Müller et al. (eds.), *Immunology and Psychiatry*, Current Topics in Neurotoxicity 8, DOI 10.1007/978-3-319-13602-8_8

Introduction

Major depressive disorder (MDD) is one of the most prevalent mental disorders, accompanied by psychosocial disability and an important impact on quality of life (Whiteford et al. 2013). In the last decades several studies have suggested activated immune-inflammatory pathways in distinct chronic diseases, including MDD (Maes et al. 1990; Leonard and Maes 2012; Maes 1995; Dantzer 2009; Raison and Miller 2011). Convergent findings suggest that MDD is accompanied by cell-mediated immune (CMI) activation and a chronic mild inflammatory response, characterized by an increased production of pro-inflammatory cytokines or their functional receptors, such as interleukin (IL)-1β, IL-2, IL-6, interferon (IFN)-γ, tumor necrosis factor (TNF)-α, soluble IL-6 receptor (sIL-6R), and IL-1 receptor antagonist (IL-1RA) (Licinio and Wong 1999; Maes 1999; Zorrilla et al. 2001; Connor and Leonard 1998; Maes et al. 1997). Abnormalities in immune-inflammatory responses may be clinically relevant since several cytokines are directly involved in pathophysiology of depressive symptoms, such as anhedonia, fatigue, and somatic symptoms (Anisman et al. 2002). Moreover, immune-inflammatory pathways are involved in neuroprogression in MDD, caused by increased neuronal apoptosis and neurotoxic effects, and reduced neurogenesis, neuroplasticity, and neurotrophic or neuroprotective factors (Catena-Dell'Osso et al. 2011; Duman and Monteggia 2006; Maes et al. 2009).

Inflammation and CMI Activation

Responses of the immune system to injury consist of two major processes, i.e. humoral and cellular immune responses. The humoral response is mainly composed of cytokines, which mediate cell-to-cell communication for an adequate immune response, and by acute-phase response proteins (APPs), e.g. C-reactive protein (PCR), haptoglobin, and complement proteins, which are the first line response against immunological insults (Silverstein 2003). We call acute "inflammation" the process of increased production of pro-inflammatory cytokines and proteins which aim to activate immune responses and eliminate possible pathogens. Usually, the process starts with activated macrophages producing TNF-α, IL-6, and IL-1β. These cytokines induce the activation of nuclear factors, like nuclear factor κB (NFκB), promoting several processes including endothelial modifications, which in turn promote leukocyte attraction, induction of APPs by hepatocytes, and T lymphocyte activation.

The cell-mediated response is initiated by the activation of T helper (Th) cells. Activated Th cells produce IL-2 that interacts with the IL-2 receptor (IL-2R or CD25) expressed by other T cells boosting immune activation. Different Th subsets (Th1, Th2, Th17, Treg) are defined by the cytokines produced by these T cell subtypes. Th1-like cells are a great source of IFN-γ and IL-2, participating in viral control. Th2 are main sources of IL-4 and IL-5, controlling parasite infections and allergic reactions. Th17-like immune cells produce IL-17 and IL-22, which fight bacterial and fungal infections, and play a role in autoimmunity (Lee et al. 2006; Bettelli et al. 2008). The regulation of immunological processes is mainly done by Treg cells producing IL-10 and transforming growth factor beta (TGF-β), which are capable to inhibit the production of pro-inflammatory cytokines and exhibit negative immunoregulatory responses. This inhibitory Treg response plays a role in the compensatory (anti-) inflammatory reflex system (CIRS) that controls acute inflammation and limits an overzealous acute inflammatory response (Maes et al. 2012a). But there are other mechanisms like the production of an inactive molecule homologue to IL-1 that acts like an antagonist of the IL-1 receptor (IL-1RA), which may inhibit pro-inflammatory IL-1-signalling. Another CIRS mechanism is the secretion of soluble receptors of TNF and IL-2 (sTNFR and sIL-2R), which damp cytokines and compete with membrane receptors of these cytokines, diminishing cytokine activity (Sims and Smith 2010; Campbell and Koch 2011; Xanthoulea et al. 2004; Lantz et al. 1990; Zorn et al. 1994).

The presence of abnormalities in immune-inflammatory pathways in MDD has been largely described in several studies in both humans and animal models (Mesquita et al. 2008; Song and Wang 2011), including case–control, longitudinal, and *post-mortem* studies (Dean et al. 2010). The association of pro-inflammatory cytokines and their receptors or receptor antagonists, such as IL-1β, IL-2, IL-6, IFN-γ, TNF-α, IL-6R, and IL-1RA, has been described in MDD (Licinio and Wong 1999; Maes 1999; Zorrilla et al. 2001; Dowlati et al. 2010). Three recent meta-analyses demonstrated higher concentrations of TNF-α, IL-6, IL-1, sIL-2R, and/or PCR in patients with MDD (Dowlati et al. 2010; Liu et al. 2012; Howren et al. 2009).

Other inflammatory signs have been found in MDD. Activated complement cascades (increased C3 and C4) and acute-phase response (increased haptoglobin, lower albumin) have been described, as well as cell-mediated immune activation (increased levels of neopterin, soluble CD8 molecule and lower levels of plasma tryptophan and increased tryptophan catabolites) (Leonard and Maes 2012). More recently, increased oxidative and nitrosative stress (O&NS) have been explored in MDD. Several studies showed lower levels of antioxidants, such as Coenzyme Q10, vitamins E, and glutathione in MDD (Leonard and Maes 2012; Maes et al. 2011). These abnormalities might be driven by inflammation and may lead to lipid peroxidation and DNA damage, further augmenting the immune-inflammatory response (Leonard and Maes 2012; Maes et al. 2011).

Cytokines and Depressive Symptoms

MDD is characterized by a combination of several symptoms, such as anhedonia, fatigue or low energy, disturbed sleep, loss of appetite, poor concentration, hopelessness, worthlessness, and suicidal ideation. Pro-inflammatory cytokines may contribute to the pathophysiology of these symptoms and the administration of cytokines may provoke them (Dantzer et al. 2008).

Different cytokines seem to play a role in the onset of depressive symptoms. The administration of compounds that provoke inflammation, e.g. lipopolysaccharides (LPS) and pro-inflammatory cytokines such as IL-1β, IL-6, and TNF-α causes a behavioral symptom complex "sickness behavior" (Song and Wang 2011; Dantzer et al. 2008; Kubera et al. 1996; Ohgi et al. 2013). This is a group of behavioral and psychological responses to acute injuries (e.g., infections) triggered by pro-inflammatory cytokines and consequent neuroinflammation (Dantzer 2009). Sickness behavior includes behaviors such as reduced psychomotor activity, diminished physical and social activities, fatigue, lethargy, reduced food and water intake, sleep abnormalities, and altered cognition. Indeed, immune challenges in rodents are able to induce depression-like syndromes in the rodent, such as anhedonia, anorexia, body weight loss, and reduced locomotor, exploratory, and social behaviors (Stepanichev et al. 2014). More specifically, administration of IL-6 contributes to behavioral alterations similar to depression, whereas IL-1β increases sickness behavior, anxiety, fatigue, memory impairment, and melancholic characteristics (Kubera et al. 1996; Anisman et al. 2008; Song et al. 2006), and TNF-α is associated with anxiety and fatigue (Jiang et al. 2008).

Evidence of the association of pro-inflammatory cytokines and MDD is the high prevalence of depression in patients under immunotherapy treatments for infectious diseases, such as hepatitis C, and cancer using IFN-α, IL-2, and IL-12 (Capuron et al. 2001; Little et al. 2006; Udina et al. 2012). IFN-α is a potent inducer of pro-inflammatory cytokines, and 20–40 % of the patients under this treatment will develop MDD, the majority of them in the first 12 weeks of treatment (Udina et al. 2012; Raison et al. 2005). The mean neurobiological hypothesis of IFN-induced depression is activation of pro-inflammatory cytokines and consequent neuroinflammation, leading to alterations in neurotransmission and neuroprogression (Udina et al. 2012). There is also evidence that the imbalance in those cytokines may interfere in noradrenaline and 5-HT levels and activate the indoleamine 2,3-dioxygenase (IDO), which is involved in the catabolism of tryptophan (Leonard and Maes 2012; Walker 2013).

Another important finding about the association of MDD and inflammation is the presence of different cytokine profiles in patients with distinct clinical features. Individuals with MDD and melancholic characteristics present more pronounced immune-inflammatory abnormalities than those without this characteristic, with higher levels of TNF-alpha levels, CRP, and IL-6 (Lamers et al. 2012). Furthermore, individuals with MDD diagnosis have different levels of IL-6 and IL-10 when compared to those with only few depressive symptoms, and those recruited from an inpatient unit show more severe profile than those recruited from the community (Hiles et al. 2012). Recently, Serafini et al. in a systematic review and meta-analysis found that suicidal ideation/attempts in MDD was associated with differences in inflammatory cytokines profile, characterized by altered levels of IL2, IL-6, IL-8, TNF-α, and vascular endothelial growth factor (VEGF) (Serafini et al. 2013).

Effects of Antidepressants in Inflammation

There is a well-documented body of evidence indicating that several effective treatments for MDD also have an impact by attenuating activated immune-inflammatory pathways, which is probably related to their clinical efficacy. Interventions like antidepressants, exercise, electroconvulsive therapy, and psychotherapy have anti-inflammatory or negative immunoregulatory effects, suggesting that a reduction in activated immune-inflammatory responses is associated with treatment responsivity (Raison et al. 2006). There are some studies showing a consistent anti-inflammatory activity of fluoxetine, sertraline, and paroxetine. In addition, the peripheral effect of these medications also seems to reflect a central modulatory effect on microglia (Hashioka et al. 2009). Consistently, anti-inflammatory effects of antidepressants have been reported (Hannestad et al. 2011). Studies in both animal and humans have been performed, showing suppression in the production and/or release of pro-inflammatory and Th1-like cytokines, such as IFN-γ, while increasing IL-10, a negative immunoregulatory cytokine (Kenis and Maes 2002). In vitro studies have demonstrated these immunosuppressive or negative immunoregulatory effects of several antidepressants classes, including tricyclic antidepressants (TCAs), selective serotonin reuptake inhibitors (SSRIs), and serotonin and norepinephrine reuptake inhibitors (SNRIs) (Liu et al. 2011; Horikawa et al. 2010; Xia et al. 1996). In animal model, fluoxetine or paroxetine reduces LPS-induced inflammation and depressive-like behaviors (Ohgi et al. 2013).

Studies in humans have shown that several SSRIs and TCAs inhibit the release of TNF-α, IL-1β, IL-6, and IFN-γ (Xia et al. 1996; Maes et al. 1999). SSRIs have anti-inflammatory effects on T-lymphocytes, dendritic cells, and neutrophils (Walker 2013). Nevertheless, a meta-analysis of studies with antidepressants showed that, overall, while pharmacological antidepressant treatment reduced depressive symptoms, it did not reduce serum levels of TNF-α. On the other hand, antidepressant treatment reduced levels of IL-1 β and possibly those of IL-6. Stratified subgroup analysis by class of antidepressant indicated that serotonin reuptake inhibitors might reduce levels of IL-6 and TNF-α. Other antidepressants, while efficacious for depressive symptoms, did not appear to reduce cytokine levels (Hannestad et al. 2011). Pro-inflammatory cytokines might also be a predictor of antidepressants responses. Patients with a poor response to amitriptyline have increased levels of TNF-α (Lanquillon et al. 2000), and early responders or non-responders to duloxetine presented distinct cytokines profiles (Fornaro et al. 2013).

Regarding electroconvulsive therapy, a recent systematic review involving a few studies suggested that a single session of ECT induces an acute, transient immune activation, whereas repetitive ECT treatment results in long-term down-regulation of immune activation (Guloksuz et al. 2014). However, inconsistency in findings and methodological issues, including sample size and lack of consideration of confounding factors affecting cytokine concentrations, preclude definitive conclusions. Psychotherapy, especially those highly structured modalities such as cognitive-behavioral therapy (CBT), also has been postulated as effective to reduce inflammation, although the evidence for this effect is still scarce.

Antidepressant Effects of Anti-Inflammatory Agents

Since MDD could be considered an immune-inflammatory disorder, the use of anti-inflammatory agents in monotherapy or as add-on agents to antidepressants has emerged as a possible therapeutic target (Maes et al. 2012b). Several agents have been tested, including celecoxib, naproxen, aspirin, and TNF-alpha antagonists. There is evidence that augmentation of antidepressants by celecoxib, a cyclooxygenase (COX)-2 inhibitor, significantly improved depressive symptoms compared to the placebo group in a 6-week randomized controlled trial (RCT) (Muller et al. 2006). The association of celecoxib and fluoxetine also seems superior to fluoxetine monotherapy (Akhondzadeh et al. 2009). Another anti-inflammatory drug, the acetylsalicylic acid, seems to accelerate the response to antidepressants when combined with SSRIs (Mendlewicz et al. 2006) and when used as add-on presented additional improvement compared to antidepressant alone (Galecki et al. 2009). Different meta-analyses found evidence that supplementation with eicosapentaenoic acid (EPA), an omega-3 polyunsaturated fatty acids (PUFAs), has a significant antidepressant activity (Appleton et al. 2010; Lin and Su 2007; Martins et al. 2012), possibly mediated by suppression of pro-inflammatory Th-1-like cytokines (Song and Wang 2011). Finally, TNF-alpha antagonists also exert positive effects on depressed mood. In a recent RCT with patients moderately resistant to previous antidepressants, infliximab was shown to be more effective than placebo in individuals with high-sensitivity C-reactive protein, a pro-inflammatory biomarker (Raison et al. 2013). In other RCT for psoriasis treatment, use of etanercept was associated with improvement of more than 50 % self-reported depression (Krishnan et al. 2007; Tyring et al. 2006).

Adoption of other substances with anti-inflammatory properties has also been emerged as possible target in MDD treatment. Minocycline is an antibiotic with anti-inflammatory and anti-oxidant actions (Soczynska et al. 2012) and has been successfully tested as adjunctive therapy for patients with unipolar psychotic depression (Miyaoka et al. 2012). N-acetyl cysteine (NAC) is a glutathione precursor with anti-oxidative and anti-inflammatory functions (Asevedo et al. 2012); it has antidepressant efficacy in bipolar disorder, suggesting a potential role in MDD (Berk et al. 2008; Magalhaes et al. 2011). Curcumin is a substance extracted from a plant used in Asian culinary with anti-inflammatory and anti-oxidative effects (Brietzke et al. 2013) and shows promising results in animal models (Kulkarni et al. 2009; Xu et al. 2005) and depressed people as well (Lopresti et al. 2014). Finally, ketamine and zinc are glutamate-NMDA antagonists, with actions on neurotransmitter circuits and the immune-inflammatory system as well. Both of them have antidepressant and anti-inflammatory effects (Lai et al. 2012; Loix et al. 2011). Although initial studies seem promising, most of these agents are still in pre-clinical phase studies (Kulkarni et al. 2009; Xu et al. 2005; Molina-Hernandez et al. 2008a; Molina-Hernandez et al. 2008b; Smaga et al. 2012).

General Medical Comorbidities and MDD

An important clinical aspect of immune-inflammatory dysregulation in MDD is its association with general medical comorbidities (GMC). Epidemiological studies have reported a high incidence of medical conditions in individuals with MDD, including but not limited to, several inflammatory (or with an inflammatory component) conditions, such as auto-immune diseases, cardiovascular diseases, diabetes, obesity, and metabolic syndrome, as well as asthma and allergies (Schoepf et al. 2014; Walker et al. 2011). Conversely, chronic medical conditions also significantly increase the risk of developing MDD (Egede 2007). Importantly from a public health perspective, both MDD and chronic medical diseases are associated with a staggering illness-associated burden, and the co-occurrence of depression with medical disorders has been linked to high levels of role disability and health-related costs (Alonso et al. 2011; Merikangas et al. 2007). Moreover, the high prevalence underscores the excess mortality documented in individuals with MDD (Cuijpers et al. 2014), that is largely a consequence of natural causes (e.g., cardiovascular disease, diabetes mellitus) rather than unnatural causes (e.g., suicide) (Cuijpers et al. 2014; Laursen et al. 2007).

Rheumatoid arthritis has been strongly associated with MDD, and improvements in mood symptoms were correlated with anti-inflammatory treatments (Sato et al. 2013). Psoriasis, an auto-immune and T-cell mediated disease of the skin, has also been shown to significantly increase the rates of depression (Weigle and McBane 2013). Moreover, anti-TNF-α therapy has been shown to lower depression scores independent of the effect on psoriasis severity (Krishnan et al. 2007). Inflammatory bowel disease, another auto-immune disorder mainly affecting the gastrointestinal system, is also well known for being strongly associated with MDD, especially during flare-ups (Graff et al. 2009).

Activation of immune-inflammatory pathways has also been recognized as a key component of cardiovascular disease (Ghattas et al. 2013). Overall, the prevalence of MDD is elevated in patients with cardiovascular disease (Fenton and Stover 2006). Patients who suffered an acute myocardial infarction (AMI) have a prevalence of depression three times higher compared to the general community (Thombs et al. 2006). Remarkably, an MDD diagnosis is present in 15–20 % of AMI patients (Thombs et al. 2006). On the other hand, depression increases the risk of cardiovascular events. A depressive mood is a predictor of coronary heart disease, with a relative risk (RR) of 1.49 (95 % CI = 1.16 − 1.92), wherein a full diagnosis of MD is an even stronger predictor, with an RR of 2.69 (95 % CI = 1.63 − 4.43) (Rugulies 2002).

Metabolic conditions (e.g., obesity, type 2 diabetes mellitus, and metabolic syndrome) are also overrepresented in the MDD population. Individuals with MDD have an approximately 50 % higher risk of developing obesity, when compared to the general population (Luppino et al. 2010). Conversely, overweight and obesity have been reported to increase the risk of onset of significant depressive symptoms (Luppino et al. 2010; Vogelzangs et al. 2010). Moreover, individuals with MDD have been shown to have increased waist circumference, higher proportion of visceral adiposity, raised levels of serum lipids, and higher incidence of hypertension, as well as, consequently, to meet criteria for metabolic syndrome more frequently (Vancampfort et al. 2013). The rate of mood disorders is increased in type 2 diabetes mellitus and vice versa, independently of the presence of overweight or obesity (Ali et al. 2006; Barnard et al. 2006). The risk of developing T2DM is increased by approximately 60 % in individuals with MDD (Rotella and Mannucci 2013a) and individuals with diabetes display a significantly elevated risk for developing MDD, independently of other risk factors (e.g., alcohol or drug use, smoking, medical comorbidities, and physical limitations) (Rotella and Mannucci 2013b).

Adipose tissue is an important source of cytokines and the role of adiposity as a moderator of immune-inflammatory dysregulation has been increasingly highlighted. Similarly to MDD, obesity has been associated with a persistent, low-grade pro-inflammatory state and meta-analytical work has reported elevated levels of C-reactive protein (CRP) in various populations with obesity (Choi et al. 2013). Therefore, the highly frequent association of MDD and metabolic conditions is of extreme relevance to the study of immunology in psychiatry. In fact, evidence indicates that obesity has a mediating/moderating effect on inflammatory dysregulation in MDD. Studies that contrasted obese and non-obese depressed individuals indicate the obese patients display significantly higher levels of inflammatory markers, when compared to normal weight patients (Schmidt et al. 2014). A meta-analysis reported that studies that adjusted for body mass index reported much lower, although still significant, effect sizes for inflammatory abnormalities in MDD (Howren et al. 2009).

Conclusions

Among the last decades, robust evidence pointed to the immune-inflammatory and neuroprogressive hypothesis of MDD. It has been well established that MDD is accompanied by activation of immune-inflammatory pathways, which in turn may cause neurotoxicity and neuroprogression. The link between MDD and inflammation offers new insights for new interventions in the treatment of this disorder, opening a new venue to new drug targets, with a promising role for negative immunoregulatory and anti-oxidant agents.

References

Akhondzadeh S, Jafari S, Raisi F, Nasehi AA, Ghoreishi A, Salehi B, Mohebbi-Rasa S, Raznahan M, Kamalipour A. Clinical trial of adjunctive celecoxib treatment in patients with major depression: a double blind and placebo controlled trial. Depress Anxiety. 2009;26:607–11.

Ali S, Stone MA, Peters JL, Davies MJ, Khunti K. The prevalence of co-morbid depression in adults with type 2 diabetes: a systematic review and meta-analysis. Diabet Med. 2006;23:1165–73.

Alonso J, Petukhova M, Vilagut G, Chatterji S, Heeringa S, Ustun TB, Alhamzawi AO, Viana MC, Angermeyer M, Bromet E, Bruffaerts R, de Girolamo G, Florescu S, Gureje O, Haro JM, Hinkov H, Hu CY, Karam EG, Kovess V, Levinson D, Medina-Mora ME, Nakamura Y, Ormel J, Posada-Villa J, Sagar R, Scott KM, Tsang A, Williams DR, Kessler RC. Days out of role due to common physical and mental conditions: results from the who world mental health surveys. Mol Psychiatry. 2011;16:1234–46.

Anisman H, Hayley S, Turrin N, Merali Z. Cytokines as a stressor: implications for depressive illness. Int J Neuropsychopharmacol. 2002;5:357–73.

Anisman H, Gibb J, Hayley S. Influence of continuous infusion of interleukin-1beta on depression-related processes in mice: corticosterone, circulating cytokines, brain monoamines, and cytokine mRNA expression. Psychopharmacology (Berl). 2008;199:231–44.

Appleton KM, Rogers PJ, Ness AR. Updated systematic review and meta-analysis of the effects of n-3 long-chain polyunsaturated fatty acids on depressed mood. Am J Clin Nutr. 2010;91:757–70.
Asevedo E, Cunha GR, Zugman A, Mansur RB, Brietzke E. N-acetylcysteine as a potentially useful medication to prevent conversion to schizophrenia in at-risk individuals. Rev Neurosci. 2012;23:353–62.
Barnard KD, Skinner TC, Peveler R. The prevalence of co-morbid depression in adults with type 1 diabetes: systematic literature review. Diabet Med. 2006;23:445–8.
Berk M, Copolov DL, Dean O, Lu K, Jeavons S, Schapkaitz I, Anderson-Hunt M, Bush AI. N-acetyl cysteine for depressive symptoms in bipolar disorder—a double-blind randomized placebo-controlled trial. Biol Psychiatry. 2008;64:468–75.
Bettelli E, Korn T, Oukka M, Kuchroo VK. Induction and effector functions of T(h)17 cells. Nature. 2008;453:1051–7.
Brietzke E, Mansur RB, Zugman A, Carvalho AF, Macedo DS, Cha DS, Abilio VC, McIntyre RS. Is there a role for curcumin in the treatment of bipolar disorder? Med Hypotheses. 2013;80:606–12.
Campbell DJ, Koch MA. Phenotypical and functional specialization of foxp3+ regulatory t cells. Nat Rev Immunol. 2011;11:119–30.
Capuron L, Ravaud A, Gualde N, Bosmans E, Dantzer R, Maes M, Neveu PJ. Association between immune activation and early depressive symptoms in cancer patients treated with interleukin-2-based therapy. Psychoneuroendocrinology. 2001;26:797–808.
Catena-Dell'Osso M, Bellantuono C, Consoli G, Baroni S, Rotella F, Marazziti D. Inflammatory and neurodegenerative pathways in depression: a new avenue for antidepressant development? Curr Med Chem. 2011;18:245–55.
Choi J, Joseph L, Pilote L. Obesity and C-reactive protein in various populations: a systematic review and meta-analysis. Obes Rev. 2013;14:232–44.
Connor TJ, Leonard BE. Depression, stress and immunological activation: the role of cytokines in depressive disorders. Life Sci. 1998;62:583–606.
Cuijpers P, Vogelzangs N, Twisk J, Kleiboer A, Li J, Penninx BW. Comprehensive meta-analysis of excess mortality in depression in the general community versus patients with specific illnesses. Am J Psychiatry. 2014;171:453–62.
Dantzer R. Cytokine, sickness behavior, and depression. Immunol Allergy Clin North Am. 2009;29:247–64.
Dantzer R, O'Connor JC, Freund GG, Johnson RW, Kelley KW. From inflammation to sickness and depression: when the immune system subjugates the brain. Nat Rev Neurosci. 2008; 9:46–56.
Dean B, Tawadros N, Scarr E, Gibbons AS. Regionally-specific changes in levels of tumour necrosis factor in the dorsolateral prefrontal cortex obtained postmortem from subjects with major depressive disorder. J Affect Disord. 2010;120:245–8.
Dowlati Y, Herrmann N, Swardfager W, Liu H, Sham L, Reim EK, Lanctot KL. A meta-analysis of cytokines in major depression. Biol Psychiatry. 2010;67:446–57.
Duman RS, Monteggia LM. A neurotrophic model for stress-related mood disorders. Biol Psychiatry. 2006;59:1116–27.
Egede LE. Major depression in individuals with chronic medical disorders: prevalence, correlates and association with health resource utilization, lost productivity and functional disability. Gen Hosp Psychiatry. 2007;29:409–16.
Fenton WS, Stover ES. Mood disorders: cardiovascular and diabetes comorbidity. Curr Opin Psychiatry. 2006;19:421–7.
Fornaro M, Rocchi G, Escelsior A, Contini P, Martino M. Might different cytokine trends in depressed patients receiving duloxetine indicate differential biological backgrounds. J Affect Disord. 2013;145:300–7.
Galecki P, Szemraj J, Bienkiewicz M, Zboralski K, Galecka E. Oxidative stress parameters after combined fluoxetine and acetylsalicylic acid therapy in depressive patients. Hum Psychopharmacol. 2009;24:277–86.

Ghattas A, Griffiths HR, Devitt A, Lip GY, Shantsila E. Monocytes in coronary artery disease and atherosclerosis: where are we now? J Am Coll Cardiol. 2013;62:1541–51.

Graff LA, Walker JR, Bernstein CN. Depression and anxiety in inflammatory bowel disease: a review of comorbidity and management. Inflamm Bowel Dis. 2009;15:1105–18.

Guloksuz S, Rutten BP, Arts B, van Os J, Kenis G. The immune system and electroconvulsive therapy for depression. J ECT. 2014;30:132–7.

Hannestad J, DellaGioia N, Bloch M. The effect of antidepressant medication treatment on serum levels of inflammatory cytokines: a meta-analysis. Neuropsychopharmacology. 2011;36:2452–9.

Hashioka S, McGeer PL, Monji A, Kanba S. Anti-inflammatory effects of antidepressants: possibilities for preventives against Alzheimer's disease. Cent Nerv Syst Agents Med Chem. 2009;9:12–9.

Hiles SA, Baker AL, de Malmanche T, Attia J. A meta-analysis of differences in IL-6 and IL-10 between people with and without depression: exploring the causes of heterogeneity. Brain Behav Immun. 2012;26:1180–8.

Horikawa H, Kato TA, Mizoguchi Y, Monji A, Seki Y, Ohkuri T, Gotoh L, Yonaha M, Ueda T, Hashioka S, Kanba S. Inhibitory effects of SSRIs on IFN-gamma induced microglial activation through the regulation of intracellular calcium. Prog Neuropsychopharmacol Biol Psychiatry. 2010;34:1306–16.

Howren MB, Lamkin DM, Suls J. Associations of depression with C-reactive protein, IL-1, and IL-6: a meta-analysis. Psychosom Med. 2009;71:171–86.

Jiang Y, Deacon R, Anthony DC, Campbell SJ. Inhibition of peripheral TNF can block the malaise associated with CNS inflammatory diseases. Neurobiol Dis. 2008;32:125–32.

Kenis G, Maes M. Effects of antidepressants on the production of cytokines. Int J Neuropsychopharmacol. 2002;5:401–12.

Krishnan R, Cella D, Leonardi C, Papp K, Gottlieb AB, Dunn M, Chiou CF, Patel V, Jahreis A. Effects of etanercept therapy on fatigue and symptoms of depression in subjects treated for moderate to severe plaque psoriasis for up to 96 weeks. Br J Dermatol. 2007;157:1275–7.

Kubera M, Symbirtsev A, Basta-Kaim A, Borycz J, Roman A, Papp M, Claesson M. Effect of chronic treatment with imipramine on interleukin 1 and interleukin 2 production by splenocytes obtained from rats subjected to a chronic mild stress model of depression. Pol J Pharmacol. 1996;48:503–6.

Kulkarni S, Dhir A, Akula KK. Potentials of curcumin as an antidepressant. Sci World J. 2009;9:1233–41.

Lai J, Moxey A, Nowak G, Vashum K, Bailey K, McEvoy M. The efficacy of zinc supplementation in depression: systematic review of randomised controlled trials. J Affect Disord. 2012;136:e31–9.

Lamers F, Vogelzangs N, Merikangas KR, de Jonge P, Beekman AT, Penninx BW. Evidence for a differential role of HPA-axis function, inflammation and metabolic syndrome in melancholic versus atypical depression. Mol Psychiatry. 2012;18:692–9.

Lanquillon S, Krieg JC, Bening-Abu-Shach U, Vedder H. Cytokine production and treatment response in major depressive disorder. Neuropsychopharmacology. 2000;22:370–9.

Lantz M, Malik S, Slevin ML, Olsson I. Infusion of tumor necrosis factor (TNF) causes an increase in circulating TNF-binding protein in humans. Cytokine. 1990;2:402–6.

Laursen TM, Munk-Olsen T, Nordentoft M, Mortensen PB. Increased mortality among patients admitted with major psychiatric disorders: a register-based study comparing mortality in unipolar depressive disorder, bipolar affective disorder, schizoaffective disorder, and schizophrenia. J Clin Psychiatry. 2007;68:899–907.

Lee GR, Kim ST, Spilianakis CG, Fields PE, Flavell RA. T helper cell differentiation: regulation by cis elements and epigenetics. Immunity. 2006;24:369–79.

Leonard B, Maes M. Mechanistic explanations how cell-mediated immune activation, inflammation and oxidative and nitrosative stress pathways and their sequels and concomitants play a role in the pathophysiology of unipolar depression. Neurosci Biobehav Rev. 2012;36:764–85.

Licinio J, Wong ML. The role of inflammatory mediators in the biology of major depression: central nervous system cytokines modulate the biological substrate of depressive symptoms, regulate stress-responsive systems, and contribute to neurotoxicity and neuroprotection. Mol Psychiatry. 1999;4:317–27.

Lin PY, Su KP. A meta-analytic review of double-blind, placebo-controlled trials of antidepressant efficacy of omega-3 fatty acids. J Clin Psychiatry. 2007;68:1056–61.
Little RF, Pluda JM, Wyvill KM, Rodriguez-Chavez IR, Tosato G, Catanzaro AT, Steinberg SM, Yarchoan R. Activity of subcutaneous interleukin-12 in aids-related Kaposi sarcoma. Blood. 2006;107:4650–7.
Liu D, Wang Z, Liu S, Wang F, Zhao S, Hao A. Anti-inflammatory effects of fluoxetine in lipopolysaccharide (LPS)-stimulated microglial cells. Neuropharmacology. 2011;61:592–9.
Liu Y, Ho RC, Mak A. Interleukin (IL)-6, tumour necrosis factor alpha (TNF-alpha) and soluble interleukin-2 receptors (SIL-2r) are elevated in patients with major depressive disorder: a meta-analysis and meta-regression. J Affect Disord. 2012;139:230–9.
Loix S, De Kock M, Henin P. The anti-inflammatory effects of ketamine: state of the art. Acta Anaesthesiol Belg. 2011;62:47–58.
Lopresti AL, Maes M, Maker GL, Hood SD, Drummond PD. Curcumin for the treatment of major depression: a randomised, double-blind, placebo controlled study. J Affect Disord. 2014;167:368–75.
Luppino FS, de Wit LM, Bouvy PF, Stijnen T, Cuijpers P, Penninx BW, Zitman FG. Overweight, obesity, and depression: a systematic review and meta-analysis of longitudinal studies. Arch Gen Psychiatry. 2010;67:220–9.
Maes M. Evidence for an immune response in major depression: a review and hypothesis. Prog Neuropsychopharmacol Biol Psychiatry. 1995;19:11–38.
Maes M. Major depression and activation of the inflammatory response system. Adv Exp Med Biol. 1999;461:25–46.
Maes M, Bosmans E, Suy E, Vandervorst C, De Jonckheere C, Raus J. Immune disturbances during major depression: upregulated expression of interleukin-2 receptors. Neuropsychobiology. 1990;24:115–20.
Maes M, Bosmans E, De Jongh R, Kenis G, Vandoolaeghe E, Neels H. Increased serum IL-6 and IL-1 receptor antagonist concentrations in major depression and treatment resistant depression. Cytokine. 1997;9:853–8.
Maes M, Song C, Lin AH, Bonaccorso S, Kenis G, De Jongh R, Bosmans E, Scharpe S. Negative immunoregulatory effects of antidepressants: inhibition of interferon-gamma and stimulation of interleukin-10 secretion. Neuropsychopharmacology. 1999;20:370–9.
Maes M, Yirmyia R, Noraberg J, Brene S, Hibbeln J, Perini G, Kubera M, Bob P, Lerer B, Maj M. The inflammatory & neurodegenerative (I&ND) hypothesis of depression: leads for future research and new drug developments in depression. Metab Brain Dis. 2009;24:27–53.
Maes M, Galecki P, Chang YS, Berk M. A review on the oxidative and nitrosative stress (O&NS) pathways in major depression and their possible contribution to the (neuro)degenerative processes in that illness. Prog Neuropsychopharmacol Biol Psychiatry. 2011;35:676–92.
Maes M, Berk M, Goehler L, Song C, Anderson G, Galecki P, Leonard B. Depression and sickness behavior are janus-faced responses to shared inflammatory pathways. BMC Med. 2012a;10:66.
Maes M, Fisar Z, Medina M, Scapagnini G, Nowak G, Berk M. New drug targets in depression: inflammatory, cell-mediated immune, oxidative and nitrosative stress, mitochondrial, antioxidant, and neuroprogressive pathways. And new drug candidates—NRF2 activators and GSK-3 inhibitors. Inflammopharmacology. 2012b;20:127–50.
Magalhaes PV, Dean OM, Bush AI, Copolov DL, Malhi GS, Kohlmann K, Jeavons S, Schapkaitz I, Anderson-Hunt M, Berk M. N-acetylcysteine for major depressive episodes in bipolar disorder. Rev Bras Psiquiatr. 2011;33:374–8.
Martins JG, Bentsen H, Puri BK. Eicosapentaenoic acid appears to be the key omega-3 fatty acid component associated with efficacy in major depressive disorder: a critique of Bloch and Hannestad and updated meta-analysis. Mol Psychiatry. 2012;17:1144–9. discussion 1163-1147.
Mendlewicz J, Kriwin P, Oswald P, Souery D, Alboni S, Brunello N. Shortened onset of action of antidepressants in major depression using acetylsalicylic acid augmentation: a pilot open-label study. Int Clin Psychopharmacol. 2006;21:227–31.
Merikangas KR, Ames M, Cui L, Stang PE, Ustun TB, Von Korff M, Kessler RC. The impact of comorbidity of mental and physical conditions on role disability in the US adult household population. Arch Gen Psychiatry. 2007;64:1180–8.

Mesquita AR, Correia-Neves M, Roque S, Castro AG, Vieira P, Pedrosa J, Palha JA, Sousa N. IL-10 modulates depressive-like behavior. J Psychiatr Res. 2008;43:89–97.

Miyaoka T, Wake R, Furuya M, Liaury K, Ieda M, Kawakami K, Tsuchie K, Taki M, Ishihara K, Araki T, Horiguchi J. Minocycline as adjunctive therapy for patients with unipolar psychotic depression: an open-label study. Prog Neuropsychopharmacol Biol Psychiatry. 2012;37: 222–6.

Molina-Hernandez M, Tellez-Alcantara NP, Perez-Garcia J, Olivera-Lopez JI, Jaramillo-Jaimes MT. Desipramine or glutamate antagonists synergized the antidepressant-like actions of intra-nucleus accumbens infusions of minocycline in male Wistar rats. Prog Neuropsychopharmacol Biol Psychiatry. 2008a;32:1660–6.

Molina-Hernandez M, Tellez-Alcantara NP, Perez-Garcia J, Olivera-Lopez JI, Jaramillo-Jaimes MT. Antidepressant-like actions of minocycline combined with several glutamate antagonists. Prog Neuropsychopharmacol Biol Psychiatry. 2008b;32:380–6.

Muller N, Schwarz MJ, Dehning S, Douhe A, Cerovecki A, Goldstein-Muller B, Spellmann I, Hetzel G, Maino K, Kleindienst N, Moller HJ, Arolt V, Riedel M. The cyclooxygenase-2 inhibitor celecoxib has therapeutic effects in major depression: results of a double-blind, randomized, placebo controlled, add-on pilot study to reboxetine. Mol Psychiatry. 2006;11: 680–4.

Ohgi Y, Futamura T, Kikuchi T, Hashimoto K. Effects of antidepressants on alternations in serum cytokines and depressive-like behavior in mice after lipopolysaccharide administration. Pharmacol Biochem Behav. 2013;103:853–9.

Raison CL, Miller AH. Is depression an inflammatory disorder? Curr Psychiatry Rep. 2011;13:467–75.

Raison CL, Borisov AS, Broadwell SD, Capuron L, Woolwine BJ, Jacobson IM, Nemeroff CB, Miller AH. Depression during pegylated interferon-alpha plus ribavirin therapy: prevalence and prediction. J Clin Psychiatry. 2005;66:41–8.

Raison CL, Capuron L, Miller AH. Cytokines sing the blues: inflammation and the pathogenesis of depression. Trends Immunol. 2006;27:24–31.

Raison CL, Rutherford RE, Woolwine BJ, Shuo C, Schettler P, Drake DF, Haroon E, Miller AH. A randomized controlled trial of the tumor necrosis factor antagonist infliximab for treatment-resistant depression: the role of baseline inflammatory biomarkers. JAMA Psychiatry. 2013;70:31–41.

Rotella F, Mannucci E. Depression as a risk factor for diabetes: a meta-analysis of longitudinal studies. J Clin Psychiatry. 2013a;74:31–7.

Rotella F, Mannucci E. Diabetes mellitus as a risk factor for depression. A meta-analysis of longitudinal studies. Diabetes Res Clin Pract. 2013b;99:98–104.

Rugulies R. Depression as a predictor for coronary heart disease. A review and meta-analysis. Am J Prev Med. 2002;23:51–61.

Sato E, Nishimura K, Nakajima A, Okamoto H, Shinozaki M, Inoue E, Taniguchi A, Momohara S, Yamanaka H. Major depressive disorder in patients with rheumatoid arthritis. Mod Rheumatol. 2013;23:237–44.

Schmidt FM, Lichtblau N, Minkwitz J, Chittka T, Thormann J, Kirkby KC, Sander C, Mergl R, Fasshauer M, Stumvoll M, Holdt LM, Teupser D, Hegerl U, Himmerich H. Cytokine levels in depressed and non-depressed subjects, and masking effects of obesity. J Psychiatr Res. 2014;55:29–34.

Schoepf D, Uppal H, Potluri R, Chandran S, Heun R. Comorbidity and its relevance on general hospital based mortality in major depressive disorder: a naturalistic 12-year follow-up in general hospital admissions. J Psychiatr Res. 2014;52:28–35.

Serafini G, Pompili M, Elena Seretti M, Stefani H, Palermo M, Coryell W, Girardi P. The role of inflammatory cytokines in suicidal behavior: a systematic review. Eur Neuropsychopharmacol. 2013;23:1672–86.

Silverstein AM. Cellular versus humoral immunology: a century-long dispute. Nat Immunol. 2003;4:425–8.

Sims JE, Smith DE. The IL-1 family: regulators of immunity. Nat Rev Immunol. 2010;10: 89–102.

Smaga I, Pomierny B, Krzyzanowska W, Pomierny-Chamiolo L, Miszkiel J, Niedzielska E, Ogorka A, Filip M. N-acetylcysteine possesses antidepressant-like activity through reduction of oxidative stress: behavioral and biochemical analyses in rats. Prog Neuropsychopharmacol Biol Psychiatry. 2012;39:280–7.

Soczynska JK, Mansur RB, Brietzke E, Swardfager W, Kennedy SH, Woldeyohannes HO, Powell AM, Manierka MS, McIntyre RS. Novel therapeutic targets in depression: minocycline as a candidate treatment. Behav Brain Res. 2012;235:302–17.

Song C, Wang H. Cytokines mediated inflammation and decreased neurogenesis in animal models of depression. Prog Neuropsychopharmacol Biol Psychiatry. 2011;35:760–8.

Song C, Horrobin DF, Leonard BE. The comparison of changes in behavior, neurochemistry, endocrine, and immune functions after different routes, doses and durations of administrations of IL-1beta in rats. Pharmacopsychiatry. 2006;39:88–99.

Stepanichev M, Dygalo NN, Grigoryan G, Shishkina GT, Gulyaeva N. Rodent models of depression: neurotrophic and neuroinflammatory biomarkers. Biomed Res Int. 2014;2014:932757.

Thombs BD, Bass EB, Ford DE, Stewart KJ, Tsilidis KK, Patel U, Fauerbach JA, Bush DE, Ziegelstein RC. Prevalence of depression in survivors of acute myocardial infarction. J Gen Intern Med. 2006;21:30–8.

Tyring S, Gottlieb A, Papp K, Gordon K, Leonardi C, Wang A, Lalla D, Woolley M, Jahreis A, Zitnik R, Cella D, Krishnan R. Etanercept and clinical outcomes, fatigue, and depression in psoriasis: double-blind placebo-controlled randomised phase III trial. Lancet. 2006;367:29–35.

Udina M, Castellvi P, Moreno-Espana J, Navines R, Valdes M, Forns X, Langohr K, Sola R, Vieta E, Martin-Santos R. Interferon-induced depression in chronic hepatitis C: a systematic review and meta-analysis. J Clin Psychiatry. 2012;73:1128–38.

Vancampfort D, Correll CU, Mitchell AJ, De Herdt A, Sienaert P, De Hert M. Metabolic syndrome and metabolic abnormalities in bipolar disorder: a meta-analysis of prevalence rates and moderators. Am J Psychiatry. 2013;170:265–74.

Vogelzangs N, Kritchevsky SB, Beekman AT, Brenes GA, Newman AB, Satterfield S, Yaffe K, Harris TB, Penninx BW, Health ABCS. Obesity and onset of significant depressive symptoms: results from a prospective community-based cohort study of older men and women. J Clin Psychiatry. 2010;71:391–9.

Walker FR. A critical review of the mechanism of action for the selective serotonin reuptake inhibitors: do these drugs possess anti-inflammatory properties and how relevant is this in the treatment of depression? Neuropharmacology. 2013;67:304–17.

Walker JR, Graff LA, Dutz JP, Bernstein CN. Psychiatric disorders in patients with immune-mediated inflammatory diseases: prevalence, association with disease activity, and overall patient well-being. J Rheumatol Suppl. 2011;88:31–5.

Weigle N, McBane S. Psoriasis. Am Fam Physician. 2013;87:626–33.

Whiteford HA, Degenhardt L, Rehm J, Baxter AJ, Ferrari AJ, Erskine HE, Charlson FJ, Norman RE, Flaxman AD, Johns N, Burstein R, Murray CJ, Vos T. Global burden of disease attributable to mental and substance use disorders: findings from the global burden of disease study 2010. Lancet. 2013;382:1575–86.

Xanthoulea S, Pasparakis M, Kousteni S, Brakebusch C, Wallach D, Bauer J, Lassmann H, Kollias G. Tumor necrosis factor (TNF) receptor shedding controls thresholds of innate immune activation that balance opposing TNF functions in infectious and inflammatory diseases. J Exp Med. 2004;200:367–76.

Xia Z, DePierre JW, Nassberger L. Tricyclic antidepressants inhibit IL-6, IL-1 beta and TNF-alpha release in human blood monocytes and IL-2 and interferon-gamma in T cells. Immunopharmacology. 1996;34:27–37.

Xu Y, Ku BS, Yao HY, Lin YH, Ma X, Zhang YH, Li XJ. The effects of curcumin on depressive-like behaviors in mice. Eur J Pharmacol. 2005;518:40–6.

Zorn U, Dallmann I, Grosse J, Kirchner H, Poliwoda H, Atzpodien J. Soluble interleukin 2 receptors abrogate IL-2 induced activation of peripheral mononuclear cells. Cytokine. 1994;6:358–64.

Zorrilla EP, Luborsky L, McKay JR, Rosenthal R, Houldin A, Tax A, McCorkle R, Seligman DA, Schmidt K. The relationship of depression and stressors to immunological assays: a meta-analytic review. Brain Behav Immun. 2001;15:199–226.

Chapter 9
Immune System Related Markers: Changes in childhood Neuropsychiatry Disorders Cause and Consequence

Tatiana Falcone and Kathleen Franco

Abstract There has been great interest in the role of inflammation in psychiatric disorders of children and adolescents. This chapter begins by describing the normal development of the immune system from fetal life and earlier through infancy, childhood, and beyond. The role of genetics, environmental stressors, and trauma will be described and the sequelae that may follow. A focus will be the transition from normal to altered immunity and how that may later be expressed in the child. Links between inflammation and mood disorders are well documented, as are those associated with psychotic disorders. The latter group has been of interest for decades with the description of schizophrenia associated with infection. This chapter will trace earlier findings as they have evolved in the more recent past with technologically advanced evidence. Post-traumatic stress disorder (PTSD), increasingly more important for mental health professionals to understand, has diverse links to the immune system. Autism spectrum disorder (ASD) is another exciting area with multiple associations to inflammation and obsessive compulsive spectrum disorders have multiple alterations of interleukins. New evidence points to immune changes in children with attention deficit hyperactive disorder, as well as those with disruptive behavior. Critical to the complete understanding of suicide are the new reports of the neuroimmune system of suicidal patients. The chapter concludes with prominent bio-markers that have been identified in recent literature having connections with child and adolescent psychopathology. Current evidence points researchers toward an exciting new frontier that may offer new and more effective treatments in the future.

T. Falcone (✉)
Departments of Psychiatry and Neurology, Neurologic Institute, Cleveland Clinic, Cleveland, OH, USA

Epilepsy Center, Neurologic Institute, Cleveland Clinic, Cleveland, OH, USA
e-mail: Falcont1@ccf.org

K. Franco
Cleveland Clinic Lerner College of Medicine, NA 2-21, 9500 Euclid Avenue, Cleveland, OH 44195, USA
e-mail: Francok@ccf.org

N. Müller et al. (eds.), *Immunology and Psychiatry*, Current Topics in Neurotoxicity 8, DOI 10.1007/978-3-319-13602-8_9

Keywords Children • Adolescents • Neuroinflammation • Psychoneuroimmunology • Trauma

Introduction

Multiple lines of evidence point to inflammation playing a role in the development of certain psychiatric disorders in some patients (Najjar et al. 2013; Zorrilla et al. 2001). There is an important link between psychiatric symptoms and the immune response. Patients with different autoimmune disorders have greater probability of psychiatric symptoms (Eaton et al. 1992), and neuroimmunological abnormalities are frequently observed in several psychiatric disorders (schizophrenia, bipolar disorder, major depression disorder, among others) (Ganguli et al. 1993; Henderson 2005; Rothermundt and Arolt 2007; Arolt and Rothermundt 2005; Birmaher et al. 1994; Capuron et al. 2007; Drexhage et al. 2008, 2010). Important correlations exist between several medical conditions such as diabetes, cancer, lupus, epilepsy, obesity, rheumatoid arthritis and the development of psychiatric psychopathology (Drexhage et al. 2008; Fellerhoff et al. 2007; Maes et al. 1991; Wichers and Maes 2002; Caplan et al. 2005; Pompili et al. 2005; Carlton et al. 2008; Bonaccorso et al. 2001; Crespel et al. 2002; Granata et al. 2011; Harrison et al. 2009; Marchi et al. 2009, 2011; Vezzani et al. 2008, 2011; Vezzani and Granata 2005). This suggests a connection between underlying inflammatory factors in several organs including the brain. The inflammatory link has also been described in pediatric patients, making the case for early intervention (Gabbay et al. 2009a; Falcone 2008, 2012; Falcone et al. 2007; Carlton et al. 2009). In certain neurologic disorders, the inflammatory changes appear much earlier than brain changes observed in MRI, and up to 10 years earlier than cognitive symptoms (Insel 2013; Insel et al. 2013). The future of psychiatry might lie on identifying early changes in children and adolescents to potentially prevent long-term psychopathology for which we have limited medication effectiveness.

Developmental Period and Immunity

In recent years, there has been great interest in determining how environmental conditions and responses to these alter a child's susceptibility to mental or physical illness. The interplay of the nervous and endocrine systems with the immune system has a central role. How these interact in the mother even before she becomes pregnant can have an impact on the child (Johnson et al. 2013). Certainly these influences are significant throughout the entire pregnancy. After birth, challenges in the environment producing "toxic stress" can affect brain plasticity and how a child's body will later respond to illnesses as an adult, such as diabetes, asthma, mental illness, or cardiovascular disease (Caserta et al. 2008; Dube et al. 2001, 2009; Dunn et al. 2012; Johnson et al. 2002; Miller and Chen 2007; Miller et al. 2009a, b; Miller and Cole 2012). There are likely periods when the new human has greater vulnerability as well as times when interventions might be more successful (McGorry et al. 2005; Phillips et al. 2005). These are of great interest to pediatricians, psychologists, and child psychiatrists. We will begin this chapter with a very basic review of these changes over time. Toxic stress occurs when prolonged or excessive stress presents in a child's life without protective buffering by stable, committed adults (Johnson et al. 2013). It's not the presence of adversity, but rather inadequate or absent protection that leads to poor outcomes (Cohen 2003; Garner et al. 2012). Unhealthy relationships with caretakers lead to unhealthy adaptations.

There is no question that the mother's physical, mental, and environmental status can impact how she feels about the pregnancy and the baby after delivery. Chronic maternal stress can impact the baby's immune system leading to more illness. Trauma over the mother's lifetime prior to pregnancy is linked to alterations of immunoglobulin E antibody levels in the neonate (Sternthal et al. 2009). Stressors can activate glucocorticoids and cytokines that circulate in the mother's body, not only to the hypothalamus and the pituitary in the brain, but also the cardiovascular and many other systems. These chemical messengers responding to stress or physical illness can play a role in fighting off illness but can also lead to "sickness behavior" causing withdrawal, fatigue, depression, and irritability (Dantzer et al. 2008). These changes can impact the fetus leading to hyperactive responses in the immune system. The functioning fetal cytokines normally interact with the mother's blood through the cord to reduce the chance of rejection. The mother's immunoglobulin G antibodies pass through the placenta. After birth, there is up-regulation of T helper 1 cellular immunity and down-regulation T helper 2 cellular immunity (Hertz-Picciotto et al. 2008; Rowe et al. 2007). If T helper 2 cellular response is strong, the child may later acquire allergies or asthma (Gaffin and Phipatanakul 2009). Generally it is believed that there is great vulnerability for allergies between infancy and 8 years of age (Hurtado and Johnson 2005).

Human infants require a great deal of care in their early years. Some mammals have longer gestation for additional development. Nine months of gestation allows the human baby to grow without its head becoming so large it cannot pass through the birth canal. Once delivered, the mother infant bond must be very strong for the infant to get the care it needs. Endocrine changes through the hypothalamus pituitary axis normally foster the mother's ability to nurture her child. There are multiple studies describing differences of infants who have grown up in an institution versus those that have a grown up with a nurturing mother (Shirtcliff et al. 2009). Infants who have had less nurturance may have hyperactive responses to stress via the hypothalamus pituitary axis. Mothers successfully breastfeeding pass along helpful immunoglobulin A antibodies and lactoferrin through breast milk (Johnson and Versalovic 2012).

Teran et al. (2011) identified important changes in immune responses from infancy to 5 years of age. They studied children in tropical Ecuador and concurred with others that young infants have greater innate immunity to microbial pathogens. In their sample, there were "increasing percentages of memory CD4 and CD8 T cells."

As children age, there is “down regulation of proinflammatory innate immune responses” with a decline in both regulatory T cells and IL-10. Significant levels of IFN-gamma in later infancy occur as increasing memory T cells appear and decreasing naïve T cells are noted. In Teran’s cohort they found this to be true in both rural and urban populations. However there was a stronger down regulation of IL-10 and a delay in INF gamma in rural children. There was a dramatic increase in memory CD 8 T cells in rural children. Urban children demonstrated more rapid down regulation of IL8 responses to TLR3 and TLR6. Figueiredo et al. (2009) found an increase in SEB-induced IL-10 in an urban population of children ages 4–11, demonstrating changes continue to occur as child ages.

Exposure to microbes early on may be met well by the developing innate immunity. If the innate immunity is hyper-responsive, immunopathology may result. A respiratory infection might precipitate the onset of asthma.

An interesting report on adolescents raised in orphanages as young children points to one window of opportunity that might already have closed. In Shirtcliff et al., adolescence who had been adopted into stable families were compared to adolescents in dysfunctional settings with continued disruption. Both groups were unable to keep herpes simplex virus dormant even in those adolescents who now had healthy protective families (Shirtcliff et al. 2009).

Although we believe the innate immune system is more responsible for disorders that show themselves early in life, like autism, the adaptive immune system may also be involved.

Continued neuroinflammation can be seen in brain tissue of these patients with autism on postmortem evaluations (Kumar et al. 2012). Cellular and humoral responses continue to morph throughout all of childhood and for that matter, throughout all of life (Pardo et al. 2005). Not only do infections, trauma, and emotional stress cause alterations of the immune system, but literally 100’s of toxins in the environment (Fellerhoff et al. 2007; Unuvar and Buyukgebiz 2012; Hagberg et al. 2012; Smith et al. 2007; Torrey et al. 2006; Zuckerman and Weiner 2005; Yolken et al. 2001).

Microorganisms living inside and outside human beings prompt changes in our immunity (Gronlund et al. 2000). It has been said that these bacteria outnumber human cells by a factor of 10. Infants are born with 3 phyla of bacteria in their gastrointestinal track, gaining more on their skin as they exit the birth canal, and additional ones from breast milk, all on their first day of life. Infants delivered by cesarean section have a higher incidence of atopic diseases like food allergies and asthma, possibly because they were not challenged by organisms in the birth canal (Gill et al. 2006). Hospitalization and gestational age also impact the microbiota. Later the infant’s diet and home setting lead to more changes pushing the immune system to adapt yet again (Negele et al. 2004). In protecting the immune system, evolution has created complex interactions affected by infections that can impact growth (Piwoz et al. 2012).

Delays or atypical changes in the infant core microbes can alter development of the immune system and produce unwelcome outcomes much later in life.

The Role of Trauma in Children

Adverse experiences during childhood, such as child abuse (CA), are associated with poor health over the lifespan (Shonkoff et al. 2012; Slopen et al. 2012a, 2013a). *The impact of CA during critical developmental periods may have longstanding consequences* (Johnson et al. 2013). CA is a mental health problem of epidemic proportions. In 2009, three million cases of CA were reported in the USA. The prevalence rate of CA exposure in adults has been reported to be 25–45 % (Johnson et al. 2002, 2013; Greeson et al. 2011).

There is an important link between CA and the development of psychopathology. Traumatic experiences during childhood have been identified as particularly strong risk factors for mood disorders, anxiety disorders [including post-traumatic stress disorder (PTSD), generalized anxiety disorder (GAD) and phobias], and psychosis (Caserta et al. 2008; Dube et al. 2009; Jumper 1995; American Academy of Pediatrics et al. 2008; Bremner 1999, 2006; Brown and Anderson 1991; Coogan et al. 2013; Dowd et al. 2010). The severity of the CA has been correlated with the number of suicide attempts during childhood or adolescence (Johnson et al. 2002).

Multiple studies report the increased expression of several inflammatory markers in patients exposed to CA (Slopen et al. 2010, 2012a, b, 2013b). Victims of childhood abuse are more susceptible to develop: allergies and asthma, autoimmune disorders, osteoarthritis, cardiovascular disorders, and metabolic disorders (Dube et al. 2001, 2009; Dunn et al. 2012; Dowd et al. 2010; Brown et al. 2009; Carrion and Wong 2012; Danese and McEwen 2012; Danese et al. 2007, 2008a; Fuller-Thomson et al. 2009, 2012). Chronic stress can activate pathways in the brain that secrete cortisol and trigger the inflammatory cascade. There is an important relationship between early life adversity and elevated levels of inflammatory markers in childhood and adulthood. Patients with a known history of childhood abuse are more likely than non-victims to have elevated inflammatory markers at an earlier age and tend to die younger (Johnson et al. 2002, 2013; Dube et al. 2001, 2009; Dunn et al. 2012; Danese et al. 2007, 2008a).

Many studies of adults that were exposed to CA have reported increased serum levels of C-reactive protein, fibrinogen, white blood cells, IL-1β, IL-6, IL-8, and TNFα among others (Dube et al. 2009; Bremner 1999; Danese and McEwen 2012; Danese et al. 2007, 2008a, 2009). Additional evidence from animal studies has also demonstrated that acute and chronic CA can trigger biological stress reactions, alter the feedback of the hypothalamic–pituitary–adrenal (HPA) axis, and trigger a sympathetic nervous system (SNS) response. These changes may increase the production of glucocorticoids, blunt the HPA axis, and produce excessive chronic inflammation as measured by C-reactive protein, IL-6 and fibrinogen (Danese et al. 2008a; Dennison et al. 2012; Vaccarino et al. 2008).

CA can severely impair learning, behavior, physical, and emotional well-being. The impact of chronic traumatic stress affects several organs systems, especially the brain. According to the technical report on the impact of toxic stress by the American Academy of Pediatrics, multiple traumatic stressors can trigger long-term alterations on brain development. These alterations can produce anatomic changes and physiologic dysregulation that are the precursors of physical and mental illness (Johnson et al. 2013).

Two important studies in humans have associated the glucocorticoid receptor gene with mood disorders. Mc Gowan et al. examined the epigenetic differences in a neuron specific glucocorticoid receptor promoter NR3C1 between the hippocampus of suicide victims with a history of childhood abuse and those suicide victims without childhood abuse. Researchers found decreased levels of glucocorticoid receptor mRNA as well as increased methylation of an NR3C1 promoter and concluded that there is an important epigenetic regulation of hippocampal glucocorticoid receptor expression in patients with mood disorders who were victims of childhood abuse (McGowan et al. 2009). Oberlander et al. studied the relationship between prenatal exposure to maternal mood and the methylation status of a CPG region in the promoter of and exon 1 F of the NR3C1 in newborns and the HPA stress reactivity as measured by multiple salivary cortisol levels. Researchers reported prenatal exposure to mood disorders in the mother during the third trimester was associated with methylation of NR3C1 at the NGF1-A binding site. This finding was also associated with increased salivary cortisol at 3 months. No effects were observed in the maternal methylation status of the mothers who were treated with SSRIs and those who were also depressed but not treated with SSRIs. Investigators concluded that the methylation status of the human NR3C1 gene in newborns is sensitive to depression during the third trimester and can possibly offer a potential epigenetic process that links maternal mood during the pregnancy and altered HPA stress reactivity during childhood (Oberlander et al. 2008).

Psychopathology in Children and Inflammation

Mood Disorders

The lifetime prevalence of depression is 15–20 % in adolescents and adults, suggesting that *depression in adults often begins in adolescence* (Brent et al. 2003). *In children and adolescents the point prevalence is about 1–2.5 % in school age children* and 2–8.3 % in adolescents (Insel and Charney 2003; Birmaher et al. 1996; Brent et al. 1998). *Fourteen to 25 % of adolescents experience at least one episode of depression before adulthood* (Nock et al. 2013). Individuals born in the twentieth century are at a greater risk to develop depression and these disorders are manifesting at a younger age in children (Brent et al. 2013).

Depression in youth on average lasts 7–9 months and recurs in up to 70 % of patients. Multiple negative sequelae are related to depression including impairment in school performance and interpersonal relationships. Suicidal ideation and attempts are very frequent in adolescents; a survey of adolescents found 19 % had seriously considered suicide in the last year, 15 % had made a plan, and 8 % made an attempt. Since 1950, the suicide rate of adolescents has quadrupled and represents 12 % of the mortality rate in this age group (Nock et al. 2013).

The link between inflammation and depression: There is strong evidence linking major depressive disorder (MDD) with immunological abnormalities, inflammation and cytokine production, that perhaps may contribute to the pathophysiology of mood disorders (Capuron and Miller 2004; Capuron et al. 2004; Segerstrom and Miller 2004). In the following paragraphs we will present evidence linking inflammation and the psychoneuroimmunological abnormalities observed in patients with mood disorders. Adult patients suffering from depression often have immunological alterations that can be detected in blood samples by clinical detection methods (Miller et al. 2003, 2009a; Capuron and Miller 2004; Capuron et al. 2004; Andreoli et al. 1993). A recent meta-analysis of over 180 studies and more than 40 immune measures has provided overwhelming evidence that immunological abnormalities are associated with mood disorders in adult patients (Dowlati et al. 2010). This evidence points to activation of the innate immune inflammatory response and alteration in the ability of the immune cells to express inflammatory cytokines. Inflammation is an important biological factor that might increase the risk of major depression (Zorrilla et al. 2001; Dowlati et al. 2010). Astrocytes and microglia are the major immunocompetent cells in the brain. Their activation points toward inflammation and immune reactions indicated by the release of cytokines and S100, an astrocytic protein at increased peripheral blood levels in adults and youth with severe mood disorders (Falcone et al. 2010a; Schroeter et al. 2013; Schroeter and Steiner 2009; Rothermundt et al. 2001a, b; Arolt et al. 2003).

Although the exact pathophysiology of mood disorders has not been unraveled, multiple studies point to increasing evidence that there is alteration of the immune system in patients with MDD (Zorrilla et al. 2001; Rothermundt et al. 2001c). One of the inflammatory models for mood disorders is the "sickness behavior theory," the organism's response to inflammation. Cytokine therapy has been used to study the pathophysiology of sickness behavior. Interferon alpha and IL2 have been used for the treatment of some immune-mediated illnesses like cancer, chronic hepatitis C, and AIDS. Some of these therapies are effective, but 30–50 % of the patients report "sickness behavior" characterized by anhedonia, anorexia, decreased libido, psychomotor retardation, fatigue, weight loss, sleep disturbance, hyperalgesia, social isolation, sad mood, worthlessness, guilt and suicidal ideation—the same symptom constellation reported in MDD (Miller et al. 2003, 2009a; Rothermundt et al. 2001a; Raison et al. 2006; Anisman et al. 1999; Motivala et al. 2005; Pittenger and Duman 2008; Smith 1991). This symptomatology is mediated by cytokine production, specifically IL1, IL6, TNF alpha, and IFN gamma. There is an important relationship between the impact of stress on the activation of the mononuclear cells, converting psychosocial stress as a trigger of inflammation and subsequently activating innate inflammatory signaling pathways and ultimately producing depressive symptoms in humans (Capuron and Miller 2004; Buter et al. 1993; Clerici et al. 2009). The opposite effect of major depression augmenting the inflammatory response to stress has also been proposed (Zorrilla et al. 2001; Birmaher et al. 1994;

Danese et al. 2007, 2008b; Buter et al. 1993; Miller et al. 2009c). The link between Nf-κB, inflammation, and psychosocial stress producing depression and worsening other chronic illnesses such as cancer, diabetes, asthma, AIDS, and osteoporosis has frequently been reported (Danese et al. 2007, 2008b).

In a recent systematic review of inflammatory markers in youth with several neuropsychiatric disorders, 28 studies were identified evaluating peripheral inflammatory markers (PIM) in youth with MDD (Mitchell and Goldstein 2014). Henje Blom et al. reported increased levels of Il-1B, IL-2, and IL-10 in a group of 60 patients with MDD after controlling for medication levels IL-6, IL-1, and IL-1B compared to a control group. Levels of IL-6 were correlated with severity of anxiety and depression (Henje Blom et al. 2012). In a study evaluating the role of childhood adversity and the probability of developing MDD, Miller and Cole (2012) reported 40 patients who developed MDD with a past history of childhood adversity among a group of 147 patients at risk with increased levels of IL-6 and C-reactive protein CRP. When 24 adolescents with mood disorders were compared to 33 adolescent controls, S100B was elevated in patients with severe depression (Falcone et al. 2010a). Alterations in the kyneurinic pathway were described in a group of 20 adolescents with MDD compared with 30 adolescents with non-MDD (Gabbay et al. 2009b). Additionally, adolescents with melancholic features had decreased blood levels of tryptophan. Gabbay et al. in a study of 30 male adolescents with MDD found increased levels of IFN-gamma in patients with MDD vs. controls. There was also a trend for increased levels of IL-6 in the group with MDD (Gabbay et al. 2009b). Brambilla et al. reported increased levels of IL-1B in patients with dysthymia compared to controls. They found decreased levels of TNF-alpha in patients with dysthymia and an important correlation between TNF-alpha and the severity of depressive symptoms (Brambilla et al. 2004).

In another systematic review of 18 studies of the role of cytokines in depression in adolescents, the authors concluded that adolescents have age-specific characteristics such as the natural killer cell activity and certain proinflammatory cytokines (IL1B and TNF alpha). They concluded that the role of cytokines in adolescents with depression might be influenced by developmental changes, puberty, stress, and trauma, important to understand when formulating treatment (Mills et al. 2013).

A recent series of articles dissects the inflammatory differences in patients with mood disorder. The Mood Inflame Group reported the presence of a proinflammatory state of the circulating monocytes of patients with mood disorders. Researchers were able to identify 19 inflammation-related genes in the circulating monocytes of patients with mood disorders (Drexhage et al. 2010; Padmos et al. 2008, 2009). This inflammatory signature was also found in the offspring of patients with mood disorders especially those who developed childhood depression. This finding supports the theory that activated monocytes precede the appearance of adult mood symptoms. The same group replicated their findings in a twin study to determine the contribution of individual differences in the gene–environment interaction. The data show that the inflammatory response system (IRS) is activated in mood disorders and that activation of the IRS preceded the onset of first mood episodes in offspring of parents with mood disorders (Padmos et al. 2008, 2009).

Psychotic Disorders

Hundreds of studies of schizophrenic illness in adults have documented immunological abnormalities (Rothermundt and Arolt 2007; Fellerhoff et al. 2007; Torrey et al. 2006; Yolken et al. 2001; Ledgerwood et al. 2003; Maes 1997, 1998; Maino et al. 2007; Potvin et al. 2008; Falcone et al. 2008a). An early study by Solov'eva and Orlovskaia (1979) described how embryos of schizophrenic mothers showed signs of microglia activity. Different inflammatory factors and immune reactions directly influence neuronal proliferation, differentiation, migration, and apoptosis. In response to brain pathology, microglia become activated (Rothermundt and Arolt 2007; Rothermundt et al. 2004; Steiner et al. 2006, 2009; Hanson and Gottesman 2005). Immune cell function allows microglia react to brain injury, either with tissue repair or induction of immune responses: phagocytosis, secretion of cytokines, neuronal growth factors, and antigen presentation. Microglia activation may sustain chronic inflammation of the brain.

Evidence from animal models of schizophrenia has reported important contributions to this field of research (Zuckerman and Weiner 2005; Meyer et al. 2005, 2006a; Boksa 2008; Patterson 2002, 2007; Romero et al. 2007). Three different groups have created immunological animal models of schizophrenia. Their findings, which examine cytokine levels, support those we observed in our preliminary studies of children and adolescents with psychosis (Falcone et al. 2008b). Meyer et al. described how the maternal cytokine-associated inflammatory response to infection might be a crucial link for the development of schizophrenia regardless of the pathogen. In this experiment researchers gave Poly I:C to pregnant mice C57BL6/J on day 9. Cytokines were examined in both the fetal and adult brain of the Poly I:C mice; Abnormalities in IL-1 β ($p<0.001$), IL-6 ($p<0.001$), IL-10 ($p<0.001$), and TNFα ($p<0.05$) were identified when compared to control fetal and adult mice. From these findings, the group concluded that cytokine-related inflammatory processes in prenatal life are potential causal agents of postnatal brain histopathology (Meyer et al. 2006a, 2008a).

Microglia seem to play an important role on the pathophysiology of psychosis. Microglia are the macrophages of brain and have several roles that are central to the immune response, brain development, and synaptic regulation. Microglia, when activated, produce inflammatory cytokines (Rothermundt and Arolt 2007; Rothermundt et al. 2004). Some of these can alter the permeability of the blood–brain barrier (BBB) and allow increased influx of monocytes and other inflammatory cells into the brain (Oby and Janigro 2006). In brain development, microglia produce some neurotrophic factors that promote synaptic function and also regulate pruning (Neuwelt et al. 2011). Postmortem studies provide evidence of the role of activated microglia in schizophrenia. Two PET studies in patients with first episode psychosis with the PK11195 marker have found activated microglia in vivo in these patients (Radewicz et al. 2000; Hirsch 2004).

D'Mello et al. (2009) proposed that during systemic inflammation microglia can recruit monocytes into the brain. In the case of psychosis, Fellerhoff et al. hypothesized that Chlamydiae stimulated the transmigration of the monocytes through the

BBB in 40 % of psychotic patients. This proposes a possible mechanism for monocyte activation subsequent production of cytokines, and initiation of the inflammatory cascade. Alteration in the BBB might lead to the transmigration of monocytes from the vessels to the brain (Fellerhoff et al. 2007).

There are few studies in children of inflammatory changes during psychosis. Sporn et al. in a study of youth with childhood onset schizophrenia reported that schizophrenic children had an increase of TNF-alpha compared to controls. This association was also correlated with the body mass index (Sporn et al. 2005). In a group of children with schizophrenia, Mittleman et al. measured levels of cytokines in the CSF and compared them with a control group of children with ADHD. Children with schizophrenia had higher levels of IL-4 (Mittleman et al. 1997). In a retrospective study of youth with first episode psychosis, our group found that children with psychosis compared to other children with behavioral disorders had an increased number of monocytes (Falcone et al. 2013).

Studies have reported increased monocytes in the blood of schizophrenic patients (Drexhage et al. 2010; Fellerhoff et al. 2007; Falcone et al. 2008a; Zorrilla et al. 1996; Nikkila et al. 1999). In a longitudinal cohort study of the progeny of schizophrenic patients, activated monocytes were described in youth who later developed psychosis (Drexhage et al. 2010; Padmos et al. 2009). In a prospective case–control study of adolescents with psychosis and suicidal ideation, patients with psychosis ($n=40$) had increased serum levels of Il-1B and IL-8 compared to controls (Falcone et al. 2013).

In a population based study of atopic disorders, inflammatory markers, and the relationship with childhood psychotic experiences, Khandaker et al. concluded that atopic disorders are associated with the risk of developing psychosis during adolescence. Psychotic experiences were more common in patients with asthma, eczema, or both. In this population based study of 3,850 adolescents, 498 of those patients experienced psychotic symptoms (Khandaker et al. 2014).

The two- hit hypothesis is derived from inflammation models during the first trimester (Smith et al. 2007; Zuckerman and Weiner 2005; Meyer et al. 2005, 2006a, b, 2008a, b; Boksa 2008; Patterson 2002, 2007; Romero et al. 2007; Ashdown et al. 2006; Brown et al. 2004; Dammann and Leviton 1997; Zuckerman et al. 2003). This model suggests that maternal stress in the first trimester either by infection, abuse or severe stress increased the production of inflammatory cells and the activation of monocytes and microglia (Zuckerman and Weiner 2005; Patterson 2002, 2007; Zuckerman et al. 2003). Subsequent production of cytokines can lead to alterations in brain development in the children who later become schizophrenic have soft neurological signs, or develop disorders like ADHD (Gogtay and Rapoport 2007). Later in life after a second hit, such as stress, bullying, infection or family dysfunction, can activated the HPA axis, and enhance the production of monocytes, the activation of monocytes and microglial inflammation of the brain (Falcone et al. 2008a, b, 2013; Bergink et al. 2014) (Fig. 9.1). In a group of children with first episode psychosis, we observed monocytosis and high levels of proinflammatory cytokines (Falcone et al. 2013).

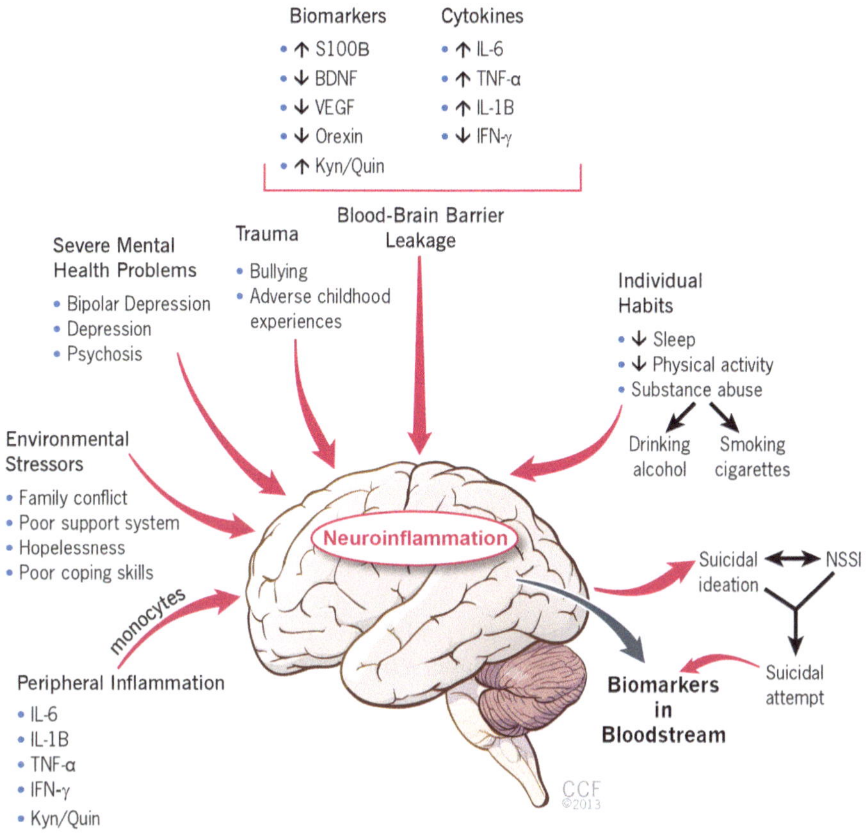

Fig. 9.1 Immune system and the environmental impact on the brain

Post-Traumatic Stress Disorder

Several studies have evaluated levels of cortisol in children exposed to trauma (Johnson et al. 2013; Shonkoff et al. 2012; McGowan et al. 2009; Carrion et al. 2002; McCrory et al. 2010). In a study of maltreated children with PTSD, 24 h levels of cortisol were increased (De Bellis et al. 2010). Goenjian et al. (1996) reported lower levels of basal cortisol in children 5 years after an earthquake. Gunnar et al. (2001) reported higher diurnal levels of cortisol after 6.5 years post trauma. In a study of girls who had attempted suicide. Girls with early sexual abuse had lower salivary levels of morning cortisol were reported by King et al. (2001). Carrion et al. reported higher levels of cortisol in 33 youth ages 10–16 and, and combined with changes in brain MRI. Youth with post-traumatic stress symptoms (PTSS) ($n = 45$) had decreased total brain tissue in comparison with healthy controls ($n = 15$). There was a significant negative association between pre-bedtime cortisol levels and left ventral prefrontal cortex gray volumes (Carrion et al. 2010a). Studies in adults report similar findings in patients with PTSD (Vythilingam et al. 2002). Some developmental differences have been reported in levels of IL-6 in children with PTSD. In both adults and children, IL-6 is elevated in individuals with PTSD. In adults the levels of IL-6 continue to be elevated chronically, while in children the IL-6 levels tend to normalize over time (Pervanidou et al. 2007).

The pathophysiological mechanisms underlying the development of neurobiological consequences of childhood trauma have not been fully elucidated. Disruption of the HPA axis with subsequent cortisol induced toxicity has been proposed as one mechanism for neuroimaging differences (Bremner 1999, 2006; McCrory et al. 2010; Marin et al. 2007). Another proposed mechanism is the induction of an inflammatory response leading to neuronal injury (Miller et al. 2009c; Herberth et al. 2008). A more recent and novel hypothesis is that brain injury could be caused by disruption of the BBB leading to influx of inflammatory cells, as well as toxic compounds which could lead to damage to several brain areas (Neuwelt et al. 2011; Falcone et al. 2015; Andrews and Neises 2012; Angelov et al. 2009; Cucullo et al. 2011; Esposito et al. 2001, 2002, 2003; Janigro 2012; Marchi et al. 2004; Rapoport et al. 1972; Reiber 2001). The last hypothesis has not been investigated as thoroughly due to the of lack of easily available tools to measure the integrity of the BBB until recently (Cucullo et al. 2011; Marchi et al. 2003, 2004; Kapural et al. 2002). Studies in soldiers before and after a parachute jump describe how chronic stress can activate peripheral blood mononuclear cells (PBMC) and increase several inflammatory markers such as nerve growth factor, cortisol, and adrenocorticotropic hormone (Aloe et al. 1994).

Animal model studies suggest that activation of PBMCs can potentially exacerbate CNS damage via inflammation (Ajmo et al. 2008). Lewitus et al. (2008) described the role of T lymphocyte infiltration at the level of the BBB and an increase of intercellular adhesion molecules (ICAM-1) on the BBB. ICAM-1 has also been associated with increased migration of inflammatory cells to the BBB. de Pablos et al. (2006) described increased levels of inflammatory markers in animals secondary to stress that were associated with a decreased number of neurons. A leaky BBB can permit further accumulation of PBMCs, that potentially will produce more inflammation and production of matrix metalloproteinases, further increasing the BBB permeability (Rosenberg 2009).

Studies in animal models have demonstrated that stressful situations, such as immobilization, forced swimming, and training in a water maze produced BBB breakdown (Sharma 2011; Sharma et al. 1990, 1991, 1992, 1995, 2007a, b, 2009, 2010, 2012; Sharma and Dey 1981, 1984, 1986, 1988; Sharma and Johanson 2007). Sharma et al. concluded after multiple investigations on the role of the BBB under stress that specific opening of the BBB occurs under different kinds of stressful situations: "emotional stress," "heat stress," "cold stress," "swimming stress," "immobilization stress," and "surgical stress." Adaptation or resilience is what makes a difference between how these stressors continue to impact the brain overtime. After the initial stress the glucocorticoid system is activated, increasing levels of ACTH that help the organism cope with the increased demands of stress (Sharma

2011; Sharma et al. 1990, 1991, 1992, 1995, 2007b, 2009, 2010, 2012, 2013; Sharma and Dey 1981, 1984, 1986, 1988; Sharma and Johanson 2007; Sharma and Cervos-Navarro 1991). Catecholamines (adrenaline, noradrenaline) are released and alter the BBB function. The initial mild to moderate stress can be an adaptive response that helps the organism to cope with the increasing demands while prolonged, chronic, severe stress can lead to brain changes in selective areas of the brain and potentially later on contributing to the development of PTSD (Shalev et al. 2009) (Fig. 9.2).

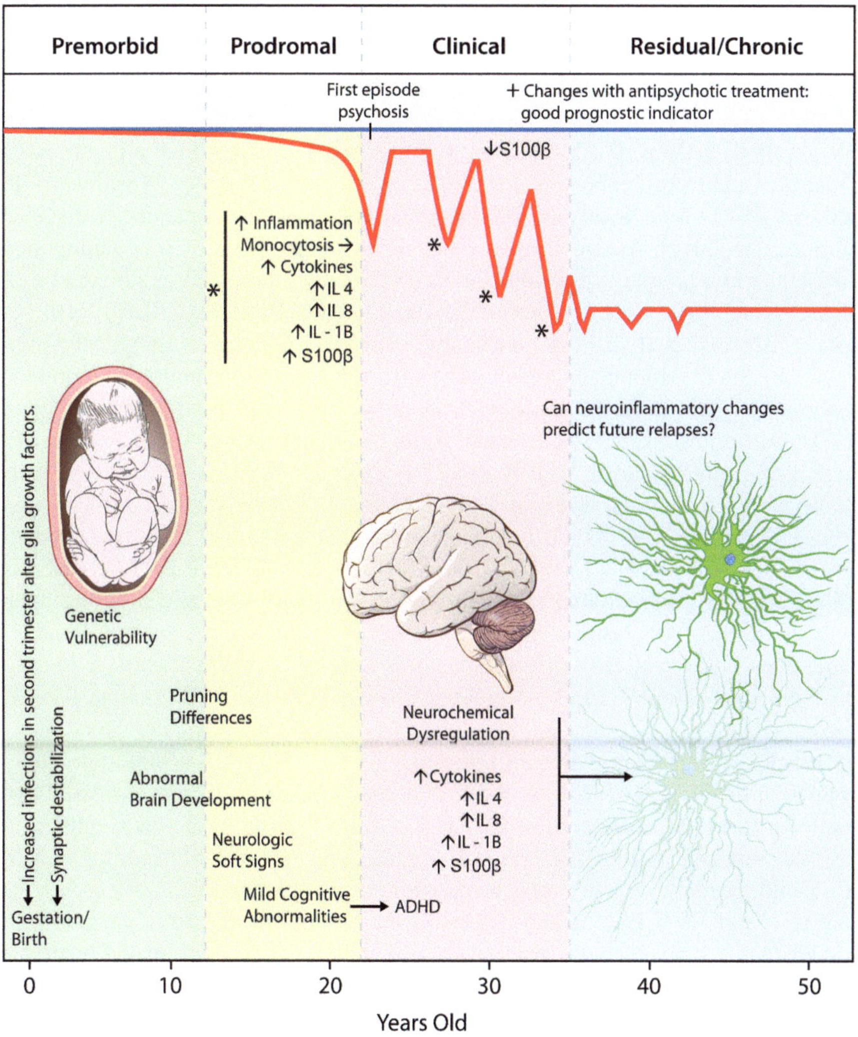

Fig. 9.2 Inflammatory changes in patients with psychosis- Please move the figure next to the psychosis section

Autism

There is very strong evidence for the role of inflammation in patients with autism. In children with autism spectrum disorder (ASD), 40 studies reported data on PIM. Al-Ayadhi and Mostafa (2012) in a study of 45 patients with ASD reported increases in levels of IL-17A in a group of ASD patients compared to controls. Manzardo et al. (2012) described 99 ASD patients vs. 40 controls and found decreased levels of IL-1alpha and IL-6 in ASD compared to controls. In a very small sample of ten patients, Bent et al. reported no changes in several cytokines but no control group was available. Also, there was no data on the gender of the participants (Bent et al. 2012). In a study in Saudi Arabia comparing 20 male patients with ASD with 19 controls, levels of IFN gamma were significantly elevated in patients with ASD (El-Ansary et al. 2011). Jyonuchi et al. identified elevated levels of IL-1, IL-6, IL12p40, IL-17, IL-23, IFN gamma, and TNF alpha after stimulation of PBMC. In study of 47 patients with ASD vs. 37 controls, there was decreased production of IL-1, IL-6, IL12p40, IL17, IL23, IFN gamma, and TNF alpha in ASD patients vs. controls (Jyonouchi et al. 2012). In a Brazilian study of 24 patients with ASD vs. 24 controls, levels of IFN gamma were increased in patients with ASD (Tostes et al. 2012). Ashwood et al. reported increased levels of IFN gamma and TNF alpha in a group of 66 patients with ASD vs. 77 controls after stimulation of PBMC. No changes were observed in IL-1B, IL-2, IL-4, IL-5, IL-6, IL8, IL-10, and IL-13 (Ashwood et al. 2011a). One study included 97 patients with ASD, 87 controls, and 39 developmentally delayed patients after plasma multiplex luminex, analysis of levels of IL-1B, IL6, IL-8, IL-12p40 were increased in patients with ASD (Ashwood et al. 2011b). In Saudi Arabia 25 males ages 4–12 vs. 16 controls, there were decreased levels of IL-6 and TNF alpha in ASD. In a study of 6 patients with ASD vs. 6 controls, levels of TNF alpha and IL-6 were increased in the ASD group (El-Ansary et al. 2011). Jyonouchi et al. (2011) reported decreased production of IL-6, IL beta, IL 12, IL 23, IL 10, sTNFRII, and TGF B in 30 ASD patients compared to 28 controls after stimulation of PBMC. Suzuki et al. (2011) in Japan reported levels of 48 plasma inflammatory markers increased in a group of 28 ASD patients vs. 28 controls. While Tobiasova in a study of 21 ASD patients vs. 15 controls, reported serum of IL-13 decreased in ASD patients. ASD patients ($n=77$) were treated with risperidone and no effect was observed in the inflammatory markers vs. placebo in ASD patients (Tobiasova et al. 2011). In a postmortem study evaluating expression of IL-6, in the frozen cerebellum of 6 patients of ASD vs 6 controls, increased expression of IL-6 in the cerebellum in ASD was identified in patients compared to controls (Wei et al. 2011). Patients with ASD and fragile X in ASD with fragile X ($n=40$) vs. patients with fragile X and no ASD ($n=64$) vs. 19 controls, levels of IL1 alpha and IL12p40 were reported increased in patients with ASD/fragile X patients vs. controls (Ashwood et al. 2010). In a group of 17 patients with ASD vs. 16 controls, levels of IL-1 B, IL-6, TNF alpha, were increased after stimulation of the Peripheral Blood Monocytes (PBMC) (Enstrom et al. 2010). In a study of 19 patients with ASD, levels of IL-6 were decreased if their PBMC were treated with BDE-47. Levels of IL-8 were increased in the ASD patients (Ashwood et al. 2009). IL-6, TNF alpha, and IFN gamma were increased in the frontal cortex of ASD patients postmortem compared to controls (Li et al. 2009). Onore et al. (2009) reported decreased IL-23 in 34 patients with ASD. Multiple infections and food allergies are frequently reported in children with ASD. Jyonouchi et al. (2008) studied levels of inflammatory markers in this group and found decreased IL-1 B and IL10 in ASD and increased IL-23 in ASD. Chez et al. evaluated levels of CSF TNF alpha in ASD and found levels increased in 10 patients. No controls were recruited in this study (Chez et al. 2007).

Hyperbaric oxygen treatment is an alternative treatment (non-FDA approved) for ASD. In a group of 18 patients with ASD without controls, levels of high sensitive CRP were elevated in three patients with ASD; after treatment with hyperbaric oxygen, the levels of CRP normalized (Rossignol et al. 2007). Molloy et al. (2006) reported elevated levels of IL-4, IL-5, and IL-13 in 20 patients with ASD vs. 20 controls. Jyonouchi et al. (2005a) analyzed the impact of gastro intestinal (GI) symptoms in children with ASD and the impact on inflammatory markers compared to controls and found increases in TNF alpha and IL-12 in ASD, but without impact on the GI symptoms. The same group in a dietary intervention study evaluated the levels of TNF alpha after stimulation of PBMC, patients with ASD had increased TNF alpha. The restricted diet had no impact on other inflammatory markers (Jyonouchi et al. 2005b). Two studies evaluated intestinal biopsy samples in children with ASD, De Felice found no difference in inflammatory markers in patients vs. controls, and Ashwood found increased CD3 TNF, CD3 IL2, CD3 IFN gamma, and CD3 IL 4 in patients with ASD (Ashwood et al. 2004). Zimmerman et al. (2005) evaluated levels of CSF and serum of several inflammatory markers in patients with ASD and found decreased levels of IL1Ra and IL 6 in patients vs. controls. In a study evaluating brain tissue from area-Brodman 19 and cerebellar cortex, in 9 males with ASD vs. 9 male controls, the inflammatory pathway gene dysregulation was significant in patients with ASD (Ginsberg et al. 2012).

Looking at these studies, there is consistent evidence for increased TNF alpha in patients with ASD. Contradictory findings in patients with ASD on the levels of IL-6. One of the important anti-inflammatory cytokines was decreased in nine studies of children with ASD (Mitchell and Goldstein 2014).

Different animal models of maternal inflammation during the first trimester have been proposed in the pathophysiology of autism (Smith et al. 2007; Patterson 2002). The role of infections during pregnancy has been extensively discussed, especially from the perspective on how early inflammatory changes can potentially alter neural circuitry and pruning. Maternal immune activation can later cause cerebellar abnormalities in the offspring of mice. Similar characteristics are shared in the immunological animal models of schizophrenia and autism (Smith et al. 2007; Zuckerman and Weiner 2005; Meyer et al. 2005, 2006a, b, 2008a, b; Romero et al. 2007; Dammann and Leviton 1997; Zuckerman et al. 2003; Depino 2006; Gilmore et al. 2004; Nawa and Takei 2006).

Obsessive Compulsive Disorder, Tourette's Disorder, and PANDAS

Several studies report increased inflammatory markers in patients with anxiety disorders. They reported an important relationship between the development of de-novo obsessive symptoms or tic disorder after the exposure to upper respiratory infection (Mitchell and Goldstein 2014).

Tourette Disorder

A study in China of 40 patients ages 4–19 with Tourette disorder (TD) compared to 40 controls reported elevated levels of IL-1, IL-6, and IL-17, in TD (Cheng et al. 2012). In a German study of 15 patients with TD vs. 10 controls, there were increased levels of TNF alpha and IL-1Ra in controls when compared to patients with TD (Matz et al. 2012). There was no differences in polymorphisms of children with TD compared to controls (Liu et al. 2011a, b; Li et al. 2013). Bos-Veneman et al. in 66 patients with TD measured serum levels of IL-2, IL-4, IL-5, IL-10, IL-12, IFN-gamma, and TNF-alpha to compare to 71 controls. There were no significant differences in serum concentrations in patients with TD (Bos-Veneman et al. 2010). In Taiwan a study of 159 patients of TD and 175 controls, polymorphisms in the IL1RN were associated with TD (OR of 7.65) (Chou et al. 2010).

Obsessive Compulsive Disorder

One study evaluated plasma levels of IL-1B, IL-2, IL-6, IL-12, and TNF alpha, in three groups: patients with TD ($n=32$), patients with TD and OCD ($n=17$) and controls ($n=14$). Levels of IL-12 were increased in TD-OCD group compared to controls, and to those with TD only. In a similar study Leckman et al. reported increased levels of IL-12 and TNF alpha in patients with TD+OCD ($n=34$) vs controls ($n=31$). Levels of IL-12 and TNF were increased during worsening symptoms of TD+OCD (Leckman et al. 2005). Levels of TNF alpha correlated inversely with obsessive compulsive disorder (OCD) symptom severity (Gabbay et al. 2009c). To the contrary levels of TNF alpha correlated inversely with OCD. Luo et al. evaluated serum levels of high sensitive CRP (hsCRP) in a group of 47 TD+OCD patients and 19 controls. There were no differences in the levels of hsCRP between the groups or after worsening of OCD symptoms (Luo et al. 2004).

Pediatric Autoimmune Neuropsychiatric Disorder

A 1990 study by the NIMH in youth with OCD described a group of children with severe symptoms and sudden onset of psychopathology (Swedo et al. 2004). Researchers noted an association between the sudden onset of symptoms of obsessive thoughts and other neuropsychiatric symptoms: irritability, anxiety, mood lability, and regression. Symptoms occurred chronologically near infections from influenza, varicella, streptococcal infections, Lyme disease, and Mycoplasma infections. Streptococcal infections were the ones most frequently recognized in association with the neuropsychiatric symptoms. In 2010 the diagnosis changed from PANDAS to PANS Pediatric Acute Onset neuropsychiatric syndrome (Singer et al. 2012). Different treatments have been recommended to treat the acute symptoms of PANDAS such as antibiotics in the absence of confirmed strept infection. There are many case reports, but there are no randomized controlled trials for PANS (Murphy et al. 2012). There are studies evaluating the use of intravenous immunoglobulin (IVIG) vs placebo, with improvement of symptoms by 40 % for IVIG and 60 % for plasmapheresis (Perlmutter et al. 1999). Two trials evaluating prophylaxis for strept infections were beneficial in reducing the rate of recurrence in PANDAS (Snider and Swedo 2004). But further research is needed.

In some prepubertal patients with anorexia, there is an antecedent of streptococcal infection. These patients have increased neurological abnormalities such as motor hyperactivity, fidgeting, or pacing (Vincenzi et al. 2010). In a recent review of the literature from 1990 to 2008 authors suggest measuring a CBC, an ESR, CRP, neopterine, a throat swab for strep and ASO titers, AntiDNAse B, AnAb, and D8/17. PANDAS-AN continues to be a controversial diagnosis in the literature (Vincenzi et al. 2010; Puxley et al. 2008).

Externalizing Disorders

ADHD

Conflicting evidence has been reported in PIM in children with ADHD. There are three German studies by the same group on this topic. One study with a group of 35 patients with ADHD and controls, levels of IFN gamma in association with lower birth weight and prematurity were reported (Oades 2011; Oades et al. 2010a, b). The second study reported levels of IL-1 were elevated in patients with a history of paternal smoking (Oades 2011). Patients with increased IL-16 and symptoms of hyperactivity were associated with errors of commission. Levels of IL-13 were associated with symptoms of omission in patients with ADHD (Oades et al. 2010a). In the last study by the same group, no differences were observed between non-medicated patients with ADHD vs. medicated patients vs. controls. There was a trend towards elevated levels of IFN gamma in patients with ADHD (Oades 2011).

Trios (n=958) of children with ADHD and their parents were compared genetically to 816 patients and three SNPs vs. 816 patients reported 3 SNPs rs7172689, rs7172689, rs4128767 in the IL-16 gene in chromosome 15 were associated with inattention in patients with ADHD (Lasky-Su et al. 2008).

In a genetic study evaluating 220 patients with ADHD, there was no association with the IL1Ra gene polymorphism (Misener et al. 2004). Lasky-Su et al. (2008)

analyzed 930 parents and patients and found IL-3 was associated with earlier onset ADHD. Segman et al. (2002) studied 86 patients with ADHD and parents, the IL-1Ra 4 repeat allele transmission was increased in patients with ADHD.

Disruptive Behavior

In an adult study examining 153 healthy adults, hostility was examined using the Cook-Medley Hostility Scale (CMHS), an association between hostility and cytokine production was described. Levels of IFN gamma, TNF alpha, and IL-5 were higher in patients with higher hostility (Janicki-Deverts et al. 2010). One study evaluated a sample of 113 children with a profile of behavioral dysregulation as measured by higher scores on the Child Behavior Check List (CBCL) scale for anxiety/depression, aggressive behavior, and attention problems. Fifty-one children scored 2.5 standard deviations above the norm on the CBCL. These children had increased CRP as compared to children with normal scores ($n = 82$) on the CBCL DP profile (Holtmann et al. 2013).

Suicide

Since not all suicidal behavior is associated with mood disorder, this section highlights suicide as a separate topic. Suicide rates are high among young people; it is the third leading cause of death in 15- to 24-year-olds and accounts for 13 % of all adolescent deaths annually (Falcone 2008; Nock et al. 2013). Suicidal ideation and attempts are very frequent in adolescents. A survey of adolescents showed that 19 % had seriously considered suicide in the last year, 15 % had made a plan and 8 % reported that they had made an attempt. Since 1950, the suicide rate of adolescents has quadrupled and represents 12 % of the mortality rate in this age group (Brent 2011). Multiple studies evaluating peripheral biological makers have been reported in suicidal patients and postmortem brains of those who have committed suicide (Birmaher et al. 1996). Different systems have been reported to show abnormalities in suicidal patients: the serotoninergic system, the HPA axis, neurotrophin factors, and the neuroimmune system (Serafini et al. 2013a). In the quest for biological markers for different mental illnesses, goals morphed based on genetic studies. If similar genes are involved in different psychiatric disorders, similar specific symptoms may have in common some specific biological pathways. Also independent environmental factors such as stress and adverse childhood experiences can alter the neuroimmune network in youth (Shonkoff et al. 2012). These alterations can predispose to suicide later in life (Dube et al. 2001, 2009; Danese et al. 2007). Special traits such as impulsivity and aggression might be more frequent in adolescent suicides (Nock et al. 2013; Nock 2009).

Although the exact pathophysiology of suicidal behavior has not been unraveled, several studies point to increasing evidence that there is an alteration in the immune system. There are at least 21 published studies of peripheral inflammatory makers in patients after a suicide attempt to date (Serafini et al. 2013a). Although the role of cytokines in suicidal patients has been investigated since the early 1980s and remains ambiguous, the results look more compelling (Johnson et al. 2002; Falcone et al. 2010a, b; Gabbay et al. 2009b; Bayard-Burfield et al. 1996; Carlborg et al. 2013; Dwivedi 2010; Dwivedi et al. 2003; Erhardt et al. 2013; Isung et al. 2012a; Janelidze et al. 2011; Kim et al. 2007; Lee and Kim 2010; Lindqvist et al. 2008, 2011; Serafini et al. 2013b; Steiner et al. 2008; Sublette et al. 2011). Melancholic patients have been consistently reported as having low monocyte counts and cytokine abnormalities (Rothermundt et al. 2001a). Due to the crucial role in the acute inflammatory reaction the monocytes–macrophage system has received the most attention (Serafini et al. 2013a). Cytokines produced by the monocytes like, IL-1beta, activate the complement system.

In adolescents, levels of several inflammatory markers have been identified in patients with suicidal ideation and behavior. Studies have reported alterations of IL-1B, IL-6, TNF-alpha, IFN-gamma, VEGF, S100B, orexin, BDNF, and Kyneurine (Falcone et al. 2010a; Gabbay et al. 2009b; Serafini et al. 2013a; Janelidze et al. 2011).

Individual Inflammatory Markers

Cytokines

The role of stress is important to help us understand the psychosocial and neurobiological underpinnings of suicidal patients (Slopen et al. 2012a; Segerstrom and Miller 2004). Stress activates the inflammatory cascade, and as a result of this reaction, microglia become activated (Slopen et al. 2012a; Bremner 1999; Serafini et al. 2013a; Singareddy et al. 2013; Lucas et al. 2013; Young 2013). These cells produce proinflammatory cytokines such as IL-1B, IL-2, IL-6, TNF; IL-1B and IL-6, impacting the permeability of the BBB (Oby and Janigro 2006; Janigro 1999, 2012). These increases have been identified in patients who have attempted suicide. Dysregulation of the HPA axis has been reported in some studies evaluating suicidal behavior. Alteration in the HPA axis is directly related to the production of cytokines. When monocytes are activated, they can cross the BBB and activate microglia, consistent with evidence from postmortem studies of patients who have committed suicide (Rothermundt et al. 2001c; Motivala et al. 2005; Steiner et al. 2008, 2013).

IL1 B

Levels of IL1 have been shown to potentiate the brain concentrations of serotonin and norepinephrine. This cytokine has also been associated with the activation of the HPA axis and is reported to increase the secretion of ACTH and cortisol (Steiner et al. 2008, 2013). Pandey et al. in a postmortem study of teenage suicide victims reported increased expression of IL1B in the area of Broadman 10 (Pandey and Dwivedi 2012; Pandey et al. 2012).

IL 6

Levels of IL6 are reported to be elevated in patients with depression, exposure to childhood trauma, and suicide. Evidence from several meta-analysis from patients with mood disorder and schizophrenia has reported increased levels of IL-6. One third of the circulating levels of IL-6 come from adipose tissue. Smoking has also been linked to increased levels of IL-6. Three studies reported an increase in the levels of IL 6 in patients with suicide (Zorrilla et al. 2001; Dowlati et al. 2010; Steiner et al. 2008).

TNF Alpha

Increased levels of norepinephrine in patients who are receiving TNF alpha have been described (Dowlati et al. 2010). Animal studies found changes on levels of TNF alpha after treatment with antidepressants. Some theories involve the effect of antidepressants effect on the glucocorticoid receptor. Increased levels of TNF alpha have been reported in five studies with suicidal patients (Zorrilla et al. 2001; Dowlati et al. 2010; Serafini et al. 2013a; Pandey and Dwivedi 2012; Pandey et al. 2012), one of them in adolescents (Pandey et al. 2012).

IFN Gamma

This cytokine has been decreased after treatment in vivo or in vitro with antidepressants (Kubera et al. 1996). Two studies reported altered levels of IFN gamma in suicidal patients (Gabbay et al. 2009b; Janelidze et al. 2011), one of them in adolescents reporting a different biological pattern compared to the adults. Decreased levels of this cytokine were found in suicidal adolescents, although sample size was small (Gabbay et al. 2009b). IFN Gamma is one of the first steps on the Kyneurinic pathway (Hughes et al. 2012).

Kyneurinic/Quinolinic Acid

Inflammation increased the production of quinolinic acid (QUIN) and kyneurinic acid (KYNA). The Kyneurinic pathway is altered in patients with mood disorder and schizophrenia (Gabbay et al. 2009b; Hughes et al. 2012). Some preliminary studies on youth with ADHD and those exposed to trauma may also be implicated by this pathway. Longitudinal studies of patients undergoing treatment for mood disorders (MoodInflame Group) have demonstrated decreased levels of kyneurinic acid in patients who are improving. Evidence from postmortem studies of mood disorder patients has reported alteration in certain areas of the brain on the Kyneurinic pathway in a group of depressed patients. Two studies have reported alterations on the Kyneurinic pathway in those attempting suicide (Steiner et al. 2013;

Hughes et al. 2012; Claes et al. 2011; Miura et al. 2008). Sublette compared levels between healthy controls and depressed patients who had attempted suicide and those who had not (Sublette et al. 2011). Erhardt et al. (2013) reported levels of increased QUIN acid in the CSF of 64 medication free suicide attempters. The authors hypothesize that alteration of the KYN pathway in suicide might be related to the glutamatergic transmission, since QUIN is an NMDA receptor agonist. The fact that ketamine has shown promising results in improving suicidality in four small studies might suggest that the NMDA receptor could be an important pathway to further study in patients with suicidal behavior (DiazGranados et al. 2010; Larkin and Beautrais 2011; Price et al. 2009; Zarate et al. 2012).

S100B

S100B is a glial and trophic marker, located primarily on the cytoplasm of oligodendrocytes and astrocytes (Arolt et al. 2003; Andreazza et al. 2007). S100B stays in the brain, unless there are alterations of the BBB. At which time this protein is identified in serum (Marchi et al. 2004). In a study on adolescents presenting with suicidal behavior, levels of S100B were increased in patients compared to controls despite a diagnosis of MDD, bipolar disorder, or schizophrenia (Falcone et al. 2010a). This study was later replicated by our group using the Columbia Suicidality Screen Rating Scale (CSSRS) as the measure for suicidal ideation in adolescents. Levels of S100B were increased in those adolescents who were at greater risk for suicide and may be possibly a tool to consider for assesment of risk (Falcone et al. 2010a).

Brain Derived Neurotrophic Factor

Brain derived neurotrophic factor (BDNF) is a neurotrophic classified as a nerve growth factor, affecting growth, plasticity, and neurogenesis. Some suggest that suicidal behavior may be related with alterations in neural plasticity. There is evidence from postmortem studies reporting a decreased number of glial cells, and glial density in the frontal cortex and hippocampus (Oby and Janigro 2006; Janigro 1999). Five studies have reported alteration in the BDNF levels of suicidal patients (Gabbay et al. 2009b; Li et al. 2013; Erhardt et al. 2013; Janelidze et al. 2011; Sublette et al. 2011; Pandey et al. 2012; Boehm et al. 2010). Additionally the mRNA expression of BDNF was decreased in the pre-frontal cortex and hippocampus of suicide victims (Dwivedi et al. 2003).

Vascular Endothelial Growth Factor

Vascular endothelial growth factor (VEGF) is an angiogenic protein with an important role in neuroprotection and neurogenesis. Viikki et al. (2010) reported that patients with a VEGF polymorphism were resistant to depression treatment.

Two studies reported a negative correlation of VEGF plasma levels proportionate to depression severity (Isung et al. 2012a, b). VEGF is reported to be low in the plasma of suicide attempters. Additionally, VEGF is a key factor in the integrity of the BBB and studies have reported alteration in the permeability of the BBB in patients with suicidal behavior (Serafini et al. 2013b).

Orexin

One of the hypothalamic neuropeptides that modulates the HPA axis has been linked to the regulation of sleep. Two studies have reported decreased CSF levels of Orexin in patients who attempted suicide. After patients received treatment (6 and 12 months) levels of orexin increased (Kim et al. 2007; Karege et al. 2005).

Conclusions

There are important advances in the field of psychoneuroimmunology in the last 25 years; however, most of the studies on inflammatory markers and psychiatric disorders come from adult populations (Zorrilla et al. 2001; Drexhage et al. 2010; Dowlati et al. 2010; Padmos et al. 2008; Shalev et al. 2009; Serafini et al. 2013a). In this chapter we focus on studies in children and adolescents (Carlton et al. 2009; Falcone et al. 2008a, 2010a, c, 2013; Gabbay et al. 2009b; Ginsberg et al. 2012; Janelidze et al. 2011; Falcone and Janigro 2008). Pediatric populations are key in our process to understand the development and progression of mental illness and in the case of finding novel treatments for those patients who have comorbid increase of inflammatory markers and different psychiatric disorders. The evidence in adult studies points to a developmental advantage when the treatment is started early (Akhondzadeh et al. 2007; Bresee et al. 2006; Muller et al. 2002, 2005; Rapaport et al. 2005; Riedel et al. 2005). The study of inflammatory markers in children is a complex field, important confounding factors have to be taken into account, such as age, Tanner stage, weight, use of medications, sleep, variability of the marker during the day, life stressors, trauma history, and family history of psychiatric disorder (Mitchell and Goldstein 2014; Lucas et al. 2013; Bahn and Schwarz 2011; Gottschalk et al. 2013; Lopresti et al. 2014). Most studies in pediatric populations are cross-sectional and retrospective. This limits our ability to identify pathogenic cause for mental illness, and only permits discussion of an association between inflammation and diverse psychiatric disorders reviewed. Despite this limitation, there is important preliminary evidence for the role of inflammation in psychiatric disorders in pediatric populations (Pardo et al. 2005; Falcone et al. 2010a, 2013; Mitchell and Goldstein 2014; Gabbay et al. 2009b, c; Padmos et al. 2009; Carrion et al. 2002, 2007, 2008, 2010a, b; Chez et al. 2007; Oades 2011; Oades et al. 2010a, b; Pandey and Dwivedi 2012; Carrion and Wong 2012; De Bellis and Kuchibhatla 2006). Taking into account the high incidence of comorbidities in child psychiatry, it is very unlikely that inflammatory markers can be used solely for diagnostic purposes. However they could be very helpful to evaluate treatment response, disease progression and help us monitor high-risk patients that may potentially benefit from preventive treatment (Mitchell and Goldstein 2014; Goldstein and Young 2013). There is progress in identifying the onset, course, and recurrence of behavioral disorders in children and adolescents; however, further research is needed to understand the role of inflammation and the relationship with psychosocial and psychobiological factors.

References

Ajmo Jr CT, Vernon DO, Collier L, Hall AA, Garbuzova-Davis S, Willing A, et al. The spleen contributes to stroke-induced neurodegeneration. J Neurosci Res. 2008;86(10):2227–34. doi:10.1002/jnr.21661.

Akhondzadeh S, Tabatabaee M, Amini H, Ahmadi Abhari SA, Abbasi SH, Behnam B. Celecoxib as adjunctive therapy in schizophrenia: a double-blind, randomized and placebo-controlled trial. Schizophr Res. 2007;90(1–3):179–85.

Al-Ayadhi LY, Mostafa GA. Elevated serum levels of interleukin-17A in children with autism. J Neuroinflammation. 2012;9:158. doi:10.1186/1742-2094-9-158.

Aloe L, Bracci-Laudiero L, Alleva E, Lambiase A, Micera A, Tirassa P. Emotional stress induced by parachute jumping enhances blood nerve growth factor levels and the distribution of nerve growth factor receptors in lymphocytes. Proc Natl Acad Sci U S A. 1994;91(22):10440–4.

American Academy of Pediatrics, Stirling J, Jr, Committee on Child Abuse and Neglect and Section on Adoption and Foster Care, American Academy of Child and Adolescent Psychiatry, Amaya-Jackson L, National Center for Child Traumatic Stress, et al. Understanding the behavioral and emotional consequences of child abuse. Pediatrics. 2008;122(3):667–73. doi:10.1542/peds.2008-1885.

Andreazza AC, Cassini C, Rosa AR, Leite MC, de Almeida LM, Nardin P, et al. Serum S100B and antioxidant enzymes in bipolar patients. J Psychiatr Res. 2007;41(6):523–9. doi:10.1016/j.jpsychires.2006.07.013.

Andreoli AV, Keller SE, Rabaeus M, Marin P, Bartlett JA, Taban C. Depression and immunity: age, severity, and clinical course. Brain Behav Immun. 1993;7(4):279–92. doi:10.1006/brbi.1993.1028.

Andrews JA, Neises KD. Cells, biomarkers, and post-traumatic stress disorder: evidence for peripheral involvement in a central disease. J Neurochem. 2012;120(1):26–36. doi:10.1111/j.1471-4159.2011.07545.x.

Angelov L, Doolittle ND, Kraemer DF, Siegal T, Barnett GH, Peereboom DM, et al. Blood–brain barrier disruption and intra-arterial methotrexate-based therapy for newly diagnosed primary CNS lymphoma: a multi-institutional experience. J Clin Oncol. 2009;27(21):3503–9. doi:10.1200/JCO.2008.19.3789.

Anisman H, Ravindran AV, Griffiths J, Merali Z. Endocrine and cytokine correlates of major depression and dysthymia with typical or atypical features. Mol Psychiatry. 1999;4(2):182–8.

Arolt V, Rothermundt M. The immunology of psychiatric disorders. Psychother Psychosom Med Psychol. 2005;55(1):36–48.

Arolt V, Peters M, Erfurth A, Wiesmann M, Missler U, Rudolf S, et al. S100B and response to treatment in major depression: a pilot study. Eur Neuropsychopharmacol. 2003;13(4):235–9.

Ashdown H, Dumont Y, Ng M, Poole S, Boksa P, Luheshi GN. The role of cytokines in mediating effects of prenatal infection on the fetus: implications for schizophrenia. Mol Psychiatry. 2006;11(1):47–55.

Ashwood P, Anthony A, Torrente F, Wakefield AJ. Spontaneous mucosal lymphocyte cytokine profiles in children with autism and gastrointestinal symptoms: mucosal immune activation and reduced counter regulatory interleukin-10. J Clin Immunol. 2004;24(6):664–73. doi:10.1007/s10875-004-6241-6.

Ashwood P, Schauer J, Pessah IN, Van de Water J. Preliminary evidence of the in vitro effects of BDE-47 on innate immune responses in children with autism spectrum disorders. J Neuroimmunol. 2009;208(1–2):130–5. doi:10.1016/j.jneuroim.2008.12.012.

Ashwood P, Nguyen DV, Hessl D, Hagerman RJ, Tassone F. Plasma cytokine profiles in Fragile X subjects: is there a role for cytokines in the pathogenesis? Brain Behav Immun. 2010;24(6):898–902. doi:10.1016/j.bbi.2010.01.008.

Ashwood P, Krakowiak P, Hertz-Picciotto I, Hansen R, Pessah I, Van de Water J. Elevated plasma cytokines in autism spectrum disorders provide evidence of immune dysfunction and are associated with impaired behavioral outcome. Brain Behav Immun. 2011a;25(1):40–5. doi:10.1016/j.bbi.2010.08.003.

Ashwood P, Krakowiak P, Hertz-Picciotto I, Hansen R, Pessah IN, Van de Water J. Altered T cell responses in children with autism. Brain Behav Immun. 2011b;25(5):840–9. doi:10.1016/j.bbi.2010.09.002.

Bahn S, Schwarz E. Serum-based biomarkers for psychiatric disorders. Nervenarzt. 2011;82(11):1395–6, 1398, 1400 passim. doi:10.1007/s00115-011-3346-z.

Bayard-Burfield L, Alling C, Blennow K, Jonsson S, Traskman-Bendz L. Impairment of the blood-CSF barrier in suicide attempters. Eur Neuropsychopharmacol. 1996;6(3):195–9.

Bent S, Bertoglio K, Ashwood P, Nemeth E, Hendren RL. Brief report: hyperbaric oxygen therapy (HBOT) in children with autism spectrum disorder: a clinical trial. J Autism Dev Disord. 2012;42(6):1127–32. doi:10.1007/s10803-011-1337-3.

Bergink V, Gibney SM, Drexhage HA. Autoimmunity, inflammation, and psychosis: a search for peripheral markers. Biol Psychiatry. 2014;75(4):324–31. doi:10.1016/j.biopsych.2013.09.037.

Birmaher B, Rabin BS, Garcia MR, Jain U, Whiteside TL, Williamson DE, et al. Cellular immunity in depressed, conduct disorder, and normal adolescents: role of adverse life events. J Am Acad Child Adolesc Psychiatry. 1994;33(5):671–8. doi:10.1097/00004583-199406000-00008.

Birmaher B, Ryan ND, Williamson DE, Brent DA, Kaufman J. Childhood and adolescent depression: a review of the past 10 years. Part II. J Am Acad Child Adolesc Psychiatry. 1996;35(12):1575–83. doi:10.1097/00004583-199612000-00008.

Boehm J, Fischer K, Bohnert M. Putative role of TNF-alpha, interleukin-8 and ICAM-1 as indicators of an early inflammatory reaction after burn: a morphological and immunohistochemical study of lung tissue of fire victims. J Clin Pathol. 2010;63(11):967–71. doi:10.1136/jcp.2010.079863.

Boksa P. Maternal infection during pregnancy and schizophrenia. J Psychiatry Neurosci. 2008;33(3):183–5.

Bonaccorso S, Puzella A, Marino V, Pasquini M, Biondi M, Artini M, et al. Immunotherapy with interferon-alpha in patients affected by chronic hepatitis C induces an intercorrelated stimulation of the cytokine network and an increase in depressive and anxiety symptoms. Psychiatry Res. 2001;105(1–2):45–55.

Bos-Veneman NG, Bijzet J, Limburg PC, Minderaa RB, Kallenberg CG, Hoekstra PJ. Cytokines and soluble adhesion molecules in children and adolescents with a tic disorder. Prog Neuropsychopharmacol Biol Psychiatry. 2010;34(8):1390–5. doi:10.1016/j.pnpbp.2010.06.028.

Brambilla F, Monteleone P, Maj M. Interleukin-1beta and tumor necrosis factor-alpha in children with major depressive disorder or dysthymia. J Affect Disord. 2004;78(3):273–7. doi:10.1016/S0165-0327(02)00315-4.

Bremner JD. Does stress damage the brain? Biol Psychiatry. 1999;45(7):797–805.

Bremner JD. Traumatic stress: effects on the brain. Dialogues Clin Neurosci. 2006;8(4):445–61.

Brent DA. Preventing youth suicide: time to ask how. J Am Acad Child Adolesc Psychiatry. 2011;50(8):738–40. doi:10.1016/j.jaac.2010.09.017.

Brent DA, Kolko DJ, Birmaher B, Baugher M, Bridge J, Roth C, et al. Predictors of treatment efficacy in a clinical trial of three psychosocial treatments for adolescent depression. J Am Acad Child Adolesc Psychiatry. 1998;37(9):906–14. doi:10.1097/00004583-199809000-00010.

Brent DA, Oquendo M, Birmaher B, Greenhill L, Kolko D, Stanley B, et al. Peripubertal suicide attempts in offspring of suicide attempters with siblings concordant for suicidal behavior. Am J Psychiatry. 2003;160(8):1486–93.

Brent DA, McMakin DL, Kennard BD, Goldstein TR, Mayes TL, Douaihy AB. Protecting adolescents from self-harm: a critical review of intervention studies. J Am Acad Child Adolesc Psychiatry. 2013;52(12):1260–71. doi:10.1016/j.jaac.2013.09.009.

Bresee CJ, Delrahim K, Maddux RE, Dolnak D, Ahmadpour O, Rapaport MH. The effects of celecoxib augmentation on cytokine levels in schizophrenia. Int J Neuropsychopharmacol. 2006;9(3):343–8.

Brown GR, Anderson B. Psychiatric morbidity in adult inpatients with childhood histories of sexual and physical abuse. Am J Psychiatry. 1991;148(1):55–61.

Brown AS, Hooton J, Schaefer CA, Zhang H, Petkova E, Babulas V, et al. Elevated maternal interleukin-8 levels and risk of schizophrenia in adult offspring. Am J Psychiatry. 2004;161(5):889–95.

Brown DW, Anda RF, Tiemeier H, Felitti VJ, Edwards VJ, Croft JB, et al. Adverse childhood experiences and the risk of premature mortality. Am J Prev Med. 2009;37(5):389–96. doi:10.1016/j.amepre.2009.06.021.

Buter J, de Vries EG, Sleijfer DT, Willemse PH, Mulder NH. Neuropsychiatric symptoms during treatment with interleukin-2. Lancet. 1993;341(8845):628.

Caplan R, Siddarth P, Gurbani S, Hanson R, Sankar R, Shields WD. Depression and anxiety disorders in pediatric epilepsy. Epilepsia. 2005;46(5):720–30.

Capuron L, Miller AH. Cytokines and psychopathology: lessons from interferon-alpha. Biol Psychiatry. 2004;56(11):819–24. doi:10.1016/j.biopsych.2004.02.009.

Capuron L, Ravaud A, Miller AH, Dantzer R. Baseline mood and psychosocial characteristics of patients developing depressive symptoms during interleukin-2 and/or interferon-alpha cancer therapy. Brain Behav Immun. 2004;18(3):205–13. doi:10.1016/j.bbi.2003.11.004.

Capuron L, Miller A, Irwin MR. Psychoneuroimmunology of depressive disorder: mechanisms and clinical implications. In: Ader R, editor. Psychoneuroimmunology, vol. I. Paris: Elsevier; 2007. p. 509–30. ISBN: 978-0-12-088576-3.

Carlborg A, Jokinen J, Jonsson EG, Erhardt S, Nordstrom P. CSF kynurenic acid and suicide risk in schizophrenia spectrum psychosis. Psychiatry Res. 2013;205(1–2):165–7. doi:10.1016/j.psychres.2012.08.021.

Carlton E, Falcone T, Tuxhorn I, Janigro D. Psychosis and epilepsy in childhood – the inflammatory hypothesis. Schizophr Res. 2008;98(Suppl 1):80.

Carrion VG, Wong SS. Can traumatic stress alter the brain? Understanding the implications of early trauma on brain development and learning. J Adolesc Health. 2012;51(2 Suppl):S23–8. doi:10.1016/j.jadohealth.2012.04.010.

Carrion VG, Wong SS. Can traumatic stress alter the brain? Understanding the implications of early trauma on brain development and learning. J Adolesc Health. 2012;51(2 Suppl):S23–8. doi:10.1016/j.jadohealth.2012.04.010.

Carrion VG, Weems CF, Ray RD, Glaser B, Hessl D, Reiss AL. Diurnal salivary cortisol in pediatric posttraumatic stress disorder. Biol Psychiatry. 2002;51(7):575–82.

Carrion VG, Weems CF, Reiss AL. Stress predicts brain changes in children: a pilot longitudinal study on youth stress, posttraumatic stress disorder, and the hippocampus. Pediatrics. 2007;119(3):509. doi:10.1542/peds. –516, 2006-2028.

Carrion VG, Garrett A, Menon V, Weems CF, Reiss AL. Posttraumatic stress symptoms and brain function during a response-inhibition task: an fMRI study in youth. Depress Anxiety. 2008;25(6):514–26. doi:10.1002/da.20346.

Carrion VG, Weems CF, Richert K, Hoffman BC, Reiss AL. Decreased prefrontal cortical volume associated with increased bedtime cortisol in traumatized youth. Biol Psychiatry. 2010a;68(5):491–3. doi:10.1016/j.biopsych.2010.05.010.

Carrion VG, Haas BW, Garrett A, Song S, Reiss AL. Reduced hippocampal activity in youth with posttraumatic stress symptoms: an FMRI study. J Pediatr Psychol. 2010b;35(5):559–69. doi:10.1093/jpepsy/jsp112.

Caserta MT, O'Connor TG, Wyman PA, Wang H, Moynihan J, Cross W, et al. The associations between psychosocial stress and the frequency of illness, and innate and adaptive immune function in children. Brain Behav Immun. 2008;22(6):933–40. doi:10.1016/j.bbi.2008.01.007.

Cheng YH, Zheng Y, He F, Yang JH, Li WB, Wang ML, et al. Detection of autoantibodies and increased concentrations of interleukins in plasma from patients with Tourette's syndrome. J Mol Neurosci. 2012;48(1):219–24. doi:10.1007/s12031-012-9811-8.

Chez MG, Dowling T, Patel PB, Khanna P, Kominsky M. Elevation of tumor necrosis factor-alpha in cerebrospinal fluid of autistic children. Pediatr Neurol. 2007;36(6):361–5. doi:10.1016/j.pediatrneurol.2007.01.012.

Chou IC, Lin HC, Wang CH, Lin WD, Lee CC, Tsai CH, et al. Polymorphisms of interleukin 1 gene IL1RN are associated with Tourette syndrome. Pediatr Neurol. 2010;42(5):320–4. doi:10.1016/j.pediatrneurol.2010.01.006.

Claes S, Myint AM, Domschke K, Del-Favero J, Entrich K, Engelborghs S, et al. The kynurenine pathway in major depression: haplotype analysis of three related functional candidate genes. Psychiatry Res. 2011;188(3):355–60. doi:10.1016/j.psychres.2011.03.012.

Clerici M, Arosio B, Mundo E, Cattaneo E, Pozzoli S, Dell'osso B, et al. Cytokine polymorphisms in the pathophysiology of mood disorders. CNS Spectr. 2009;14(8):419–25.

Cohen JA. Treating acute posttraumatic reactions in children and adolescents. Biol Psychiatry. 2003;53(9):827–33.

Coogan PF, Wise LA, O'Connor GT, Brown TA, Palmer JR, Rosenberg L. Abuse during childhood and adolescence and risk of adult-onset asthma in African American women. J Allergy Clin Immunol. 2013;131(4):1058–63. doi:10.1016/j.jaci.2012.10.023.

Crespel A, Coubes P, Rousset M, Brana C, Rougier A, Rondouin G, et al. Inflammatory reactions in human medial temporal lobe epilepsy with hippocampal sclerosis. Brain Res. 2002;952(2):159–69.

Cucullo L, Hossain M, Puvenna V, Marchi N, Janigro D. The role of shear stress in blood–brain barrier endothelial physiology. BMC Neurosci. 2011;12:40. doi:10.1186/1471-2202-12-40.

D'Mello C, Le T, Swain MG. Cerebral microglia recruit monocytes into the brain in response to tumor necrosis factoralpha signaling during peripheral organ inflammation. J Neurosci. 2009;29(7):2089–102. doi:10.1523/JNEUROSCI.3567-08.2009.

Dammann O, Leviton A. Maternal intrauterine infection, cytokines, and brain damage in the preterm newborn. Pediatr Res. 1997;42(1):1–8.

Danese A, McEwen BS. Adverse childhood experiences, allostasis, allostatic load, and age-related disease. Physiol Behav. 2012;106(1):29–39. doi:10.1016/j.physbeh.2011.08.019.

Danese A, Pariante CM, Caspi A, Taylor A, Poulton R. Childhood maltreatment predicts adult inflammation in a life-course study. Proc Natl Acad Sci U S A. 2007;104(4):1319–24. doi:10.1073/pnas.0610362104.

Danese A, Moffitt TE, Pariante CM, Ambler A, Poulton R, Caspi A. Elevated inflammation levels in depressed adults with a history of childhood maltreatment. Arch Gen Psychiatry. 2008;65(4):409–15. doi:10.1001/archpsyc.65.4.409.

Danese A, Moffitt TE, Pariante CM, Ambler A, Poulton R, Caspi A. Elevated inflammation levels in depressed adults with a history of childhood maltreatment. Arch Gen Psychiatry. 2008b;65(4):409–15.

Danese A, Moffitt TE, Harrington H, Milne BJ, Polanczyk G, Pariante CM, et al. Adverse childhood experiences and adult risk factors for age-related disease: depression, inflammation, and clustering of metabolic risk markers. Arch Pediatr Adolesc Med. 2009;163(12):1135–43. doi:10.1001/archpediatrics.2009.214.

Dantzer R, O'Connor JC, Freund GG, Johnson RW, Kelley KW. From inflammation to sickness and depression: when the immune system subjugates the brain. Nat Rev Neurosci. 2008;9(1):46–56. doi:10.1038/nrn2297.

De Bellis MD, Kuchibhatla M. Cerebellar volumes in pediatric maltreatment-related posttraumatic stress disorder. Biol Psychiatry. 2006;60(7):697–703. doi:10.1016/j.biopsych.2006.04.035.

De Bellis MD, Hooper SR, Woolley DP, Shenk CE. Demographic, maltreatment, and neurobiological correlates of PTSD symptoms in children and adolescents. J Pediatr Psychol. 2010;35(5):570–7. doi:10.1093/jpepsy/jsp116.

de Pablos RM, Villaran RF, Arguelles S, Herrera AJ, Venero JL, Ayala A, et al. Stress increases vulnerability to inflammation in the rat prefrontal cortex. J Neurosci. 2006;26(21):5709–19. doi:10.1523/JNEUROSCI.0802-06.2006.

Dennison U, McKernan D, Cryan J, Dinan T. Schizophrenia patients with a history of childhood trauma have a pro-inflammatory phenotype. Psychol Med. 2012;42(9):1865–71. doi:10.1017/S0033291712000074.

Depino AM. Maternal infection and the offspring brain. J Neurosci. 2006;26(30):7777–8.

DiazGranados N, Ibrahim LA, Brutsche NE, Ameli R, Henter ID, Luckenbaugh DA, et al. Rapid resolution of suicidal ideation after a single infusion of an *N*-methyl-D-aspartate antagonist in patients with treatment-resistant major depressive disorder. J Clin Psychiatry. 2010;71(12):1605–11. doi:10.4088/JCP.09m05327blu.

Dowd JB, Zajacova A, Aiello AE. Predictors of inflammation in U.S. children aged 3–16 years. Am J Prev Med. 2010;39(4):314–20. doi:10.1016/j.amepre.2010.05.014.

Dowlati Y, Herrmann N, Swardfager W, Liu H, Sham L, Reim EK, et al. A meta-analysis of cytokines in major depression. Biol Psychiatry. 2010;67(5):446–57. doi:10.1016/j.biopsych.2009.09.033.

Drexhage RC, Padmos RC, de Wit H, Versnel MA, Hooijkaas H, van der Lely AJ, et al. Patients with schizophrenia show raised serum levels of the pro-inflammatory chemokine CCL2: association with the metabolic syndrome in patients? Schizophr Res. 2008;102(1–3):352–5.

Drexhage RC, Knijff EM, Padmos RC, Heul-Nieuwenhuijzen L, Beumer W, Versnel MA, et al. The mononuclear phagocyte system and its cytokine inflammatory networks in schizophrenia and bipolar disorder. Expert Rev Neurother. 2010;10(1):59–76.

Dube SR, Anda RF, Felitti VJ, Chapman DP, Williamson DF, Giles WH. Childhood abuse, household dysfunction, and the risk of attempted suicide throughout the life span: findings from the Adverse Childhood Experiences Study. JAMA. 2001;286(24):3089–96.

Dube SR, Fairweather D, Pearson WS, Felitti VJ, Anda RF, Croft JB. Cumulative childhood stress and autoimmune diseases in adults. Psychosom Med. 2009;71(2):243–50. doi:10.1097/PSY.0b013e3181907888.

Dunn EC, Gilman SE, Willett JB, Slopen NB, Molnar BE. The impact of exposure to interpersonal violence on gender differences in adolescent-onset major depression: results from the national comorbidity survey replication (NCS-R). Depress Anxiety. 2012;29(5):392–9. doi:10.1002/da.21916.

Dwivedi Y. Brain-derived neurotrophic factor and suicide pathogenesis. Ann Med. 2010;42(2):87–96. doi:10.3109/07853890903485730.

Dwivedi Y, Rizavi HS, Conley RR, Roberts RC, Tamminga CA, Pandey GN. Altered gene expression of brain-derived neurotrophic factor and receptor tyrosine kinase B in postmortem brain of suicide subjects. Arch Gen Psychiatry. 2003;60(8):804–15. doi:10.1001/archpsyc.60.8.804.

Eaton WW, Hayward C, Ram R. Schizophrenia and rheumatoid arthritis: a review. Schizophr Res. 1992;6(3):181–92.

El-Ansary AK, Ben Bacha AG, Al-Ayadhi LY. Proinflammatory and proapoptotic markers in relation to mono and di-cations in plasma of autistic patients from Saudi Arabia. J Neuroinflammation. 2011;8:142. doi:10.1186/1742-2094-8-142.

Enstrom AM, Onore CE, Van de Water JA, Ashwood P. Differential monocyte responses to TLR ligands in children with autism spectrum disorders. Brain Behav Immun. 2010;24(1):64–71. doi:10.1016/j.bbi.2009.08.001.

Erhardt S, Lim CK, Linderholm KR, Janelidze S, Lindqvist D, Samuelsson M, et al. Connecting inflammation with glutamate agonism in suicidality. Neuropsychopharmacology. 2013;38(5):743–52. doi:10.1038/npp.2012.248.

Esposito P, Gheorghe D, Kandere K, Pang X, Connolly R, Jacobson S, et al. Acute stress increases permeability of the blood–brain-barrier through activation of brain mast cells. Brain Res. 2001;888(1):117–27.

Esposito P, Chandler N, Kandere K, Basu S, Jacobson S, Connolly R, et al. Corticotropin-releasing hormone and brain mast cells regulate blood–brain-barrier permeability induced by acute stress. J Pharmacol Exp Ther. 2002;303(3):1061–6. doi:10.1124/jpet.102.038497.

Esposito P, Basu S, Letourneau R, Jacobson S, Theoharides TC. Corticotropin-releasing factor (CRF) can directly affect brain microvessel endothelial cells. Brain Res. 2003;968(2):192–8.
Falcone T. Biomarkers in youth with major depressive disorder. Neurosci Pathways. 2012;1:28–9.
Falcone T. Increased suicidality among children and adolescents with epilepsy and depression. Neuropsychiatry Rev. 2008;19(2).
Falcone T, Janigro D. Inflammation in children with epilepsy may be a key to psychosis. Neuropsychiatry Rev. 2008;9(2).
Falcone T, Carlton E, Tuxhorn I, Janigro D. Psychosis and epilepsy in childhood how strong are the ties. Epilepsia. 2007;48(S6):60.
Falcone T, Carlton E, Franco K, Simon B, Janigro D. 259 – Monocytosis in a group of children with first episode psychosis. Schizophr Res. 2008a;98:140.
Falcone T, Janigro D, Franco K, Fattal O. Inflammation, psychosis and the brain. What do we really know? Schizophr Res. 2008b;98:30.
Falcone T, Janigro D, Lovell R, Simon B, Brown CA, Herrera M, Mynt AM, Anand A, S100B Blood Levels and Childhood Trauma in Adolescent Inpatients, *Journal of Psychiatric Research* (2015), doi: 10.1016/ j.jpsychires.2014.12.002.
Falcone T, Fazio V, Lee C, Simon B, Franco K, Marchi N, et al. Serum S100B: a potential biomarker for suicidality in adolescents? PLoS One. 2010a;5(6):e11089. doi:10.1371/journal.pone.0011089.
Falcone T, Mishra L, Carlton E, Butler RS, Simon B, Janigro D, et al. Suicidal behavior in children and adolescents with first episode psychosis. Clin Schizophr Relat Psychoses. 2010b;4(1):34–40. doi:10.3371/CSRP.4.1.2.
Falcone T, Mishra L, Carlton E, Lee C, Butler R, Janigro D, et al. Suicidal behavior in adolescents with first-episode psychosis. Clin Schizophr Relat Psychoses. 2010c;4(1):34–40.
Falcone T, Carlton E, Lee C, Janigro M, Fazio V, Forcen FE, et al. Does systemic inflammation play a role in pediatric psychosis? Clin Schizophr Relat Psychoses. 2013;14:1–43. doi:10.3371/CSRP.FACA.030813.
Fellerhoff B, Laumbacher B, Mueller N, Gu S, Wank R. Associations between Chlamydophila infections, schizophrenia and risk of HLA-A10. Mol Psychiatry. 2007;12(3):264–72.
Figueiredo CA, Alcantara-Neves NM, Veiga R, Amorim LD, Dattoli V, Mendonca LR, et al. Spontaneous cytokine production in children according to biological characteristics and environmental exposures. Environ Health Perspect. 2009;117(5):845–9. doi:10.1289/ehp.0800366.
Fuller-Thomson E, Stefanyk M, Brennenstuhl S. The robust association between childhood physical abuse and osteoarthritis in adulthood: findings from a representative community sample. Arthritis Rheum. 2009;61(11):1554–62. doi:10.1002/art.24871.
Fuller-Thomson E, Bejan R, Hunter JT, Grundland T, Brennenstuhl S. The link between childhood sexual abuse and myocardial infarction in a population-based study. Child Abuse Negl. 2012;36(9):656–65. doi:10.1016/j.chiabu.2012.06.001.
Gabbay V, Klein RG, Guttman LE, Babb JS, Alonso CM, Nishawala M, Katz Y, Gaite MR, Gonzalez CJ. A preliminary study of cytokines in suicidal and nonsuicidal adolescents with major depression. J Child Adolesc Psychopharmacol. 2009a;19(4):423–30.
Gabbay V, Klein RG, Guttman LE, Babb JS, Alonso CM, Nishawala M, et al. A preliminary study of cytokines in suicidal and nonsuicidal adolescents with major depression. J Child Adolesc Psychopharmacol. 2009b;19(4):423–30. doi:10.1089/cap.2008.0140.
Gabbay V, Coffey BJ, Guttman LE, Gottlieb L, Katz Y, Babb JS, et al. A cytokine study in children and adolescents with Tourette's disorder. Prog Neuropsychopharmacol Biol Psychiatry. 2009c;33(6):967–71. doi:10.1016/j.pnpbp.2009.05.001.
Gaffin JM, Phipatanakul W. The role of indoor allergens in the development of asthma. Curr Opin Allergy Clin Immunol. 2009;9(2):128–35.
Ganguli R, Brar JS, Chengappa KN, Yang ZW, Nimgaonkar VL, Rabin BS. Autoimmunity in schizophrenia: a review of recent findings. Ann Med. 1993;25(5):489–96.
Garner AS, Shonkoff JP, Committee on Psychosocial Aspects of Child and Family Health, Committee on Early Childhood, Adoption, and Dependent Care, Section on Developmental

and Behavioral Pediatrics. Early childhood adversity, toxic stress, and the role of the pediatrician: translating developmental science into lifelong health. Pediatrics. 2012;129(1):e224–31. doi:10.1542/peds.2011-2662.
Gill SR, Pop M, Deboy RT, Eckburg PB, Turnbaugh PJ, Samuel BS, et al. Metagenomic analysis of the human distal gut microbiome. Science. 2006;312(5778):1355–59. pii: 312/5778/1355.
Gilmore JH, Jarskog LF, Vadlamudi S, Lauder JM. Prenatal infection and risk for schizophrenia: IL-Ibeta, IL-6, and TNFalpha inhibit cortical neuron dendrite development. Neuropsychopharmacology. 2004;29(7):1221–9.
Ginsberg MR, Rubin RA, Falcone T, Ting AH, Natowicz MR. Brain transcriptional and epigenetic associations with autism. PLoS One. 2012;7(9):e44736. doi:10.1371/journal.pone.0044736.
Goenjian AK, Yehuda R, Pynoos RS, Steinberg AM, Tashjian M, Yang RK, et al. Basal cortisol, dexamethasone suppression of cortisol, and MHPG in adolescents after the 1988 earthquake in Armenia. Am J Psychiatry. 1996;153(7):929–34.
Gogtay N, Rapoport J. Childhood onset schizophrenia and other early-onset psychotic disorders. In: Martin A, Volkmar FR, editors. Lewis's child and adolescent psychiatry, a comprehensive textbook. 4th ed. Colombia: Lippincott Williams & Wilkins; 2007. p. 493.
Goldstein BI, Young LT. Toward clinically applicable biomarkers in bipolar disorder: focus on BDNF, inflammatory markers, and endothelial function. Curr Psychiatry Rep. 2013;15(12):425. doi:10.1007/s11920-013-0425-9.
Gottschalk MG, Schwarz E, Bahn S. Biomarker research in neuropsychiatry: challenges and potential. Fortschr Neurol Psychiatr. 2013;81(5):243–9. doi:10.1055/s-0033-1335235.
Granata T, Cross H, Theodore W, Avanzini G. Immune-mediated epilepsies. Epilepsia. 2011;52(3): 5–11.
Greeson JK, Briggs EC, Kisiel CL, Layne CM, Ake 3rd GS, Ko SJ, et al. Complex trauma and mental health in children and adolescents placed in foster care: findings from the National Child Traumatic Stress Network. Child Welfare. 2011;90(6):91–108.
Gronlund MM, Arvilommi H, Kero P, Lehtonen OP, Isolauri E. Importance of intestinal colonisation in the maturation of humoral immunity in early infancy: a prospective follow up study of healthy infants aged 0–6 months. Arch Dis Child Fetal Neonatal Ed. 2000;83(3):F186–92.
Gunnar MR, Morison SJ, Chisholm K, Schuder M. Salivary cortisol levels in children adopted from Romanian orphanages. Dev Psychopathol. 2001;13(3):611–28.
Hagberg H, Gressens P, Mallard C. Inflammation during fetal and neonatal life: implications for neurological and neuropsychiatric disease in children and adults. Ann Neurol. 2012;71:444–57.
Hanson DR, Gottesman II. Theories of schizophrenia: a genetic-inflammatory-vascular synthesis. BMC Med Genet. 2005;6:7.
Harrison NA, Brydon L, Walker C, Gray MA, Steptoe A, Critchley HD. Inflammation causes mood changes through alterations in subgenual cingulate activity and mesolimbic connectivity. Biol Psychiatry. 2009;66(5):407–14. doi:10.1016/j.biopsych.2009.03.015.
Henderson DC. Schizophrenia and comorbid metabolic disorders. J Clin Psychiatry. 2005;66 Suppl 6:11–20.
Henje Blom E, Lekander M, Ingvar M, Asberg M, Mobarrez F, Serlachius E. Pro-inflammatory cytokines are elevated in adolescent females with emotional disorders not treated with SSRIs. J Affect Disord. 2012;136(3):716–23. doi:10.1016/j.jad.2011.10.002.
Herberth G, Weber A, Roder S, Elvers HD, Kramer U, Schins RP, et al. Relation between stressful life events, neuropeptides and cytokines: results from the LISA birth cohort study. Pediatr Allergy Immunol. 2008;19(8):722–9. doi:10.1111/j.1399-3038.2008.00727.x.
Hertz-Picciotto I, Park HY, Dostal M, Kocan A, Trnovec T, Sram R. Prenatal exposures to persistent and non-persistent organic compounds and effects on immune system development. Basic Clin Pharmacol Toxicol. 2008;102(2):146–54. doi:10.1111/j.1742-7843.2007.00190.x.
Hirsch S. Glial changes measured by [11C](R)-PK11195 PET in patients with psychosis and cognitive decline are associated with impaired event potential mismatch negativity. Schizophr Res. 2004;67 Suppl 1:103.

Holtmann M, Poustka L, Zepf FD, Banaschewski T, Priller J, Bolte S, et al. Severe affective and behavioral dysregulation in youths is associated with a proinflammatory state. Z Kinder Jugendpsychiatr Psychother. 2013;41(6):393–9. doi:10.1024/1422-4917/a000255.

Hughes MM, Carballedo A, McLoughlin DM, Amico F, Harkin A, Frodl T, et al. Tryptophan depletion in depressed patients occurs independent of kynurenine pathway activation. Brain Behav Immun. 2012;26(6):979–87. doi:10.1016/j.bbi.2012.05.010.

Hurtado A, Johnson RJ. Hygiene hypothesis and prevalence of glomerulonephritis. Kidney Int Suppl. 2005;(97):S62–7. doi:10.1111/j.1523-1755.2005.09711.x.

Insel TR. The case for preemption. Director's Blog, 16 January 2013.

Insel TR, Charney DS. Research on major depression: strategies and priorities. JAMA. 2003;289(23):3167–8. doi:10.1001/jama.289.23.3167.

Insel TR, Landis SC, Collins FS. Research priorities. The NIH BRAIN initiative. Science. 2013;340(6133):687–8. doi:10.1126/science.1239276.

Isung J, Mobarrez F, Nordstrom P, Asberg M, Jokinen J. Low plasma vascular endothelial growth factor (VEGF) associated with completed suicide. World J Biol Psychiatry. 2012a;13(6): 468–73. doi:10.3109/15622975.2011.624549.

Isung J, Aeinehband S, Mobarrez F, Martensson B, Nordstrom P, Asberg M, et al. Low vascular endothelial growth factor and interleukin-8 in cerebrospinal fluid of suicide attempters. Transl Psychiatry. 2012b;2:e196. doi:10.1038/tp.2012.123.

Janelidze S, Mattei D, Westrin A, Traskman-Bendz L, Brundin L. Cytokine levels in the blood may distinguish suicide attempters from depressed patients. Brain Behav Immun. 2011;25(2): 335–9. doi:10.1016/j.bbi.2010.10.010.

Janicki-Deverts D, Cohen S, Doyle WJ. Cynical hostility and stimulated Th1 and Th2 cytokine production. Brain Behav Immun. 2010;24(1):58–63. doi:10.1016/j.bbi.2009.07.009.

Janigro D. Blood–brain barrier, ion homeostatis and epilepsy: possible implications towards the understanding of ketogenic diet mechanisms. Epilepsy Res. 1999;37(3):223–32.

Janigro D. Are you in or out? Leukocyte, ion, and neurotransmitter permeability across the epileptic blood–brain barrier. Epilepsia. 2012;53 Suppl 1:26–34. doi:10.1111/j.1528-1167.2012.03472.x.

Johnson CL, Versalovic J. The human microbiome and its potential importance to pediatrics. Pediatrics. 2012;129(5):950–60. doi:10.1542/peds.2011-2736.

Johnson JG, Cohen P, Gould MS, Kasen S, Brown J, Brook JS. Childhood adversities, interpersonal difficulties, and risk for suicide attempts during late adolescence and early adulthood. Arch Gen Psychiatry. 2002;59(8):741–9.

Johnson SB, Riley AW, Granger DA, Riis J. The science of early life toxic stress for pediatric practice and advocacy. Pediatrics. 2013;131(2):319–27. doi:10.1542/peds.2012-0469.

Jumper SA. A meta-analysis of the relationship of child sexual abuse to adult psychological adjustment. Child Abuse Negl. 1995;19(6):715–28.

Jyonouchi H, Geng L, Ruby A, Reddy C, Zimmerman-Bier B. Evaluation of an association between gastrointestinal symptoms and cytokine production against common dietary proteins in children with autism spectrum disorders. J Pediatr. 2005a;146(5):605–10. doi:10.1016/j.jpeds.2005.01.027.

Jyonouchi H, Geng L, Ruby A, Zimmerman-Bier B. Dysregulated innate immune responses in young children with autism spectrum disorders: their relationship to gastrointestinal symptoms and dietary intervention. Neuropsychobiology. 2005b;51(2):77–85. doi:10.1159/000084164.

Jyonouchi H, Geng L, Cushing-Ruby A, Quraishi H. Impact of innate immunity in a subset of children with autism spectrum disorders: a case control study. J Neuroinflammation. 2008;5:52. doi:10.1186/1742-2094-5-52.

Jyonouchi H, Geng L, Streck DL, Toruner GA. Children with autism spectrum disorders (ASD) who exhibit chronic gastrointestinal (GI) symptoms and marked fluctuation of behavioral symptoms exhibit distinct innate immune abnormalities and transcriptional profiles of peripheral blood (PB) monocytes. J Neuroimmunol. 2011;238(1–2):73–80. doi:10.1016/j.jneuroim.2011.07.001.

Jyonouchi H, Geng L, Streck DL, Toruner GA. Immunological characterization and transcription profiling of peripheral blood (PB) monocytes in children with autism spectrum disorders

(ASD) and specific polysaccharide antibody deficiency (SPAD): case study. J Neuroinflammation. 2012;9:4. doi:10.1186/1742-2094-9-4.

Kapural M, Krizanac-Bengez L, Barnett G, Perl J, Masaryk T, Apollo D, et al. Serum S-100beta as a possible marker of blood–brain barrier disruption. Brain Res. 2002;940(1–2):102–4.

Karege F, Vaudan G, Schwald M, Perroud N, La Harpe R. Neurotrophin levels in postmortem brains of suicide victims and the effects of antemortem diagnosis and psychotropic drugs. Brain Res Mol Brain Res. 2005;136(1–2):29–37. doi:10.1016/j.molbrainres.2004.12.020.

Khandaker GM, Zammit S, Lewis G, Jones PB. A population-based study of atopic disorders and inflammatory markers in childhood before psychotic experiences in adolescence. Schizophr Res. 2014;152(1):139–45. doi:10.1016/j.schres.2013.09.021.

Kim YK, Lee HP, Won SD, Park EY, Lee HY, Lee BH, et al. Low plasma BDNF is associated with suicidal behavior in major depression. Prog Neuropsychopharmacol Biol Psychiatry. 2007;31(1):78–85. doi:10.1016/j.pnpbp.2006.06.024.

King JA, Mandansky D, King S, Fletcher KE, Brewer J. Early sexual abuse and low cortisol. Psychiatry Clin Neurosci. 2001;55(1):71–4. doi:10.1046/j.1440-1819.2001.00787.x.

Kubera M, Symbirtsev A, Basta-Kaim A, Borycz J, Roman A, Papp M, et al. Effect of chronic treatment with imipramine on interleukin 1 and interleukin 2 production by splenocytes obtained from rats subjected to a chronic mild stress model of depression. Pol J Pharmacol. 1996;48(5):503–6.

Kumar B, Prakash A, Sewal RK, Medhi B, Modi M. Drug therapy in autism: a present and future perspective. Pharmacol Rep. 2012;64(6):1291–304.

Larkin GL, Beautrais AL. A preliminary naturalistic study of low-dose ketamine for depression and suicide ideation in the emergency department. Int J Neuropsychopharmacol. 2011;14(8):1127–31. doi:10.1017/S1461145711000629.

Lasky-Su J, Neale BM, Franke B, Anney RJ, Zhou K, Maller JB, et al. Genome-wide association scan of quantitative traits for attention deficit hyperactivity disorder identifies novel associations and confirms candidate gene associations. Am J Med Genet B Neuropsychiatr Genet. 2008;147B(8):1345–54. doi:10.1002/ajmg.b.30867.

Leckman JF, Katsovich L, Kawikova I, Lin H, Zhang H, Kronig H, et al. Increased serum levels of interleukin-12 and tumor necrosis factor-alpha in Tourette's syndrome. Biol Psychiatry. 2005;57(6):667–73. doi:10.1016/j.biopsych.2004.12.004.

Ledgerwood LG, Ewald PW, Cochran GM. Genes, germs, and schizophrenia: an evolutionary perspective. Perspect Biol Med. 2003;46(3):317–48.

Lee BH, Kim YK. BDNF mRNA expression of peripheral blood mononuclear cells was decreased in depressive patients who had or had not recently attempted suicide. J Affect Disord. 2010;125(1–3):369–73. doi:10.1016/j.jad.2010.01.074.

Lewitus GM, Cohen H, Schwartz M. Reducing post-traumatic anxiety by immunization. Brain Behav Immun. 2008;22(7):1108–14. doi:10.1016/j.bbi.2008.05.002.

Li X, Chauhan A, Sheikh AM, Patil S, Chauhan V, Li XM, et al. Elevated immune response in the brain of autistic patients. J Neuroimmunol. 2009;207(1–2):111–6. doi:10.1016/j.jneuroim.2008.12.002.

Li Z, Qi D, Chen J, Zhang C, Yi Z, Yuan C, et al. Venlafaxine inhibits the upregulation of plasma tumor necrosis factor-alpha (TNF-alpha) in the Chinese patients with major depressive disorder: a prospective longitudinal study. Psychoneuroendocrinology. 2013;38(1):107–14. doi:10.1016/j.psyneuen.2012.05.005.

Lindqvist D, Isaksson A, Traskman-Bendz L, Brundin L. Salivary cortisol and suicidal behavior—a follow-up study. Psychoneuroendocrinology. 2008;33(8):1061–8.

Lindqvist D, Janelidze S, Erhardt S, Traskman-Bendz L, Engstrom G, Brundin L. CSF biomarkers in suicide attempters—a principal component analysis. Acta Psychiatr Scand. 2011;124(1):52–61. doi:10.1111/j.1600-0447.2010.01655.x.

Liu S, Yi M, Qi F, Che F, Ma X. Lack of association between polymorphism -592A/C in the promoter region of the IL10 gene and Tourette's syndrome in a family-based association study in the Chinese Han population. Genet Test Mol Biomarkers. 2011a;15(10):733–5. doi:10.1089/gtmb.2010.0272.

Liu S, Yi M, Wang M, Sun Y, Che F, Ma X. Association of IL8–251A/T, IL12B–1188A/C and TNF-alpha -238A/G polymorphisms with Tourette syndrome in a family-based association study in a Chinese Han population. Neurosci Lett. 2011b;495(2):155–8. doi:10.1016/j.neulet.2011.03.060.

Lopresti AL, Maker GL, Hood SD, Drummond PD. A review of peripheral biomarkers in major depression: the potential of inflammatory and oxidative stress biomarkers. Prog Neuropsychopharmacol Biol Psychiatry. 2014;48:102–11. doi:10.1016/j.pnpbp.2013.09.017.

Lucas M, O'Reilly EJ, Mirzaei F, Okereke OI, Unger L, Miller M, et al. Cigarette smoking and completed suicide: results from 3 prospective cohorts of American adults. J Affect Disord. 2013;151(3):1053–8. doi:10.1016/j.jad.2013.08.033.

Luo F, Leckman JF, Katsovich L, Findley D, Grantz H, Tucker DM, et al. Prospective longitudinal study of children with tic disorders and/or obsessive-compulsive disorder: relationship of symptom exacerbations to newly acquired streptococcal infections. Pediatrics. 2004;113(6):e578–85.

Maes M. Cytokines in schizophrenia. Biol Psychiatry. 1997;42(4):308–9.

Maes M. Interleukin-2 and schizophrenia. Psychiatry Res. 1998;77(1):63–4.

Maes M, Bosmans E, Suy E, Vandervorst C, Dejonckheere C, Raus J. Antiphospholipid, antinuclear, Epstein–Barr and cytomegalovirus antibodies, and soluble interleukin-2 receptors in depressive patients. J Affect Disord. 1991;21(2):133–40.

Maino K, Gruber R, Riedel M, Seitz N, Schwarz M, Muller N. T- and B-lymphocytes in patients with schizophrenia in acute psychotic episode and the course of the treatment. Psychiatry Res. 2007;152(2–3):173–80.

Manzardo AM, Henkhaus R, Dhillon S, Butler MG. Plasma cytokine levels in children with autistic disorder and unrelated siblings. Int J Dev Neurosci. 2012;30(2):121–7. doi:10.1016/j.ijdevneu.2011.12.003.

Marchi N, Fazio V, Cucullo L, Kight K, Masaryk T, Barnett G, et al. Serum transthyretin monomer as a possible marker of blood-to-CSF barrier disruption. J Neurosci. 2003;23(5):1949–55.

Marchi N, Cavaglia M, Fazio V, Bhudia S, Hallene K, Janigro D. Peripheral markers of blood–brain barrier damage. Clin Chim Acta. 2004;342(1–2):1–12. doi:10.1016/j.cccn.2003.12.008.

Marchi N, Fan Q, Ghosh C, Fazio V, Bertolini F, Betto G, et al. Antagonism of peripheral inflammation reduces the severity of status epilepticus. Neurobiol Dis. 2009;33(2):171–81. doi:10.1016/j.nbd.2008.10.002.

Marchi N, Granata T, Freri E, Ciusani E, Ragona F, Puvenna V, et al. Efficacy of anti-inflammatory therapy in a model of acute seizures and in a population of pediatric drug resistant epileptics. PLoS One. 2011;6(3):e18200.

Marin TJ, Martin TM, Blackwell E, Stetler C, Miller GE. Differentiating the impact of episodic and chronic stressors on hypothalamic–pituitary–adrenocortical axis regulation in young women. Health Psychol. 2007;26(4):447–55. doi:10.1037/0278-6133.26.4.447.

Matz J, Krause DL, Dehning S, Riedel M, Gruber R, Schwarz MJ, et al. Altered monocyte activation markers in Tourette's syndrome: a case–control study. BMC Psychiatry. 2012;12:29. doi:10.1186/1471-244X-12-29.

McCrory E, De Brito SA, Viding E. Research review: the neurobiology and genetics of maltreatment and adversity. J Child Psychol Psychiatry. 2010;51(10):1079–95. doi:10.1111/j.1469-7610.2010.02271.x.

McGorry P, Nordentoft M, Simonsen E. Introduction to 'Early psychosis: a bridge to the future'. Br J Psychiatry. 2005;187(48):s1–3. doi:10.1192/bjp.187.48.s1.

McGowan PO, Sasaki A, D'Alessio AC, Dymov S, Labonte B, Szyf M, et al. Epigenetic regulation of the glucocorticoid receptor in human brain associates with childhood abuse. Nat Neurosci. 2009;12(3):342–8. doi:10.1038/nn.2270.

Meyer U, Feldon J, Schedlowski M, Yee BK. Towards an immuno-precipitated neurodevelopmental animal model of schizophrenia. Neurosci Biobehav Rev. 2005;29(6):913–47.

Meyer U, Nyffeler M, Engler A, Urwyler A, Schedlowski M, Knuesel I, et al. The time of prenatal immune challenge determines the specificity of inflammation-mediated brain and behavioral pathology. J Neurosci. 2006a;26(18):4752–62.

Meyer U, Feldon J, Schedlowski M, Yee BK. Immunological stress at the maternal-foetal interface: a link between neurodevelopment and adult psychopathology. Brain Behav Immun. 2006b;20(4):378–88.

Meyer U, Nyffeler M, Yee BK, Knuesel I, Feldon J. Adult brain and behavioral pathological markers of prenatal immune challenge during early/middle and late fetal development in mice. Brain Behav Immun. 2008a;22(4):469–86.

Meyer U, Nyffeler M, Schwendener S, Knuesel I, Yee BK, Feldon J. Relative prenatal and postnatal maternal contributions to schizophrenia-related neurochemical dysfunction after in utero immune challenge. Neuropsychopharmacology. 2008b;33(2):441–56.

Miller G, Chen E. Unfavorable socioeconomic conditions in early life presage expression of proinflammatory phenotype in adolescence. Psychosom Med. 2007;69(5):402–9. doi:10.1097/PSY.0b013e318068fcf9.

Miller GE, Cole SW. Clustering of depression and inflammation in adolescents previously exposed to childhood adversity. Biol Psychiatry. 2012;72(1):34–40. doi:10.1016/j.biopsych.2012.02.034.

Miller GE, Freedland KE, Carney RM, Stetler CA, Banks WA. Pathways linking depression, adiposity, and inflammatory markers in healthy young adults. Brain Behav Immun. 2003;17(4):276–85.

Miller GE, Rohleder N, Cole SW. Chronic interpersonal stress predicts activation of pro- and anti-inflammatory signaling pathways 6 months later. Psychosom Med. 2009a;71(1):57–62. doi:10.1097/PSY.0b013e318190d7de.

Miller JM, Kinnally EL, Ogden RT, Oquendo MA, Mann JJ, Parsey RV. Reported childhood abuse is associated with low serotonin transporter binding in vivo in major depressive disorder. Synapse. 2009b;63(7):565–73. doi:10.1002/syn.20637.

Miller AH, Maletic V, Raison CL. Inflammation and its discontents: the role of cytokines in the pathophysiology of major depression. Biol Psychiatry. 2009c;65(9):732–41. doi:10.1016/j.biopsych.2008.11.029.

Mills NT, Scott JG, Wray NR, Cohen-Woods S, Baune BT. Research review: the role of cytokines in depression in adolescents: a systematic review. J Child Psychol Psychiatry. 2013;54(8):816–35.

Misener VL, Schachar R, Ickowicz A, Malone M, Roberts W, Tannock R, et al. Replication test for association of the IL-1 receptor antagonist gene, IL1RN, with attention-deficit/hyperactivity disorder. Neuropsychobiology. 2004;50(3):231–4. doi:10.1159/000079976.

Mitchell RH, Goldstein BI. Inflammation in children and adolescents with neuropsychiatric disorders: a systematic review. J Am Acad Child Adolesc Psychiatry. 2014;53(3):274–96. doi:10.1016/j.jaac.2013.11.013.

Mittleman BB, Castellanos FX, Jacobsen LK, Rapoport JL, Swedo SE, Shearer GM. Cerebrospinal fluid cytokines in pediatric neuropsychiatric disease. J Immunol. 1997;159(6):2994–9.

Miura H, Ozaki N, Sawada M, Isobe K, Ohta T, Nagatsu T. A link between stress and depression: shifts in the balance between the kynurenine and serotonin pathways of tryptophan metabolism and the etiology and pathophysiology of depression. Stress. 2008;11(3):198–209. doi:10.1080/10253890701754068.

Molloy CA, Morrow AL, Meinzen-Derr J, Schleifer K, Dienger K, Manning-Courtney P, et al. Elevated cytokine levels in children with autism spectrum disorder. J Neuroimmunol. 2006;172(1–2):198–205. doi:10.1016/j.jneuroim.2005.11.007.

Motivala SJ, Sarfatti A, Olmos L, Irwin MR. Inflammatory markers and sleep disturbance in major depression. Psychosom Med. 2005;67(2):187–94. doi:10.1097/01.psy.0000149259.72488.09.

Muller N, Riedel M, Scheppach C, Brandstatter B, Sokullu S, Krampe K, et al. Beneficial antipsychotic effects of celecoxib add-on therapy compared to risperidone alone in schizophrenia. Am J Psychiatry. 2002;159(6):1029–34.

Muller N, Riedel M, Schwarz MJ, Engel RR. Clinical effects of COX-2 inhibitors on cognition in schizophrenia. Eur Arch Psychiatry Clin Neurosci. 2005;255(2):149–51.

Murphy TK, Storch EA, Lewin AB, Edge PJ, Goodman WK. Clinical factors associated with pediatric autoimmune neuropsychiatric disorders associated with streptococcal infections. J Pediatr. 2012;160(2):314–9. doi:10.1016/j.jpeds.2011.07.012.

Najjar S, Pearlman DM, Devinsky O, Najjar A, Zagzag D. Neurovascular unit dysfunction with blood–brain barrier hyperpermeability contributes to major depressive disorder: a review of clinical and experimental evidence. J Neuroinflammation. 2013;10:142. doi:10.1186/1742-2094-10-142.

Nawa H, Takei N. Recent progress in animal modeling of immune inflammatory processes in schizophrenia: implication of specific cytokines. Neurosci Res. 2006;56(1):2–13.

Negele K, Heinrich J, Borte M, von Berg A, Schaaf B, Lehmann I, et al. Mode of delivery and development of atopic disease during the first 2 years of life. Pediatr Allergy Immunol. 2004;15(1):48–54.

Neuwelt EA, Bauer B, Fahlke C, Fricker G, Iadecola C, Janigro D, et al. Engaging neuroscience to advance translational research in brain barrier biology. Nat Rev Neurosci. 2011;12(3): 169–82.

Nikkila HV, Muller K, Ahokas A, Miettinen K, Rimon R, Andersson LC. Accumulation of macrophages in the CSF of schizophrenic patients during acute psychotic episodes. Am J Psychiatry. 1999;156(11):1725–9.

Nock MK. Suicidal behavior among adolescents: correlates, confounds, and (the search for) causal mechanisms. J Am Acad Child Adolesc Psychiatry. 2009;48(3):237–9. doi:10.1097/CHI.0b013e318196b944.

Nock MK, Green JG, Hwang I, McLaughlin KA, Sampson NA, Zaslavsky AM, et al. Prevalence, correlates, and treatment of lifetime suicidal behavior among adolescents: results from the National Comorbidity Survey Replication Adolescent Supplement. JAMA Psychiatry. 2013;70(3):300–10. doi:10.1001/2013.jamapsychiatry.55.

Oades RD. An exploration of the associations of pregnancy and perinatal features with cytokines and tryptophan/kynurenine metabolism in children with attention-deficit hyperactivity disorder (ADHD). Atten Defic Hyperact Disord. 2011;3(4):301–18. doi:10.1007/s12402-011-0062-2.

Oades RD, Dauvermann MR, Schimmelmann BG, Schwarz MJ, Myint AM. Attention-deficit hyperactivity disorder (ADHD) and glial integrity: S100B, cytokines and kynurenine metabolism—effects of medication. Behav Brain Funct. 2010a;6:29. doi:10.1186/1744-9081-6-29.

Oades RD, Myint AM, Dauvermann MR, Schimmelmann BG, Schwarz MJ. Attention-deficit hyperactivity disorder (ADHD) and glial integrity: an exploration of associations of cytokines and kynurenine metabolites with symptoms and attention. Behav Brain Funct. 2010b;6:32. doi:10.1186/1744-9081-6-32.

Oberlander TF, Weinberg J, Papsdorf M, Grunau R, Misri S, Devlin AM. Prenatal exposure to maternal depression, neonatal methylation of human glucocorticoid receptor gene (NR3C1) and infant cortisol stress responses. Epigenetics. 2008;3(2):97–106.

Oby E, Janigro D. The blood–brain barrier and epilepsy. Epilepsia. 2006;47(11):1761–74.

Onore C, Enstrom A, Krakowiak P, Hertz-Picciotto I, Hansen R, Van de Water J, et al. Decreased cellular IL-23 but not IL-17 production in children with autism spectrum disorders. J Neuroimmunol. 2009;216(1–2):126–9. doi:10.1016/j.jneuroim.2009.09.005.

Padmos RC, Hillegers MH, Knijff EM, Vonk R, Bouvy A, Staal FJ, et al. A discriminating messenger RNA signature for bipolar disorder formed by an aberrant expression of inflammatory genes in monocytes. Arch Gen Psychiatry. 2008;65(4):395–407.

Padmos RC, Van Baal GC, Vonk R, Wijkhuijs AJ, Kahn RS, Nolen WA, et al. Genetic and environmental influences on pro-inflammatory monocytes in bipolar disorder: a twin study. Arch Gen Psychiatry. 2009;66(9):957–65.

Pandey GN, Dwivedi Y. Neurobiology of teenage suicide. In: Dwivedi Y, editor. The neurobiological basis of suicide. Boca Raton: Taylor & Francis Group, LLC; 2012.

Pandey GN, Rizavi HS, Ren X, Fareed J, Hoppensteadt DA, Roberts RC, et al. Proinflammatory cytokines in the prefrontal cortex of teenage suicide victims. J Psychiatr Res. 2012;46(1):57–63. doi:10.1016/j.jpsychires.2011.08.006.

Pardo CA, Vargas DL, Zimmerman AW. Immunity, neuroglia and neuroinflammation in autism. Int Rev Psychiatry. 2005;17(6):485–95. doi:10.1080/02646830500381930.

Patterson PH. Maternal infection: window on neuroimmune interactions in fetal brain development and mental illness. Curr Opin Neurobiol. 2002;12(1):115–8.

Patterson PH. Neuroscience. Maternal effects on schizophrenia risk. Science. 2007;318(5850): 576–7.

Perlmutter SJ, Leitman SF, Garvey MA, Hamburger S, Feldman E, Leonard HL, et al. Therapeutic plasma exchange and intravenous immunoglobulin for obsessive-compulsive disorder and tic disorders in childhood. Lancet. 1999;354(9185):1153–8. doi:10.1016/S0140-6736(98)12297-3.

Pervanidou P, Kolaitis G, Charitaki S, Margeli A, Ferentinos S, Bakoula C, et al. Elevated morning serum interleukin (IL)-6 or evening salivary cortisol concentrations predict posttraumatic stress disorder in children and adolescents six months after a motor vehicle accident. Psychoneuroendocrinology. 2007;32(8–10):991–9. pii: S0306-4530(07)00177-1.

Phillips LJ, McGorry PD, Yung AR, McGlashan TH, Cornblatt B, Klosterkotter J. Prepsychotic phase of schizophrenia and related disorders: recent progress and future opportunities. Br J Psychiatry. 2005;187(48):s33–44. doi:10.1192/bjp.187.48.s33.

Pittenger C, Duman RS. Stress, depression, and neuroplasticity: a convergence of mechanisms. Neuropsychopharmacology. 2008;33(1):88–109.

Piwoz E, Sundberg S, Rooke J. Promoting healthy growth: what are the priorities for research and action? Adv Nutr. 2012;3(2):234–41. doi:10.3945/an.111.001164.

Pompili M, Girardi P, Ruberto A, Tatarelli R. Suicide in the epilepsies: a meta-analytic investigation of 29 cohorts. Epilepsy Behav. 2005;7(2):305–10.

Potvin S, Stip E, Sepehry AA, Gendron A, Bah R, Kouassi E. Inflammatory cytokine alterations in schizophrenia: a systematic quantitative review. Biol Psychiatry. 2008;63(8):801–8. doi:10.1016/j.biopsych.2007.09.024.

Price RB, Nock MK, Charney DS, Mathew SJ. Effects of intravenous ketamine on explicit and implicit measures of suicidality in treatment-resistant depression. Biol Psychiatry. 2009;66(5):522–6. doi:10.1016/j.biopsych.2009.04.029.

Puxley F, Midtsund M, Iosif A, Lask B. PANDAS anorexia nervosa—endangered, extinct or nonexistent? Int J Eat Disord. 2008;41(1):15–21. doi:10.1002/eat.20462.

Radewicz K, Garey LJ, Gentleman SM, Reynolds R. Increase in HLA-DR immunoreactive microglia in frontal and temporal cortex of chronic schizophrenics. J Neuropathol Exp Neurol. 2000;59(2):137–50.

Raison CL, Capuron L, Miller AH. Cytokines sing the blues: inflammation and the pathogenesis of depression. Trends Immunol. 2006;27(1):24–31.

Rapaport MH, Delrahim KK, Bresee CJ, Maddux RE, Ahmadpour O, Dolnak D. Celecoxib augmentation of continuously ill patients with schizophrenia. Biol Psychiatry. 2005;57(12): 1594–6.

Rapoport SI, Hori M, Klatzo I. Testing of a hypothesis for osmotic opening of the blood–brain barrier. Am J Physiol. 1972;223(2):323–31.

Reiber H. Dynamics of brain-derived proteins in cerebrospinal fluid. Clin Chim Acta. 2001;310(2):173–86.

Riedel M, Strassnig M, Schwarz MJ, Muller N. COX-2 inhibitors as adjunctive therapy in schizophrenia: rationale for use and evidence to date. CNS Drugs. 2005;19(10):805–19.

Romero E, Ali C, Molina-Holgado E, Castellano B, Guaza C, Borrell J. Neurobehavioral and immunological consequences of prenatal immune activation in rats. Influence of antipsychotics. Neuropsychopharmacology. 2007;32(8):1791–804.

Rosenberg GA. Matrix metalloproteinases and their multiple roles in neurodegenerative diseases. Lancet Neurol. 2009;8(2):205–16. doi:10.1016/S1474-4422(09)70016-X.

Rossignol DA, Rossignol LW, James SJ, Melnyk S, Mumper E. The effects of hyperbaric oxygen therapy on oxidative stress, inflammation, and symptoms in children with autism: an open-label pilot study. BMC Pediatr. 2007;7:36. doi:10.1186/1471-2431-7-36.

Rothermundt M, Arolt V. Schizophrenia and immunity. In: Ader R, editor. Psychoneuroimmunology. 4th ed. Boston: Elsevier; 2007. p. 563–77.

Rothermundt M, Arolt V, Fenker J, Gutbrodt H, Peters M, Kirchner H. Different immune patterns in melancholic and non-melancholic major depression. Eur Arch Psychiatry Clin Neurosci. 2001a;251(2):90–7.

Rothermundt M, Arolt V, Wiesmann M, Missler U, Peters M, Rudolf S, et al. S-100B is increased in melancholic but not in non-melancholic major depression. J Affect Disord. 2001b;66(1):89–93.

Rothermundt M, Arolt V, Peters M, Gutbrodt H, Fenker J, Kersting A, et al. Inflammatory markers in major depression and melancholia. J Affect Disord. 2001c;63(1–3):93–102.

Rothermundt M, Falkai P, Ponath G, Abel S, Burkle H, Diedrich M, et al. Glial cell dysfunction in schizophrenia indicated by increased S100B in the CSF. Mol Psychiatry. 2004;9(10):897–9.
Rowe J, Kusel M, Holt BJ, Suriyaarachchi D, Serralha M, Hollams E, et al. Prenatal versus postnatal sensitization to environmental allergens in a high-risk birth cohort. J Allergy Clin Immunol. 2007;119(5):1164–73. doi:10.1016/j.jaci.2007.02.016.
Schroeter ML, Steiner J. Elevated serum levels of the glial marker protein S100B are not specific for schizophrenia or mood disorders. Mol Psychiatry. 2009;14(3):235–7. doi:10.1038/mp.2008.85.
Schroeter ML, Sacher J, Steiner J, Schoenknecht P, Mueller K. Serum S100B represents a new biomarker for mood disorders. Curr Drug Targets. 2013;14(11):1237–48.
Segerstrom SC, Miller GE. Psychological stress and the human immune system: a meta-analytic study of 30 years of inquiry. Psychol Bull. 2004;130(4):601–30. doi:10.1037/0033-2909.130.4.601.
Segman RH, Meltzer A, Gross-Tsur V, Kosov A, Frisch A, Inbar E, et al. Preferential transmission of interleukin-1 receptor antagonist alleles in attention deficit hyperactivity disorder. Mol Psychiatry. 2002;7(1):72–4. doi:10.1038/sj/mp/4000919.
Serafini G, Pompili M, Elena Seretti M, Stefani H, Palermo M, Coryell W, et al. The role of inflammatory cytokines in suicidal behavior: a systematic review. Eur Neuropsychopharmacol. 2013a;23(12):1672–86. doi:10.1016/j.euroneuro.2013.06.002.
Serafini G, Pompili M, Lindqvist D, Dwivedi Y, Girardi P. The role of neuropeptides in suicidal behavior: a systematic review. Biomed Res Int. 2013;2013:687575. doi:10.1155/2013/687575.
Shalev H, Serlin Y, Friedman A. Breaching the blood–brain barrier as a gate to psychiatric disorder. Cardiovasc Psychiatry Neurol. 2009;2009:278531. doi:10.1155/2009/278531.
Sharma HS. Blood-CNS barrier, neurodegeneration and neuroprotection: recent therapeutic advancements and nano-drug delivery. J Neural Transm. 2011;118(1):3–6. doi:10.1007/s00702-010-0542-0.
Sharma HS, Cervos-Navarro J. Role of histamine in pathophysiology of heat stress in rats. Agents Actions Suppl. 1991;33:97–102.
Sharma HS, Dey PK. Impairment of blood–brain barrier (BBB) in rat by immobilization stress: role of serotonin (5-HT). Indian J Physiol Pharmacol. 1981;25(2):111–22.
Sharma HS, Dey PK. Role of 5-HT on increased permeability of blood–brain barrier under heat stress. Indian J Physiol Pharmacol. 1984;28(4):259–67.
Sharma HS, Dey PK. Probable involvement of 5-hydroxytryptamine in increased permeability of blood–brain barrier under heat stress in young rats. Neuropharmacology. 1986;25(2):161–7.
Sharma HS, Dey PK. EEG changes following increased blood–brain barrier permeability under long-term immobilization stress in young rats. Neurosci Res. 1988;5(3):224–39.
Sharma HS, Johanson CE. Blood-cerebrospinal fluid barrier in hyperthermia. Prog Brain Res. 2007;162:459–78. doi:10.1016/S0079-6123(06)62023-2.
Sharma HS, Olsson Y, Dey PK. Changes in blood–brain barrier and cerebral blood flow following elevation of circulating serotonin level in anesthetized rats. Brain Res. 1990;517(1–2):215–23.
Sharma HS, Cervos-Navarro J, Dey PK. Increased blood–brain barrier permeability following acute short-term swimming exercise in conscious normotensive young rats. Neurosci Res. 1991;10(3):211–21.
Sharma HS, Kretzschmar R, Cervos-Navarro J, Ermisch A, Ruhle HJ, Dey PK. Age-related pathophysiology of the blood–brain barrier in heat stress. Prog Brain Res. 1992;91:189–96.
Sharma HS, Westman J, Navarro JC, Dey PK, Nyberg F. Probable involvement of serotonin in the increased permeability of the blood–brain barrier by forced swimming. An experimental study using Evans blue and 131I-sodium tracers in the rat. Behav Brain Res. 1995;72(1–2):189–96.
Sharma HS, Patnaik R, Patnaik S, Mohanty S, Sharma A, Vannemreddy P. Antibodies to serotonin attenuate closed head injury induced blood brain barrier disruption and brain pathology. Ann N Y Acad Sci. 2007a;1122:295–312. doi:10.1196/annals.1403.022.
Sharma HS, Sjoquist PO, Ali SF. Drugs of abuse-induced hyperthermia, blood–brain barrier dysfunction and neurotoxicity: neuroprotective effects of a new antioxidant compound H-290/51. Curr Pharm Des. 2007b;13(18):1903–23.

Sharma HS, Muresanu D, Sharma A, Patnaik R. Cocaine-induced breakdown of the blood–brain barrier and neurotoxicity. Int Rev Neurobiol. 2009;88:297–334. doi:10.1016/S0074-7742(09)88011-2.
Sharma HS, Sjoquist PO, Ali SF. Alterations in blood–brain barrier function and brain pathology by morphine in the rat. Neuroprotective effects of antioxidant H-290/51. Acta Neurochir Suppl. 2010;106:61–6. doi:10.1007/978-3-211-98811-4_10.
Sharma HS, Castellani RJ, Smith MA, Sharma A. The blood–brain barrier in Alzheimer's disease: novel therapeutic targets and nanodrug delivery. Int Rev Neurobiol. 2012;102:47–90. doi:10.1016/B978-0-12-386986-9.00003-X.
Sharma A, Muresanu DF, Patnaik R, Sharma HS. Size- and age-dependent neurotoxicity of engineered metal nanoparticles in rats. Mol Neurobiol. 2013;48(2):386–96. doi:10.1007/s12035-013-8500-0.
Shirtcliff EA, Coe CL, Pollak SD. Early childhood stress is associated with elevated antibody levels to herpes simplex virus type 1. Proc Natl Acad Sci U S A. 2009;106(8):2963–7. doi:10.1073/pnas.0806660106.
Shonkoff JP, Garner AS, Committee on Psychosocial Aspects of Child and Family Health, Committee on Early Childhood, Adoption, and Dependent Care, Section on Developmental and Behavioral Pediatrics. The lifelong effects of early childhood adversity and toxic stress. Pediatrics. 2012;129(1):e232–46. doi:10.1542/peds.2011-2663.
Singareddy R, Krishnamurthy VB, Vgontzas AN, Fernandez-Mendoza J, Calhoun SL, Shaffer ML, et al. Subjective and objective sleep and self-harm behaviors in young children: a general population study. Psychiatry Res. 2013;209(3):549–53. doi:10.1016/j.psychres.2013.03.036.
Singer HS, Gilbert DL, Wolf DS, Mink JW, Kurlan R. Moving from PANDAS to CANS. J Pediatr. 2012;160(5):725–31. doi:10.1016/j.jpeds.2011.11.040.
Slopen N, Lewis TT, Gruenewald TL, Mujahid MS, Ryff CD, Albert MA, et al. Early life adversity and inflammation in African Americans and whites in the midlife in the United States Survey. Psychosom Med. 2010;72(7):694–701. doi:10.1097/PSY.0b013e3181e9c16f.
Slopen N, Koenen KC, Kubzansky LD. Childhood adversity and immune and inflammatory biomarkers associated with cardiovascular risk in youth: a systematic review. Brain Behav Immun. 2012a;26(2):239–50. doi:10.1016/j.bbi.2011.11.003.
Slopen N, Koenen KC, Kubzansky LD. Childhood adversity and immune and inflammatory biomarkers associated with cardiovascular risk in youth: a systematic review. Brain Behav Immun. 2012;26(2):239–50. doi:10.1016/j.bbi.2011.11.003.
Slopen N, Kubzansky LD, McLaughlin KA, Koenen KC. Childhood adversity and inflammatory processes in youth: a prospective study. Psychoneuroendocrinology. 2013a;38(2):188–200. doi:10.1016/j.psyneuen.2012.05.013.
Slopen N, Kubzansky LD, McLaughlin KA, Koenen KC. Childhood adversity and inflammatory processes in youth: a prospective study. Psychoneuroendocrinology. 2013;38(2):188–200. doi:10.1016/j.psyneuen.2012.05.013.
Smith RS. The immune system is a key factor in the etiology of psychosocial disease. Med Hypotheses. 1991;34(1):49–57.
Smith SEP, Li J, Garbett K, Mirnics K, Patterson PH. Maternal immune activation alters fetal brain development through interleukin-6. J Neurosci. 2007;27(40):10695–702.
Snider LA, Swedo SE. PANDAS: current status and directions for research. Mol Psychiatry. 2004;9(10):900–7. doi:10.1038/sj.mp.4001542.
Solov'eva Z, Orlovskaia DD. Microglia-type cells in normal and pathologic human embryonic brains. Zh Nevropatol Psikhiatr Im S S Korsakova. 1979;79(7):852–7.
Sporn AL, Bobb AJ, Gogtay N, Stevens H, Greenstein DK, Clasen LS, et al. Hormonal correlates of clozapine-induced weight gain in psychotic children: an exploratory study. J Am Acad Child Adolesc Psychiatry. 2005;44(9):925–33. doi:10.1097/01.chi.0000170552.15798.dd.
Steiner J, Bielau H, Bernstein HG, Bogerts B, Wunderlich MT. Increased cerebrospinal fluid and serum levels of S100B in first-onset schizophrenia are not related to a degenerative release of glial fibrillar acidic protein, myelin basic protein and neurone-specific enolase from glia or neurones. J Neurol Neurosurg Psychiatry. 2006;77(11):1284–7.

Steiner J, Bielau H, Brisch R, Danos P, Ullrich O, Mawrin C, et al. Immunological aspects in the neurobiology of suicide: elevated microglial density in schizophrenia and depression is associated with suicide. J Psychiatr Res. 2008;42(2):151–7. doi:10.1016/j.jpsychires.2006.10.013.

Steiner J, Walter M, Wunderlich MT, Bernstein HG, Pantli B, Brauner M, Jacobs R, Gos T, Rothermundt M, Bogerts B. A new pathophysiological aspect of S100B in schizophrenia: potential regulation of S100B by its scavenger soluble RAGE. Biol Psychiatry. 2009;65:1107–10.

Steiner J, Gos T, Bogerts B, Bielau H, Drexhage HA, Bernstein HG. Possible impact of microglial cells and the monocyte-macrophage system on suicidal behavior. CNS Neurol Disord Drug Targets. 2013;12(7):971–9.

Sternthal MJ, Enlow MB, Cohen S, Canner MJ, Staudenmayer J, Tsang K, et al. Maternal interpersonal trauma and cord blood IgE levels in an inner-city cohort: a life-course perspective. J Allergy Clin Immunol. 2009;124(5):954–60. doi:10.1016/j.jaci.2009.07.030.

Sublette ME, Galfalvy HC, Fuchs D, Lapidus M, Grunebaum MF, Oquendo MA, et al. Plasma kynurenine levels are elevated in suicide attempters with major depressive disorder. Brain Behav Immun. 2011;25(6):1272–8. doi:10.1016/j.bbi.2011.05.002.

Suzuki K, Matsuzaki H, Iwata K, Kameno Y, Shimmura C, Kawai S, et al. Plasma cytokine profiles in subjects with high-functioning autism spectrum disorders. PLoS One. 2011;6(5):e20470. doi:10.1371/journal.pone.0020470.

Swedo SE, Leonard HL, Rapoport JL. The pediatric autoimmune neuropsychiatric disorders associated with streptococcal infection (PANDAS) subgroup: separating fact from fiction. Pediatrics. 2004;113(4):907–11.

Teran R, Mitre E, Vaca M, Erazo S, Oviedo G, Hubner MP, et al. Immune system development during early childhood in tropical Latin America: evidence for the age-dependent down regulation of the innate immune response. Clin Immunol. 2011;138(3):299–310. doi:10.1016/j.clim.2010.12.011.

Tobiasova Z, van der Lingen KH, Scahill L, Leckman JF, Zhang Y, Chae W, et al. Risperidone-related improvement of irritability in children with autism is not associated with changes in serum of epidermal growth factor and interleukin-13. J Child Adolesc Psychopharmacol. 2011;21(6):555–64. doi:10.1089/cap.2010.0134.

Torrey EF, Leweke MF, Schwarz MJ, Mueller N, Bachmann S, Schroeder J, et al. Cytomegalovirus and schizophrenia. CNS Drugs. 2006;20(11):879–85.

Tostes MH, Teixeira HC, Gattaz WF, Brandao MA, Raposo NR. Altered neurotrophin, neuropeptide, cytokines and nitric oxide levels in autism. Pharmacopsychiatry. 2012;45(6):241–3. doi:10.1055/s-0032-1301914.

Unuvar T, Buyukgebiz A. Fetal and neonatal endocrine disruptors. J Clin Res Pediatr Endocrinol. 2012;4(2):51–60. doi:10.4274/jcrpe.569.

Vaccarino V, Brennan M, Miller AH, Bremner JD, Ritchie JC, Lindau F, et al. Association of major depressive disorder with serum myeloperoxidase and other markers of inflammation: a twin study. Biol Psychiatry. 2008;64(6):476–83. doi:10.1016/j.biopsych.2008.04.023.

Vezzani A, Granata T. Brain inflammation in epilepsy: experimental and clinical evidence. Epilepsia. 2005;46(11):1724–43.

Vezzani A, Balosso S, Ravizza T. The role of cytokines in the pathophysiology of epilepsy. Brain Behav Immun. 2008;22(6):797–803. doi:10.1016/j.bbi.2008.03.009.

Vezzani A, French J, Bartfai T, Barma T. The role of inflammation in epilepsy. Nat Rev Neurol. 2011;7(1):31. -32-40.

Viikki M, Anttila S, Kampman O, Illi A, Huuhka M, Setala-Soikkeli E, et al. Vascular endothelial growth factor (VEGF) polymorphism is associated with treatment resistant depression. Neurosci Lett. 2010;477(3):105–8. doi:10.1016/j.neulet.2010.04.039.

Vincenzi B, O'Toole J, Lask B. PANDAS and anorexia nervosa—a spotters' guide: suggestions for medical assessment. Eur Eat Disord Rev. 2010;18(2):116–23. doi:10.1002/erv.977.

Vythilingam M, Heim C, Newport J, Miller AH, Anderson E, Bronen R, et al. Childhood trauma associated with smaller hippocampal volume in women with major depression. Am J Psychiatry. 2002;159(12):2072–80.

Wei H, Zou H, Sheikh AM, Malik M, Dobkin C, Brown WT, et al. IL-6 is increased in the cerebellum of autistic brain and alters neural cell adhesion, migration and synaptic formation. J Neuroinflammation. 2011;8:52. doi:10.1186/1742-2094-8-52.

Wichers M, Maes M. The psychoneuroimmuno-pathophysiology of cytokine-induced depression in humans. Int J Neuropsychopharmacol. 2002;5(4):375–88.

Yolken RH, Bachmann S, Ruslanova I, Lillehoj E, Ford G, Torrey EF, et al. Antibodies to Toxoplasma gondii in individuals with first-episode schizophrenia. Clin Infect Dis. 2001;32(5): 842–4.

Young SN. Elevated incidence of suicide in people living at altitude, smokers and patients with chronic obstructive pulmonary disease and asthma: possible role of hypoxia causing decreased serotonin synthesis. J Psychiatry Neurosci. 2013;38(5):130002. doi:10.1503/jpn.130002.

Zarate Jr CA, Brutsche NE, Ibrahim L, Franco-Chaves J, Diazgranados N, Cravchik A, et al. Replication of ketamine's antidepressant efficacy in bipolar depression: a randomized controlled add-on trial. Biol Psychiatry. 2012;71(11):939–46. doi:10.1016/j.biopsych.2011.12.010.

Zimmerman AW, Jyonouchi H, Comi AM, Connors SL, Milstien S, Varsou A, et al. Cerebrospinal fluid and serum markers of inflammation in autism. Pediatr Neurol. 2005;33(3):195–201. doi:10.1016/j.pediatrneurol.2005.03.014.

Zorrilla EP, Cannon TD, Gur RE, Kessler J. Leukocytes and organ-nonspecific autoantibodies in schizophrenics and their siblings: markers of vulnerability or disease? Biol Psychiatry. 1996;40(9):825–33.

Zorrilla EP, Luborsky L, McKay JR, Rosenthal R, Houldin A, Tax A, et al. The relationship of depression and stressors to immunological assays: a meta-analytic review. Brain Behav Immun. 2001;15(3):199–226. doi:10.1006/brbi.2000.0597.

Zuckerman L, Weiner I. Maternal immune activation leads to behavioral and pharmacological changes in the adult offspring. J Psychiatr Res. 2005;39(3):311–23.

Zuckerman L, Rehavi M, Nachman R, Weiner I. Immune activation during pregnancy in rats leads to a postpubertal emergence of disrupted latent inhibition, dopaminergic hyperfunction, and altered limbic morphology in the offspring: a novel neurodevelopmental model of schizophrenia. Neuropsychopharmacology. 2003;28(10):1778–89.

Chapter 10
Cytokines and Related Metabolic Markers in Adult Neuropsychiatric Disorders: Possible Roles in Clinical Application

Aye-Mu Myint and Markus J. Schwarz

Abstract The importance of inflammatory response system activation indicated by abnormalities in cytokines and related metabolic markers in adult psychiatric disorders has been well studied and documented. The important aspects were reported not only in terms of pathophysiological mechanisms but also from the aspect of early diagnosis, prediction of the course of disease and prediction of response to treatment and subgrouping of clinical diagnostic groups. Future application of those markers in early diagnosis, prediction and prevention of suicide and personalized medicine for choice of treatment, either with psychotropic medications or adjuvant therapy, is necessary. The innovation effort to bring those results to application is challenging. New therapeutic developments should be carried out with great care. The past and present findings from the aspect of clinical applications and future perspectives are discussed in this chapter.

Keywords Inflammation • Cytokines • Metabolism • Tryptophan • Kynurenines • Tyrosine • Depression • Schizophrenia • Bipolar • Immunomodulation

Introduction

Among the adult neuropsychiatric disorders, depression, bipolar disorders and schizophrenia are the disorders which have highest socioeconomic impact. It is estimated that 38.2 % of the European population suffers from mental disorders

A.M. Myint
Department of Psychiatry and Psychotherapy, PsychoNeuroImmunology Research Group, Ludwig-Maximilians-University Munich, Munich, Germany

Otto von Guericke University, Magdeburg, Germany

School for Mental Health and Neuroscience, Faculty of Medicine, Maastricht University, Universiteitssingel 40, 6229 Maastricht, The Netherlands

M.J. Schwarz (✉)
Institut fuer Laboratoriumsmedizin, Klinikum der Universitaet Muenchen, Muenchen, Germany
e-mail: markus.schwarz@med.uni-muenchen.de

N. Müller et al. (eds.), *Immunology and Psychiatry*, Current Topics in Neurotoxicity 8, DOI 10.1007/978-3-319-13602-8_10

each year and, when age and comorbidity are adjusted, 164.8 million persons are affected each year (Wittchen et al. 2011). That in turn burdens society. In 2010, the total cost in Europe of brain disorders was €798 billion and, of that, €113.4 billion was spent on mood disorders and €93.9 billion for psychotic disorders (Gustavsson et al. 2011). Major depressive disorder (MDD) is by far the most important of these major psychiatric disorders. Disorders of the brain, and mental disorders in particular, contribute to 26.6 % of the total disease burden worldwide, and depression ranks highest among the most disabling diseases (Wittchen et al. 2011). In terms of prevalence, MDD is the third-highest prevalent disorder (6.9 %) in adults. The World Health Organization has estimated that depression will be the second most important cause of disability by 2030 (Mathers and Loncar 2005). It is noteworthy that approximately 50 % of people with depression do not consult a physician, and approximately 50 % of depression cases are not recognized by general practitioners. Among those who consult a physician, only approximately 30 % experience remission (Scott et al. 2003). The prevalence and incidence of bipolar mania and schizophrenia are less than for depression. The lifetime prevalence of bipolar spectrum disorder (BD) in the general US population is 6.4 %, with 0.8 % experiencing manic episodes and 0.5 % hypomanic episodes (Judd and Akiskal 2003). The point prevalence of schizophrenia is approximately 5 per thousand in the population, and the lifetime prevalence is 4 per thousand, ranging from 1.6 to 12.1 per thousand (Messias et al. 2007; Saha et al. 2005). The incidence of schizophrenia is approximately 0.20 per thousand per year, with regional variations at a magnitude of approximately 0.04–0.58 per thousand (Messias et al. 2007). A report on the 12-month prevalence rates for bipolar disorder and psychotic disorders in Europe, including schizophrenia, describes values of 0.9 and 0.8 %, respectively (Wittchen and Jacobi 2005).

For many decades, there have been two schools of thought when psychiatric disorders are considered. One considers that psychiatric disorders are states of mind that are purely non-biological and due to the psychological stress and the state of mind of a person. The other considers psychiatric disorders to be the result of some neurochemical changes in the brain. This latter is known as biological psychiatry. Research in biological psychiatry has advanced our knowledge more and more and there are well-documented data showing that psychiatric disorders are not only because of the neurochemical changes in the brain but also because of the biochemical changes in the body that further induce the changes of neurochemistry in the brain. In fact, the involvement of the biochemical state of the body in mood change was considered even in the time of the ancient Greeks. Hippocrates (c. 460–370 BC) is considered the first physician to describe the clinical picture of depression in the form of the term "melancholia". Hippocratic medicine was based on the concept of the four humours (phlegm, blood, yellow bile and black bile) and the belief that disease or ailment was caused by an imbalance in one of the four basic bodily liquids. According to Hippocrates, a person with a preponderance of black bile tended to have a melancholic disposition with particular mental and physical symptoms such as repulsion to eating, despondency, insomnia, irritability and restlessness (Sartorius 2001). The involvement of the genetic element is also quite well documented, although genes alone are not sufficient to induce the biological changes that result in psychiatric disorders, and the environmental factor is an important part in the aetiological mechanism. Environmental factors can even be the cause of changes in the genes through epigenetic mechanisms and involvement of epigenetics in psychiatric disorders is documented. Documented evidence shows that, among the environmental factors, lifetime experiences and childhood trauma, natural disasters, wars and epidemic infections composed of both physical and biological traumatic experiences are associated with the development of psychiatric disorders. Taking all factors together, psychiatric disorders in aetiology are quite heterogeneous, since the aetiological mechanism involves the interaction between environmental factors and the biological changes in the body and brain, including genetics and the state of mind. The latter is partly influenced by societal or family support or work environment. It could be stated that the psychiatric disorders are disorders of human life, meaning not just of the brain or of the body or of the state of mind, but of all those factors plus life time experiences.

Biological Factors in Aetiology of Psychiatric Disorders

There are several theories in terms of biological mechanisms involved in different psychiatric disorders. However, there is some overlap in the mechanisms and no one molecule is specific for one particular disorder.

Monoamines

Among the neurotransmitters involved in the aetiology of the above-mentioned adult psychiatric disorders, serotonin, dopamine and noradrenaline are the earliest neurotransmitters discovered in the middle of the twentieth century. The deficiencies in monoamines during depression became of interest in research focusing on the neurobiology of depression. The antidepressant actions of the monoamine oxidase inhibitors (MAOIs) have supported the theory of a serotonin deficiency during depression. Selective serotonin reuptake inhibitors (SSRIs) have been developed and are widely used, based on this theory. However, there are still controversies. Why do substances such as cocaine and amphetamines, which also promote serotonergic neurotransmission, not work as antidepressants (Hindmarch 2001)? Why do other drugs, such as tianeptine, which actually enhance serotonin reuptake, have antidepressant activities (Loo and Saiz-Ruiz 1999; Pineyro and Blier 1999; Wilde et al. 1995)? Why is it necessary to wait approximately 2–3 weeks before experiencing antidepressant effects following SSRI treatment, although the medication is able to increase serotonin availability within 1–2 days? The involvement of serotonergic abnormalities in schizophrenia has also long been suspected. The hallucinogenic drug lysergic acid diethylamide (LSD), which has structural similarity to serotonin and thus acts on the receptor serving as a serotonergic antagonist, could provoke symptoms similar to those of schizophrenia (i.e., hallucinations and paranoid delusions) (Fischman 1983). Some second-generation antipsychotics also have effects on serotonin receptors, particularly 5HT2A. However, pure serotonergic medications do not have antipsychotic effects. It has been suggested that serotonin may not be directly involved in schizophrenia but rather may act through interactions with other neurotransmitter systems (Alex and Pehek 2007). Due to the involvement of other monoamines, such as catecholamine deficiencies, noradrenaline also became of interest in the research of the pathophysiology of depression at around the same time as did the serotonergic mechanisms. Indeed, noradrenaline reuptake inhibitors (NRIs), such as venlafaxine, have been developed and are currently used in the treatment of depression. In schizophrenia, dopamine (DA), which is the precursor of catecholamine, was shown to be involved, and there may be an imbalance between the subcortical and cortical DA systems. Whereas the subcortical systems may be hyperactive due to the hyperstimulation of DR2 and result in positive symptoms, negative symptoms and cognitive impairment may be due to cortical projection, which is hypoactive (hypostimulation of DR1) (Guillin et al. 2007).

GABA and Glutamate

The reduced inhibitory action of the neurotransmitter γ-aminobutyric acid (GABA), which is involved in the synchronization of neuronal activity underlying working memory (Lewis et al. 2005), is also part of the pathophysiological mechanism of schizophrenia. Several post-mortem studies revealed indirect evidence of this, including the reduced expression of some isoforms of the enzyme glutamic acid decarboxylase (GAD) that converts glutamate into GABA (Lewis et al. 2005; Akbarian and Huang 2006). The GABAergic interneurons play important roles in regulating glutamatergic excitatory activity (Marty et al. 2000). The hypothesis of the involvement of glutamatergic neurotransmission was confirmed by a pharmacological observation. Non-competitive glutamate *N*-methyl-D-aspartate (NMDA) receptor antagonists, such as phencyclidine (PCP) and ketamine, have been shown to have psychotomimetic effects (Javitt and Zukin 1991). In contrast, patients with MDD showed significantly higher levels of the NMDA receptor subunit genes, GRIN2B and GRIN2C expression, and the metabotropic receptor genes, GRM4 and GRM5 expression, in noradrenergic neurons from locus coeruleus brain region. That finding indicated disrupted glutamatergic–noradrenergic interactions at the level of locus coeruleus in MDD. That supported the fact that glutamate antagonists may provide rapid antidepressant effects and may prevent suicide (Chandley et al. 2014). Indeed, the NMDA receptor antagonist ketamine showed efficient antidepressant effects in patients with depression (McGirr et al. 2014). In terms of GABAergic neurotransmission, similar to psychosis, reduction in GABA was also observed in the cerebrospinal fluid (CSF) of the patients with depression compared to patients with bipolar disorders or healthy controls (Mann et al. 2014).

Stress Hormones

Elevated level of plasma cortisol was observed in depressed patients compared to those who are not depressed (Board et al. 1957; Sachar et al. 1970). The lack of negative feedback mechanism was shown in those patients as a negative Dexamethasone Suppression Test (DST) which became a diagnostic marker of melancholic depression (Carroll et al. 1968a, b). It was also reported that Arginine Vasopressin (AVP) is the main activator of the HPA axis due to chronic stress and major depression (Scott and Dinan 1998). A systematic review on cortisol concluded that, in general the stress hormone axis is enhanced in schizophrenia, regardless of the heterogeneity in outcome of cortisol level and DST test studies (Bradley and Dinan 2010). Elevated cortisol metabolism in terms of enhanced urinary 5-α-reductase, 5-β-reductase and 11β-hydroxysteroid dehydrogenase activities was observed in both schizophrenia and bipolar patients (Steen et al. 2011). The results of the AVP studies in schizophrenia patients are also heterogenic. Even from the same research group, one study reported increased AVP in the serum of schizophrenia patients compared with controls and another reported decreased AVP in the serum of patients with schizophrenia (Rubin et al. 2013, 2014).

Immune System and Related Metabolic Pathways

Inflammatory Response System

Studies have shown activated inflammatory response system (IRS) in some patients with one of the adult psychiatric disorders such as major depression, bipolar disorder and schizophrenia. Although transient immune reactions to danger signals are an important part of essential body defence mechanisms, if these become chronic, these reactions can be deleterious to the organism. In addition, the mediators secreted or produced by the immune cells along these reactions are not 100 % positive or harmless to the surrounding cells and tissues. Moreover, these molecules induce not only effects directly but also through interaction with other endocrine and metabolic pathways. This network makes attempts to keep the body systems in balance. Any changes that disturb the balance can be detrimental to the organism.

Evidence implying a role for the pro-inflammatory cytokines in the aetiology of depression has been provided by studies on the changes in Interleukin (IL)-1, IL-6 and tumour necrosis factor (TNF) α in depressed patients (Maes 1994; Myint et al. 2005; Rothermundt et al. 2001) and also by the effects of interferon (IFN) α on non-depressed individuals being treated for hepatitis C (Hosoda et al. 2000) or malignant melanoma (Schafer et al. 1999). Those neuropsychiatric changes in IFNα treatment appear to be a consequence of the neurotransmitter and endocrine changes induced by the cytokines, rather than the pathological condition of the diseases (Valentine et al. 1998). Further evidence of immune activation in depressed patients is provided by the studies showing raised plasma concentrations of IL-1, IL-6, IFNγ, soluble IL-6 and IL-2 receptors, and the IL-1 receptor antagonist. These changes are correlated with a rise in plasma positive acute phase proteins (Sluzewska et al. 1996). Effective antidepressant treatments largely attenuated such immune changes (Sluzewska et al. 1995; Maes et al. 1999). In addition to the increases in pro-inflammatory cytokines, there is also evidence in the blood of depressed patients of an increased number of T-helper (Th), T-memory, activated T-cells and B-cells that act as a source of the plasma cytokines (Birmaher et al. 1994; Maes et al. 1994). The imbalance between the Th1 and Th2 was reported in depressed patients (Myint et al. 2005). It is also well known that microglia produce inflammatory mediators such as PGE2, cyclo-oxygenase-1 and -2 (COX-1 and -2), pro-inflammatory cytokines and inducible nitric oxide synthase (iNOS) (Chen et al. 2005; Campuzano et al. 2008; Depboylu et al. 2011) and the PGE2 content of the saliva, serum and CSF of depressed patients is reported to be increased (Ohishi et al. 1988; Linnoila et al. 1983).

Bipolar disorder is also associated with increased plasma levels of pro-inflammatory cytokines, such as IL-6, TNFα and IL-1β (Berk et al. 2011; Drexhage et al. 2010). The signalling molecules of IL-1, Nuclear factor-κB (NFκB) and IL-1 receptor antagonist (IL-1RA) protein and mRNA levels are increased in the post-mortem frontal cortex of bipolar patients (Rao et al. 2010). Increased acute phase protein level such as C-reactive protein (CRP) (Dickerson et al. 2007) is also associated with bipolar disorder. Regarding the balances between Th1, Th2 and Th3 cytokines, in bipolar mania patients, the ratio between Th1 to Th3 cytokines (IFNγ/TGFβ1) and the ratio between Th2 to Th3 cytokines (IL-4/TGFβ1) are significantly higher than controls (Kim et al. 2004a). Since Th3 cytokine regulates the balance between Th1 and Th2 cells, reduction in Th3 is considered unfavourable.

In schizophrenia, most of the cytokines changes are not very consistent. The most consistent findings are with IL-6. Most of the studies reported increased blood levels of IL-6 (Maes et al. 1995; Lin et al. 1998; Garcia-Miss Mdel et al. 2010) although there are some contradictory findings (Kim et al. 2000; Singh et al. 2009). The soluble IL-6 receptor levels in both serum and CSF were associated with severity of symptoms in paranoid schizophrenia (Muller et al. 1997). The findings in IL-2 are also controversial (Kim et al. 1998, 2000; Licinio et al. 1993). Regarding IL-1, the serum levels and mRNA expression in peripheral blood mononuclear cells are increased in drug-naïve first-episode patients. The increased mRNA levels of IL-1 are associated with increased NFκB expression (Song et al. 2009). Regarding the balance between Th1, Th2 and Th3, the ratio Th1/Th2 is increased in the plasma of drug-naïve and drug-free patients, but Th3 hardly changes (Kim et al. 2004b).

Some studies have reported on certain chemokines and chemokine receptors in psychiatric disorders, although no clear functional relationships have been reported. In this area, only non-specific association of chemokines such as macrophage chemotactic protein-1, macrophage inhibitory protein-1 and regulation and activation factors secreted by normal T-cells (RANTES) were reported in psychiatric disorders (Stuart and Baune 2014). No clear association with clinical symptoms and response to treatment has yet been clearly reported. The relationship of chemokines to adult psychiatric disorders needs to be researched considerably more. Advances in measurement technologies make this a promising research area.

Immune System Related Metabolic Pathways

There are two main metabolic pathways which are related to immune activation as well as to the network of neurochemicals, namely: tryptophan metabolism and tyrosine metabolisms (Fig. 10.1).

Tryptophan metabolism in psychiatric depression has been studied since 1960s, since tryptophan is the essential amino acid from which serotonin is synthesized. Increased degradation of tryptophan into kynurenine (KYN) could induce depressive mood (Lapin and Oxenkrug 1969). Normally, tryptophan is metabolized to kynurenine in the liver by the tryptophan 2,3-dioxygenase (TDO) (Watanabe et al. 1980). The activity is mainly controlled by the tryptophan level itself, resulting in a stable metabolism. The central availability of tryptophan mainly depends on the competition by the large amino acids at the transport across BBB and partially depends on the cerebral demand (Fernstrom 1977). KYN is further catabolized into 3-hydroxy-kynurenine (3HK) by the kynurenine-3-mono-oxygenase (KMO) enzyme, and then to 3-hydroxyanthranilic acid (HAA) through the action of kynureninase.

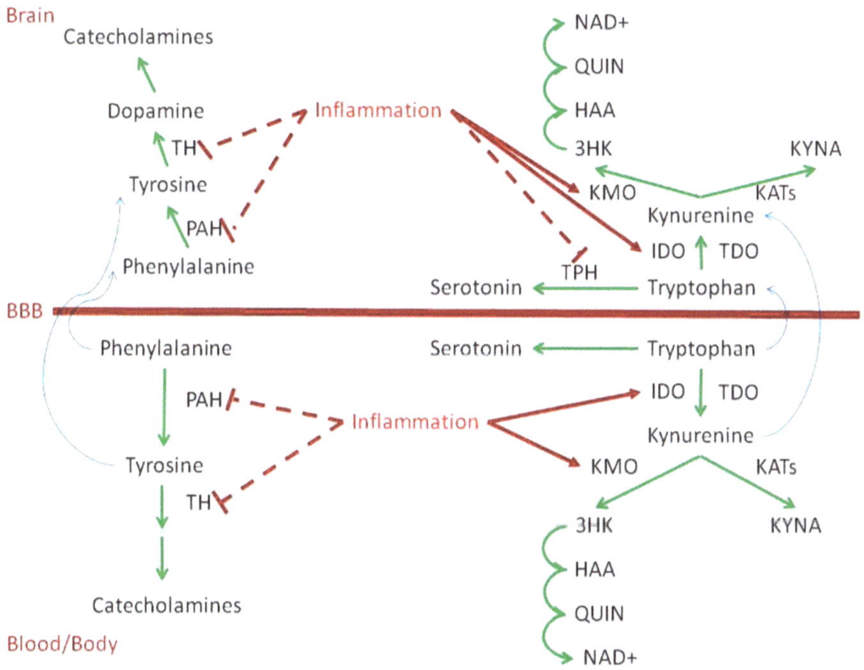

Fig. 10.1 Effects of inflammation on related metabolisms. The lines in *green* colour = metabolic process, the lines in *blue* colour = transport across blood brain barrier, the lines in *red* colour (complete) = activation, the lines in *red* (interrupted) = inhibitory effect, *IDO* indoleamine 2,3 dioxygenase, *TDO* tryptophan 2,3 dioxygenase, *KMO* kynurenine-3 mono-oxygenase, *KATs* kynurenine aminotransferases, *PAH* phenylalanine 5-hydroxylase, *TH* tyrosine 5-hydroxylase, *TPH* tryptophan hydroxylase, *KYNA* kynurenic acid, *3HK* 3-hydroxy-kynurenine, *HAA* 3-hydroxyanthranilic acid, *QUIN* quinolinic acid, *NAD* nicotinamide adenine dinucleotide, *BBB* blood–brain barrier

After that, the catabolism continues either into the complete oxidation pathway and forms adenosine triphosphate (ATP) which occurs mainly in the liver or into quinolinic acid (QUIN) which is finally degraded into nicotinamide adenine dinucleotide (NAD). From the complete oxidation pathway, picolinic acid (PIC) is also formed in small quantity. In physiological condition, the catabolism goes mainly into ATP formation and only a minor portion goes into NAD formation (Leklem 1971). In normal state, to get a normal NAD requirement for the nervous system, QUIN synthesis occurs only transiently in the liver and it does not accumulate in the hepatocytes (Bender 1989). KYN can also be catabolized by the kynurenine aminotransferases (KATs) into kynurenic acid (KYNA). Tryptophan metabolism is generally influenced by age and gender (Leklem 1971). Since KYN itself could be transported across the blood–brain barrier, in addition to the kynurenine formed in the brain by tryptophan breakdown, extra KYN is available from the periphery for further kynurenine metabolism in the brain. Sixty percent of brain KYN was contributed from the periphery (Gal and Sherman 1980). In the brain, tryptophan catabolism occurs mainly in the astrocytes and microglia (Grant et al. 2000; Grant and Kapoor 1998; Heyes et al. 1996), although some neurons also possess indoleamine 2,3-dioxygenase (IDO) and/or TDO2 (Miller et al. 2004). The astrocytes are the main source of KYNA because of a lack of KMO enzymes, whereas microglia and macrophages are the main sources for QUIN (Guillemin et al. 2000, 2001, 2005a). The astrocytes metabolize QUIN produced by the neighbouring microglia (Guillemin et al. 2001).

In case of inflammation, the enzyme indoleamine 2,3-dioxygenase (IDO) degrades tryptophan in the extrahepatic tissues (Heyes et al. 1993; Mellor and Munn 1999). Enzyme activity is enhanced by pro-inflammatory cytokines such as interferon-γ (IFNγ) (Carlin et al. 1987; Yasui et al. 1986). Thus, the extrahepatic tryptophan metabolism is shifted away from the liver (Moffett et al. 1998). In this case, tryptophan breakdown through KYN pathway occurs mainly in the blood and lymphoid tissues (Moffett and Namboodiri 2003). The IDO activity is inhibited by the anti-inflammatory cytokine IL4 (Musso et al. 1994). In case of stress or related conditions where cortisol secretion is enhanced, TDO activity is also further enhanced by glucocorticoids (Knox 1951; Salter and Pogson 1985). In this case, the KYN formation becomes much higher than physiological condition. Since the liver-cell uptake of KYN is not efficient for extrahepatic KYN, the further KYN catabolism mainly occurs extrahepatically. The activity of KMO is also enhanced by pro-inflammatory cytokines (Mellor and Munn 1999). In case of inflammation, formation of 3HK therefore becomes enhanced much faster than KYNA formation and the balance between formation of 3HK and KYNA shifted to the arm of the metabolism with 3HK and QUIN. In the presence of inflammation, activated monocytes are also found to be the robust producers of QUIN (Chiarugi et al. 2001). The accumulation of 3HK could induce neuronal apoptosis (Okuda et al. 1998), while accumulation of QUIN, the endogenous NMDA-R agonist (Bender and McCreanor 1985), induces excitotoxic neurodegenerative changes (Schwarcz et al. 1983) and astrocytes apoptosis (Guillemin et al. 2005b). However, KYNA is the NMDA-R antagonist (Perkins and Stone 1982) and is protective against excitotoxicity of QUIN (Kim and Choi 1987), although accumulation of KYNA could induce glutamatergic hypo-functioning and might disturb cognitive function (Olney et al. 1991). Moreover, KYNA is an antagonist of α7-nicotinic acid acetylcholine receptor (α7nAchR) (Hilmas et al. 2001) and influences the expression of non-α7nAchR (Hilmas et al. 2001). This in turn disturbs the α7 and non-α7nAchR mediated release of noradrenaline (NA), dopamine (DA) and acetylcholine (Ach) (Myint 2012).

Non-depressed patients with hepatitis C who were treated with IFNα and who later developed depression showed an increase in IL-6 and a decrease in KYNA, which showed significant association with the development of depressive symptoms. Since IFNα could induce 15 kD protein, this would further increase IFNγ production and activate IDO and KMO (Wichers et al. 2005). Another study also showed increased KYN pathway with an increase in both KYNA and QUIN in the CSF of IFNα treated patients (Raison 2010). In the patients with major depression who are drug naïve or medication free for at least 4 months, an imbalance between those neuroprotective and neurotoxic pathways with lower protective metabolite has been observed (Myint et al. 2007a). The possible important role of KYNA in depression was supported by a recent genetic study which demonstrates that a single nucleotide polymorphism (SNP) from the KAT-III gene is associated with major depression and bipolar depression (Claes et al. 2011). A magnetic resonance (MR) spectroscopy study in melancholic depressed adolescents also reported that the choline levels in the brain, which indicated the turnover of cells, showed positive correlation with serum KYN and the HAA/KYN ratio (Gabbay 2010) and serum KYN and the HAA/KYN ratio were positively correlated with depression scores.

The plasma of bipolar mania patients shows reduced tryptophan levels (Myint et al. 2007b) and an increased KYN to tryptophan ratio (Reininghaus et al. 2014). A post-mortem study on TDO2 and KYN in the anterior cingulate cortex (ACC) showed increased TDO2 expression and KYN levels in the brains of bipolar patients (Miller et al. 2006). In a study on the in vitro fibroblast culture from patients with bipolar disorder and controls, an increase of both 3HK and KYNA was observed in the culture of the cells from the patients; when the culture was challenged immunologically, the cytokines IL1 and IL6 and the 3HK and 3HK/KYNA were significantly increased in the culture from the patients compared with controls (Johansson et al. 2013).

The findings in schizophrenia patients are quite controversial. A study in post-mortem brain tissue in different cortical regions revealed increased KYNA levels in schizophrenic samples compared to a control sample, particularly in the prefrontal cortex (PFC) (Schwarcz et al. 2001). Another investigation in the amygdala observed an insignificant increase of KYNA in medicated schizophrenics (Miller et al. 2006). Increased levels of KYNA were also observed in the CSF of drug-naive first-episode schizophrenic patients (Erhardt et al. 2001). It was hypothesized that accumulation of KYNA in the brain may lead to schizophrenic symptoms (Erhardt et al. 2003). However, not only the positive symptoms and cognitive impairment but also the negative symptoms and loss of brain volume (Takahashi et al. 2009) are components of the psychopathology of schizophrenia. Yao and group have reported on 3HK levels in schizophrenia (Yao 2010), demonstrating the positive association between OHK and the total symptoms score at the time of recruitment. Our recent finding in medication-naïve schizophrenia patients indicated increased plasma 3HK and decreased plasma KYNA compared to healthy controls (Myint et al. 2011a). Part of the same study mentioned above, which tested the in vitro fibroblast culture from patients with schizophrenia and controls, also observed the increase of both 3HK and KYNA in the culture of the cells from the schizophrenia patients; when the culture was challenged immunologically, the cytokines IL1 and IL6, and the 3HK and 3HK/KYNA were significantly increased in the culture from schizophrenia patients compared with controls (Johansson et al. 2013).

Another metabolic pathway that links the immune system and neurochemicals is the tyrosine metabolism, from which dopamine and catecholamines are synthesized. In this metabolism the tetrahydrobiopterin (BH4) is the cofactor of phenylalanine 4-hydroxylase (PAH), tyrosine 5-hydroxylase (TH), tryptophan 5-hydroxylase (TPH), nitric oxide synthases (NOS) (Werner-Felmayer et al. 2002) and alkylglycerol mono-oxygenase (Werner-Felmayer et al. 2002). PAH converts phenylalanine to tyrosine and subsequently TH initiates the production of DA, adrenaline and noradrenaline through the formation of L-3,4-dihydroxyphenylalanine (L-DOPA). The enzyme TPH is also important for synthesis of serotonin from tryptophan. Since BH4 is a labile molecule, the oxidation of BH4 by molecular oxygen is irreversible (Connor et al. 1979). Under conditions of oxidative stress, one can expect that metabolites such as BH4 are rapidly destroyed (Fuchs et al. 2001; Widner et al. 2001) and the enzymes dependent on this metabolite lose their activity. In this case, the activities of PAH and TH are reduced, leading to an increase in phenylalanine/tyrosine ratio and a decrease in dopamine synthesis from tyrosine. This mechanism was proposed as part of the pathophysiology of depression (Sperner-Unterweger et al. 2014), based on the finding in the study in an elderly population that increased tryptophan catabolism was associated with the depressive symptoms of lassitude, reduced motivation, anorexia and pessimistic thoughts, whereas phenylalanine/tyrosine alterations showed a more pronounced correlation to neurovegetative symptoms, including sleep disturbance, digestive symptoms, fatigue, sickness and motor symptoms (Capuron et al. 2011).

Oxidative Stress Related Biomarkers

The evidence of increased oxidative stress, which is also related to activation of IRS, is also well documented in psychiatric disorders. A recent post-mortem brain study has reported that the gene expression levels of oxidative defence enzymes superoxide dismutases, catalase and glutathione peroxidase in association with telomere length were significantly lower in white matter oligodendrocytes from patients with MDD as compared to controls (Szebeni et al. 2014). The increased levels of an early component of the peroxidation chain in the plasma samples of euthymic patients with bipolar disorder were also reported (Andreazza et al. 2015). The redox dysbalance, which is the imbalance between anti-oxidants and pro-oxidants (reactive oxygen species and reactive nitrogen species) was also well documented in schizophrenia (Steullet et al. 2014). However, there is no report from in-depth biomarker studies on oxidative stress markers in relation to clinical course and clinical symptoms or treatment response in adult psychiatric disorders.

Clinical Aspects of Cytokines and Related Metabolic Biomarkers

From the late twentieth century and the beginning of the twenty-first century to date, the field of research in immune changes and its related metabolic changes in psychiatric disorders developed significantly in terms of translational and clinical studies. Evidences of the associations of changes in cytokines and the metabolites with clinical features, response to treatment and choice of treatment in different adult psychiatric disorders were documented in clinical settings.

Association with Clinical Symptoms and Response to Psychotrophic Medication

Major Depressive Disorders

The role of high Th1 and Th2 cytokines in the blood of the patients with major depression has been well documented (Maes 1994; Musselman et al. 2001; Seidel et al. 1996). It has also been found that antidepressants can decrease the Th1/Th2 or pro-inflammatory/anti-inflammatory cytokine ratio (Kubera et al. 2001). A study on Th1, Th2 and Th3 balance has shown that there was an increase in Th1 cytokine IFNγ positivity in the depressed patients, and the Th3 cytokine TGFβ1, which regulates the balance between Th1 and Th2, was increased in depressed patients following effective antidepressant treatment (Myint et al. 2005). This study has shown that the Th1 and Th2 balance was dependent on age and gender. The concentrations of the blood inflammatory markers at baseline are also related to symptom severity and outcome of the treatment. In our recent study, increase in pro-inflammatory markers such as IFNγ, TNFα and CRP is associated with severity of the symptoms such as lack of interest and psychomotor retardation (Halaris et al. 2015). In a study on IFNα-treated hepatitis C patients, concentrations of the baseline soluble interleukin-2 receptor (sIL-2r), IL-6 and IL-10 were significantly increased in those who developed clinical depression during treatment compared with those who did not (Wichers et al. 2006). Altered plasma and CSF IL-6 (Lindqvist et al. 2009) and chemokines (Janelidze et al. 2013) are also reported to be associated with suicide attempts. Those findings indicated that the activation of IRS might be involved in the development of depressive symptoms. In addition, the activation of IRS can induce changes in the tryptophan degradation pathway.

A recent immunogenetic study on the IFNγ gene reported that the presence of IFNγ CA repeat allele 2 homozygous, which was associated with increased IFNγ secretion, had significant association with higher kynurenine concentrations in controls ($F=4.47$, $p=0.038$) as well as in patients ($F=3.79$, $p=0.045$), and also with an increase of tryptophan breakdown over time during the study period ($F=6.0$, $p=0.019$) (Myint et al. 2013). The imbalances in the tryptophan degradation pathway were also associated with symptom severity. In our recent study, the serum KYNA concentration showed significant inverse correlation with depressive symptom and paranoid symptom scores in depressed patients (Halaris et al. 2015). A recent study on the polymorphism of the genes of the enzymes involved in the tryptophan degradation pathway reported that KATIII enzyme polymorphism is associated with unipolar and bipolar depression and associated somatic anxiety symptoms (Claes et al. 2011). A study on people who have attempted suicide showed that they have significantly higher concentrations of QUIN in their plasma and CSF, regardless of the associated psychiatric disorders, including schizophrenia (Erhardt et al. 2013). The increase in QUIN immune-positive microglia cell density in the anterior mid cingulate cortex and subgenual anterior cingulate cortex areas of the ACC of post-mortem brain tissues was also reported in those patients with unipolar and bipolar depression who committed suicide compared with the control brain tissues from those who died from other causes (Steiner et al. 2011). These findings pointed out the fact that neuronal excitotoxicity during depression played an important role in suicide and raised a question as to whether the reduction of NMDA-mediated neurotoxicity using ketamine could be a preventive therapy for suicide in patients with high serum or plasma QUIN concentration. Unlike in the ACC, the QUIN immune-positive microglia cell density was reduced in the right CA1 region of the hippocampus area in both unipolar and bipolar depressed suicidal brains (Busse et al. 2014). The confounding effect of minor hippocampal degeneration could not be excluded, although there was no gross volume loss. These findings indicated that changes in the brain are area-specific and cannot be generalized as in the periphery.

Fortunately, currently available antidepressants could help to a certain degree in terms of biochemical changes. The 6-week treatment with SSRIs could increase the ratio of KYNA to KYN, indicating the rebalancing the shift of kynurenine pathway (Myint et al. 2007a). In another study, monotherapy with escitalopram could reduce inflammatory molecule CRP and the neurotoxic QUIN around the 8th week of therapy, although those rose again around the 12th week, whereas 3HK was significantly reduced in the 12th week (Halaris et al. 2015). In this study, it was reported that those patients who responded to the therapy had higher initial serum CRP, although those patients did not have a reduction in anti-inflammatory cytokines and imbalance in the kynurenines. In another study on the changes in kynurenines in the CSF, it was observed that the higher the protective metabolite KYNA in relation to the toxic 3HK, the better the response to antidepressants and mood stabilizers in patients with affective disorders (Schwarz et al. in press). Those findings together indicated that activated IRS is not always a negative indicator as long as it is not yet associated with imbalance in the whole immune-neurochemical network.

Bipolar Disorder

The studies in bipolar disorder reported a somewhat different but similar direction to that of major depression. In terms of cytokines changes, a meta-analysis on 18 studies with a total of 761 bipolar disorder patients and 919 healthy controls reported that concentrations of soluble IL-2 receptor (sIL2R), TNFα, soluble tumour necrosis factor receptor type 1 (sTNFR1) ($p<0.001$ each), sIL6R ($p=0.01$) and IL-4 ($p=0.04$) were significantly higher in bipolar patients compared with healthy controls, whereas there was no significant difference in other cytokine (Munkholm and Brauner 2013). Another meta-analysis on 30 studies reported the difference between phases such as manic and euthymic for sIL2R, sIL6R, IL4 and TNFα (Modabbernia et al. 2013). In general, maniac phase showed more inflammatory changes compared with euthymic phase.

In a study on tryptophan metabolites in bipolar mania, it was reported that bipolar manic patients have a significantly lower tryptophan index than normal controls ($F=9.779$, $p=0.004$) (Myint et al. 2007b) and the mean tryptophan breakdown index was increased significantly after a 6-week treatment. The reduction in plasma tryptophan and reduction in tryptophan index showed significant negative correlation with reduction in YMRS score ($r=-0.577$, $p=0.019$ and $r=-0.520$, $p=0.039$, respectively). In another study, both blood kynurenine concentrations and the KYN-to-tryptophan ratio were significantly higher in the total sample of euthymic patients with bipolar disorder, with greater increases noted in both parameters in the subsample of overweight patients. When compared to controls, peripheral neopterin concentrations were significantly lower. Within the patient group, there were also significant between-group differences in neopterin concentrations, with higher levels in those who were overweight and in patients in the later stages of illness compared to earlier stages (Reininghaus et al. 2014).

Schizophrenia

The results of the studies in schizophrenia are still controversial. However, involvement of infection and immune changes in schizophrenia has been well accepted. Prenatal maternal infection of immune changes that programmed the brain changes in the foetus in the early neurodevelopmental stage which predisposes the development of schizophrenia in adulthood was a well-recognized hypothesis (Meyer et al. 2006).

The plasma and serum levels of tryptophan metabolites, which are related to immune activation, showed significant associations with clinical symptom scores. A study on serum 3HK in schizophrenia discussed that there is an association between initial 3HK concentrations and (1) total positive and negative symptoms score (PANSS) at the time of recruitment, and (2) delayed response of positive symptoms at 4-week neuroleptic treatment in first-episode neuroleptic-naïve schizophrenia patients (Condray et al. 2011). In another study, it was reported that the baseline plasma tryptophan levels in schizophrenia patients showed negative correlations with the PANSS positive symptoms scores (Kim et al. 2009). It was also reported that the initial plasma KYNA concentrations were associated with a reduced PANSS positive symptoms score and a reduced depressive symptoms score upon discharge. In contrast, the higher the KYNA/KYN ratio on admission, the higher the positive symptoms score on admission, although the higher the KYNA/KYN ratio on admission the better the reduction in depressive symptoms scores at discharge. Moreover, the increase in this ratio after 6-week medication was associated with better reduction in the positive symptoms, general symptoms and depressive symptoms scores. Initial plasma 3HK concentrations were negatively associated with the admission PANSS positive symptoms score. However, this study reported that, the higher the plasma 3HK on admission, the lower the reduction in general symptom scores at the time of discharge (Myint et al. 2011b). In another study on CSF kynurenines and response to treatment, it was observed that the higher the KYNA in the CSF in relation to 3HK, the better the response to treatment (Schwarz et al. in press). Taking those findings together could lead to the interpretation that an increase in KYNA concentration might induce positive symptoms and cognitive symptoms, although the lower concentration of KYNA could result in poor response to treatment.

A post-mortem brain study demonstrated the increase in KYNA levels in schizophrenic samples compared to those of controls, particularly in the PFC (Schwarcz et al. 2001). Another post-mortem brain study on the amygdale region also showed an insignificant increase of KYNA in medicated schizophrenics (Miller et al. 2006). A recent immunohistochemical study on a post-mortem brain showed reduction in QUIN positive microglia cell density in the CA1 region in schizophrenia patients compared with controls (Gos et al. 2014). Since a similar finding was reported in the study for major depression and bipolar depression (Busse et al. 2014) as well, it could be hypothesized that reduction in NMDA-R agonist QUIN in the hippocampus area might be related to impaired cognitive function in psychiatric disorders.

Cytokines and Related Metabolism in Immunomodulation Therapy

The proof of concept regarding the involvement of immune activation in the pathophysiology of psychiatric disorders is that the anti-inflammatory medications improved the symptoms of those disorders. Several medications which are neither antidepressants nor psychotropic drugs but which modulate the immune system directly and indirectly are reported to give positive outcomes in treatment of adult psychiatric disorders.

The most studied anti-inflammatory medication is the COX2 inhibitor, celecoxib. Celecoxib is used as an add-on to the currently used antidepressants or antipsychotics. Clinical studies have suggested that the COX2 inhibitor celecoxib has positive effects on cognitive function in depressed patients (Muller et al. 2006; Chen et al. 2010). In a double-blind controlled trial in which celecoxib was given as an add-on to an antidepressant, reboxetine, those depressed patients with higher KYN-to-tryptophan ratios responded to the add-on treatment better than those with lower KYN-to-tryptophan ratios (N. Müller, Biomarkers in celecoxib trial, personal communication). Since this ratio indicates the degradation of tryptophan to kynurenine, which indirectly indicates the activation of IRS, it could be interpreted that those with an inflammatory condition or an activated IRS in association with depression responded better to COX2 inhibitor add-on medication. In the same study, it was also observed in the placebo+reboxetine group that those who responded to reboxetine had a lower initial QUI/KYNA ratio. In addition, in overall combined group, the responders had a lower initial QUIN/KYN ratio. Those findings indicated that the general response to treatment might be better if the neurotoxic QUIN formation is lower, whereas a celecoxib add-on is necessary for those with activated IRS. To date, there is only one study of celecoxib in treatment-resistant bipolar disorder (Nery et al. 2008) and the significant reduction in Hamilton Depression Score was noted at the first week time-point but not afterward. No biomarker was measured in this study. In another study on celecoxib add-on therapy in treatment-resistant depression, no clinical data was reported; only a reduction in blood TNFα in the celecoxib group was mentioned (Kargar et al. 2014). The suppression of TNF, which is a pro-inflammatory cytokine, could be one of the mechanisms involved in better improvement of depressive symptoms in celecoxib add-on therapy. Treatment with an anti-TNF antibody, infliximab, did not show a clinical advantage in overall analyses, but a baseline hs-CRP>5 mg/L was found to be the point at which infliximab-treated patients began to exhibit a greater decrease in HAM-D-17 scores than placebo-treated subjects (Raison et al. 2013). Further study continued with analyses of transcriptome for response to treatment with infliximab revealed that 148 transcripts were predictive of infliximab response and were associated with gluconeogenesis and cholesterol transport, and those were enriched in a network regulated by the hepatocyte nuclear factor (HNF)4-alpha, a transcription factor involved in gluconeogenesis and cholesterol and lipid homeostasis (Mehta et al. 2013). Interestingly, tryptophan metabolite QUIN, which is associated with response to treatment, is also associated with suppression of gluconeogenesis (Lardy 1971). Therefore, gluconeogenesis might be considered involved in the response to antidepressant treatment. Moreover, the cytokines and related metabolic biomarkers should be applied to the treatment with anti-inflammatory therapy since these medications will work if only IRS activation is involved. There was only one open-labelled trial of aspirin in depression (Galecki et al. 2009) and it did not show a satisfactory advantage over fluoxetine, although some population-based studies showed advantages over mood and cognitive function (Dinnerstein and Halm 1970; Lieberman et al. 1987; Kang et al. 2007).

In schizophrenia, unlike in affective disorders, the meta-analysis study on the randomized controlled trials with celecoxib and aspirin reported only a marginal effect on the PANSS total scores and that was mainly due to the effect of aspirin (Nitta et al. 2013). In the placebo-controlled trial with aspirin 100 mg/day as adjuvant to antipsychotics, total psychopathology was reduced significantly by aspirin in those patients who had the lowest initial value of Th1:Th2 ratio (Laan et al. 2010).

It was reported that only those patients with early stage of schizophrenia responded to anti-inflammatory therapy (Nitta et al. 2013).

Another medication which is not an anti-inflammatory, although it induces anti-inflammatory-like action in the brain by suppression of microglia activation, is a second-generation tetracycline, minocycline. The earliest case report was by a Japanese group who reported the antipsychotic effect especially on negative symptoms in two patients with schizophrenia treated with minocycline as adjunct to haloparidol (Miyaoka et al. 2007). That report was followed by another double-blind randomized study on 54 schizophrenia patients who were treated with minocycline as an adjunct to atypical antipsychotics, which delivered a beneficial outcome in terms of negative symptoms, general symptoms and cognitive function, especially in the early stage patients (Levkovitz et al. 2010). Another case report in which minocycline was used as an adjunct therapy to clozapine in two patients with treatment-resistant schizophrenia also reported the beneficial effect on the negative symptoms (Kelly et al. 2011). Another three more randomized double-blind placebo-controlled trials in schizophrenia also reported the beneficial effect of minocycline on the negative symptoms (Fekadu et al. 2013; Khodaie-Ardakani et al. 2014; Liu et al. 2014). To date, only an open-label study of minocycline therapy in combination with SSRIs on patients with psychotic depression has reported that 6-week therapy could reduce HDRS, Brief Psychiatric Rating Scale and psychotic symptoms (Miyaoka et al. 2012). None of the studies have mentioned the role of cytokines and related metabolic biomarkers in terms of prediction of treatment response. Markers related to microglia activation, such as QUIN, or imaging of glia activity could be useful markers to predict the response to treatment. This area still needs to be investigated, since no medication will give 100 % response, and treatment-response indicators are important in future personalized psychiatry.

Another medication which indirectly suppresses the IRS activation is polyunsaturated fatty acid (PUFA) which is widely studied and has shown beneficial effects, especially in schizophrenia. Meta-analysis of 11 placebo-controlled trials conducted on patients with a DSM-defined diagnosis of MDD reported significant clinical benefits of omega-3 PUFA treatment compared to placebo (Grosso et al. 2014). Earlier, several placebo-controlled trials of omega-3 fatty acid as either mono- or add-on therapy in schizophrenia in the beginning of the twenty-first century showed advantage of PUFA over the placebo (Peet 2008; Müller 2013), although a meta-analysis showed no advantage (Fusar-Poli and Berger 2012). However, a recent study from Norway in which PUFA or vitamins C and E add on to antipsychotics in acute psychotic patients with schizophrenia showed that, in patients with low erythrocyte PUFA, both add-on medications worsen the psychotic symptoms, but not if PUFA was combined with vitamins (Bentsen et al. 2013). This negative effect could be due to drug interaction of those add-on medications with antipsychotics. Another possibility is that those patients with low erythrocyte PUFA might also have vitamin C and E deficiency, and that the positive effects of vitamin C and E will occur only if PUFA levels are normal and vice versa. Therefore, it was reported that it was safe to give those add-on medications in combination.

Those studies indicated the role of biomarkers in choosing an anti-inflammatory therapy to predict the outcome. In addition it should be noted that it is not that simple to conclude that low blood level of a biological compound indicates the need for that compound as an add-on therapy. Since biological compounds or metabolites or cytokines do not work simply as a single molecule but through the interaction with other chemicals, careful interpretation considering the possible interactions is essential.

Future Perspectives and Conclusion

Based on the above findings, there are several options open for future research studies related to clinical application. From the diagnostic aspects, studies for early diagnosis and for choice of therapy will be the promising approach in terms of prevention and personalized medicine. However, in terms of therapeutic research, there should be a cautious approach, since careless manipulation of the system can induce long-term deleterious effect that could not be easily detected in the short-term animal experiments.

Diagnostic Perspective

The studies on the role of cytokines and related biomarkers have clearly indicated that, in a certain percentage of patients with adult psychiatric disorders, there are an activated IRS and disturbances in the metabolism related to immune function. Based on those findings and also on findings from the studies of the anti-inflammatory treatment trials, it could be concluded that activated IRS and dysbalance in immune function and related metabolism are part of the aetiology of adult psychiatric disorders. Therefore, those disturbances are expected to take place before the disturbances in brain neurochemicals in some patients. In this case, those early changes could be considered as early detection of abnormal changes in the system. To detect those early disturbances, it is necessary to know the normal profile of those markers in each individual, since immune system related metabolic profile can be influenced by many individual factors. If those early disturbances persist for a longer period than regular infection or inflammation should, it is necessary to consider this as detection of early pathological changes and those changes should be corrected in time by using simple anti-inflammatory medications or vitamins. That would prevent the unnecessary use of psychotropic medications, which not only could induce untoward effects but also could increase the cost of treatment and socioeconomic burden.

In addition to the early diagnostic perspectives, some of the markers such as QUIN in the blood and CSF could be applied for a predictive marker for suicide.

By giving medications that could block the NMDA-R and induce antidepressant effects such as ketamine to those with high QUIN in the blood, we might prevent untimely death of the patients and could improve the quality of life of the patients as well as of the family members.

Moreover, by stratifying the whole group of patients with mood disorders and psychoses, those markers might be able to predict the response to treatment of particular psychotropic medications. Future research should be carried out in this direction. In this way, we might be able to give the medication that better fits with the patient from the beginning of the illness. This might in turn prevent the development of treatment resistance and chronicity of the disorders which result in considerable socioeconomic burdens.

However, to reach this applicability, further innovation in terms of the development of a diagnostic test system which could be applied at the community health-care level is necessary. Since most of the molecules are small molecules, the development of a sensitive and specific test system is a challenge to the scientific community. In addition, standardization among the available methods is also important to establish a reference system for new developments.

Therapeutic Aspect

Regarding the therapeutic aspects use of simpler medications with low possible side effects or application of life-style intervention, such as exercise and mindfulness practice for prevention and as adjuvant therapy, should be encouraged. However, it is important to realize that life-style intervention studies do not always work as we expect. This could be due to genetics and sociocultural predisposition that would make a person get benefit from a particular way of intervention. One way of life-style change cannot be expected to produce a similar effect in the whole population. Future studies that cover this aspect should be considered to make possible providing personalized advice.

Regarding the development of new compounds, it should be considered with great care. The results from the previous studies have clearly indicated that the pathophysiology of the adult psychiatric disorders is not due to one gene or one molecule. Manipulation based on the change of a molecule without proper monitoring could cause severe untoward effect. For example, in the kynurenine pathway, shift to 3HK can induce toxic changes but use of KMO inhibitor to block the pathway can induce deficiency and energy metabolism and can induce some psychiatric symptoms which we may not easily be able to detect in the acute animal models. Another example is the inhibitor of the KAT II enzyme to reduce the formation of KYNA with the intention of treating the cognitive symptoms in schizophrenia. This could lead the metabolic pathway to increased formation of QUIN which can induce not only excitotoxicity but also suicidal symptoms which are impossible to detect in the acute animal models.

Conclusion

Future research studies are needed to bring the previously collected, well-documented findings to clinical application, and innovation is needed to improve the clinical management. At the same time, some more clinical, translational and basic studies are required to bridge the gaps still open in terms of understanding these disorders.

Acknowledgements The work of Myint AM was funded be European Collaborative Project, Moodinflame (Grant No. 22963) and Marie-Curie IAPP Project, Psych-aid (Grant No. 286334).

References

Akbarian S, Huang HS. Molecular and cellular mechanisms of altered GAD1/GAD67 expression in schizophrenia and related disorders. Brain Res Rev. 2006;52(2):293–304.

Alex KD, Pehek EA. Pharmacologic mechanisms of serotonergic regulation of dopamine neurotransmission. Pharmacol Ther. 2007;113(2):296–320.

Andreazza AC, Gildengers A, Rajji TK, Zuzarte PM, Mulsant BH, Young LT. Oxidative stress in older patients with bipolar disorder. Am J Geriatr Psychiatry. 2015;23(3):314–9.

Bender DA. Effects of a dietary excess of leucine and of the addition of leucine and 2-oxo-isocaproate on the metabolism of tryptophan and niacin in isolated rat liver cells. Br J Nutr. 1989;61(3):629–40.

Bender DA, McCreanor GM. Kynurenine hydroxylase: a potential rate-limiting enzyme in tryptophan metabolism. Biochem Soc Trans. 1985;13(2):441–3.

Bentsen H, Osnes K, Refsum H, Solberg DK, Bohmer T. A randomized placebo-controlled trial of an omega-3 fatty acid and vitamins E+C in schizophrenia. Transl Psychiatry. 2013;3:e335.

Berk M, Kapczinski F, Andreazza AC, Dean OM, Giorlando F, Maes M, Yucel M, Gama CS, Dodd S, Dean B, et al. Pathways underlying neuroprogression in bipolar disorder: focus on inflammation, oxidative stress and neurotrophic factors. Neurosci Biobehav Rev. 2011;35(3):804–17.

Birmaher B, Rabin BS, Garcia MR, Jain U, Whiteside TL, Williamson DE, al-Shabbout M, Nelson BC, Dahl RE, Ryan ND. Cellular immunity in depressed, conduct disorder, and normal adolescents: role of adverse life events. J Am Acad Child Adolesc Psychiatry. 1994;33(5):671–78.

Board F, Wadeson R, Persky H. Depressive affect and endocrine functions; blood levels of adrenal cortex and thyroid hormones in patients suffering from depressive reactions. AMA Arch Neurol Psychiatry. 1957;78(6):612–20.

Bradley AJ, Dinan TG. A systematic review of hypothalamic–pituitary–adrenal axis function in schizophrenia: implications for mortality. J Psychopharmacol. 2010;24(4 Suppl):91–118.

Busse M, Busse S, Myint AM, Gos T, Dobrowolny H, Müller UJ, Bogerts B, Bernstein HG, Steiner J. Decreased quinolinic acid in the hippocampus of depressive patients: evidence for local anti-inflammatory and neuroprotective responses? Eur Arch Psychiatry Clin Neurosci. 2014 (submitted).

Campuzano O, Castillo-Ruiz MM, Acarin L, Castellano B, Gonzalez B. Distinct pattern of microglial response, cyclooxygenase-2, and inducible nitric oxide synthase expression in the aged rat brain after excitotoxic damage. J Neurosci Res. 2008;86(14):3170–83.

Capuron L, Schroecksnadel S, Feart C, Aubert A, Higueret D, Barberger-Gateau P, Laye S, Fuchs D. Chronic low-grade inflammation in elderly persons is associated with altered tryptophan and tyrosine metabolism: role in neuropsychiatric symptoms. Biol Psychiatry. 2011;70(2):175–82.

Carlin JM, Borden EC, Sondel PM, Byrne GI. Biologic-response-modifier-induced indoleamine 2,3-dioxygenase activity in human peripheral blood mononuclear cell cultures. J Immunol. 1987;139(7):2414–8.

Carroll BJ, Martin FI, Davies B. Resistance to suppression by dexamethasone of plasma 11-O.H.C.S. levels in severe depressive illness. Br Med J. 1968a;3(5613):285–7.

Carroll BJ, Martin FI, Davies B. Pituitary–adrenal function in depression. Lancet. 1968b; 1(7556):1373–4.

Chandley MJ, Szebeni A, Szebeni K, Crawford JD, Stockmeier CA, Turecki G, Kostrzewa RM, Ordway GA. Elevated gene expression of glutamate receptors in noradrenergic neurons from the locus coeruleus in major depression. Int J Neuropsychopharmacol. 2014:17(10):1569–78.

Chen S, Averett NT, Manelli A, Ladu MJ, May W, Ard MD. Isoform-specific effects of apolipoprotein E on secretion of inflammatory mediators in adult rat microglia. J Alzheimers Dis. 2005;7(1):25–35.

Chen CY, Tzeng NS, Chen YC. Maintenance therapy of celecoxib for major depression with mimicking neuropsychological dysfunction. Gen Hosp Psychiatry. 2010;32(6):647. e647–649.

Chiarugi A, Calvani M, Meli E, Traggiai E, Moroni F. Synthesis and release of neurotoxic kynurenine metabolites by human monocyte-derived macrophages. J Neuroimmunol. 2001; 120(1–2):190–8.

Claes S, Myint AM, Domschke K, Del-Favero J, Entrich K, Engelborghs S, De Deyn P, Mueller N, Baune B, Rothermundt M. The kynurenine pathway in major depression: haplotype analysis of three related functional candidate genes. Psychiatry Res. 2011;188(3):355–60.

Condray R, Dougherty Jr GG, Keshavan MS, Reddy RD, Haas GL, Montrose DM, Matson WR, McEvoy J, Kaddurah-Daouk R, Yao JK. 3-Hydroxykynurenine and clinical symptoms in first-episode neuroleptic-naive patients with schizophrenia. Int J Neuropsychopharmacol. 2011; 14(6):756–67.

Connor MJ, Pheasant AE, Blair JA. The identification of p-acetamidobenzoate as a folate degradation product in rat urine. Biochem J. 1979;178(3):795–7.

Depboylu C, Weihe E, Eiden LE. COX1 and COX2 expression in non-neuronal cellular compartments of the rhesus macaque brain during lentiviral infection. Neurobiol Dis. 2011;42(1): 108–15.

Dickerson F, Stallings C, Origoni A, Boronow J, Yolken R. Elevated serum levels of C-reactive protein are associated with mania symptoms in outpatients with bipolar disorder. Prog Neuropsychopharmacol Biol Psychiatry. 2007;31(4):952–5.

Dinnerstein AJ, Halm J. Modification of placebo effects by means of drugs: effects of aspirin and placebos on self-rated moods. J Abnorm Psychol. 1970;75(3):308–14.

Drexhage RC, Knijff EM, Padmos RC, Heul-Nieuwenhuijzen L, Beumer W, Versnel MA, Drexhage HA. The mononuclear phagocyte system and its cytokine inflammatory networks in schizophrenia and bipolar disorder. Expert Rev Neurother. 2010;10(1):59–76.

Erhardt S, Blennow K, Nordin C, Skogh E, Lindstrom LH, Engberg G. Kynurenic acid levels are elevated in the cerebrospinal fluid of patients with schizophrenia. Neurosci Lett. 2001; 313(1–2):96–8.

Erhardt S, Schwieler L, Engberg G. Kynurenic acid and schizophrenia. Adv Exp Med Biol. 2003;527:155–65.

Erhardt S, Lim CK, Linderholm KR, Janelidze S, Lindqvist D, Samuelsson M, Lundberg K, Postolache TT, Traskman-Bendz L, Guillemin GJ, et al. Connecting inflammation with glutamate agonism in suicidality. Neuropsychopharmacology. 2013;38(5):743–52.

Fekadu A, Mesfin M, Medhin G, Alem A, Teferra S, Gebre-Eyesus T, Seboxa T, Assefa A, Hussein J, Lemma MT, et al. Adjuvant therapy with minocycline for schizophrenia (The MINOS Trial): study protocol for a double-blind randomized placebo-controlled trial. Trials. 2013;14:406.

Fernstrom JD. Effects on the diet on brain neurotransmitters. Metabolism. 1977;26(2):207–23.

Fischman LG. Dreams, hallucinogenic drug states, and schizophrenia: a psychological and biological comparison. Schizophr Bull. 1983;9(1):73–94.

Fuchs D, Jaeger M, Widner B, Wirleitner B, Artner-Dworzak E, Leblhuber F. Is hyperhomocysteinemia due to the oxidative depletion of folate rather than to insufficient dietary intake? Clin Chem Lab Med. 2001;39(8):691–4.

Fusar-Poli P, Berger G. Eicosapentaenoic acid interventions in schizophrenia: meta-analysis of randomized, placebo-controlled studies. J Clin Psychopharmacol. 2012;32(2):179–85.

Gabbay V, Liebes L, Katz Y, Liu S, Mendoza S, Babb JS, Klein RG, Gonen O. The kynurenine pathway in adolescent depression: preliminary findings from a proton MR spectroscopy study. Prog Neuropsychopharmacol Biol Psychiatry. 2010;34(1):37–44.

Gal EM, Sherman AD. L-kynurenine: its synthesis and possible regulatory function in brain. Neurochem Res. 1980;5(3):223–39.

Galecki P, Szemraj J, Bienkiewicz M, Zboralski K, Galecka E. Oxidative stress parameters after combined fluoxetine and acetylsalicylic acid therapy in depressive patients. Hum Psychopharmacol. 2009;24(4):277–86.

Garcia-Miss Mdel R, Perez-Mutul J, Lopez-Canul B, Solis-Rodriguez F, Puga-Machado L, Oxte-Cabrera A, Gurubel-Maldonado J, Arankowsky-Sandoval G. Folate, homocysteine, interleukin-6, and tumor necrosis factor alfa levels, but not the methylenetetrahydrofolate reductase C677T polymorphism, are risk factors for schizophrenia. J Psychiatr Res. 2010;44(7):441–6.

Gos T, Myint AM, Schiltz K, Meyer-Lotz G, Dobrowolny H, Busse S, Muller UJ, Mawrin C, Bernstein HG, Bogerts B, et al. Reduced microglial immunoreactivity for endogenous NMDA receptor agonist quinolinic acid in the hippocampus of schizophrenia patients. Brain Behav Immun. 2014;41:59–64.

Grant RS, Kapoor V. Murine glial cells regenerate NAD, after peroxide-induced depletion, using either nicotinic acid, nicotinamide, or quinolinic acid as substrates. J Neurochem. 1998; 70(4):1759–63.

Grant RS, Naif H, Espinosa M, Kapoor V. IDO induction in IFN-gamma activated astroglia: a role in improving cell viability during oxidative stress. Redox Rep. 2000;5(2–3):101–4.

Grosso G, Pajak A, Marventano S, Castellano S, Galvano F, Bucolo C, Drago F, Caraci F. Role of omega-3 fatty acids in the treatment of depressive disorders: a comprehensive meta-analysis of randomized clinical trials. PLoS One. 2014;9(5):e96905.

Guillemin GJ, Smith DG, Kerr SJ, Smythe GA, Kapoor V, Armati PJ, Brew BJ. Characterisation of kynurenine pathway metabolism in human astrocytes and implications in neuropathogenesis. Redox Rep. 2000;5(2–3):108–11.

Guillemin GJ, Kerr SJ, Smythe GA, Smith DG, Kapoor V, Armati PJ, Croitoru J, Brew BJ. Kynurenine pathway metabolism in human astrocytes: a paradox for neuronal protection. J Neurochem. 2001;78(4):842–53.

Guillemin GJ, Smythe G, Takikawa O, Brew BJ. Expression of indoleamine 2,3-dioxygenase and production of quinolinic acid by human microglia, astrocytes, and neurons. Glia. 2005a; 49(1):15–23.

Guillemin GJ, Wang L, Brew BJ. Quinolinic acid selectively induces apoptosis of human astrocytes: potential role in AIDS dementia complex. J Neuroinflammation. 2005b;2:16.

Guillin O, Abi-Dargham A, Laruelle M. Neurobiology of dopamine in schizophrenia. Int Rev Neurobiol. 2007;78:1–39.

Gustavsson A, Svensson M, Jacobi F, Allgulander C, Alonso J, Beghi E, Dodel R, Ekman M, Faravelli C, Fratiglioni L, et al. Cost of disorders of the brain in Europe 2010. Eur Neuropsychopharmacol. 2011;21(10):718–79.

Halaris A, Myint AM, Savant V, Meresh E, Lim C, Hoppensteadt D, Fareed J, Sinacore J. Does Escitalopram reduce neurotoxic effects in major depression? Psychiatry Research. 2015. Accepted.

Heyes MP, Saito K, Major EO, Milstien S, Markey SP, Vickers JH. A mechanism of quinolinic acid formation by brain in inflammatory neurological disease. Attenuation of synthesis from L-tryptophan by 6-chlorotryptophan and 4-chloro-3-hydroxyanthranilate. Brain. 1993;116(Pt 6): 1425–50.

Heyes MP, Achim CL, Wiley CA, Major EO, Saito K, Markey SP. Human microglia convert l-tryptophan into the neurotoxin quinolinic acid. Biochem J. 1996;320(Pt 2):595–7.

Hilmas C, Pereira EF, Alkondon M, Rassoulpour A, Schwarcz R, Albuquerque EX. The brain metabolite kynurenic acid inhibits alpha7 nicotinic receptor activity and increases non-alpha7 nicotinic receptor expression: physiopathological implications. J Neurosci. 2001;21(19):7463–73.

Hindmarch I. Expanding the horizons of depression: beyond the monoamine hypothesis. Hum Psychopharmacol. 2001;16(3):203–18.

Hosoda S, Takimura H, Shibayama M, Kanamura H, Ikeda K, Kumada H. Psychiatric symptoms related to interferon therapy for chronic hepatitis C: clinical features and prognosis. Psychiatry Clin Neurosci. 2000;54(5):565–72.

Janelidze S, Ventorp F, Erhardt S, Hansson O, Minthon L, Flax J, Samuelsson M, Traskman-Bendz L, Brundin L. Altered chemokine levels in the cerebrospinal fluid and plasma of suicide attempters. Psychoneuroendocrinology. 2013;38(6):853–62.

Javitt DC, Zukin SR. Recent advances in the phencyclidine model of schizophrenia. Am J Psychiatry. 1991;148(10):1301–8.

Johansson AS, Owe-Larsson B, Asp L, Kocki T, Adler M, Hetta J, Gardner R, Lundkvist GB, Urbanska EM, Karlsson H. Activation of kynurenine pathway in ex vivo fibroblasts from patients with bipolar disorder or schizophrenia: cytokine challenge increases production of 3-hydroxykynurenine. J Psychiatr Res. 2013;47(11):1815–23.

Judd LL, Akiskal HS. The prevalence and disability of bipolar spectrum disorders in the US population: re-analysis of the ECA database taking into account subthreshold cases. J Affect Disord. 2003;73(1–2):123–31.

Kang JH, Cook N, Manson J, Buring JE, Grodstein F. Low dose aspirin and cognitive function in the women's health study cognitive cohort. BMJ. 2007;334(7601):987.

Kargar M, Yousefi A, Mojtahedzadeh M, Akhondzadeh S, Artounian V, Abdollahi A, Ahmadvand A, Ghaeli P. Effects of celecoxib on inflammatory markers in bipolar patients undergoing electroconvulsive therapy: a placebo-controlled, double-blind, randomised study. Swiss Med Wkly. 2014;144:w13880.

Kelly DL, Vyas G, Richardson CM, Koola M, McMahon RP, Buchanan RW, Wehring HJ. Adjunct minocycline to clozapine treated patients with persistent schizophrenia symptoms. Schizophr Res. 2011;133(1–3):257–8.

Khodaie-Ardakani MR, Mirshafiee O, Farokhnia M, Tajdini M, Hosseini SM, Modabbernia A, Rezaei F, Salehi B, Yekehtaz H, Ashrafi M, et al. Minocycline add-on to risperidone for treatment of negative symptoms in patients with stable schizophrenia: randomized double-blind placebo-controlled study. Psychiatry Res. 2014;215(3):540–6.

Kim JP, Choi DW. Quinolinate neurotoxicity in cortical cell culture. Neuroscience. 1987;23(2): 423–32.

Kim YK, Lee MS, Suh KY. Decreased interleukin-2 production in Korean schizophrenic patients. Biol Psychiatry. 1998;43(9):701–4.

Kim YK, Kim L, Lee MS. Relationships between interleukins, neurotransmitters and psychopathology in drug-free male schizophrenics. Schizophr Res. 2000;44(3):165–75.

Kim YK, Myint AM, Lee BH, Han CS, Lee SW, Leonard BE, Steinbusch HW. T-helper types 1, 2, and 3 cytokine interactions in symptomatic manic patients. Psychiatry Res. 2004a; 129(3):267–72.

Kim YK, Myint AM, Lee BH, Han CS, Lee HJ, Kim DJ, Leonard BE. Th1, Th2 and Th3 cytokine alteration in schizophrenia. Prog Neuropsychopharmacol Biol Psychiatry. 2004b;28(7):1129–34.

Kim YK, Myint AM, Verkerk R, Scharpe S, Steinbusch H, Leonard B. Cytokine changes and tryptophan metabolites in medication-naive and medication-free schizophrenic patients. Neuropsychobiology. 2009;59(2):123–9.

Knox WE. Two mechanisms which increase in vivo the liver tryptophan peroxidase activity: specific enzyme adaptation and stimulation of the pituitary adrenal system. Br J Exp Pathol. 1951;32(5):462–9.

Kubera M, Lin AH, Kenis G, Bosmans E, van Bockstaele D, Maes M. Anti-inflammatory effects of antidepressants through suppression of the interferon-gamma/interleukin-10 production ratio. J Clin Psychopharmacol. 2001;21(2):199–206.

Laan W, Grobbee DE, Selten JP, Heijnen CJ, Kahn RS, Burger H. Adjuvant aspirin therapy reduces symptoms of schizophrenia spectrum disorders: results from a randomized, double-blind, placebo-controlled trial. J Clin Psychiatry. 2010;71(5):520–7.

Lapin IP, Oxenkrug GF. Intensification of the central serotoninergic processes as a possible determinant of the thymoleptic effect. Lancet. 1969;1(7586):132–6.

Lardy HA. The role of tryptophan metabolites in regulating gluconeogenesis. Am J Clin Nutr. 1971;24(7):764–5.

Leklem JE. Quantitative aspects of tryptophan metabolism in humans and other species: a review. Am J Clin Nutr. 1971;24(6):659–72.

Levkovitz Y, Mendlovich S, Riwkes S, Braw Y, Levkovitch-Verbin H, Gal G, Fennig S, Treves I, Kron S. A double-blind, randomized study of minocycline for the treatment of negative and cognitive symptoms in early-phase schizophrenia. J Clin Psychiatry. 2010;71(2):138–49.

Lewis DA, Hashimoto T, Volk DW. Cortical inhibitory neurons and schizophrenia. Nat Rev Neurosci. 2005;6(4):312–24.

Licinio J, Seibyl JP, Altemus M, Charney DS, Krystal JH. Elevated CSF levels of interleukin-2 in neuroleptic-free schizophrenic patients. Am J Psychiatry. 1993;150(9):1408–10.

Lieberman HR, Wurtman RJ, Emde GG, Coviella IL. The effects of caffeine and aspirin on mood and performance. J Clin Psychopharmacol. 1987;7(5):315–20.

Lin A, Kenis G, Bignotti S, Tura GJ, De Jong R, Bosmans E, Pioli R, Altamura C, Scharpe S, Maes M. The inflammatory response system in treatment-resistant schizophrenia: increased serum interleukin-6. Schizophr Res. 1998;32(1):9–15.

Lindqvist D, Janelidze S, Hagell P, Erhardt S, Samuelsson M, Minthon L, Hansson O, Bjorkqvist M, Traskman-Bendz L, Brundin L. Interleukin-6 is elevated in the cerebrospinal fluid of suicide attempters and related to symptom severity. Biol Psychiatry. 2009;66(3):287–92.

Linnoila M, Whorton AR, Rubinow DR, Cowdry RW, Ninan PT, Waters RN. CSF prostaglandin levels in depressed and schizophrenic patients. Arch Gen Psychiatry. 1983;40(4):405–6.

Liu F, Guo X, Wu R, Ou J, Zheng Y, Zhang B, Xie L, Zhang L, Yang L, Yang S, et al. Minocycline supplementation for treatment of negative symptoms in early-phase schizophrenia: a double blind, randomized, controlled trial. Schizophr Res. 2014;153(1–3):169–76.

Loo H, Saiz-Ruiz J, Costa e Silva J, Ansseau M, Herrington R, Vaz-Serra A, Dilling H, de Risio S. Efficacy and safety of tianeptine in the treatment of depressive disorders in comparison with fluoxetine. J Affect Disord. 1999;56(2–3):109–18.

Maes M. Cytokines in major depression. Biol Psychiatry. 1994;36(7):498–9.

Maes M, Lambrechts J, Suy E, Vandervorst C, Bosmans E. Absolute number and percentage of circulating natural killer, non-MHC-restricted T cytotoxic, and phagocytic cells in unipolar depression. Neuropsychobiology. 1994;29(4):157–63.

Maes M, Bosmans E, Calabrese J, Smith R, Meltzer HY. Interleukin-2 and interleukin-6 in schizophrenia and mania: effects of neuroleptics and mood stabilizers. J Psychiatr Res. 1995;29(2): 141–52.

Maes M, Song C, Lin AH, Bonaccorso S, Kenis G, De Jongh R, Bosmans E, Scharpe S. Negative immunoregulatory effects of antidepressants: inhibition of interferon-gamma and stimulation of interleukin-10 secretion. Neuropsychopharmacology. 1999;20(4):370–9.

Mann JJ, Oquendo MA, Watson KT, Boldrini M, Malone KM, Ellis SP, Sullivan G, Cooper TB, Xie S, Currier D. Anxiety in major depression and cerebrospinal fluid free gamma-aminobutyric acid. Depress Anxiety. 2014;31(10):814–21.

Marty S, Wehrle R, Sotelo C. Neuronal activity and brain-derived neurotrophic factor regulate the density of inhibitory synapses in organotypic slice cultures of postnatal hippocampus. J Neurosci. 2000;20(21):8087–95.

Mathers C, Loncar D. Updated projections of global mortality and burden of disease, 2002–2030: data sources, methods and results. Geneva: WHO; 2005.

McGirr A, Berlim MT, Bond DJ, Fleck MP, Yatham LN, Lam RW. A systematic review and meta-analysis of randomized, double-blind, placebo-controlled trials of ketamine in the rapid treatment of major depressive episodes. Psychol Med. 2015;45(4):693–704.

Mehta D, Raison CL, Woolwine BJ, Haroon E, Binder EB, Miller AH, Felger JC. Transcriptional signatures related to glucose and lipid metabolism predict treatment response to the tumor necrosis factor antagonist infliximab in patients with treatment-resistant depression. Brain Behav Immun. 2013;31:205–15.

Mellor AL, Munn DH. Tryptophan catabolism and T-cell tolerance: immunosuppression by starvation? Immunol Today. 1999;20(10):469–73.

Messias EL, Chen CY, Eaton WW. Epidemiology of schizophrenia: review of findings and myths. Psychiatr Clin North Am. 2007;30(3):323–38.
Meyer U, Feldon J, Schedlowski M, Yee BK. Immunological stress at the maternal–foetal interface: a link between neurodevelopment and adult psychopathology. Brain Behav Immun. 2006;20(4):378–88.
Miller CL, Llenos IC, Dulay JR, Barillo MM, Yolken RH, Weis S. Expression of the kynurenine pathway enzyme tryptophan 2,3-dioxygenase is increased in the frontal cortex of individuals with schizophrenia. Neurobiol Dis. 2004;15(3):618–29.
Miller CL, Llenos IC, Dulay JR, Weis S. Upregulation of the initiating step of the kynurenine pathway in postmortem anterior cingulate cortex from individuals with schizophrenia and bipolar disorder. Brain Res. 2006;1073–1074:25–37.
Miyaoka T, Yasukawa R, Yasuda H, Hayashida M, Inagaki T, Horiguchi J. Possible antipsychotic effects of minocycline in patients with schizophrenia. Prog Neuropsychopharmacol Biol Psychiatry. 2007;31(1):304–7.
Miyaoka T, Wake R, Furuya M, Liaury K, Ieda M, Kawakami K, Tsuchie K, Taki M, Ishihara K, Araki T, et al. Minocycline as adjunctive therapy for patients with unipolar psychotic depression: an open-label study. Prog Neuropsychopharmacol Biol Psychiatry. 2012;37(2):222–6.
Modabbernia A, Taslimi S, Brietzke E, Ashrafi M. Cytokine alterations in bipolar disorder: a meta-analysis of 30 studies. Biol Psychiatry. 2013;74(1):15–25.
Moffett JR, Namboodiri MA. Tryptophan and the immune response. Immunol Cell Biol. 2003;81(4):247–65.
Moffett JR, Blinder KL, Venkateshan CN, Namboodiri MA. Differential effects of kynurenine and tryptophan treatment on quinolinate immunoreactivity in rat lymphoid and non-lymphoid organs. Cell Tissue Res. 1998;293(3):525–34.
Müller N. The role of anti-inflammatory treatment in psychiatric disorders. Psychiatr Danub. 2013;25(3):292–8.
Muller N, Dobmeier P, Empl M, Riedel M, Schwarz M, Ackenheil M. Soluble IL-6 receptors in the serum and cerebrospinal fluid of paranoid schizophrenic patients. Eur Psychiatry. 1997;12(6):294–9.
Muller N, Schwarz MJ, Dehning S, Douhe A, Cerovecki A, Goldstein-Muller B, Spellmann I, Hetzel G, Maino K, Kleindienst N, et al. The cyclooxygenase-2 inhibitor celecoxib has therapeutic effects in major depression: results of a double-blind, randomized, placebo controlled, add-on pilot study to reboxetine. Mol Psychiatry. 2006;11(7):680–4.
Munkholm K, Brauner JV, Kessing LV, Vinberg M. Cytokines in bipolar disorder vs. healthy control subjects: a systematic review and meta-analysis. J Psychiatr Res. 2013;47(9):1119–33.
Musselman DL, Miller AH, Porter MR, Manatunga A, Gao F, Penna S, Pearce BD, Landry J, Glover S, McDaniel JS, et al. Higher than normal plasma interleukin-6 concentrations in cancer patients with depression: preliminary findings. Am J Psychiatry. 2001;158(8):1252–7.
Musso T, Gusella GL, Brooks A, Longo DL, Varesio L. Interleukin-4 inhibits indoleamine 2,3-dioxygenase expression in human monocytes. Blood. 1994;83(5):1408–11.
Myint AM. Kynurenines: from the perspective of major psychiatric disorders. FEBS J. 2012; 279(8):1375–85.
Myint AM, Leonard BE, Steinbusch HW, Kim YK. Th1, Th2, and Th3 cytokine alterations in major depression. J Affect Disord. 2005;88(2):167–73.
Myint AM, Kim YK, Verkerk R, Scharpe S, Steinbusch H, Leonard B. Kynurenine pathway in major depression: evidence of impaired neuroprotection. J Affect Disord. 2007a;98(1–2):143–51.
Myint AM, Kim YK, Verkerk R, Park SH, Scharpe S, Steinbusch HW, Leonard BE. Tryptophan breakdown pathway in bipolar mania. J Affect Disord. 2007b;102(1–3):65–72.
Myint AM, Schwarz M, Verkerk R, Scharpe S, Steinbusch HWM, Leonard BE, Kim YK. Imbalance of kynurenine metabolites in drug naive schizophrenia. Brain Behav Immun. 2011a;25(8):6.
Myint AM, Schwarz MJ, Verkerk R, Mueller HH, Zach J, Scharpe S, Steinbusch HW, Leonard BE, Kim YK. Reversal of imbalance between kynurenic acid and 3-hydroxykynurenine by antipsychotics in medication-naive and medication-free schizophrenic patients. Brain Behav Immun. 2011b;25(8):1576–81.

Myint AM, Bondy B, Baghai TC, Eser D, Nothdurfter C, Schule C, Zill P, Muller N, Rupprecht R, Schwarz MJ. Tryptophan metabolism and immunogenetics in major depression: a role for interferon-gamma gene. Brain Behav Immun. 2013;31:128–33.
Nery FG, Monkul ES, Hatch JP, Fonseca M, Zunta-Soares GB, Frey BN, Bowden CL, Soares JC. Celecoxib as an adjunct in the treatment of depressive or mixed episodes of bipolar disorder: a double-blind, randomized, placebo-controlled study. Hum Psychopharmacol. 2008;23(2):87–94.
Nitta M, Kishimoto T, Muller N, Weiser M, Davidson M, Kane JM, Correll CU. Adjunctive use of nonsteroidal anti-inflammatory drugs for schizophrenia: a meta-analytic investigation of randomized controlled trials. Schizophr Bull. 2013;39(6):1230–41.
Ohishi K, Ueno R, Nishino S, Sakai T, Hayaishi O. Increased level of salivary prostaglandins in patients with major depression. Biol Psychiatry. 1988;23(4):326–34.
Okuda S, Nishiyama N, Saito H, Katsuki H. 3-Hydroxykynurenine, an endogenous oxidative stress generator, causes neuronal cell death with apoptotic features and region selectivity. J Neurochem. 1998;70(1):299–307.
Olney JW, Labruyere J, Wang G, Wozniak DF, Price MT, Sesma MA. NMDA antagonist neurotoxicity: mechanism and prevention. Science. 1991;254(5037):1515–8.
Peet M. Omega-3 polyunsaturated fatty acids in the treatment of schizophrenia. Isr J Psychiatry Relat Sci. 2008;45:19–25.
Perkins MN, Stone TW. An iontophoretic investigation of the actions of convulsant kynurenines and their interaction with the endogenous excitant quinolinic acid. Brain Res. 1982;247(1):184–7.
Pineyro G, Blier P. Autoregulation of serotonin neurons: role in antidepressant drug action. Pharmacol Rev. 1999;51(3):533–91.
Raison CL, Dantzer R, Kelley KW, Lawson MA, Woolwine BJ, Vogt G, Spivey JR, Saito K, Miller AH. CSF concentrations of brain tryptophan and kynurenines during immune stimulation with IFN-alpha: relationship to CNS immune responses and depression. Mol Psychiatry. 2010;15(4):393–403.
Raison CL, Rutherford RE, Woolwine BJ, Shuo C, Schettler P, Drake DF, Haroon E, Miller AH. A randomized controlled trial of the tumor necrosis factor antagonist infliximab for treatment-resistant depression: the role of baseline inflammatory biomarkers. JAMA Psychiatry. 2013;70(1):31–41.
Rao JS, Harry GJ, Rapoport SI, Kim HW. Increased excitotoxicity and neuroinflammatory markers in postmortem frontal cortex from bipolar disorder patients. Mol Psychiatry. 2010;15(4):384–92.
Reininghaus EZ, McIntyre RS, Reininghaus B, Geisler S, Bengesser SA, Lackner N, Hecht K, Birner A, Kattnig F, Unterweger R, et al. Tryptophan breakdown is increased in euthymic overweight individuals with bipolar disorder: a preliminary report. Bipolar Disord. 2014;16(4):432–40.
Rothermundt M, Arolt V, Peters M, Gutbrodt H, Fenker J, Kersting A, Kirchner H. Inflammatory markers in major depression and melancholia. J Affect Disord. 2001;63(1–3):93–102.
Rubin LH, Carter CS, Bishop JR, Pournajafi-Nazarloo H, Harris MS, Hill SK, Reilly JL, Sweeney JA. Peripheral vasopressin but not oxytocin relates to severity of acute psychosis in women with acutely-ill untreated first-episode psychosis. Schizophr Res. 2013;146(1–3):138–43.
Rubin LH, Carter CS, Bishop JR, Pournajafi-Nazarloo H, Drogos LL, Hill SK, Ruocco AC, Keedy SK, Reilly JL, Keshavan MS, et al. Reduced levels of vasopressin and reduced behavioral modulation of oxytocin in psychotic disorders. Schizophr Bull. 2014;40(6):1374–84.
Sachar EJ, Hellman L, Fukushima DK, Gallagher TF. Cortisol production in depressive illness. A clinical and biochemical clarification. Arch Gen Psychiatry. 1970;23(4):289–98.
Saha S, Chant D, Welham J, McGrath J. A systematic review of the prevalence of schizophrenia. PLoS Med. 2005;2(5):e141.
Salter M, Pogson CI. The role of tryptophan 2,3-dioxygenase in the hormonal control of tryptophan metabolism in isolated rat liver cells. Effects of glucocorticoids and experimental diabetes. Biochem J. 1985;229(2):499–504.

Sartorius N. Concepts of depression: sporadic revolutions of continuous evolution. Hum Psychopharmacol. 2001;16(S1):S3–6.
Schafer M, Messer T, Wegner U, Schmid-Wendtner MH, Volkenandt M. Psychiatric side effects during adjuvant therapy with interferon-alphain patients with malignant melanoma. Clinical evaluation as well as diagnostic and therapeutic possibilities. Hautarzt. 1999;50(9):654–8.
Schwarcz R, Whetsell Jr WO, Mangano RM. Quinolinic acid: an endogenous metabolite that produces axon-sparing lesions in rat brain. Science. 1983;219(4582):316–8.
Schwarcz R, Rassoulpour A, Wu HQ, Medoff D, Tamminga CA, Roberts RC. Increased cortical kynurenate content in schizophrenia. Biol Psychiatry. 2001;50(7):521–30.
Schwarz MJ, Myint AM, Holdenrieder S, Seemüller F, Tischinger M, Riedel M, Müller N. The association between peripheral and CSF kynurenine metabolites in patients with affective disorders and schizophrenia. Brain Behav Imm. Under review.
Scott LV, Dinan TG. Vasopressin and the regulation of hypothalamic–pituitary–adrenal axis function: implications for the pathophysiology of depression. Life Sci. 1998;62(22):1985–98.
Scott J, Palmer S, Paykel E, Teasdale J, Hayhurst H. Use of cognitive therapy for relapse prevention in chronic depression. Cost-effectiveness study. Br J Psychiatry. 2003;182:221–7.
Seidel A, Arolt V, Hunstiger M, Rink L, Behnisch A, Kirchner H. Increased CD56+ natural killer cells and related cytokines in major depression. Clin Immunol Immunopathol. 1996;78(1): 83–5.
Singh B, Bera NK, Nayak CR, Chaudhuri TK. Decreased serum levels of interleukin-2 and interleukin-6 in Indian Bengalee schizophrenic patients. Cytokine. 2009;47(1):1–5.
Sluzewska A, Rybakowski JK, Laciak M, Mackiewicz A, Sobieska M, Wiktorowicz K. Interleukin-6 serum levels in depressed patients before and after treatment with fluoxetine. Ann N Y Acad Sci. 1995;762:474–6.
Sluzewska A, Rybakowski J, Bosmans E, Sobieska M, Berghmans R, Maes M, Wiktorowicz K. Indicators of immune activation in major depression. Psychiatry Res. 1996;64(3):161–7.
Song XQ, Lv LX, Li WQ, Hao YH, Zhao JP. The interaction of nuclear factor-kappa B and cytokines is associated with schizophrenia. Biol Psychiatry. 2009;65(6):481–8.
Sperner-Unterweger B, Kohl C, Fuchs D. Immune changes and neurotransmitters: possible interactions in depression? Prog Neuropsychopharmacol Biol Psychiatry. 2014;48:268–76.
Steen NE, Methlie P, Lorentzen S, Hope S, Barrett EA, Larsson S, Mork E, Almas B, Lovas K, Agartz I, et al. Increased systemic cortisol metabolism in patients with schizophrenia and bipolar disorder: a mechanism for increased stress vulnerability? J Clin Psychiatry. 2011;72(11):1515–21.
Steiner J, Walter M, Gos T, Guillemin GJ, Bernstein HG, Sarnyai Z, Mawrin C, Brisch R, Bielau H, Meyer zu Schwabedissen L, et al. Severe depression is associated with increased microglial quinolinic acid in subregions of the anterior cingulate gyrus: evidence for an immune-modulated glutamatergic neurotransmission? J Neuroinflammation. 2011;8:94.
Steullet P, Cabungcal JH, Monin A, Dwir D, O'Donnell P, Cuenod M, Do KQ. Redox dysregulation, neuroinflammation, and NMDA receptor hypofunction: a "central hub" in schizophrenia pathophysiology? Schizophr Res. 2014;1–11. In press [Epub ahead of print].
Stuart MJ, Baune BT. Chemokines and chemokine receptors in mood disorders, schizophrenia, and cognitive impairment: a systematic review of biomarker studies. Neurosci Biobehav Rev. 2014;42:93–115.
Szebeni A, Szebeni K, DiPeri T, Chandley MJ, Crawford JD, Stockmeier CA, Ordway GA. Shortened telomere length in white matter oligodendrocytes in major depression: potential role of oxidative stress. Int J Neuropsychopharmacol. 2014;17(10):1579–1589.
Takahashi T, Wood SJ, Soulsby B, McGorry PD, Tanino R, Suzuki M, Velakoulis D, Pantelis C. Follow-up MRI study of the insular cortex in first-episode psychosis and chronic schizophrenia. Schizophr Res. 2009;108(1–3):49–56.
Valentine AD, Meyers CA, Kling MA, Richelson E, Hauser P. Mood and cognitive side effects of interferon-alpha therapy. Semin Oncol. 1998;25(1 Suppl 1):39–47.

Watanabe Y, Fujiwara M, Yoshida R, Hayaishi O. Stereospecificity of hepatic L-tryptophan 2,3-dioxygenase. Biochem J. 1980;189(3):393–405.

Werner-Felmayer G, Golderer G, Werner ER. Tetrahydrobiopterin biosynthesis, utilization and pharmacological effects. Curr Drug Metab. 2002;3(2):159–73.

Wichers MC, Koek GH, Robaeys G, Verkerk R, Scharpe S, Maes M. IDO and interferon-alpha-induced depressive symptoms: a shift in hypothesis from tryptophan depletion to neurotoxicity. Mol Psychiatry. 2005;10(6):538–44.

Wichers MC, Kenis G, Leue C, Koek G, Robaeys G, Maes M. Baseline immune activation as a risk factor for the onset of depression during interferon-alpha treatment. Biol Psychiatry. 2006;60(1):77–9.

Widner B, Fuchs D, Leblhuber F, Sperner-Unterweger B. Does disturbed homocysteine and folate metabolism in depression result from enhanced oxidative stress? J Neurol Neurosurg Psychiatry. 2001;70(3):419.

Wilde MI, Benfield P. Tianeptine. A review of its pharmacodynamic and pharmacokinetic properties, and therapeutic efficacy in depression and coexisting anxiety and depression. Drugs. 1995;49(3):411–39.

Wittchen HU, Jacobi F. Size and burden of mental disorders in Europe—a critical review and appraisal of 27 studies. Eur Neuropsychopharmacol. 2005;15(4):357–76.

Wittchen HU, Jacobi F, Rehm J, Gustavsson A, Svensson M, Jonsson B, Olesen J, Allgulander C, Alonso J, Faravelli C, et al. The size and burden of mental disorders and other disorders of the brain in Europe 2010. Eur Neuropsychopharmacol. 2011;21(9):655–79.

Yao JK, Dougherty Jr GG, Reddy RD, Keshavan MS, Montrose DM, Matson WR, Rozen S, Krishnan RR, McEvoy J, Kaddurah-Daouk R. Altered interactions of tryptophan metabolites in first-episode neuroleptic-naive patients with schizophrenia. Mol Psychiatry. 2010; 15(9):938–53.

Yasui H, Takai K, Yoshida R, Hayaishi O. Interferon enhances tryptophan metabolism by inducing pulmonary indoleamine 2,3-dioxygenase: its possible occurrence in cancer patients. Proc Natl Acad Sci U S A. 1986;83(17):6622–6.

Chapter 11
Depression, the Metabolic Syndrome and Neurodegeneration

Brian E. Leonard

Abstract The review summarises the evidence that chronic low grade inflammation triggers a number of metabolic changes that are ultimately responsible for the physical ill-health (such as type 2 diabetes, heart disease and cancer) which frequently characterise major depression.

The possible mechanisms involve a dysfunction of glucose metabolism due to an insensitivity of insulin receptor signalling which results from the action of superoxide radicals produced by intermediates of the tryptophan–kynurenine pathway. Additional metabolic changes occur as a result of a decrease in mitochondrial activity.

Increasing epidemiological and clinical evidence suggests that chronic depression is often a prelude to dementia in later life. Possible metabolic mechanisms whereby this occurs involve the neurodegenerative effects of pro-inflammatory cytokines, the consequences of oxidative stress and the action of the neurotoxins formed by the tryptophan–kynurenine pathway. The review concludes with a summary of some of the nutritional factors that may have a neuroprotective action, largely due to their anti-inflammatory effects.

Keywords Insulin insensitivity • Diabetes type 2 • Pro-inflammatory cytokines • Oxidative stress • Endogenous neurotoxins • Neurodegeneration • Alzheimer's disease • B-vitamins • Anti inflammatory lipids

Introduction

In recent years, depression research has extended from the consideration of the consequences of neurotransmitter dysfunction to the role that the endocrine and immune systems may contribute to the pathophysiology of the disorder. There are a number of reasons for these changes. Thus the occurrence of low grade

B.E. Leonard (✉)
National University of Ireland, Galway, Ireland
e-mail: belucg@iol.ie

N. Müller et al. (eds.), *Immunology and Psychiatry*, Current Topics in Neurotoxicity 8, DOI 10.1007/978-3-319-13602-8_11

inflammation has been shown to impact on neurotransmitter function and shown to be associated with many of the metabolic changes associated with depression (Leonard 2010; Maes 1995; Smith 1991; Leonard 2014). These observations have led to the identification of a link between the consequences of chronic inflammation and an increase in the frequency of type 2 diabetes and heart disease in chronically depressed patients and also with the increased possibility of dementia in the elderly patient (McIntyre et al. 2012; Leonard 2013).

The question therefore arises whether such disparate changes are epiphenomena of a chronic psychiatric condition reflecting life-style changes involving nutrition and poor diet, lack of self-care associated with exposure to pathogens, etc., or more fundamentally that the metabolic changes are initiated by cytokines and other inflammatory mediators that are ultimately responsible for the psychopathological and chronic physical ill-health which characterise many chronic psychiatric disorders.

The purpose of this review is to consider how the metabolic changes caused by chronic low grade inflammation may initiate physical and mental changes associated with depression.

The Tryptophan–Kynurenine Pathway and the Metabolic Syndrome

Over 40 years ago, Lapin and colleagues, from the Bekhterev Psychoneurological Research Institute in Leningrad, published a series of experimental studies demonstrating the importance of the tryptophan–kynurenine pathway in the action of imipramine and suggested that depression resulted as a consequence of the adverse effects of the metabolic products of this pathway (Lapin and Oxenkrug 1969; Lapin 1973). As such metabolic products included quinolinic acid and 3-hydroxykynurenine that were shown experimentally to cause anxiety and stress-like changes in rodents, Lapin further characterised the "neurokynurenines" as the neurochemical link between depression and anxiety states (Lapin 1973).

Since that time, there have been a plethora of experimental and clinical studies that demonstrate changes in the tryptophan–kynurenine pathway (Stone 1993; Myint and Kim 2003; Oxenkrug 2011) and which implicate quinolinic acid and 3-hydroxykynurenine as important neurotoxins which compromise neuronal function by enhancing oxidative stress (3-hydroxyanthranilic acid and 3-hydroxykynurenine) and by activating the *N*-methyl-D-aspartate glutamate receptor thereby causing neuronal apoptosis (Guillemin and Brew 2002).

A summary of the main metabolic stages in the tryptophan–kynurenine pathway is shown in Fig. 11.1.

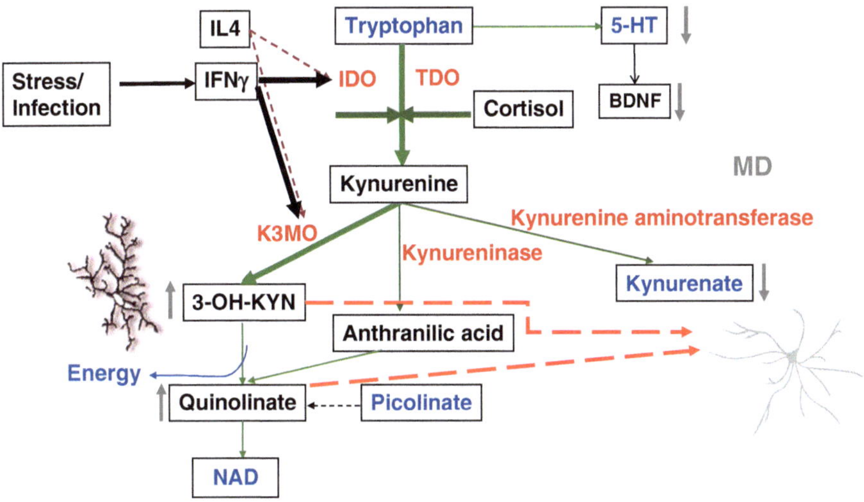

Fig. 11.1 Summary of main steps in the tryptophan–kynurenine pathway. In depression, the neurodegenerative pathway, leading to the synthesis of the neurotoxins quinolinic and picolinic acids, is increased while the neuroprotective pathway, in which kynurenic acid is the main end product, is decreased. The synthesis of nicotinamide adenine dinucleotide (NAD+) from quinolinic acid is an important co-factor for the electron transport system in mitochondria. Xantheuric and 3-hydroxyanthranilic acid are sources of reactive oxygen species which contribute to insulin insensitivity and further compromise energy metabolism (see Myint and Kim 2014; Oxenkrug 2013). *IDO* indoleamine 2,3 dioxygenase, *TDO* tryptophan dioxygenase, *KAT* kynurenine aminotransferase, *KMO* kynurenine monooxygenase, *ICD* isocitric dehydrogenase

The Tryptophan–Kynurenine Pathway and Deficits in Intermediary Metabolism

In recent years, attention has centred on the neurotoxic consequences of the increase in quinolinic acid and the intermediates formed from kynurenine in the tryptophan–kynurenine pathway. While such neurotoxins undoubtedly play a critical role in the neurodegenerative changes associated with chronic psychiatric disorders such as depression and schizophrenia, it is often overlooked that quinolinic acid is also an important substrate for the formation of nicotinamide adenine dinucleotide (NAD+). As NAD+ is a key component of the respiratory chain, chronic pathological changes that reduce its formation are liable to have adverse consequences for intermediary metabolism particularly in neurons that are critically dependent on high energy sources. It is estimated that approximately 99 % of tryptophan that is not used for protein and serotonin synthesis is metabolised to NAD+ via the tryptophan–kynurenine pathway, and therefore this pathway is important for the synthesis of this vital co-factor (Gal and Sherman 1980; Han et al. 2010). This situation would be compounded by a reduction in the availability of insulin, a key factor in the transport of glucose into neurons (Oxenkrug 2013). As there is evidence that insulin receptor resistance is a frequent feature of depression, and other major psychiatric disorders and with age-related pathology associated with the dementias (Lee et al. 2013), it seems reasonable to conclude that a chronic decrease in high energy substrates resulting from a deficit in glucose and essential co-factors may be of crucial importance in understanding the causes of increased neuronal apoptosis (Lee et al. 2013). This situation is further complicated by mitochondrial dysfunction in depression which results in a decrease in the synthesis of adenosine triphosphate (ATP) and related high energy molecules, combined with an increase in oxidative damage, while the synthesis of superoxide radicals resulting from a decrease in the respiratory chain increases the damage to the mitochondrial membranes by opening the permeability transition pores (Sas et al. 2007). In addition, oxygen free radical synthesis is enhanced by xanthenuric acid and 3-hydroxykynurenine which are formed in the brain as a result of the inflammation enhanced tryptophan–kynurenine pathway. This will be further discussed later.

An essential co-factor in the control of many of the intermediates in the tryptophan–kynurenine pathway is pyridoxal-5-phosphate (P5P), the active form of vitamin B 6. It is well established that vitamin B6, together with vitamin B12 and folate, is involved in the methylation reactions that contribute to the synthesis of the monoamine neurotransmitters, phospholipids and nucleotides, all of which are functionally compromised in depression. Thus a deficiency of dietary vitamin B6 could have an impact on depression and recent studies have demonstrated that low plasma P5P levels are inversely correlated with the severity of depressive symptoms particularly in the elderly (Merete et al. 2008). Other investigators have reported that the B vitamins reduced the symptoms of major depression in post stroke patients (Almeida et al. 2010) while, in a Japanese study, a higher vitamin B6 status was associated with a decreased risk of depression (Nanri et al. 2013). It should be noted however that not all epidemiological studies on the vitamin status have reported the beneficial effects of vitamin B 6 (Sanchez-Villegas et al. 2009).

In the tryptophan–kynurenine pathway, P5P is a key co-factor for kynureninase, kynurenine aminotransferase and kynurenine monooxygenase. A deficit in P5P would therefore result in a shift in the synthesis of 3-hydroxykynurenine from NAD+ synthesis to the synthesis of xanthenuric acid. Xanthenuric acid contributes to insulin insensitivity by forming a chelation complex with insulin, thereby resulting in a reduction in the function of insulin (Oxenkrug 2013). Other changes caused by xanthenuric acid also occur in the pancreatic islet cells (Hattori and Kotake 1984). Support for these findings has been provided by the early clinical studies in which the urinary excretion of xanthenuric acid was shown to be increased in patients with type 2 diabetes (Hattori and Kotake 1984), and by experimental studies which demonstrated that xanthenuric acid induced diabetes in rats (Kotaki et al. 1975). Such observations are of relevance to major depression where it has been reported that there is a 65 % increased risk of diabetes (Campayo et al. 2010).

The reduction in kynureninase and kynurenine monooxygenase associated with reduced P5P would lead to an increase in the kynurenine–kynurenic acid component of the pathway and thereby further contribute to insulin insensitivity. This is due to the fact that kynurenic acid, which acts as an agonist at *N*-methyl-D-aspartate glutamate receptors, has been shown experimentally to inhibit glucose metabolism (Lam et al. 2010).

The Metabolic Syndrome and Chronic Low Grade Inflammation

The metabolic syndrome in major psychiatric disorders such as depression and schizophrenia is characterised by central obesity, dyslipidaemia, hypertension and type 2 diabetes, changes which are linked to insulin insensitivity. Vascular induced cognitive impairment and late-life depression are also associated with the metabolic and inflammatory status that commonly occur in chronic depression (Dilman et al. 1979; Oxenkrug 2011). A common feature of all these conditions is chronic, low grade inflammation, in which the initiating factors are the pro-inflammatory cytokine and inflammatory chemokines. There are a substantial number of publications which demonstrate the importance of the cytokines in major depression (Kim et al. 2007; Leonard 2010; Maes 1995; Maes et al. 1999; Raison et al. 2006) and of these interferon-gamma appears to play a critical role by activating indoleamine dioxygenase (IDO) and thereby decreasing the synthesis of brain serotonin by increasing the activity of the tryptophan–kynurenine pathway (see Fig. 11.1). This results in an increase in the end products of the pathway, quinolinic and picolinic acids. These neurotoxins activate inducible nitric oxide synthase, thereby increasing the concentration of nitric oxide and the nitrosylation of macromolecules including nucleic acids (Cuzzocrea et al. 2001). This process would be enhanced by a deficiency of P5P.

The activity of interferon-gamma is reflected in the concentration of neopterin by the activation of guanosine triphosphate (GTP) cyclohydrolase to form dihydrobiopterin, the precursor of neopterin (Sucher and Schroecksnadel 2010; Oxenkrug 2011; Fuchs and Avanzas 2009).

The relationship between inflammation, tetrahydrobiopterin and nitric oxide synthesis is complex. Interferon-gamma, released from Th-I helper cells, natural killer cells and cytotoxic T-cells, induces GTP-cyclohydrolase, which facilitates nitric oxide synthesis via tetrahydrobiopterin (Wirleitner et al. 2005). In the presence of adequate tissue tetrahydrobiopterin, nitric oxide is the main end product. However, should there be a deficiency of tetrahydrobiopterin, which may occur in depression and contribute to a deficiency in monoamine neurotransmitters for which it is an essential co-factor (Moens and Kass 2007; Werner-Falmayer et al. 2002), nitric oxide synthase mainly produces superoxide anions which react with any nitric oxide present to form cytotoxic peroxynitrite (Sperner-Unterweger et al. 2014). The importance of tetrahydrobiopterin in inflammation and the consequences of its deficiency in the neurotoxic changes associated with chronic depression have been reviewed by Sperner-Unterweger et al. (2014) and serve to emphasise the importance of dietary co-factors as causative factors in major depression.

Nitric oxide plays a critical role in inflammation not only by activating the glutamatergic system, thereby contributing to neuronal apoptosis, but also by increasing the mobilisation of arachidonic acid in the neuronal membranes (Stewart 2007). An important inflammatory product of arachidonic acid is prostaglandin E2 (PGE2), formed by the action of cyclooxygenase-2 (COX-2).

COX-2 is induced by inflammation but also contributes to the inflammatory cascade by activating the EP2 receptors located on microglia (Jiang et al. 2011). Thus the activation of the tryptophan–kynurenine pathway and IDO by inflammatory mediators contributes directly to neuronal apoptosis and neurodegeneration. These effects are further complicated by the consequences of insulin insensitivity and mitochondrial dysfunction (Sas et al. 2007).

Depression, Inflammation and Dementia: The Metabolic Link?

While a history of depression, particularly in the elderly, is an important risk factor for dementia in later life (Jorm 2001; Chen et al. 1999; Olazaran et al. 2013), the precise mechanism linking depression with dementia is unclear. From the clinical perspective, chronic depression could be a prodrome of dementia and it seems possible that, as a consequence of low grade inflammation, damage to neuronal networks ultimately contributes to the onset of dementia. This has been reviewed by Leonard (2001a, b) and by Leonard and Myint (2006).

Several large epidemiological and clinical studies have demonstrated an association between chronic depression and Alzheimer's disease. There is also evidence that depression increases the risk of Alzheimer's disease (Byers and Yaffe 2011; Devanand et al. 1996; Ownby et al. 2006; Wilson et al. 2002). While the relationship between the pathology of depression and Alzheimer's disease is undoubtedly multi-factorial, there is evidence from the Rotterdam Scan Study of Geerlings et al. (2000) that the risk of Alzheimer's disease increased from 2.3-fold for late onset depression to 3.8-fold for early onset depression. Thus the longer the duration of the disorder, the greater the increased risk of Alzheimer's disease. It is possible that while cerebrovascular complications may increase the risk of late-life depression (Lesser et al. 1996) this is less likely to be a precipitating factor in early life onset depression. One possible explanation for the association of early life depression with dementia involves an increase in the synthesis of beta amyloid plaque synthesis, as suggested by the study of Rapp et al. (2006) who showed that the post mortem beta amyloid plaque density of patients with Alzheimer's disease was significantly increased in those who had chronic depression before they developed dementia. Other researchers have demonstrated that the plasma beta amyloid levels are increased in depressed patients which may provide a possible prodromal marker of Alzheimer's disease (Sun et al. 2008). Further support for this view has been provided by Namekawa et al. (2013) who demonstrated that the serum beta amyloid (Abeta) 40/Abeta 42 ratio in early onset and in elderly depressed patients was significantly higher in depressed patients than in elderly non-depressed and that the ratio was negatively correlated with the age of onset of the depression. This not only demonstrates an important link between depression and Alzheimer's disease but also identifies a possible blood marker as a risk factor for the disease in later life.

The association between physical ill-health and chronic depression also plays a role in predisposing the patient to dementia. Cerebrovascular disease, hypertension and type 2 diabetes are just some of the important changes that contribute to cognitive impairment in dementia and chronic depression in the elderly (Launer 2002; Barnes et al. 2006). Dotson et al. (2010) have also shown that recent depressive symptoms and the incidence of dementia are associated with mild cognitive impairment, changes which may be associated with hippocampal and pre-frontal cortical dysfunction. The link between cognitive decline and the metabolic syndrome has been demonstrated recently in the cohort study of participants in the Action to Control Cardiovascular Risk in Diabetes (ACCORD-MIND) trial in which it was shown that depression in patients with type-2 diabetes was associated with greater cognitive decline (Sullivan et al. 2013).

Inflammatory mechanisms undoubtedly play an important role in the pathophysiology of Alzheimer's disease, and inflammatory cytokines in the vicinity of the beta amyloid deposits and neurofibrillary tangles are suggestive of a causal link between inflammation and the pathology of Alzheimer's disease. Of the various pro-inflammatory cytokines that are increased in depression and Alzheimer's disease, tumour necrosis factor (TNF)-alpha is known to be associated with amyloid plaques and to stimulate beta amyloid synthesis (Park and Bowers 2010). The activity of TNF-alpha is mediated through its two membrane bound receptors, TNF R1&2, which are detached from the cell surface by the TNF-alpha converting enzyme (TACE) into their soluble forms. The activity of TACE has been shown to increase in the CSF of patients with Alzheimer's disease and also in those with mild cognitive impairment, a condition which is often a prelude to Alzheimer's (Jiang et al. 2011); Zhang et al. (2013) also reported that TACE activity and the concentration of TNF receptors in the plasma are significantly increased in both these groups when compared to healthy controls. Zhang et al. (2013) suggest that TNF-alpha signalling may drive the pathology of Alzheimer's disease and contribute to the neurotoxic changes induced by beta amyloid. In addition, the changes in TACE and the soluble TNF-alpha receptors correlate with the plasma apolipoprotein (apo) E4 allele concentration, an association which may be linked to higher systemic and brain concentrations of TNF-alpha and IL-6 (Lynch et al. 2003); both these pro-inflammatory cytokines are also elevated in patients with major depression (Maes 1999).

From these clinical observations, it seems reasonable to conclude that the metabolic changes associated with chronic inflammation in depression not only contribute to the physical disorders that are co-morbid with chronic depression but are also causally related to the neurodegenerative changes that are causally linked to dementia. While specific neurotoxins, such as the products of the tryptophan–kynurenine pathway, no doubt play a critical role in these changes, it is possible that the underlying metabolic changes associated with insulin insensitivity and mitochondrial dysfunction are responsible for the fundamental changes underlying dementia. Perhaps it is now time to consider ways in which these metabolic changes could be reversed in the early stages of cognitive decline before the neurodegenerative changes become dominant.

Could Diet Play a Role in Attenuating the Metabolic Changes in Depression?

The relationship between life-threatening infections (such as meningitis, sepsis and cerebral malaria) that involve the dramatic release of pro-inflammatory cytokines and other inflammatory mediators from central and peripheral immune cells are well known to cause profound metabolic changes (Douglas and Shaw 1989; Balkwill et al. 1990; Waage et al. 1989). However, the question arises if similar metabolic changes occur in patients in which chronic low grade inflammation occurs and which are not associated with life-threatening external causes of infection such as pathogenic viruses or bacteria. The chronicity of inflammation in major psychiatric disorders such as depression and schizophrenia, and the dementias, raises the possibility that the dietary content of anti-oxidants, anti-inflammatory lipids and vitamin co-factors which reduce inflammation may have a role to play in modulating low grade inflammation. There are a number of publications that have considered this possibility and the topic will only be briefly covered here (Gimble 1992, 1998; Warren et al. 1987; Grunfeld and Feingold 1992).

Role of anti-oxidants. As free radicals of oxygen and nitrogen are produced as part of the inflammatory cascade that are an important means of defence against invading microorganisms, endogenous anti-oxidants (such as vitamin E and other tocopherols, vitamin C, beta carotene, glutathione, caeruloplasmin and metallothiones) are also produced to counteract the damaging effects of the free radicals. In addition, there are numerous enzymes which destroy intracellular free radicals which include superoxide didmutase, catalase, glutathione peroxidise and reductase. These enzymes require selenium, copper and zinc as co-factors (Gimble 1998). Thus an adequate dietary source of anti-oxidant vitamins and trace metals to counteract any deficits induced by chronic inflammation could be an important adjunctive treatment to the psychotropic drugs that are required to treat the symptoms.

Role of polyunsaturated fatty acids. The lipid components of the neuronal membranes determine the types of inflammatory mediators that are released by an inflammatory response. The n-3 and n-6 fatty acids are important components of the neuronal membranes and the ratio of these fatty acids affects the relative proportion of inflammatory to the anti-inflammatory components. Thus the relative excess of n-6 fatty acids promotes the formation of arachidonic acid which acts as precursor for the prostaglandins and other eicosanoids involved in the inflammatory response (Calder 2008). By contrast, the eicosanoids produced from the n-3 fatty acids, such as eicosanopentaenoic acid (EPA) and docosahexaenoic acid (DHA), exhibit anti-inflammatory effects and inhibit the actions of pro-inflammatory cytokines such as IL-1 beta; IL-1 synthesis is increased in the ageing brain and even more extensively in the brain of the patient with Alzheimer's disease (Griffin et al. 1985; Song et al. 2003, 2004). Such results suggest that the composition of dietary polyunsaturated fatty acids could influence the synthesis of PGE2 and related inflammatory mediators that are derived from arachidonic acid.

The potential importance of diet in inflammation is reflected in the evolutionary changes in human diet whereby the ratio of n-6 to n-3 fatty acids changed from approximately 1:1 in the early diet of mankind to 10:1 in the modern diet (Simopoulos 2003). In addition to the reduction in endogenous anti-inflammatory eicosanoids in depression (Maes et al. 1999), there is also an increase in pro-inflammatory mediators associated with many psychiatric and neurological disorders. In chronic depression, Alzheimer's disease, chronic epilepsy and Parkinson's disease, the over-expression of IL-1 has been implicated in the initiation and progression of the neuropathological changes, changes that are linked to the activation of microglia and to the synthesis of beta amyloid (Griffin et al. 1985, 1995). Furthermore, the increase in IL-1 is linked to a decrease in synaptic plasticity which in the hippocampus is reflected in a decrease in long-term potentiation and learning (Murray and Lynch 1998; Song and Horrobin 2004). Such changes may help to explain the cognitive impairment associated with chronic depression, ageing and dementia; in rats chronic EPA treatment prevents the impairment of memory caused by IL-1 (Song et al. 2004). While the precise mechanism for this anti-inflammatory effects of EPA is uncertain, there is evidence that the n-3 fatty acid, DHA, is converted by microglia to the potent anti-inflammatory 17-resolvins by cyclooxygenase-2 (Serhan et al. 2000) The resolvins block the synthesis of pro-inflammatory cytokines in microglia by inhibiting the activation of nuclear factor kappa-beta (NFκB) (Marcheselli et al. 2000). In addition to the effects of EPA on microglia induced inflammation, it is also evident that the fatty acids increase the concentration of the anti-inflammatory cytokine IL-10, which acts as an antagonist of inflammatory cytokines in vivo (Bluthe et al. 1999). These experimental observations of the effects of EPA have been extended by Song et al. (2009) to demonstrate that the subcellular inflammatory changes initiated by bilateral olfactory bulbectomy in the rat (a chronic model of depression; Leonard and Song 2002) were attenuated by chronic treatment with EPA.

Epidemiological studies support the experimental observations that polyunsaturated fatty acids protect the brain from the detrimental effects of age-associated over-expression of pro-inflammatory cytokines. Thus elderly people who consume fish or other types of sea food, diets rich in n-3 fatty acids, on a weekly basis are at a lower risk of suffering from Alzheimer's disease and other dementias (Barber-Gateau et al. 2002; Kalmijn et al. 2004). Thus it would be anticipated that by improving the diet by including components with an anti-inflammatory effect would impact beneficially on the mental and physical state of health of the patient. Thus the addition of anti-oxidants, vitamins of the B complex and fish oil to the diet appears to be a reasonable strategy, particularly in the middle-aged and elderly.

Conclusion

The purpose of this review was to illustrate how the metabolic changes in major depression occur as part of the cascade of changes initiated by chronic low grade inflammation. These changes are exacerbated by deficiencies in a number of dietary co-factors that contribute to insulin insensitivity and the impairment of glucose metabolism. Energy metabolism in the brain is further compromised by a reduction in mitochondrial function.

It is reasoned that, by increasing the B-vitamin, anti-oxidant and polyunsaturated fatty acids in the diet in those prone to depression may improve mental and physical health and possibly reduced the risk of dementia. Clearly, more clinical and experimental research needs to be undertaken to define more precisely how these dietary factors produce their beneficial effects and how significant these effects are in reducing the onset of age-associated memory deficits and consequent neurodegenerative changes.

References

Almeida OP, Marsh K, Alfonso H, Flicker L, et al. B vitamins reduce the long-term risk of depression after stroke: the VITATOPS-DEP trial. Ann Neurol. 2010;68:503–10.

Balkwill F, Burke F, Talbot D. Evidence of TNF/cachectin production in cancer. Lancet. 1990; 2:1229–32.

Barber-Gateau P, Latenneur L, Deschamps V, et al. Fish, meat and risk of dementia: cohort study. Br Med J. 2002;325:932–3.

Barnes DE, Alexopoulos GS, Lopez OL, et al. Depressive symptoms, vascular disease and mild cognitive impairment: findings from the cardiovascular health study. Arch Gen Psychiatry. 2006;63:273–9.

Bluthe RM, Castanon N, Pousset F, et al. Central injection of IL-10 antagonises the behavioural effects of lipopolysaccharide in rats. Psychoneuroendocrinology. 1999;24:301–11.

Byers AL, Yaffe K. Depression and risk of developing dementia. Nat Rev Neurol. 2011;7: 323–31.

Calder PC. The relationship between the fatty acid composition of immune cells and their function. Prostaglandins Leukot Essent Fatty Acids. 2008;79:101–8.

Campayo A, de Jonge P, Roy JF, Saz P, et al. Depressive disorder and incident diabetes mellitus: the effect of characteristics of depression. Am J Psychiatry. 2010;167:580–8.

Chen P, Ganguli M, Mubant BH, De Korsky ST. The temporal relationship between depressive symptoms and dementia: a community based prospective study. Arch Gen Psychiatry. 1999;56: 261–6.

Cuzzocrea S, Riley DP, Capute AP, Salvemini D. Antioxidant therapy: a new approach to shock, inflammation and ischaemia/reperfusion injury. Pharmacol Rev. 2001;53:135–59.

Devanand DP, Sano M, Tang MX, et al. Depressed mood and the incidence of Alzheimer's disease in elderly living in the community. Arch Gen Psychiatry. 1996;53:175–82.

Dilman M, Lapin IP, Oxenkrug GF. Serotonin and ageing. In: Essman W, editor. Serotonin in health and disease, vol. 5. London: Spectrum Press; 1979. p. 111–23.

Dotson VM, Beydoun MA, Zonderman AB. Recurrent depressive symptoms and the incidence of dementia and mild cognitive impairment. Neurology. 2010;75:27–34.

Douglas RG, Shaw JHF. Metabolic response to sepsis and trauma. Br J Surg. 1989;76:115–20.

Fuchs D, Avanzas P. The role of neopterin in atherosclerosis and cardiovascular risk assessment. Curr Med Chem. 2009;16:4644–53.

Gal EM, Sherman AD. L-Kynurenine: its synthesis and possible regulatory function in brain. Neurochem Res. 1980;5:223–39.

Geerlings MI, Schoevers RA, Bechman AT, et al. Depression and the risk of cognitive decline in Alzheimer's disease: results of 2 prospective community based studies in the Netherlands. Br J Psychiat. 2000;176:568–75.

Gimble RF. Dietary manipulation of the inflammatory cascade. Proc Nutr Soc. 1992;51:285–93.
Gimble RF. Nutritional modulation of cytokine biology. Nutrition. 1998;14:634–40.
Griffin DE, Moser HW, Mendoza Q, Moench TR, et al. Identification of the inflammatory cells in the CNS of patients with adrenoleukodystrophy. Ann Neurol. 1985;18:660–4.
Grunfeld C, Feingold KR. TNF, interleukin and interferon induce changes in lipid metabolism as part of host defence. Proc Soc Exp Biol Med. 1992;200:224–30.
Guillemin GJ, Brew BJ. Implications of the kynurenine pathway and quinolinic acid in Alzheimer's disease. Redox Rep. 2002;7:199–206.
Han Q, Tao C, Tagle DA, Li J. Structure, expression and function of kynurenine aminotransferase in human and rodent brains. Cell Mol Life Sci. 2010;67:353–68.
Hattori M, Kotake Y. Studies on the urinary excretion of xanthenuric acid in diabetics. Acta Vitaminol Enzymol. 1984;16:221–8.
Jiang H, Hempel D, Prvulovic D, et al. Elevated CSF levels of TACE activity and soluble TNF receptors in subjects with mild cognitive impairment and patients with Alzheimer's disease. Mol Neurodegener. 2011;6:69–77.
Jorm AF. History of depression as a risk factor for dementia: an update. Aust N Z J Psychiatry. 2001;35:776–81.
Kalmijn S, van Boxtel MP, Ocke M, et al. Dietary intake of fatty acids and fish in relationship to cognitive performance at middle-age. Neurology. 2004;62:275–80.
Kim Y-K, Na KS, Shin KH, Jung HY, et al. Cytokine imbalance in the pathophysiology of major depressive disorder. Prog Neuropsychopharmacol Biol Psychiatry. 2007;31:1044–53.
Kotaki Y, Ueda T, Mori T, Igali S, et al. Abnormal tryptophan metabolism and experimental diabetes by xanthenuric acid. Acta Vitaminol Enzymol. 1975;29:2369.
Lam CK, Chari M, Su BB, Cheung GW, et al. Activation of *N*-methyl-D-aspartate receptors in the dorsal vagal complex lowers glucose production. J Biol Chem. 2010;285:21913–21.
Lapin IP. Kynurenines as probable participants of depression. Pharmakopsychiatr Neurop sychopharmakol. 1973;6:273–9.
Lapin IP, Oxenkrug GF. Intensification of the central serotonergic processes as possible determinants of the thymoleptic effect. Lancet. 1969;1:32–9.
Launer LJ. Demonstrating the case that Alzheimer's disease is a vascular disease: epidemiologic evidence. Ageing Res Rev. 2002;1:61–77.
Lee S, Tong M, Hang S, et al. Cerebrospinal fluid and brain indices in insulin resistance, oxidative stress and neuroinflammation in early and late Alzheimer's disease. Alzheimer's disease and Parkinsonism. 2013. doi:10.4172/2161-0460.1000128.
Leonard BE. Changes in the immune system in depression and dementia. Int J Dev Neurosci. 2001a;19:305–21.
Leonard BE. Brain cytokines and the psychopathology of depression. In: Leonard BE, editor. Antidepressants. Basel: Birkhauser Verlag; 2001b. p. 109–20.
Leonard BE. The concept of depression as a dysfunction of the immune system. Curr Immunol Rev. 2010;6:205–12.
Leonard BE. Inflammation as the cause of the metabolic syndrome in depression. Mod Trends Pharmacopsychiatri. 2013;28:117–26.
Leonard BE. Impact of inflammation on neurotransmitter changes in major depression: an insight into the action of antidepressants. Prog Neuropsychopharmacol Biol Psychiatry. 2014; 48:261–7.
Leonard BE, Myint A-M. Inflammation and depression: is there a causal connection with dementia? Neurotox Res. 2006;10:149–60.
Leonard BE, Song C. Changes in the immune system in rodent models of depression. Int J Neuropsychopharmacol. 2002;5:345–56.
Lesser IM, Boone KB, Mehringer CM, et al. Cognitive and white matter hyperintensities in older depressed patients. Am J Psychiat. 1996;153:1280–87.
Lynch J, Tang W, Wang H, et al. APOE genotype and APOEmimetic peptide modify the systemic and CNS inflammatory response. J Biol Chem. 2003;278:48529–33.
Maes M. Evidence for an immune response in major depression: a review and hypothesis. Prog Neuropsychopharmacol Biol Psychiatry. 1995;19:11–38.

Maes M. Major depression and the activation of the inflammatory response system. Adv Exp Med Biol. 1999;461:25–45.
Maes M, Christophe A, Delanghe J, et al. Lowered omega 3 polyunsaturated fatty acids in serum phospholipid and cholesteryl esters in depressed patients. Psychiatry Res. 1999;85:275–91.
Marcheselli VL, Hong S, Lukiw WJ, et al. Novel docoanoids inhibit brain ischaemia-reperfusion mediated leukocyte infiltration and proinflammatory gene expression. J Physiol Pharmacol. 2000;51:643–54.
McIntyre RS, Rosenbluth M, Ramasulbbu R, et al. Managing medical and psychiatricco-morbidity in individuals with major depression and bipolar disorder. Ann Clin Psychiatry. 2012;24: 163–9.
Merete C, Falcon LM, Tucker KL. Vitamin B6 is associated with depressive symptomatology in Massachusetts Elders. J Am Coll Nutr. 2008;27:421–7.
Moens AL, Kass DA. Therapeutic potential and tetrahydrobiopterin for treating vascular cardiac disease. J Cardiovasc Pharmacol. 2007;50:238–46.
Murray CA, Lynch MA. Evidence that increased hippocampal expression of cytokine IL-1beta is a common trigger for age- and stress impairments in LTP. J Neurosci. 1998;18:2974–81.
Myint A-M, Kim Y-K. Cytokine-serotonin interaction through IDO: a neurodegeneration hypothesis of depression. Med Hypotheses. 2003;61:519–25.
Myint A-M, Kim Y-K. Network beyond IDO in psychiatric disorders: revisiting the neurodegeneration hypothesis. Prog Neuropsychopharmacol Biol Psychiatry. 2014;48:304–13.
Namekawa Y, Baba H, Maeshima H, et al. Heterogeneity of elderly depressives at increased risk of Alzheimer's disease and amyloid beta protein metabolism. Prog Neuropsychopharmacol Biol Psychiatry. 2013;43:203–8.
Nanri A, Pham NM, Kurotani K, Kume A, et al. Serum pyridoxal concentrations and depressive symptoms among Japanese adults: results of a prospective study. Eur J Clin Nutr. 2013;67: 1060–5.
Olazaran J, Trincado R, Bermyo-Pareja F. Cumulative effect of depression and dementia risk. Int J Alzheimers Dis. 2013. doi:10.1155/2013/457175.
Ownby RL, Crocco E, Acevedo A, et al. Depression and risk of Alzheimer's disease: systematic review, meta-analysis and meta-regression analysis. Arch Gen Psychiatry. 2006;63:530–8.
Oxenkrug GF. Interferon gamma inducible kynurenines/pteridines inflammation cascade: implications for ageing-associated psychiatric and medical disorders. J Neural Transm. 2011;118: 75–85.
Oxenkrug GF. Insulin resistance and dysregulation of tryptophan-kynurenine-nicotinamide adenine dinucleotide pathways. Mol Neurobiol. 2013;48:294–301.
Park KM, Bowers WT. TNF-alpha mediated signaling in neuronal homeostasis and dysfunction. Cell Signal. 2010;22:977–83.
Raison CL, Capuron L, Miller AH. Cytokines sing the blues: inflammation and the pathogenesis of depression. Trends Immunol. 2006;27:24–31.
Rapp MA, Sahnaider-Beeri M, Grossman HT, et al. Increased hippocampal plaques and tangles in patients with Alzheimer's disease with a life-time history of major depression. Arch Gen Psychiatry. 2006;63:161–7.
Sanchez-Villegas A, Poreste J, Schlatter J, Pla J, et al. Association between folate, vitamin B6 and vitamin B12 intake and depression in the SUN cohort study. J Hum Nutr Diet. 2009;22: 122–33.
Sas K, Robotka H, Toldi J, Vecsei L. Mitochondria, metabolic disturbances, oxidative stress and the kynurenine system, with a focus on neurodegenerative disorders. J Neurol Sci. 2007;257: 221–39.
Serhan CN, Clish CB, Brannon J, et al. Antimicroinflammatory lipid signals generated from dietary n-3 fatty acids via cyclooxygenase 2 and transcellular processing: a novel mechanism for NSAID and n-3 PUFA therapeutic actions. J Physiol Pharmacol. 2000;51:643–54.
Simopoulos AP. Importance of the ratio of omega 6/omega 3 essential fatty acids: evolutionary aspects. World Rev Nutr Diet. 2003;92:1–23.
Smith RS. The macrophage theory of depression. Med Hypotheses. 1991;35:298–306.

Song C, Horrobin DF. Omega-3 fatty acidethyl-eicosapentaenoic acid, but not soya bean oil, attenuates memory impairment induced by central il-1 beta administration. J Lipid Res. 2004; 45:1112–21.

Song C, Li X, Leonard BE, Horrobin DF. Effects of dietary n-3 or n-6 fatty acids on IL-1 beta induced anxiety, stress and inflammation responses in rats. J Lipid Res. 2003;44:1984–91.

Song C, Phillips AG, Leonard BE, Horrobin DF. Ethyl-eicosapentaenoic acid ingestion prevents corticosterone – mediated memory impairment induced by central administration of IL-1 beta in rats. Mol Psychiatry. 2004;9:630–8.

Song C, Zhang XY, Manku M. Increased PLA2 activity and inflammatory responses but decreased NGF expression in the olfactory bulbectomised rat model of depression: effects of chronic ethyl-eicosapentaenoic acid treatment. J Neurosci. 2009;29:14–22.

Sperner-Unterweger B, Kohl C, Fuchs D. Immune changes and neurotransmitters: possible interactions in depression. Prog Neuropsychopharmacol Biol Psychiatry. 2014;48:268–76.

Stewart JC. Negative emotions and 3 year progression of subclinical atherosclerosis. Arch Gen Psychiatry. 2007;64:225–33.

Stone TW. Neuropharmacology of quinolinic and kynurenic acids. Pharmacol Rev. 1993;45: 310–79.

Sucher R, Schroecksnadel K. Neopterin, a prognostic marker in human malignancies. Cancer Lett. 2010;287:13–22.

Sullivan MD, Katon WJ, Lovato LC, et al. Association of depression with accelerated cognitive decline among patients with type-2 diabetes in the ACCORD-Mind trial. JAMA Psychiatry. 2013;70:1041–7.

Sun X, Steffens DC, Au R, et al. Amyloid associated depression: a prodromal depression of Alzheimer's disease? Arch Gen Psychiatry. 2008;65:542–50.

Waage A, Brandtzoeg P, Halstensen P, et al. Complex pattern of cytokines in serum of patients with meningococcal septic shock. J Exp Med. 1989;169:333–40.

Warren RS, Starnes EF, Gabrilov JL, et al. The acute metabolic effects of TNF administration in humans. Arch Surg. 1987;122:1396–402.

Werner-Falmayer G, Goldener G, Wernes ER. Tetrahydrobiopterin biosynthesis utilization and pharmacological effects. Curr Drug Metab. 2002;3:159–73.

Wilson RS, Barnes LL, Mendes de Leon CF, et al. Depressive symptoms, cognitive decline and risk of Alzheimer's disease in older persons. Neurology. 2002;59:364–70.

Wirleitner B, Neuranter G, Schroecksnadel K, et al. Interferon-gamma induced conversion of tryptophan immunologic and neuropsychiatric aspects. Curr Med Chem. 2005;10:1581–91.

Zhang J, Jia J, Qin W, Wang S. Combination of plasma TNF receptors signalling proteins, beta amyloid and apolipoprotein E for the detection of Alzheimer's disease. Neurosci Lett. 2013; 541:99–104.

Chapter 12
Microglia Activation, Herpes Infection, and NMDA Receptor Inhibition: Common Pathways to Psychosis?

Hans C. Klein, Janine Doorduin, Lot de Witte, and Erik F.J. de Vries

Abstract Microglia are the resident macrophages of the brain. Microglia play important housekeeping roles during brain development and during exposure to psychosocial stress, toxins, and infectious pathogens. The hippocampus is a vulnerable brain region in response to these external stressors. In patients with psychosis, we found a major activation of microglia in the hippocampus, which could explain in part the volume loss of the hippocampus after a psychotic episode of patients. Recent research suggests low-grade replication of herpes simplex virus (HSV) as a potential environmental trigger of microglia activation in schizophrenia. A common mechanism of psychosis is proposed that involves activation of microglia by toxic, auto-immune, or infectious factors, converging to a blockade of the NMDA-NR1 receptor in the hippocampus.

Keywords Herpes simplex virus • Microglia • Schizophrenia • NMDA-R

Introduction on Microglia Physiology

The adult brain consists of terminally differentiated neurons (10 %), immune cells such as microglia (10 %), and cells with important support functions such as glia and oligodendrocytes (80 %). Microglia play central roles in neurogenesis,

---

H.C. Klein (✉)
Department of Psychiatry, University of Groningen, University Medical Centre Groningen, CC44, Hanzeplein 1, 9713 GZ Groningen, The Netherlands

University of Groningen, University Medical Centre Groningen, Nuclear Medicine Molecular Imaging, Hanzeplein 1, 9713 GZ Groningen, The Netherlands
e-mail: h.c.klein@umcg.nl

J. Doorduin • E.F.J. de Vries
University of Groningen, University Medical Centre Groningen, Nuclear Medicine Molecular Imaging, Hanzeplein 1, 9713 GZ Groningen, The Netherlands

L. de Witte
Department of Psychiatry, University of Utrecht, Heidelberglaan 100, 3584 CX Utrecht, The Netherlands

N. Müller et al. (eds.), *Immunology and Psychiatry*, Current Topics in Neurotoxicity 8, DOI 10.1007/978-3-319-13602-8_12

synaptogenesis (forming of new neurons and connections respectively) (Ekdahl et al. 2009; Ziv et al. 2006), and also in apoptotic death during development of the nervous system in the embryo. The yolk sac is the major source of microglia precursors during development, and these stem cells differentiate further within the brain (Aguzzi et al. 2013; Swinnen et al. 2013). Final differentiation is dependent on the immediate cellular environment of the microglia. Close to the vascular compartment the precursors differentiate into perivascular macrophages, and within the parenchyma the typical microglia appear (Aguzzi et al. 2013). White matter and gray matter microglia develop into different microscopic shapes and show different immune responses after stimulation (Hart et al. 2012; Olah et al. 2012). Microglia are structurally and morphologically different in the different loci of the cerebral cortex, motor, sensor or association cortex (Olah et al. 2011) and are subject to aging (Damani et al. 2011; Hart et al. 2012).

When resting, microglia have the typical ramified phenotype, with slender processes extending far beyond the cell soma. In the activation state of microglia, the ramifications retract, the cell body swells, and the microglia adopt an amoeboid appearance. Microglia are sometimes functionally classified in three theatrical types, the "good," the "bad," and the "resting" (Henkel et al. 2009). The activated M2 phenotype (the "good") excretes growth factors to promote neurogenesis and synaptogenesis. The activated M1 phenotype (the "bad") excretes excess of pro-inflammatory and pro-apoptotic compounds that result in killing of neurons. The resting microglia exert basal housekeeping functions and provide neurons with precursors for energy metabolism and neurotransmitter production. It should be kept in mind that the M1 phenotype could be "bad" in one situation (e.g., auto-immune killing of neurons in multiple sclerosis), but might be "good" in another situation (e.g., clearance of superfluous neurons during embryogenesis or killing of virus infected neurons). The M2 phenotype, on the other hand, could also be "bad" in particular situations (e.g., promotion of oncogenesis).

Microglia Activation in Psychotic Patients

Positron emission tomography (PET) is a unique tool to measure the expression or activity of receptors, enzymes, or transporters in the brain. In short, a pharmaceutical of choice is tagged with a positron emitting atom, such as ^{18}F or ^{11}C. Since all organic compounds contain carbon atoms, incorporation of ^{11}C into the molecule can fully preserve the desired structure formula and physical and biological properties of the compound of interest. ^{18}F can be used to substitute a fluorine atom in, for example, a drug without structural changes, or to substitute a hydroxyl group OH or hydrogen atom in the parent compound, causing slight modification of the properties of the compound. When the radiopharmaceutical is injected intravenously in a study subject, the blood circulation distributes the pharmaceutical to all organs. When the receptor, transporter, or enzyme of interest is present in a particular organ, the compound will be bound, imported, or metabolized. These processes can result in cellular "trapping" of the radiopharmaceutical. The accumulation of radioactivity emitted by the radiopharmaceutical can be detected by the PET camera. Each positron emitted from the radioligand collides with a nearby electron and "annihilates" into two gamma ray quanta with an energy of exactly 511 keV that leave the annihilation site at an angle of 180°. The PET camera consists of a ring of detectors around the subject. The camera is equipped with a coincidence detection system, which allows only simultaneous detection of two gamma rays with the right energy by opposite detector crystals. The collective time-activity data generated from these coincidences are converted into tomographic images. The kinetic information from the PET scan can be used to quantify absolute uptake of the compound, as well as biological parameters like the binding potential, volume of distribution, or metabolic rate.

Activated microglia and associated cell types express the mitochondrial 18 kDa translocator protein TSPO (Cosenza-Nashat et al. 2009). The radiopharmaceutical [^{11}C]PK11195 binds to the TSPO and has been applied to detect activation of microglia in many neurological disorders (Doorduin et al. 2008): Increased binding of [^{11}C]PK11195 was shown to occur in, e.g. multiple sclerosis, Alzheimer dementia, and Parkinson's disease, but also in normal aging (Schuitemaker et al. 2012). To our knowledge, only three studies have been performed in psychiatric patients. A study by van Berckel et al. in recovered first episode schizophrenia patients. In this study, only non-significant differences in microglia activation between schizophrenic patients and controls were observed (van Berckel et al. 2008). We performed a study in schizophrenic patients in the psychotic phase and found profound microglia activation in the hippocampus of patients as compared to healthy controls (Doorduin et al. 2009). In a very recent study in euthymic bipolar patients, we observed a small, but significant increase in microglia activation in the left hippocampus (Haarman et al. 2014). Thus, these studies collectively demonstrate the involvement of activated microglia in psychiatric disorders.

Pathophysiology of Herpes Simplex Encephalitis

With the help of nature's examples, we can build hypotheses on the etiology of disease. A very clear example is the observation of Semmelweiss that doctors who didn't wash hands after conducting autopsy on obstetric patients were prone to inflict fevers and death in women they helped with labor afterwards. In the days of Semmelweiss the medical community had no clue of how diseases could be transmitted. In some respects nothing has changed. We still have no clue how the severe and very debilitating disease schizophrenia is caused (Yolken and Torrey 1995). Therefore we need typical examples as in the days of Semmelweiss (Bortolato and Godar 2010). One typical example of schizophrenia induced by transmission is the syndrome of herpes encephalitis. The syndrome is very rare, afflicting approximately 1 per 200,000 cases per year, but the diagnosis and therapy are very well described (Schlitt et al. 1985; Whitley and Lakeman 1995). The syndrome typically starts with psychotic experiences, such as psychomotor agitation, anxiety, auditory hallucinations, and delusions of persecution and also classical catatonia can occur (Drachman and Adams 1962; Kapur et al. 1994; Koehler and Guth 1979; Misra and Hay 1971; Raskin and Frank 1974; Schlitt et al. 1985). In this "prodromal" phase neurological deficits are rare. EEG may already reveal typical changes. Prodromal symptoms may last from days to weeks before they progress to neurological signs and symptoms. Headache is often prominent but not necessarily present. After this phase, epileptic insults occur frequently. In the last phase, patients become paralytic and comatose. Prognosis was poor before antiviral drugs were discovered: Death rate of herpes encephalitis patients was 80 % and half of the surviving patients remained severely handicapped (Whitley and Gnann 2003). The current treatment of choice is intravenous high dose aciclovir, which should be started as soon as the diagnosis is suspected. Cerebrospinal fluid (CSF) and serum examination of herpes simplex virus (HSV) of antibodies and PCR confirm the diagnosis. Survival is nowadays 90 % (Schmutzhard 2001). Typically the hippocampus and temporal cortical lobe are affected. A very interesting case study described that the hippocampus remained in an inflammatory state for a long time after herpes encephalitis (Cagnin et al. 2001). They showed with PET scans that microglia in the hippocampus remained activated after treatment and that tissue with high levels of activated microglia showed substantial atrophy on structural MRI scans at 6 months follow-up. Tissue loss of hippocampus is responsible for the severe immediate memory problems after herpes encephalitis. This experiment of nature shows that psychotic symptoms, hippocampal inflammation with loss of tissue in the temporal lobe, and cognitive decline afterwards can all be caused by a well-known and very prevalent virus: the HSV. In this line of thinking the hypothesis was formed that the origin of schizophrenic psychosis may be related to the microglia activation that is triggered by HSV, but without an obligate progression to paralysis or other severe neurological symptoms (yet). Deterioration of immediate memory and tissue loss of the hippocampus in schizophrenia might have exactly this inflammatory origin (Schretlen et al. 2010).

Modification of Herpes Encephalitis by Antipsychotics

HSV-1 is highly adapted to be carried lifelong in most humans without any symptoms or signs. HSV-1 is primarily transmitted orally by intimate (French) kissing. The majority of the adult population harbors antibodies against HSV-1. As a rule the primary infection goes unnoticed and only a minority of the population experiences typical reactivations of the virus from time to time: Cold sores. After primary transmission the virus travels anterograde from the lip region to the cell body of the sensory neuron in the ganglion of the trigeminal nerve. After a couple of replication cycles in the ganglion neuron, the virus is eventually contained by cellular and humoral immunity and forced into a latent state within the neuron. In this state, the genome of the virus is retained in an epigenome in the nucleus. After immunological, toxic, or stress challenges the virus may reactivate within the ganglion, and start to travel retrograde back to the oral region. Then the typical cold sores occur. Retrograde transport of the virus may also remain unnoticed and result in silent spreading of small amounts of the virus detectible by PCR. The virus also might travel from the trigeminal ganglion anterogradely into the brain nuclei (Johnson 1998). Anterograde transport of the virus into the brain is well documented, but the prevalence is unknown. Anterograde transport probably occurs in many individuals without any history of herpes encephalitis: Genetic material of herpes simplex is found post-mortem in the midbrain and hippocampus of healthy individuals (Baringer and Pisani 1994) and, e.g., in the brain of patients with multiple sclerosis (Gordon et al. 1996) and Alzheimer's dementia (Wozniak et al. 2009).

The human strain of HSV-1 is adapted to human life by a multi-million year co-evolution of the virus with the mammal host. The ancestry tree of human HSV-1 glycoprotein B parallels the ancestry of early mammals to homo sapiens (Maeda et al. 1992). The human strains of HSV-1 may have become generally apathogenic in man by means of natural selection. However, the human strain of HSV-1 is very pathogenic when applied in the (not adapted) rat (Buursma et al. 2005). We hypothesize that the rat model of herpes encephalitis is a copy of nature's experiment of behavioral, immunological, and neuropathological changes in herpes related schizophrenia. Therefore we evaluated the behavioral changes and microglia activation in this specific model and tested the effects of antipsychotics.

In short, animals were intranasally inoculated with HSV-1 or saline (placebo). Disease symptoms were scored daily, and, open field behavior and prepulse-inhibition was assessed at day 2 and 4 after inoculation. On day 6 after inoculation a [^{11}C]PK11195 PET scan was performed. Profound foci of microglia activation were found, which corresponded with the HSV-1 infected region (bulbus olfactorius) and its immediate connection (mesencephalon). Rats with encephalitis showed increased exploratory behavior resembling psychomotor agitation and hyperactivity in schizophrenia. The microglia activation on PET scans and increased exploratory behavior was significantly decreased by treatment with the antipsychotic clozapine and risperidone, but not by haloperidol (Doorduin et al. 2010a).

Potential Role of Herpes in Psychosis

To prove or disprove infection as cause of disease is very difficult in brain diseases in general and especially for pathogens that may remain silent in a state of latency. The classic postulates of Koch then have only limited value (Rivers 1937):

- Presence of the supposed infectious pathogen in all diseased cases,
- No presence of the supposed infectious pathogen in healthy individuals,
- Induction of the disease by the supposed pathogen after repeated culture.

The first postulate (presence of the pathogen in the brain of the diseased person) is impossible to meet, because the brain is generally inaccessible during the active phase of the disease of interest: The psychotic phase in schizophrenia disease. The second postulate is impossible to meet for all infectious agents that are capable of a latent, non-replicating state during health. Therefore an alternative way to test the hypothesis in man is proposed: Presence of a molecular signature of *replicating* virus in the brain as proof of infection (instead of presence of the pathogen itself). During latency inhibitory transcripts silence the majority of genes of the virus. The major inhibitory transcript of HSV is LAT (latency associated transcript). The LAT transcript is antisense to the most important immediate early gene ICP0 of HSV (Thompson et al. 2003) which is essential for reactivation of the virus. Upon reactivation of HSV, immediate early, early, and late mRNA transcripts are produced consecutively (Halford et al. 2001). The late transcripts code for enzymes those are essential for DNA replication of the virus. HSV encodes its own thymidine kinase enzyme, which is essential for the phosphorylation of thymidine to incorporate into DNA. The Thymidine kinase is not present in (non-dividing) neuronal cells. Therefore, the enzyme encoded by the viral transcript is essential for viral replication. Aciclovir is a compound that specifically phosphorylated by the viral thymidine kinase. After subsequent conversion into the triphosphate, aciclovir can inhibit DNA polymerases and terminate DNA replication (Hanes et al. 2007).

The selectivity of aciclovir and drugs derived from aciclovir for replicating herpes viruses was used to design two clinical studies: a diagnostic study and an intervention trial.

Primary question of the diagnostic study was whether HSV thymidine kinase (as a surrogate marker of replicating herpes viruses) might be differentially expressed in severely psychotic patients as compared to less affected patients. To determine HSV thymidine kinase activity directly inside the brain, the radiopharmaceutical [^{18}F]FHBG, derived from the antiviral drug penciclovir, was administered in psychotic patients. This radiopharmaceutical was produced by substitution of a hydroxyl group in the parent compound penciclovir with ^{18}F. After phosphorylation by the thymidine kinase of HSV, [^{18}F]FHBG remains trapped inside the infected cell. The uptake of [^{18}F]FHBG in severely psychotic and mildly psychotic patients was compared. The metabolic rate was significantly higher in the temporal lobe of severely psychotic patients as compared to mildly psychotic patients (Doorduin et al. 2010b).

Primary question of an ongoing placebo-controlled intervention trial is whether high dose treatment with valaciclovir (the oral prodrug of aciclovir) can reduce microglia activation in schizophrenia patients. The biological availability of orally ingested acyclovir is low due to a major first pass effect (80 %). Valaciclovir, the valyl ester of aciclovir, is metabolized in the liver into the active compound acyclovir. The intention of the trial is to dose patients with sufficient valaciclovir to reach the levels of aciclovir in serum similar to i.v treatment for herpes encephalitis (50 mg/kg, 4 times daily). Corresponding oral dose of valaciclovir was calculated (2 g, 4 times daily). Patients receive either valaciclovir or placebo in a double blind fashion. Patient recruitment is in progress and inclusion is nearly complete. Primary endpoint of the study is the reduction of [^{11}C]PK11195 binding potential in drug vs. placebo treated patients. Secondary outcome measures are reduction of positive psychotic symptoms and improvement of cognition in drug treated patients vs. controls.

NMDA Receptor Blockade: Common Pathway to Psychosis

Schizophrenia is a disorder with recurrent psychotic episodes as the major diagnostic criterion. As mentioned before, psychosis is associated with activation of microglia in the hippocampal region. Hippocampal inflammation could represent the biological cause of volume reduction of the hippocampus on MRI in patients after their first psychotic episode. However, to conclude about a causal relationship is premature and the cause–effect relationship might actually be the reverse: The activation of microglia during psychosis might in fact be an ongoing protective response and an ultimate effort to prevent degeneration of brain tissue. The exact phenotype of the activated microglia in psychotic patients cannot be characterized with PET imaging and might occur in two opposite directions (see Fig. 12.1): Similar to activation state of T-cells (Th1/Th2) and macrophages (M1/M2), microglia activation is hypothesized to occur in either predominant M1 or M2 responses (Durafourt et al. 2012). The M1 response is the classically activated cellular response with release of free radicals and pro-inflammatory cytokines (the "bad" phenotype). The M2 phenotype is the alternatively activated humoral response with the production of anti-inflammatory cytokines and growth factors (the "good" phenotype). The balance between M1 and M2 is tightly regulated by the cytokines released in the microenvironment. Which phenotype is dominant during psychosis is not known, but in schizophrenia the balance in the periphery is towards Th2 (Muller and Schwarz 2006; Najjar et al. 2013). The protective phenotype of microglia also expresses MHCII class antigens (Najjar et al. 2013): In schizophrenia patients kynurenic acid response in serum, CSF, and brain tissue is prevailing at the expense of quinolinic acid, which might imply that the M2 response of microglia prevails (Najjar et al. 2013) as depicted in Fig. 12.1: A balance towards kynurenic acid at the cost of quinolinic acid, which exerts opposite effects on NMDA receptors, as they block and stimulate NMDA receptor-mediated glutamate responses, respectively (Müller et al. 2012). There is ample evidence that blockade of the NMDA receptors by either antibodies (Hung et al. 2011; Kleyensteuber et al. 2010; Van de Riet and Schieveld 2013) or hallucinogenic drugs (Javitt et al. 2012) may cause psychosis: It is important to realize that experimental blockade of the NMDA receptors with drugs such as ketamine causes acute paranoid psychosis in the majority of exposed healthy volunteers, in a very similar presentation as in schizophrenia psychosis (Vollenweider et al. 1997a, b). Also important to note is that auto-antibodies against the NMDA receptors provoke classic paranoid psychotic episodes (Hung et al. 2011). The production of NMDA receptor antibodies may be a consequence of herpes encephalitis (Armangue et al. 2013; Desena et al. 2014; Leypoldt et al. 2013; Prüss et al. 2012) and can explain recurrent episodes choreatic movements long after the infection (Hacohen et al. 2014; Mohammad et al. 2014).

It is very interesting that the neurotoxic effects of herpes simplex encephalitis might be due overactivation of the NMDA receptor by the excitatory quinolinic acid (Reinhard 1998), and the more subtle behavioral changes by blockade of the NMDA receptor by the inhibitory kynurenic acid (Atlas et al. 2013).

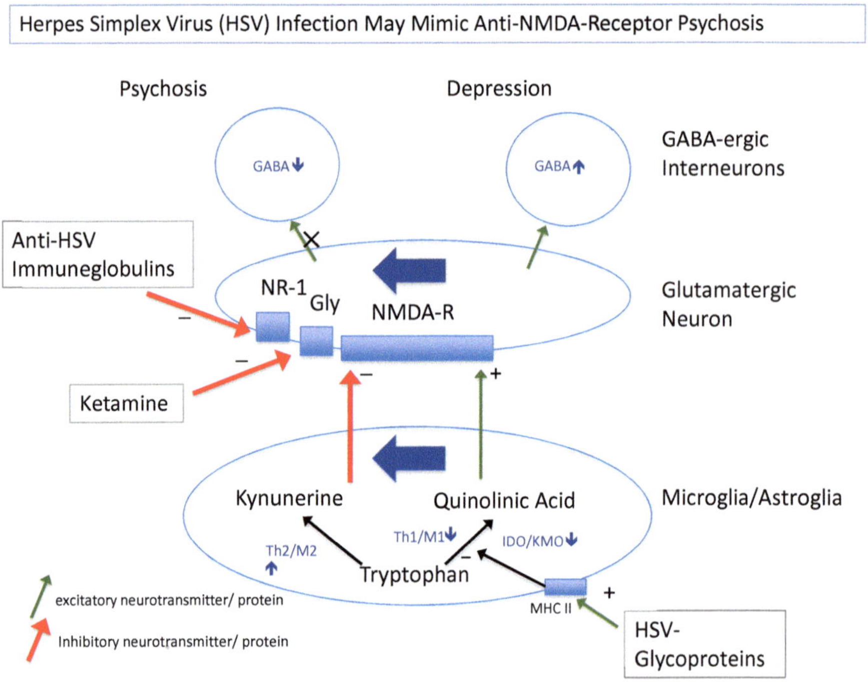

Fig. 12.1 Inflammatory pathways to psychosis, adapted from Najjar et al. (2013). In the microglia/astroglia an M2 phenotype is favored, leading to a balance shift to the left in the microglia/astroglia metabolism in the figure: The cell produces more of the NMDA-R *antagonist* kynunerine is in comparison with the NMDA-R *agonist* quinolinic acid. Exposition to the glycoproteins of HSV induces MHCII class epitope expression, which also favors the left shift in the microglia/astroglia metabolism, this occurs by inhibition of IDO (indolamine deoxygenase) and KMO (kynunerine mono oxygenase) of the cell and thereby blocking of quinolinic acid formation. Net effect of the left shift in the metabolism of the microglia/astroglia is inhibition of the NMDA-R on glutamatergic (glutamate releasing) neurons. This leads to a reduction of the release of glutamate (crossed *arrow* on the above left) of the glutamatergic neuron. The postsynaptic GABAergic neurons then become *less* stimulated. Because GABA is the major inhibitory neurotransmitter of the brain (approx. 30 % of all neurotransmission is GABA ergic!) this leads to a state of disinhibition, which is assumed to be the cause of anti-NMDA psychosis. HSV antibodies may also act directly on the NMDA (NR-1) receptors and may induce psychosis in cases indistinguishable from "regular" anti-NMDA encephalitis. Also ketamine (and PCP) act directly against the NMDA-R (glycine site) and cause psychosis in 50 % of volunteers in a study by Vollenweider et al. (1997a). In contrast, hyperactivation of the NMDA-R by quinolinic acid is not associated with psychosis but with depression, negative symptoms of schizophrenia and neurodegenerative changes and is supposed to be due to an inflammatory response with Th1/M1 character

All these clinical and pre-clinical observations on the etiology of immune or pharmacologically mediated psychoses suggest that the final common pathway to psychosis might be NMDA receptor antagonism. Drugs (ketamine), immune/neurotransmitter signaling molecules (kynurenic acid) and specific (NMDA) antibodies can all produce the classic phenotype of a paranoid psychosis with predominant positive symptoms (hallucinations and delusions).

Blockade of NMDA receptors is hypothesized to cause psychosis by the following mechanism. Glutamate is an excitatory neurotransmitter released from pyramidal cells. Glutamate is released predominantly to interneurons that possess the NMDA receptor and release the inhibitory neurotransmitter GABA. By blocking the effects of glutamate on the NMDA receptor of interneurons, these inhibitory neurons deactivate. This leads to less release of the inhibitory neurotransmitter GABA and consequent disinhibition/activation of cortical circuitry (Olney et al. 1991), which may be responsible for the psychotic phenomena.

Conclusion

It is very difficult to delineate etiology in mental disorders. The brain is not easy to approach for molecular tests. However, making use of unique molecular probes for PET imaging, we might come closer to the understanding of which external factors might be triggers for transition from perfect health to severely debilitating diseases such as schizophrenia. Psychosis of different etiology might share the pathogenesis in the common mechanism of the inhibition of NMDA signaling. Psychosis could therefore be mediated by NMDA blockade as a result of environmental factors such as social stress, hallucinatory drug use, auto-immune disease, or neurotropic viruses.

References

Aguzzi A, Barres BA, Bennett ML. Microglia: scapegoat, saboteur, or something else? Science. 2013;339(6116):156–61. doi:10.1126/science.1227901.

Armangue T, Leypoldt F, Málaga I, Raspall-Chaure M, Marti I, Nichter C, et al. Herpes simplex virus encephalitis is a trigger of brain autoimmunity. Ann Neurol. 2013;75:317–23. doi:10.1002/ana.24083.

Atlas A, Franzen-Röhl E, Söderlund J, Jönsson EG, Samuelsson M, Schwieler L, et al. Sustained elevation of kynurenic acid in the cerebrospinal fluid of patients with herpes simplex virus type 1 encephalitis. Int J Tryptophan Res. 2013;6:89–96. doi:10.4137/IJTR.S13256.

Baringer JR, Pisani P. Herpes simplex virus genomes in human nervous system tissue analyzed by polymerase chain reaction. Ann Neurol. 1994;36(6):823–9. doi:10.1002/ana.410360605.

Bortolato M, Godar SC. Animal models of virus-induced neurobehavioral sequelae: recent advances, methodological issues, and future prospects. Interdiscip Perspect Infect Dis. 2010; 2010:380456. doi:10.1155/2010/380456.

Buursma AR, de Vries EFJ, Garssen J, Kegler D, van Waarde A, Schirm J, et al. [18F]FHPG positron emission tomography for detection of herpes simplex virus (HSV) in experimental HSV encephalitis. J Virol. 2005;79(12):7721–7. doi:10.1128/JVI.79.12.7721-7727.2005.

Cagnin A, Myers R, Gunn RN, Lawrence AD, Stevens T, Kreutzberg GW, et al. In vivo visualization of activated glia by [11C] (R)-PK11195-PET following herpes encephalitis reveals projected neuronal damage beyond the primary focal lesion. Brain. 2001;124(Pt 10):2014–27. Retrieved from http://www.ncbi.nlm.nih.gov/pubmed/11571219.

Cosenza-Nashat M, Zhao M-L, Suh H-S, Morgan J, Natividad R, Morgello S, Lee SC. Expression of the translocator protein of 18 kDa by microglia, macrophages and astrocytes based on immunohistochemical localization in abnormal human brain. Neuropathol Appl Neurobiol. 2009;35(3):306–28. doi:10.1111/j.1365-2990.2008.01006.x.

Damani MR, Zhao L, Fontainhas AM, Amaral J, Fariss RN, Wong WT. Age-related alterations in the dynamic behavior of microglia. Aging Cell. 2011;10(2):263–76. doi:10.1111/j.1474-9726.2010.00660.x.

Desena A, Graves D, Warnack W, Greenberg BM. Herpes simplex encephalitis as a potential cause of anti-*N*-methyl-D-aspartate receptor antibody encephalitis: report of 2 cases. JAMA Neurol. 2014;73:344–6. doi:10.1001/jamaneurol.2013.4580.

Doorduin J, de Vries EFJ, Dierckx RA, Klein HC. PET imaging of the peripheral benzodiazepine receptor: monitoring disease progression and therapy response in neurodegenerative disorders. Curr Pharm Des. 2008;14(31):3297–315. Retrieved from http://www.ncbi.nlm.nih.gov/pubmed/19075709.

Doorduin J, de Vries EFJ, Willemsen ATM, de Groot JC, Dierckx RA, Klein HC. Neuroinflammation in schizophrenia-related psychosis: a PET study. J Nucl Med. 2009;50(11):1801–7. doi:10.2967/jnumed.109.066647.

Doorduin J, Klein HC, Dierckx RAJO, de Vries EFJ. HSV-1 induced neuroinflammation and antipsychotic treatment affect P-glycoprotein activity in the rat. Eur J Nucl Med. 2010;37:S250.

Doorduin J, de Vries EFJ, Willemsen ATM, Dierckx A, Klein HC. Chapter 10: Herpes viruses and neuroinflammation: PET imaging and implication in schizophrenia. University Groningen. 2010.

Drachman DA, Adams RD. Herpes simplex and acute inclusion-body encephalitis. Arch Neurol. 1962;7:45–63. Retrieved from http://www.ncbi.nlm.nih.gov/pubmed/13887679.

Durafourt BA, Moore CS, Zammit DA, Johnson TA, Zaguia F, Guiot M-C, et al. Comparison of polarization properties of human adult microglia and blood-derived macrophages. Glia. 2012;60(5):717–27. doi:10.1002/glia.22298.

Ekdahl CT, Kokaia Z, Lindvall O. Brain inflammation and adult neurogenesis: the dual role of microglia. Neuroscience. 2009;158(3):1021–9. doi:10.1016/j.neuroscience.2008.06.052.

Gordon L, McQuaid S, Cosby SL. Detection of herpes simplex virus (types 1 and 2) and human herpesvirus 6 DNA in human brain tissue by polymerase chain reaction. Clin Diagn Virol. 1996;6(1):33–40. Retrieved from http://www.ncbi.nlm.nih.gov/pubmed/15566888.

Haarman BC, Riemersma-Van der Lek RF, de Groot JC, Ruhé HG, Klein HC, Zandstra TE, et al. Neuroinflammation in bipolar disorder – a [(11)C]-(R)-PK11195 positron emission tomography study. Brain Behav Immun. 2014;40:219–25. doi:10.1016/j.bbi.2014.03.016.

Hacohen Y, Deiva K, Pettingill P, Waters P, Siddiqui A, Chretien P, et al. *N*-methyl-D-aspartate receptor antibodies in post-herpes simplex virus encephalitis neurological relapse. Mov Disord. 2014;29(1):90–6. doi:10.1002/mds.25626.

Halford WP, Kemp CD, Isler JA, Davido DJ, Schaffer PA. ICP0, ICP4, or VP16 expressed from adenovirus vectors induces reactivation of latent herpes simplex virus type 1 in primary cultures of latently infected trigeminal ganglion cells. J Virol. 2001;75(13):6143–53. doi:10.1128/JVI.75.13.6143-6153.2001.

Hanes JW, Zhu Y, Parris DS, Johnson KA. Enzymatic therapeutic index of acyclovir. Viral versus human polymerase gamma specificity. J Biol Chem. 2007;282(34):25159–67. doi:10.1074/jbc.M703972200.

Hart AD, Wyttenbach A, Perry VH, Teeling JL. Age related changes in microglial phenotype vary between CNS regions: grey versus white matter differences. Brain Behav Immun. 2012;26(5):754–65. doi:10.1016/j.bbi.2011.11.006.

Henkel JS, Beers DR, Zhao W, Appel SH. Microglia in ALS: the good, the bad, and the resting. J Neuroimmune Pharmacol. 2009;4(4):389–98. doi:10.1007/s11481-009-9171-5.

Hung T-Y, Foo N-H, Lai M-C. Anti-*N*-methyl-D-aspartate receptor encephalitis. Pediatr Neonatol. 2011;52(6):361–4. doi:10.1016/j.pedneo.2011.08.012.

Javitt DC, Zukin SR, Heresco-Levy U, Umbricht D. Has an angel shown the way? Etiological and therapeutic implications of the PCP/NMDA model of schizophrenia. Schizophr Bull. 2012;38(5):958–66. doi:10.1093/schbul/sbs069.

Johnson RT. Viral infections of the central nervous system. Philadelphia: Lippincott Raven; 1998.

Kapur N, Barker S, Burrows EH, Ellison D, Brice J, Illis LS, et al. Herpes simplex encephalitis: long term magnetic resonance imaging and neuropsychological profile. J Neurol Neurosurg Psychiatry. 1994;57(11):1334–42. Retrieved from http://www.pubmedcentral.nih.gov/articlerender.fcgi?artid=1073183&tool=pmcentrez&rendertype=abstract.
Kleyensteuber B, Ruterbusch V, Bennett J, Llewellyn D, Loeffler G. Limbic encephalitis presenting with seizures, anterograde amnesia, and psychosis in a patient seven weeks status post immature ovarian teratoma removal. Mil Med. 2010;175(8):616–8. Retrieved from http://www.ncbi.nlm.nih.gov/pubmed/20731268.
Koehler K, Guth W. The mimicking of mania in "benign" herpes simplex encephalitis. Biol Psychiatry. 1979;14(2):405–11. Retrieved from http://www.ncbi.nlm.nih.gov/pubmed/476227.
Leypoldt F, Titulaer MJ, Aguilar E, Walther J, Bönstrup M, Havemeister S, et al. Herpes simplex virus-1 encephalitis can trigger anti-NMDA receptor encephalitis: case report. Neurology. 2013;81(18):1637–9. doi:10.1212/WNL.0b013e3182a9f531.
Maeda K, Horimoto T, Norimine J, Kawaguchi Y, Tomonaga K, Niikura M, et al. Identification and nucleotide sequence of a gene in feline herpesvirus type 1 homologous to the herpes simplex virus gene encoding the glycoprotein B. Arch Virol. 1992;127(1–4):387–97. Retrieved from http://www.ncbi.nlm.nih.gov/pubmed/1333759.
Misra PC, Hay GG. Encephalitis presenting as acute schizophrenia. Br Med J. 1971;1(5748):532–3. Retrieved from http://www.pubmedcentral.nih.gov/articlerender.fcgi?artid=1795274&tool=pmcentrez&rendertype=abstract.
Mohammad SS, Sinclair K, Pillai S, Merheb V, Aumann TD, Gill D, et al. Herpes simplex encephalitis relapse with chorea is associated with autoantibodies to *N*-methyl-D-aspartate receptor or dopamine-2 receptor. Mov Disord. 2014;29(1):117–22. doi:10.1002/mds.25623.
Muller N, Schwarz M. Schizophrenia as an inflammation-mediated dysbalance of glutamatergic neurotransmission. Neurotox Res. 2006;10(2):131–48. Retrieved from http://www.ncbi.nlm.nih.gov/pubmed/17062375.
Müller N, Myint A-M, Schwarz MJ. Inflammation in schizophrenia. Adv Protein Chem Struct Biol. 2012;88:49–68. doi:10.1016/B978-0-12-398314-5.00003-9.
Najjar S, Pearlman DM, Alper K, Najjar A, Devinsky O. Neuroinflammation and psychiatric illness. J Neuroinflammation. 2013;10:43. doi:10.1186/1742-2094-10-43.
Olah M, Biber K, Vinet J, Boddeke HW. Microglia phenotype diversity. CNS Neurol Disord Drug Targets. 2011;10(1):108–18. Retrieved from http://www.ncbi.nlm.nih.gov/pubmed/21143141.
Olah M, Amor S, Brouwer N, Vinet J, Eggen B, Biber K, Boddeke HW. Identification of a microglia phenotype supportive of remyelination. Glia. 2012;60(2):306–21. doi:10.1002/glia.21266.
Olney JW, Labruyere J, Wang G, Wozniak DF, Price MT, Sesma MA. NMDA antagonist neurotoxicity: mechanism and prevention. Science. 1991;254(5037):1515–8.
Prüss H, Finke C, Höltje M, Hofmann J, Klingbeil C, Probst C, et al. *N*-methyl-D-aspartate receptor antibodies in herpes simplex encephalitis. Ann Neurol. 2012;72(6):902–11. doi:10.1002/ana.23689.
Raskin DE, Frank SW. Herpes encephalitis with catatonic stupor. Arch Gen Psychiatry. 1974;31(4):544–6. Retrieved from http://www.ncbi.nlm.nih.gov/pubmed/4370982.
Reinhard JF. Altered tryptophan metabolism in mice with herpes simplex virus encephalitis: increases in spinal cord quinolinic acid. Neurochem Res. 1998;23(5):661–5. Retrieved from http://www.ncbi.nlm.nih.gov/pubmed/9566604.
Rivers TM. Viruses and Koch's postulates. J Bacteriol. 1937;33(1):1–12. Retrieved from http://www.pubmedcentral.nih.gov/articlerender.fcgi?artid=545348&tool=pmcentrez&rendertype=abstract.
Schlitt M, Lakeman FD, Whitley RJ. Psychosis and herpes simplex encephalitis. South Med J. 1985;78(11):1347–50. Retrieved from http://www.ncbi.nlm.nih.gov/pubmed/4071143.
Schmutzhard E. Viral infections of the CNS with special emphasis on herpes simplex infections. J Neurol. 2001;248(6):469–77. Retrieved from http://www.ncbi.nlm.nih.gov/pubmed/11499636.
Schretlen DJ, Vannorsdall TD, Winicki JM, Mushtaq Y, Hikida T, Sawa A, et al. Neuroanatomic and cognitive abnormalities related to herpes simplex virus type 1 in schizophrenia. Schizophr Res. 2010;118(1–3):224–31. doi:10.1016/j.schres.2010.01.008.

Schuitemaker A, van der Doef TF, Boellaard R, van der Flier WM, Yaqub M, Windhorst AD, et al. Microglial activation in healthy aging. Neurobiol Aging. 2012;33:1067–72. doi:10.1016/j.neurobiolaging.2010.09.016.

Swinnen N, Smolders S, Avila A, Notelaers K, Paesen R, Ameloot M, et al. Complex invasion pattern of the cerebral cortex bymicroglial cells during development of the mouse embryo. Glia. 2013;61(2):150–63. doi:10.1002/glia.22421.

Thompson RL, Shieh MT, Sawtell NM. Analysis of herpes simplex virus ICP0 promoter function in sensory neurons during acute infection, establishment of latency, and reactivation in vivo. J Virol. 2003;77(22):12319–30. doi:10.1128/JVI.77.22.12319.

van Berckel BN, Bossong MG, Boellaard R, Kloet R, Schuitemaker A, Caspers E, et al. Microglia activation in recent-onset schizophrenia: a quantitative (R)-[11C]PK11195 positron emission tomography study. Biol Psychiatry. 2008;64(9):820–2. doi:10.1016/j.biopsych.2008.04.025.

van de Riet EH, Schieveld JN. First-onset psychosis, anti-NMDAR encephalitis, schizophrenia and Consultation-Liaison psychiatry. Gen Hosp Psychiatry. 2013;35:442–3. doi:10.1016/j.genhosppsych.2013.01.014.

Vollenweider FX, Leenders KL, Scharfetter C, Antonini A, Maguire P, Missimer J, Angst J. Metabolic hyperfrontality and psychopathology in the ketamine model of psychosis using positron emission tomography (PET) and [18F]fluorodeoxyglucose (FDG). Eur Neuropsychopharmacol. 1997a;7(1):9–24. Retrieved from http://www.ncbi.nlm.nih.gov/pubmed/9088881.

Vollenweider FX, Leenders KL, Scharfetter C, Maguire P, Stadelmann O, Angst J. Positron emission tomography and fluorodeoxyglucose studies of metabolic hyperfrontality and psychopathology in the psilocybin model of psychosis. Neuropsychopharmacology. 1997b;16(5):357–72. doi:10.1016/S0893-133X(96)00246-1.

Whitley RJ, Gnann JW. The epidemiology and clinical manifestations of herpes simplex infections. In: Roizman B, Whitley RJ, Lopez C, editors. The human herpes viruses. New York, NY: Raven; 2003. p. 69–105.

Whitley RJ, Lakeman F. Herpes simplex virus infections of the central nervous system: therapeutic and diagnostic considerations. Clin Infect Dis. 1995;20(2):414–20. Retrieved from http://www.ncbi.nlm.nih.gov/pubmed/7742450.

Wozniak MA, Mee AP, Itzhaki RF. Herpes simplex virus type 1 DNA is located within Alzheimer's disease amyloid plaques. J Pathol. 2009;217(1):131–8. doi:10.1002/path.2449.

Yolken RH, Torrey EF. Viruses, schizophrenia, and bipolar disorder. Clin Microbiol Rev. 1995;8(1):131–45. Retrieved from http://www.pubmedcentral.nih.gov/articlerender.fcgi?artid=172852&tool=pmcentrez&rendertype=abstract.

Ziv Y, Ron N, Butovsky O, Landa G, Sudai E, Greenberg N, et al. Immune cells contribute to the maintenance of neurogenesis and spatial learning abilities in adulthood. Nat Neurosci. 2006; 9(2):268–75. doi:10.1038/nn1629.

Chapter 13
Role of Autoimmunity and Infections in Tourette Syndrome

Pieter J. Hoekstra

Abstract Tourette's syndrome (TS) is characterized by the presence of both motor and vocal tics. The classically proposed autoimmunity model of TS is analogous to Sydenham's chorea. Per this model, tics and associated phenomena are thought to arise as a consequence of the immunological response to infections with Group A beta hemolytic Streptococci (GABHS). Antibodies directed against the streptococci are thought to cross-react with structures of the central nervous system, subsequently leading to damage to these structures, which eventually is thought to result in tics and associated features.

Autoimmunity in TS may be triggered by infections. Relatively good evidence is available about a possible association of streptococcal infections with TS. The available evidence seems to suggest that patients with TS are more susceptible to infections. Most evidence points to more frequent streptococcal infections. Moreover animal studies have quite convincingly demonstrated that immunization with streptococcal antigens or transfer of antibodies evoked through immunization can indeed lead to tics. Also, there is also some indication about the possible involvement of non-streptococcal infections, including common viral infections, in exacerbations of tic disorders.

Why would there be more frequent infections in patients with TS? Some studies have pointed to altered immune functioning as potential explanations, which might induce antineuronal autoantibodies. Animal models have quite convincingly pointed to the pathogenic relevance of infection-induced antineuronal antibodies. Obviously, standardization of assays in human sera should be next important steps to further our understanding of the role of antineuronal antibodies in TS.

Keywords Tourette syndrome • Tics • PANDAS • Autoimmunity • Antineuronal antibodies • IVIG • Plasmapheresis • Antibiotics • Cytokines • Streptococcal infections

---

P.J. Hoekstra (✉)
Department of Child and Adolescent Psychiatry, University of Groningen, University Medical Center Groningen, Groningen, The Netherlands
e-mail: p.hoekstra@accare.nl

N. Müller et al. (eds.), *Immunology and Psychiatry*, Current Topics in Neurotoxicity 8, DOI 10.1007/978-3-319-13602-8_13

Clinical Features

Tourette's syndrome (TS) is a neuropsychiatric disorder, characterized by the presence of both motor and vocal tics. A tic is a sudden, rapid, recurrent, nonrhythmic motor movement or vocalization. Tics vary greatly between and within individuals. Most motor tics are brief, sudden and meaningless muscle movements, such as eye blinking, nose twitching or shoulder shrugging (simple motor tics). In contrast, complex motor tics appear more purposeful and involve several muscle groups. Examples include touching other people or objects, retracing steps when walking and various complex hand gestures. Similarly, vocal tics may be subdivided into simple and complex tics, ranging from meaningless sounds such as throat clearing, sniffing and barking to the sudden utterance of words, phrases and full sentences which may in a minority of cases include echolalia and coprolalia. Tic intensity can vary substantially, ranging from barely visible or audible tics to extremely forceful or loud expressions. Quite often, tics are mild, in which case they hardly attract attention from others and do not interfere with everyday life. On the other hand, powerful and frequent tics may severely interfere with everyday activities, including speech, driving and walking. In exceptional cases, tics may lead to physical injury, including joint dislocation and other tissue damage. Also, patients who display more severe or complex tics may be stigmatized as a result of the unusual, inappropriate and bizarre character of their tics. Most patients do not experience their tics as entirely beyond their control; many individuals describe premonitory urges preceding their tics, feelings that are momentarily relieved by the performance of tics and may be temporarily ignored by suppressing the tics.

The age of onset of tics is mostly between 2 and 15 years with a median of 7 (Bruun and Budman 1997). Facial tics are normally the initial symptom. Males are more commonly affected than females (Tanner and Goldman 1997). Movements generally decrease during sleep and may be suppressed for short periods while the patient is awake (Jankovic 1997). Tics typically occur in bouts during the day alternated with relatively tic-free periods within the course of a day (Peterson and Leckman 1998). Similarly, the course of tics over a period of months to years often waxes and wanes in severity (Coffey et al. 1994). Furthermore, the type of tics in an individual patient is typically variable over time, with some tics disappearing and new ones appearing in the course of time. With increasing age, however, symptoms tend to decrease in intensity and to show less variation over time regarding both severity and type of tics (Bruun and Budman 1997). TS is not rare; population studies estimate prevalence rates for TS between around 1 % (Robertson 2008).

One important feature of TS is its well-known association with a range of behavioural disorders and comorbid psychopathology, which may be more clinically relevant than the tics themselves (Mol Debes 2013). Attention-deficit/hyperactivity disorder (ADHD) is known to affect 50 % of referred patients (Olfson et al. 2011). Obsessive-compulsive symptoms constitute another common cophenomenon of the spectrum of tic disorders (Wanderer et al. 2012). Also, many children show significant problems with social functioning (Kurlan et al. 1996).

Although genetic factors are known play a pivotal role in TS, currently, our knowledge about the pathogenesis is still inconclusive. Structural and functional neuroimaging studies point to the involvement of pathways leading from cortex through basal ganglia to thalamus (Leckman 2002). The basal ganglia facilitate the ability to effectively switch between motor and mental behaviours, required for producing novel behaviour. Failures to do so may result in the repetitive production of stereotyped movements, thoughts or behaviours as in TS.

Presumed Model of Autoimmunity for TS

Over the past two decades a significant amount of research has been conducted on the possible role of infections and immune alterations in TS. The classically proposed model of pathogenesis of tic disorders in this research is analogous to Sydenham's chorea (Swedo 1994). Per this model, tics and associated phenomena are thought to arise as a consequence of the immunological response to infections with Group A beta hemolytic Streptococci (GABHS). Antibodies directed against the streptococci are hypothesized to cross-react with structures of the central nervous system, subsequently leading to damage to these structures, which eventually is thought to result in tics and associated features. This supposed mechanism of autoimmunity is a classic model of molecular mimicry between host and microorganism. Subsequent reinfections according to the presumed mechanism would lead to symptom exacerbations.

Several clinical observations have led to the hypothesis that Sydenham's chorea might be a model for some types of TS and childhood-onset obsessive-compulsive disorder (OCD). First, it had been noted that patients with Sydenham's chorea shared certain behavioural characteristics with patients with OCD and/or tic disorders, such as emotional lability, marked irritability, but also frank obsessive-compulsive symptoms (Swedo et al. 1993). Second, a substantial number of children with OCD were reported to show choreiform movements or tics (Steingard and Dillon-Stout 1992). In addition, in some carefully studied children with TS and/or OCD, an episodic course and/or abrupt onset of their symptoms seemed to be temporally related to signs of GABHS infections (Swedo 1994). Following these observations, case studies began to appear in the literature in the mid-1990s, in which children with OCD and/or tic disorders were described who suddenly demonstrated severe forms of tics and obsessive-compulsive symptoms and in whom a temporal relationship between symptom onset or exacerbations and GABHS or viral infections seemed apparent (e.g., Allen et al. 1995; Tucker et al. 1996). In some cases, streptococcal reinfections were associated with the reinduction of neuropsychiatric symptoms (Swedo et al. 1998).

PANDAS as a Proposed Disease Entity

Researchers of the National Institute of Mental Health (NIMH) subsequently proposed criteria to identify a putatively unique subgroup of children with TS and/or OCD, in whom symptom onset and/or exacerbations were abrupt, dramatic and temporally related to GABHS infections. They designated it by the term paediatric autoimmune neuropsychiatric disorders associated with streptococcal infections (PANDAS; Swedo et al. 1998). Research criteria have been outlined for the PANDAS subgroup, which require the presence of OCD and/or tic disorder, prepubertal symptom onset, sudden onset or episodic course of symptoms, temporal association between streptococcal infections and neuropsychiatric symptom exacerbations and associated neurological abnormalities (Swedo et al. 1998). What percentage of children with TS and/or OCD would meet criteria for PANDAS has not yet been investigated.

The concept of PANDAS strongly centres on Sydenham's chorea as the putative disease model, including the suggested pathogenic role of autoantibodies that cross-react with brain antigens as a hypothesized consequence of structural homology with streptococcal antigens (Swedo 1994). Introduction of the concept of PANDAS has certainly done much to stimulate research and raise clinical awareness of the potential relevance of immune factors in tic and related disorders. Concerns regarding the validity of PANDAS have been expressed, however (Kurlan 2008; Shulman 1999; Hoekstra et al. 2002; Singer and Loiselle 2003). These include criticism about the vagueness of some of the criteria for PANDAS. For example, when should an exacerbation be considered abrupt? Or what time frame constitutes a temporal relationship between GABHS infections and symptom exacerbations? Or, what is sufficient proof of a GABHS infection? The diagnostic criteria for PANDAS are therefore not easy to apply in clinical practice as it is not straightforward to demonstrate a temporal association between GABHS infection and symptom onset and exacerbations, since streptococcal infections are fairly common in children in general, as are remissions and exacerbations in children with TS. We also lack a laboratory test for PANDAS. Case studies of patients with suspected PANDAS continue to appear in the literature (e.g., Chmelik et al. 2004; van Toorn et al. 2004; Gabbay and Coffey 2003; Storch et al. 2004; Maini et al. 2012; Das and Radhakrishnan 2012; Hachiya et al. 2013).

Some direct comparisons between PANDAS cases and non-PANDAS TS and/or OCD have also been published. In a prospective study, Kurlan et al. (2008) found that patients with PANDAS more often had exacerbations which were temporally associated (within 4 weeks) with a GABHS infection than non-PANDAS cases. Yet, the majority of the clinical exacerbations in PANDAS cases had no observable temporal relationship to GABHS infection. The authors concluded that GABHS infection is therefore not the only or even the most common antecedent event associated with exacerbations for patients fulfilling PANDAS criteria when followed over time. However, in a more recent longitudinal study, with a similar design of comparing tic exacerbation triggers between PANDAS and non-PANDAS cases, remarkably, occurrences of GABHS infections followed by a subsequent tic exacerbation were always in the *non-PANDAS* cases (even though also in the majority of non-PANDAS cases, exacerbations were not preceded by a GABHS infection). This latter finding challenges the validity of the clinical PANDAS criteria to discriminate between GABHS and non-GABHS related TS.

On the basis of concerns over the diagnostic criteria for PANDAS, a broader concept of childhood acute neuropsychiatric symptoms (CANS) has recently been proposed (Singer et al. 2012). The proposed CANS classification does not require association with a specific organism, limitation of symptoms to tics or OCD, a specific age range, or recurrence of symptoms. It does, however, require an acute dramatic onset, a comprehensive history and examination, and diagnostic evaluation. The usefulness of CANS as an entity remains to be investigated.

Association of TS with Infections

Rather than trying to identify PANDAS cases through directly implying an association of TS with infections, based on predefined PANDAS criteria, a scientifically perhaps better approach to demonstrate a relationship between TS and infections, would be to study the relevance of infections in unselected patients with TS. Interestingly, single time point studies have repeatedly found higher anti-streptococcal antibody titres (antistreptolysin O and antideoxyribonuclease B) in TS. Muller et al. (2000) found that 85 % of the subjects with TS vs 8 % of normal controls had elevated antideoxyribonuclease B levels. The same researchers subsequently showed the presence of increased titres against the streptococcal M12 and M19 proteins in patients with TS as compared with controls, while antibody titres against M1, M4 and M6 did not differ between the TS and control groups (Muller et al. 2001). Also, Cardona and Orefici (2001) reported significantly higher mean antistreptolysin titres in children with tics compared to control children, and found a positive correlation between antistreptolysin titres and severity of the tics as measured by the Yale Global Tic Severity Scale. Higher mean levels of anti-streptococcal antibody titres in patients with TS compared to controls have also been reported by others (Morer et al. 2006; Creti et al. 2004; Rizzo et al. 2006; Martino et al. 2011; Cheng et al. 2012), although there have also been negative studies, mainly in the USA, that failed to find increased single time point levels of anti-streptococcal antibodies in tic disorder subjects (Singer et al. 1998; Peterson et al. 2000; Morshed et al. 2002; Loiselle et al. 2003).

The cross-sectional design makes these findings somewhat hard to interpret, especially since antibody levels in these subjects were not assessed at the time of the first appearance of their tics, nor at a time of symptom exacerbation, but rather at an arbitrary point in time. Where could increased levels of anti-streptococcal antibodies point to? Are patients with tics more likely to encounter streptococcal infections? Or does their immune system differently respond to streptococci, which may include a more prolonged humoral response? There is indeed retrospectively collected data that indicates that patients with TS are more likely than controls to have experienced streptococcal infection prior to disease onset date (Mell et al. 2005; Leslie et al. 2008), although these findings were not confirmed in a more recent study (Schrag et al. 2009). However, we really need more large-scale prospective studies to fully understand the relevance of streptococcal infections in TS.

Overall, the available longitudinal studies to date have been somewhat mixed. A first small-scale longitudinal study involving 47 patients with TS and/or OCD found that the association between symptom exacerbations and new GABHS infections among patients was no greater than that expected on the basis of chance (Luo et al. 2004). In another longitudinal study, however, positive correlations were detected between streptococcal titres and obsessive-compulsive severity rating changes in a subgroup of patients with large symptom fluctuations over time (Murphy et al. 2004). These subjects were also more likely to have elevated streptococcal titres in comparison with patients without such dramatic fluctuations. A subsequent longitudinal study in which 693 healthy children (ages 3–12 years) were enrolled who were studied monthly during 8 months revealed a strong relationship between GABHS throat culture and subsequent motor and behaviour changes, in the form of choreiform movements and hyperactivity (Murphy et al. 2007). In another longitudinal study, newly diagnosed infections were predictive of modest increases in future tic and obsessive-compulsive symptom severity, even more so when there were high levels of psychosocial stress (Lin et al. 2010). In contrast, Martino et al. (2011) who followed 144 patients with TS at 3-month intervals found that new infections did not predict clinical exacerbations.

Despite some inconsistency in these data, overall, an association between GABHS infections and TS is noticeable. The results of currently ongoing large-scale longitudinal studies on the association between infections and tic onset and tic exacerbations will have to be awaited before definitive conclusions can be drawn about the role of GABHS infections in TS.

Support for the potential of GABHS infections to induce tic-like behaviours also stems from animal work. It has been shown that exposure of rats to GABHS antigen leads to the production of antineuronal antibodies concomitant with the development of increased compulsive-like behaviour and motor disturbances (Brimberg et al. 2012). Previously it was shown that mice that were immunized and boosted with a streptococcal homogenate in Freund's adjuvant exhibited motoric and behavioural disturbances in association with the presence of serum antibodies that were immunoreactive to several brain regions including globus pallidus and thalamus (Hoffman et al. 2004). This was not the case with mice that were boosted with Freund's adjuvant only. These results are consistent with the hypothesis that immune response to streptococci can result in motor and behaviour alterations and suggest that anti-streptococcal antibodies cross-reactive with brain components may play a role in their pathophysiology.

Interestingly, a number of studies also seem to implicate a role for non-streptococcal infections. Case reports have suggested that both Borna virus (Dietrich et al. 2005) and mycoplasma pneumoniae (Muller et al. 2004; Matsuo et al. 2004) may be associated with tics and/or OCD. In a longitudinal study, our group reported an association between virally induced common colds and tic severity exacerbations in unselected patients with TS (Hoekstra et al. 2005). Also, a close temporal relationship between upper respiratory tract infection and the subsequent onset of tics in two adult patients was described, as was the case in three patients who developed a dystonic syndrome with presence of anti-basal ganglia antibodies after an upper respiratory infection (Edwards et al. 2004). Finally, a case–control study involving 32 patients with TS and 30 matched healthy controls demonstrated a significantly higher rate of elevated antibody titres against Chlamydia trachomatis ($P=0.017$) in patients with TS as compared to controls (Krause et al. 2010). A trend toward a higher prevalence in the Tourette's group was shown for Toxoplasma. Thus, evidence for a role of infections of infections in TS is certainly not confined to streptococcal infections.

In conclusion, relatively good evidence is available about a possible association of streptococcal infections with TS. The available evidence seems to suggest that patients with TS are more susceptible to infections. Most evidence points to more frequent streptococcal infections. Moreover animal studies have quite convincingly demonstrated that immunization with streptococcal antigens or transfer of antibodies evoked through immunization can indeed lead to tics. However, there is also some indication about the possible involvement of non-streptococcal infections, including common viral infections, in exacerbations of tic disorders.

Why would there be more frequent infections in patients with TS? Some studies have pointed to altered immune functioning as potential explanations. One investigation found that the level of IgA was decreased in patients with TS/OCD compared with control subjects, possibly explaining why the children with TS are more prone to upper respiratory tract infections (Kawikova et al. 2010); however, lowered IgA levels were not confirmed in a subsequent study, which, instead, indicated decreased serum IgG3, and possibly also IgM levels in patients with TS compared to controls (Bos-Veneman et al. 2011). In a more recent case–control comparison, patients with TS appeared to have a lower receptor expression of TLR4 after stimulation with lipopolysaccharide (meant to mimic a bacterial infection) and higher levels of soluble CD14 (Weidinger et al. 2014). These findings might represent an impaired activation of the innate immune response in TS, especially with regard to bacterial infection. The impaired response to pathogens could form an explanation for the higher susceptibility to infections. Another recent study found that the number of monocytes was significantly higher in patients with TS than in healthy controls, whereas concentrations of TNF-alpha, sIL1-ra, and sCD14 were significantly lower in patients with TS (Matz et al. 2012). The monocyte dysregulation in TS along with a possible dysbalance in innate immunity could be another important lead towards explaining why patients with TS may be predisposed to infections.

Antineuronal Antibodies

Do infections induce autoimmunity, causally related to the occurrence of tics, or is the association of TS with infections just an epiphenomenon? The presence of autoantibodies that react with parts of the brain thought to be involved in tic disorders is potentially a strong line of evidence in favour of the autoimmune hypothesis of TS. Unfortunately, as yet no standardized methodology of assessing anti-basal ganglia antibodies is available. Rather, different research centres have all used different methods, which make results difficult to compare. For example, to investigate whether a subject's blood may contain antibodies that react with brain antigens, different substrates of brain tissue have been utilized across studies, ranging from immortalized neuronal cell lines to animal and human brains, which have, moreover, been prepared in different ways. Also, methods of assessing antibody binding to brain tissue differed between centres. Earlier studies used immunofluorescence microscopy, whereas later studies made use of enzyme-linked immunosorbent assays (ELISA).

So far, several research groups have reported the increased presence of these antineuronal antibodies in sera from patients with TS, compared to healthy controls (Kiessling et al. 1993; Murphy et al. 1997; Singer et al. 1998, 1999; Laurino et al. 1997; Morshed et al. 2002). Kiessling et al. (1993) were the first to assess antineuronal antibody status in children with recent onset of movement disorders (TS, motor and/or vocal tics, chorea, or choreiform movements), compared with a group of children referred for evaluation of ADHD, behaviour disorders, and learning disabilities who did not show signs of a movement disorder. They applied an indirect immunofluorescence technique, with unfixed frozen human caudate nucleus sections as antigenic substrate, using undiluted sera and fluorescein isothiocyanate-labeled secondary antibody directed against human IgG, and found 44 % of children with a movement disorder to be strongly positive for antineuronal antibodies versus 21 % of the control group. An additional study found positive antineuronal staining in 39 % of children with tic and/or obsessive compulsive disorder as compared to 24 % of healthy control children (Murphy et al. 1997). Subsequent studies using ELISA against either an immortalized neuronal cell line (Laurino et al. 1997; Singer et al. 1999), human basal ganglia (Singer et al. 1998), neuronal antigens (Cheng et al. 2012), or rat brain (Morshed et al. 2002; Yeh et al. 2006) in general confirmed the increased levels of serum antineuronal antibodies in patients with TS. Presence of these antibodies in sera of tic disorder patients fulfilling PANDAS criteria has also been found with high specificity and sensitivity when compared to control groups (children with neurological disease, with autoimmune disease, or with uncomplicated streptococcal infection) using both ELISA and Western blotting (Church et al. 2004). Common binding was observed to basal ganglia. An Italian study that used indirect tissue immunofluorescence identified anti-basal ganglia antibodies in almost two thirds of PANDAS cases and in less than 10 % of children with uncomplicated active streptococcal infection (Pavone et al. 2004). Investigation of oligoclonal bands of immunoglobulin G in cerebrospinal fluid in 21 patients with TS showed that 38 % of the patients exhibited pathological bands indicating a humoral immune response in the central nervous system and pointing to involvement of immunological mechanisms in TS pathology (Wenzel et al. 2011).

An interesting piece of evidence for the contribution of GABHS-induced antibodies to TS was provided by a study that tested the antibody response of tic patient sera to a representative panel of recombinant GABHS antigens (Bombaci et al. 2009). Sera from children with neither tic disorder nor overt GABHS infection were also analysed. The protein recognition patterns of these two sera groups were compared with those obtained using sera from children with GABHS-associated pharyngitis. This comparative analysis identified 25 antigens recognized by sera of the three patient groups and 21 antigens recognized by tic and pharyngitis sera, but poorly or not recognized by sera from children without tic. Interestingly, these antigens appeared to be, in quantitative terms, more immunogenic in tic than in pharyngitis patients. Additionally, a third group of antigens appeared to be preferentially and specifically recognized by tic sera. The authors concluded that this may be relevant in the context of one of the hypotheses proposing that GABHS antigen-dependent induction of autoantibodies in susceptible individuals may be involved in the occurrence of tic disorders.

There have, however, also been a number of negative studies, failing to find differences in antineuronal antibody status between patients with TS or PANDAS versus healthy controls (Brilot et al. 2011; Morris et al. 2009; Singer et al. 2005a); another study failed to find associations between clinical exacerbations of patients with TS and antineuronal antibodies (Singer et al. 2008).

It should be noted that as yet, no single neuronal antigenic structure has been identified as target for the putative antineuronal antibodies. Studies aimed at identifying neuronal antigens have indicated the identification of a candidate auto-antigen, pyruvate kinase by two independent research groups (Kansy et al. 2006; Dale et al. 2004), although binding to pyruvate kinase was not confirmed in a later study (Gause et al. 2009). Another study demonstrated that antibodies in patients with PANDAS reacted with the neuronal cell surface of the caudate-putamen and signal neuronal cells by inducing calcium-calmodulin dependent protein (CaM) kinase II activity. Depletion of serum IgG abrogated CaM kinase II cell signaling and reactivity of cerebrospinal fluid was blocked by streptococcal antigen *N*-acetyl-beta-D-glucosamine (GlcNAc; Kirvan et al. 2006). A recent study indicated that sera from patients with PANDAS specifically targeted dopaminergic neurons by inducing inhibitory signaling of the dopamine 2 receptor on transfected neuronal cells (Cox et al. 2013). However, binding to the dopamine receptor was not present in the sera of a series of patients with PANDAS or TS in two other studies (Dale et al. 2012; Morris-Berry et al. 2013).

Support for a pathogenic role of antineuronal autoantibodies in the disease process stems from studies in which animal models were developed to investigate whether serum or purified IgG from patients with TS can induce tic-like behaviour in rats (Hallett et al. 2000; Taylor et al. 2002). In these studies, serum or IgG was microinfused through cannulas placed in regions of the neostriatum known to induce stereotypies, after which the rats were observed for development of movements or utterances. Hallett et al. (2000) infused dilute serum from five patients with TS, with high antibody titres against human neuroblastoma, bilaterally into the ventral striatal region of the rat. Results showed a significant increase in tic-like behaviours (e.g., licks and forepaw shakes) and episodic utterances in the TS group, which was not observed when sera from healthy controls were microinfused. Taylor et al. (2002) infused serum from 12 patients with TS, who had high antibody titres against rat striatum, bilaterally into a different brain area, the ventrolateral striatal region. This led to a significant increase of oral stereotypies over a 5-day period of observation. However, these results were not confirmed in two subsequent studies aimed at replicating earlier findings about induction of stereotypic movements in rodents through passive transfer of antibodies by means of striatal microinfusions (Loiselle et al. 2004; Singer et al. 2005b). However, in a more recent study, passive transfer of GABHS-induced immunoglobulin G in mice has been shown to lead to tic-like behaviours (Yaddanapudi et al. 2010). Also, IgG from GABHS-exposed rats was demonstrated to react with D1 and D2 dopamine receptors and 5HT-2A and 5HT-2C serotonin receptors in vitro, while in vivo, IgG deposits in the striatum of infused rats colocalized with specific brain proteins such as dopamine receptors, the serotonin transporter and other neuronal proteins (Lotan et al. 2014).

In conclusion, the assessment of serum autoantibodies has appeared to be a somewhat conflicting research area in tic disorders, with a number of potentially promising leads towards characterizing antigenic structures at a molecular level, but also a fair amount of negative studies. Animal models, however, have quite convincingly pointed to the pathogenic relevance of infection-induced antineuronal antibodies. Obviously, standardization of assays in human sera should be next important steps to further our understanding of the role of antineuronal antibodies in TS.

Overactivity of Immune Responses and Association with Allergies

A number of studies have indicated increased activity of the immune system in patients with TS. One marker of cellular immune activation is the increased degradation of tryptophan via the kynurenine pathway, leading to elevated plasma levels of kynurenine and subsequent metabolites (Heyes 1996; Meyer et al. 1995). Upregulation of the kynurenine pathway can be induced through increased activity of Indoleamine 2,3-dioxygenase (IDO), which is sensitive to Interferon-gamma, a major cytokine of cellular immunity, thus, the increased levels of kynurenine may possibly be reflecting immune activation. Conversion of tryptophan to kynurenine can be triggered by GABHS, as has been shown in vitro (Murr et al. 1997). Some authors also found that kynurenine increases tic-like behaviour in an animal model of TS: in mice, head-shakes which had been induced by the 5-hydroxytryptamine receptor agonist dimethoxyiodo-phenyl-aminopropane were potentiated by administration of kynurenine (McCreary and Handley 1995).

In an initial small-scale study (involving seven patients with a tic disorder versus controls) the serum kynurenine level was found to be clearly increased in all seven patients whereas serum tryptophan was normal (Rickards et al. 1996). In a subsequent larger scale study, involving 72 patients with TS and 46 matched controls, again, plasma kynurenine levels were found to be significantly elevated (Gaynor et al. 1997). We know of one other independent report of increased plasma kynurenine (Chappel et al. 1995). Interestingly, both case–control studies (Rickards et al. 1996; Gaynor et al. 1997) reported a significant positive correlation in patients with TS between levels of kynurenine and neopterin. Neopterin is a marker of cellular immunity, which is, like IDO activity, induced by cytokines. Increased levels of neopterin therefore support possible involvement of (auto)immunity. Increased levels of neopterin in patients with TS compared to controls were confirmed in two later studies (Hoekstra et al. 2007; Matz et al. 2012). Another piece of evidence for increased immune activation in patients with TS was provided by the finding of significantly elevated serum levels of vascular cell adhesion molecule-1 and E-selectin in patients with TS, compared to controls (Martino et al. 2005).

Studies which investigated cytokines in paediatric neuropsychiatric disease have been relatively scarce. Patients with OCD were reported in one study to show a relative preponderance in cerebrospinal fluid of type 1 cytokines, notably interleukin-2, suggesting the involvement of cell-mediated immunity (Mittleman et al. 1997). This could be consistent with a role for streptococcal infection, through the involvement of streptococcal erythrogenic toxins, which can act as superantigens and are known to induce type 1 cytokines (Bhatnagar et al. 1999). In another study, serum levels of IL-2 were also significantly higher in patients with TS and comorbid OCD compared to non-OCD TS; TS patients with comorbid OCD also had significantly elevated IL-12 plasma levels compared to healthy controls (Gabbay et al. 2009). Interleukin-12 (along with tumour necrosis factor alpha concentrations) in serum had also been found to be elevated in another cohort of patients with TS compared with control subjects, whereas both of these markers were further increased during periods of symptom exacerbation (Leckman et al. 2005). In a more recent study (Cheng et al. 2012), plasma interleukin (IL)-1β, IL-6, IL-17 and soluble gp130 concentrations were significantly higher in the TS group compared with the control group, whereas the soluble IL-6 receptor concentration was significantly decreased in the TS group compared with the control group. The authors interpreted this as indication of altered immune activity in patients with TS. Finally, one small-scale study (Hsieh et al. 2010) found changes in immune markers during tic exacerbations in some patients (higher percentages of natural killer cells and memory T cells) in a small-scale longitudinal study.

Studies focusing on expression of a B-lymphocytes cell marker, D8/17 have found inconsistent results. Antibodies against D8/17 were originally prepared from fusions of spleen cells from mice that had repeatedly been immunized with isolated human B-lymphocytes obtained from rheumatic fever or rheumatic heart disease patients (Khanna et al. 1989). A supposedly D8/17-specific IgM antibody has been reported to bind to an increased percentage of B-lymphocytes in rheumatic fever patients and patients with tic disorder, OCD, PANDAS, autistic disorder, and anorexia nervosa with a high sensitivity and specificity in comparison with controls (reviewed in Hoekstra et al. 2002). These initial studies seemed to suggest that increased D8/17 B cell expression could serve as an interesting peripheral susceptibility marker for TS. Unfortunately, several problems considerably reduce the enthusiasm for D8/17. More recent studies (Weisz et al. 2004; Morer et al. 2005, 2006) that used more sophisticated methods (flow cytometry versus immunofluorescence microscopy; proper isotype control to rule out non-specific binding) failed to replicate earlier findings, perhaps caused by insufficient stability of the antibody. It is conceivable that original positive findings were entirely due to increased non-specific binding to the receptors for the constant parts of IgM molecules (Fc-mu) on B-cells (Hoekstra et al. 2004a), and that later studies failed to find increased binding of the original IgM antibody due to the instability of the IgM structure. Of interest is also that the percentage of CD19-positive B-cells appeared to be significantly elevated in rheumatic fever and patients with TS, along with that of streptococcal pharyngitis patients, possibly pointing to a role for inflammation and/or autoimmunity (Weisz et al. 2004).

The limited work that has been done on gene expression profiling fits with the notion of increased immune responses. In post-mortem brain tissue sections of the basal ganglia area of four adult patients with TS in comparison with four control subjects, significantly increased expression of monocyte chemotactic factor-1 (a marker of chronic inflammation) and interleukin-2 (a growth factor derived from T-lymphocytes) was observed in TS cases (6.5- and 2.3-fold increase, respectively; Morer et al. 2010). This supports the presence of chronic inflammatory processes in the basal ganglia in patients with TS. Interestingly, circulating leukocytes may form a reasonable alternative to post-mortem brain tissue when studying RNA expression relevant for neuropsychiatric disorders. When studying gene expression profiles in peripheral blood in 16 patients with TS and 16 control subjects, patients with tic disorders were found to be characterized by several gene expression clusters including one cluster in which six genes that were all associated with immune cell function were over-expressed (Tang et al. 2005). In a subsequent study with peripheral blood, it was shown that most persons with TS ($n=10/16$) over-expressed the Natural Killer genes (Lit et al. 2007). Natural Killer cells are a key component of the innate immune system and are primarily activated in response to viruses and cellular transformation. There may be a relationship between Natural Killer function and autoimmunity (e.g., La Gruta et al. 2001).

The notion of activated immune mechanism in TS would make it understandable that allergies would be more frequent in individuals with TS. Indeed, a number of recent studies point to an interesting association between allergies and TS. Data in a case–control study from the Taiwan National Health Insurance Research Database, comprising 845 2- to 18-year-old patients with newly diagnosed TS and 3,378 controls indicated that the presence of allergic rhinitis, atopic dermatitis, asthma and allergic conjunctivitis strongly correlated with higher risk of TS. In a model simultaneously considering all 4 allergic diseases, subjects with allergic rhinitis showed double the risk of TS (adjusted odds ratio of 2.18); adjusted odds ratios were 1.82, 1.61, and 1.33, respectively, for asthma, dermatitis and allergic conjunctivitis (Chang et al. 2011). A significant association between TS and allergic disease was also present in two smaller scale studies (Ho et al. 1999; Yuce et al. 2014). Indirectly, a link between allergies and TS is supported by the involvement of histamine related genes with TS as indicated by analysis of linkage in a two-generation pedigree leading to the identification of the involvement of a rare functional mutation in the HDC gene encoding L-histidine decarboxylase, the rate-limiting enzyme in histamine biosynthesis (Ercan-Sencicek et al. 2010).

Are There Implications for Immune-Based Treatment?

Does the possible involvement of autoimmunity have consequences for treatment? Existing clinical trials concerning immunomodulatory therapy for TS is still very limited. Most clinical evidence stems from a few case studies that have been presented in the literature. A small number of case studies have reported improvement in tic severity after immunosuppression with corticosteroids (Kondo and Kabasawa 1978; Geschwind and Kondo 1979; Matarazzo 1992). Other case studies in the literature have described dramatic symptom improvement in children fulfilling criteria of PANDAS with plasmapheresis (Allen et al. 1995; Tucker et al. 1996; Giedd et al. 1996) or intravenous immunoglobulin (IVIG) (Allen et al. 1995; Muller et al. 1997; Perlmutter et al. 1998; Zykov et al. 2009). In some of these case studies, after successful plasmapheresis, to prevent new GABHS infections penicillin prophylaxis has been provided (Allen et al. 1995).

There has been one placebo-controlled study on the efficacy of immunomodulatory therapy in patients with TS/OCD (Perlmutter et al. 1999). In that study, a total number of 30 children meeting PANDAS criteria were randomly assigned to either plasmapheresis, IVIG or a placebo condition (saline solution given in the same manner as IVIG), and subsequently each patient was given a regimen of penicillin prophylaxis. Symptom severity was rated at baseline and at 1 month and 12 months after treatment by use of standard assessment scales for OCD, tics, anxiety, depression and global functioning. At 1 month after treatment the IVIG/placebo masking was broken. At 1 month, the IVIG and plasma-exchange groups showed striking improvements in severity of OCD symptoms, anxiety and overall functioning. Tic symptoms were significantly improved by plasma exchange only. The children in the placebo condition did not show any amelioration. Interestingly, these improvements were maintained at 1 year after treatment for both plasmapheresis and IVIG. Given the low numbers of patients involved in that study and the lack of replication studies, the possible effectiveness of plasma exchange should be considered preliminary at present. The same is true for intravenously administered immune globulins (IVIG). This was also a promising treatment modality in the study of Perlmutter et al. (1999) involving PANDAS patients, but IVIG did not appear to be more effective than placebo in a study with unselected patients (Hoekstra et al. 2004b). Another treatment option could be to treat or prevent streptococcal infections with antibiotics. One year treatment with antibiotics, either azithromycin or penicillin appeared to decrease the number of neuropsychiatric symptoms in children with PANDAS when compared to the year preceding antibiotic treatment (Snider et al. 2005). In a retrospective study, otherwise healthy children with documented streptococcal throat infection who were subsequently treated with antibiotics were not at higher risk for developing tics or obsessive-compulsive symptoms (Perrin et al. 2004). The study lacked a group of children who had not been given antibiotics though. This is an area that deserves further study, though, using more controlled designs given that the study by Snider and colleagues was not placebo-controlled, and, more importantly, that the number of exacerbations in the year prior to treatment had been assessed retrospectively only, as compared to a prospective assessment during active treatment. Considering the scarce evidence, currently, immune-based treatments should not be routinely given. Antibiotic treatment may be associated with development of resistance. Also, plasma exchange and IVIG are invasive treatments that should only be considered in controlled trials. Future research should focus on the identification of patients in whom autoimmunity may be involved and who might most likely profit from immunomodulatory treatments.

Conclusion and Future Directions

A slowly but steadily growing body of research data has indicated an association of TS with common infections alongside with indications of the involvement of autoimmunity in the pathogenesis of at least a subgroup of patients with tic and related disorders. Most relevant is the progress in animal models where it has been shown that GABHS immunization and induced antibodies leads to tic-like behaviours. Indeed, to classify a disease as autoimmune, transmissibility of symptoms to animals through passive transfer of antibodies or T-lymphocytes is a prerequisite. However, although several research groups continue to report intriguing findings that appear to confirm an association between the tic/OCD spectrum and both infections and immune system alterations, this is still a research area of considerable controversy (e.g., Martino et al. 2009) and there have indeed been a number of negative studies. Furthermore, the current status of the relevance of antineuronal antibodies in human sera of patients with TS is unclear. There is no standardized assay and no serum antibody test available that can be used in clinical practice. The characterization of the brain antigenic structures recognized by the antineuronal antibodies awaits further study. Post-mortem studies in search of inflammatory alterations also warrant further attention.

So far, an association between increased anti-streptococcal antibody titres and the presence of tic disorder or OCD has mostly been investigated in cross-sectional or retrospective designs or rather small scale longitudinal designs. Prospective longitudinal findings that unequivocally link symptom onset to preceding infections are not yet available. Most urgently therefore, large-scale longitudinal data are needed to investigate the role of infections with regard to symptom onset and fluctuations. Such a study is currently being conducted as part of the European Multicentre Tics in Children Studies (EMTICS). In addition, it remains unclear whether an association between immune factors and tics and obsessive-compulsive symptoms may only refer to a subgroup of patients, so-called Paediatric Autoimmune Neuropsychiatric Disorders Associated with Streptococcal infections (PANDAS), or to tic disorders and paediatric OCD in general, or even to a broader range of post infectious neuropsychiatric symptoms. On a more general note, the validity of the clinical PANDAS criteria remains to be demonstrated. A final area in urgent need of more studies are immune-based treatment studies. All in all, tic disorders and paediatric OCD are model neuropsychiatric disorders in which to investigate the complex relationship between brain functioning, autoimmunity, and human behaviour in an interplay between genetic and environmental factors.

References

Allen AJ, Leonard HL, Swedo SE. Case study: a new infection-triggered, autoimmune subtype of pediatric OCD and Tourette's syndrome. J Am Acad Child Adolesc Psychiatry. 1995;34:307–11.

Bhatnagar A, Grover A, Ganguly NK. Superantigen-induced T cell responses in acute rheumatic fever and chronic rheumatic heart disease patients. Clin Exp Immunol. 1999;116:100–6.

Bombaci M, Grifantini R, Mora M, Reguzzi V, Petracca R, Meoni E, et al. Protein array profiling of tic patient sera reveals a broad range and enhanced immune response against Group A Streptococcus antigens. PLoS One. 2009;4:e6332.

Bos-Veneman NG, Olieman R, Tobiasova Z, Hoekstra PJ, Katsovich L, Bothwell AL, et al. Altered immunoglobulin profiles in children with Tourette syndrome. Brain Behav Immun. 2011;25:532–8.

Brilot F, Merheb V, Ding A, Murphy T, Dale RC. Antibody binding to neuronal surface in Sydenham chorea, but not in PANDAS or Tourette syndrome. Neurology. 2011;76:1508–13.

Brimberg L, Benhar I, Mascaro-Blanco A, Alvarez K, Lotan D, Winter C, et al. Behavioral, pharmacological, and immunological abnormalities after streptococcal exposure: a novel rat model of Sydenham chorea and related neuropsychiatric disorders. Neuropsychopharmacology. 2012;37:2076–87.

Bruun RD, Budman CL. The course and prognosis of Tourette syndrome. Neurol Clin. 1997;15: 291–8.

Cardona F, Orefici G. Group A streptococcal infections and tic disorders in an Italian pediatric population. J Pediatr. 2001;138:71–5.

Chang YT, Li YF, Muo CH, Chen SC, Chin ZN, Kuo HT, et al. Correlation of Tourette syndrome and allergic disease: nationwide population-based case–control study. J Dev Behav Pediatr. 2011;32:98–102.

Chappel PB, Anderson GM, Goodman WK, Price LH, Hall LM, Cohen DJ, et al. Kynurenine pathway metabolities in cerebrospinal fluid and plasma of Tourette syndrome patients. Abstr Soc Neurosci. 1995;21:1111.

Cheng YH, Zheng Y, He F, Yang JH, Li WB, Wang ML, et al. Detection of autoantibodies and increased concentrations of interleukins in plasma from patients with Tourette's syndrome. J Mol Neurosci. 2012;48:219–24.

Chmelik E, Awadallah N, Hadi FS, Quinn K, Franco K. Varied presentation of PANDAS: a case series. Clin Pediatr (Phila). 2004;43:379–82.

Church AJ, Dale RC, Giovannoni G. Anti-basal ganglia antibodies: a possible diagnostic utility in idiopathic movement disorders? Arch Dis Child. 2004;89:611–4.

Coffey BJ, Miguel EC, Savage CR, Rauch SL. Tourette's disorder and related problems: a review and update. Harv Rev Psychiatry. 1994;2:121–32.

Cox CJ, Sharma M, Leckman JF, Zuccolo J, Zuccolo A, Kovoor A, et al. Brain human monoclonal autoantibody from Sydenham chorea targets dopaminergic neurons in transgenic mice and signals dopamine D2 receptor: implications in human disease. J Immunol. 2013;191:5524–41.

Creti R, Cardona F, Pataracchia M, Hunolstein CV, Cundari G, Romano A, et al. Characterisation of group A streptococcal (GAS) isolates from children with tic disorders. Indian J Med Res. 2004;119(Suppl):174–8.

Dale RC, Candler PM, Church AJ, Wait R, Pocock JM, Giovannoni G, et al. Glycolytic enzymes on neuronal membranes are candidate auto-antigens in post-streptococcal neuropsychiatric disorders. Mov Disord. 2004;19 Suppl 9:S33.

Dale RC, Merheb V, Pillai S, Wang D, Cantrill L, Murphy TK, et al. Antibodies to surface dopamine-2 receptor in autoimmune movement and psychiatric disorders. Brain. 2012;135:3453–68.

Das A, Radhakrishnan A. A case of PANDAS with Kleine–Levin type periodic hypersomnia. Sleep Med. 2012;13:319–20.

Dietrich DE, Zhang Y, Bode L, Munte TF, Hauser U, Schmorl P, et al. Brain potential amplitude varies as a function of Borna disease virus-specific immune complexes in obsessive-compulsive disorder. Mol Psychiatry. 2005;10:515–20.

Edwards MJ, Dale RC, Church AJ, Trikouli E, Quinn NP, Lees AJ, et al. Adult-onset tic disorder, motor stereotypes, and behavioural disturbance associated with anti-basal ganglia antibodies. Mov Disord. 2004;19:1190–6.

Ercan-Sencicek AG, Stillman AA, Ghosh AK, Bilguvar K, O'Roak BJ, Mason CE, et al. L-Histidine decarboxylase and Tourette's syndrome. N Engl J Med. 2010;362:1901–8.

Gabbay V, Coffey B. Obsessive-compulsive disorder, Tourette's disorder, or pediatric autoimmune neuropsychiatric disorders associated with Streptococcus in an adolescent? Diagnostic and therapeutic challenges. J Child Adolesc Psychopharmacol. 2003;13:209–12.

Gabbay V, Coffey BJ, Guttman LE, Gottlieb L, Katz Y, Babb JS, et al. A cytokine study in children and adolescents with Tourette's disorder. Prog Neuropsychopharmacol Biol Psychiatry. 2009;33:967–71.

Gause C, Morris C, Vernekar S, Pardo-Villamizar C, Grados MA, Singer HS. Antineuronal antibodies in OCD: comparisons in children with OCD-only, OCD+chronic tics and OCD+PANDAS. J Neuroimmunol. 2009;214:118–24.

Gaynor CM, Rickards EHG, Kariyawasam SH, Sidey FM, Robertson MM, Corbett JA, et al. Increased plasma kynurenine in Tourette syndrome may be due to induction of IDO. J Psychopharmacol. 1997;(Suppl 11):A23.

Geschwind N, Kondo K. Corticosteroid therapy and Tourette syndrome. Ann Neurol. 1979;5:495.

Giedd JN, Rapoport JL, Leonard HL, Richter D, Swedo SE. Case study: acute basal ganglia enlargement and obsessive-compulsive symptoms in an adolescent boy. J Am Acad Child Adolesc Psychiatry. 1996;35:913–5.

Hachiya Y, Miyata R, Tanuma N, Hongou K, Tanaka K, Shimoda K, et al. Autoimmune neurological disorders associated with group-A beta-hemolytic streptococcal infection. Brain Dev. 2013;35:670–4.

Hallett JJ, Harling-Berg CJ, Knopf PM, Stopa EG, Kiessling LS. Anti-striatal antibodies in Tourette syndrome cause neuronal dysfunction. J Neuroimmunol. 2000;111:195–202.

Heyes MP. The kynurenine pathway and neurologic disease. Therapeutic strategies. Adv Exp Med Biol. 1996;398:125–9.

Ho CS, Shen EY, Shyur SD, Chiu NC. Association of allergy with Tourette's syndrome. J Formos Med Assoc. 1999;98:492–5.

Hoekstra PJ, Kallenberg CG, Korf J, Minderaa RB. Is Tourette's syndrome an autoimmune disease? Mol Psychiatry. 2002;7:437–45.

Hoekstra PJ, Bijzet J, Limburg PC, Kallenberg CG, Minderaa RB. Elevated binding of D8/17-specific monoclonal antibody to B lymphocytes in Tic disorder patients. Am J Psychiatry. 2004a;161:1501–2.

Hoekstra PJ, Minderaa RB, Kallenberg CG. Lack of effect of intravenous immunoglobulins on tics: a double-blind placebo-controlled study. J Clin Psychiatry. 2004b;65:537–42.

Hoekstra PJ, Manson WL, Steenhuis MP, Kallenberg CG, Minderaa RB. Association of common cold with exacerbations in pediatric but not adult patients with tic disorder: a prospective longitudinal study. J Child Adolesc Psychopharmacol. 2005;15:285–92.

Hoekstra PJ, Anderson GM, Troost PW, Kallenberg CG, Minderaa RB. Plasma kynurenine and related measures in tic disorder patients. Eur Child Adolesc Psychiatry. 2007;16 Suppl 1:71–7.

Hoffman KL, Hornig M, Yaddanapudi K, Jabado O, Lipkin WI. A murine model for neuropsychiatric disorders associated with group A beta-hemolytic streptococcal infection. J Neurosci. 2004;18:1780–91.

Hsieh MY, Lee WI, Lin KL, Hung PC, Chou ML, Chang MY, et al. Immunologic analysis and serum heavy metal levels in exacerbated Tourette syndrome. Pediatr Allergy Immunol. 2010; 21:e764–71.

Jankovic J. Tourette syndrome. Phenomenology and classification of tics. Neurol Clin. 1997; 15:267–75.

Kansy JW, Katsovich L, McIver KS, Pick J, Zabriskie JB, Lombroso PJ, et al. Identification of pyruvate kinase as an antigen associated with Tourette syndrome. J Neuroimmunol. 2006;181:165–76.

Kawikova I, Grady BP, Tobiasova Z, Zhang Y, Vojdani A, Katsovich L, et al. Children with Tourette's syndrome may suffer immunoglobulin A dysgammaglobulinemia: preliminary report. Biol Psychiatry. 2010;67(7):679–83.

Khanna AK, Buskirk DR, Williams Jr RC, Gibofsky A, Crow MK, Menon A, et al. Presence of a non-HLA B cell antigen in rheumatic fever patients and their families as defined by a monoclonal anti-body. J Clin Invest. 1989;83:1710–6.

Kiessling LS, Marcotte AC, Culpepper L. Antineuronal antibodies in movement disorders. Pediatrics. 1993;92:39–43.

Kirvan CA, Swedo SE, Snider LA, Cunningham MW. Antibody-mediated neuronal cell signaling in behavior and movement disorders. J Neuroimmunol. 2006;179:173–9.

Kondo K, Kabasawa T. Improvement in Gilles de la Tourette syndrome after corticosteroid therapy. Ann Neurol. 1978;4:387.

Krause D, Matz J, Weidinger E, Wagner J, Wildenauer A, Obermeier M, et al. Association between intracellular infectious agents and Tourette's syndrome. Eur Arch Psychiatry Clin Neurosci. 2010;260:359–63.

Kurlanrlan R, Johnson D, Kaplan EL. Tourette Syndrome Study Group. Streptococcal infection and exacerbations of childhood tics and obsessive-compulsive symptoms: a prospective blinded cohort study. Pediatrics. 2008;121(6):1188–97.

Kurlan R, Daragjati C, Como PG, McDermott MP, Trinidad KS, Roddy S, et al. Non-obscene complex socially inappropriate behavior in Tourette's syndrome. J Neuropsychiatry Clin Neurosci. 1996;8:311–7.

La Gruta NL, Van Driel IR, Toh BH, Gleeson PA. The role of natural killer cells in the induction of autoimmune gastritis. Autoimmunity. 2001;34:147–54.

Laurino JP, Hallett J, Kiessling LS, Benson M, Pelletier T, Kuhn C. An immunoassay for anti-neuronal antibodies associated with involuntary repetitive movement disorders. Ann Clin Lab Sci. 1997;27:230–5.

Leckman JF. Tourette's syndrome. Lancet. 2002;360:1577–86.

Leckman JF, Katsovich L, Kawikova I, Lin H, Zhang H, Kronig H, et al. Increased serum levels of interleukin-12 and tumor necrosis factor-alpha in Tourette's syndrome. Biol Psychiatry. 2005;57:667–73.

Leslie DL, Kozma L, Martin A, Landeros A, Katsovich L, King RA, et al. Neuropsychiatric disorders associated with streptococcal infection: a case–control study among privately insured children. J Am Acad Child Adolesc Psychiatry. 2008;47:1166–72.

Lin H, Williams KA, Katsovich L, Findley DB, Grantz H, Lombroso PJ, et al. Streptococcal upper respiratory tract infections and psychosocial stress predict future tic and obsessive-compulsive symptom severity in children and adolescents with Tourette syndrome and obsessive-compulsive disorder. Biol Psychiatry. 2010;67:684–91.

Lit L, Gilbert DL, Walker W, Sharp FR. A subgroup of Tourette's patients overexpress specific natural killer cell genes in blood: a preliminary report. Am J Med Genet B Neuropsychiatr Genet. 2007;144B:958–63.

Loiselle CR, Wendlandt JT, Rohde CA, Singer HS. Antistreptococcal, neuronal and nuclear antibodies in Tourette syndrome. Pediatr Neurol. 2003;28:119–25.

Loiselle CR, Lee O, Moran TH, Singer HS. Striatal microinfusion of Tourette syndrome and PANDAS sera: failure to induce behavioral changes. Mov Disord. 2004;19:390–6.

Lotan D, Benhar I, Alvarez K, Mascaro-Blanco A, Brimberg L, Frenkel D, et al. Behavioral and neural effects of intra-striatal infusion of anti-streptococcal antibodies in rats. Brain Behav Immun. 2014;38:249–62.

Luo F, Leckman JF, Katsovich L, Findley D, Grantz H, Tucker DM, et al. Prospective longitudinal study of children with tic disorders and/or obsessive-compulsive disorder: relationship of symptom exacerbations to newly acquired streptococcal infections. Pediatrics. 2004;113:e578–85.

Maini B, Bathla M, Dhanjal GS, Sharma PD. Pediatric autoimmune neuropsychiatric disorders after Streptococcus infection. Indian J Psychiatry. 2012;54:375–7.

Martino D, Church AJ, Defazio G, Dale RC, Quinn NP, Robertson MM, et al. Soluble adhesion molecules in Gilles de la Tourette's syndrome. J Neurol Sci. 2005;234:79–85.

Martino D, Defazio G, Giovannoni G. The PANDAS subgroup of tic disorders and childhood-onset obsessive-compulsive disorder. J Psychosom Res. 2009;67:547–57.

Martino D, Chiarotti F, Buttiglione M, Cardona F, Creti R, Nardocci N, et al. The relationship between group A streptococcal infections and Tourette syndrome: a study on a large service-based cohort. Dev Med Child Neurol. 2011;53:951–7.

Matarazzo EB. Tourette's syndrome treated with ACTH and prednisone: report of two cases. J Child Adolesc Psychopharmacol. 1992;2:215–26.

Matsuo M, Tsuchiya K, Hamasaki Y, Singer HS. Restless legs syndrome: association with streptococcal or mycoplasma infection. Pediatr Neurol. 2004;31:119–21.

Matz J, Krause DL, Dehning S, Riedel M, Gruber R, Schwarz MJ, et al. Altered monocyte activation markers in Tourette's syndrome: a case–control study. BMC Psychiatry. 2012;12:29.

McCreary AC, Handley SL. Kynurenine potentiates the DOI head-shake in mice. J Psychopharmacol. 1995;9:68–9.

Mell LK, Davis RL, Owens D. Association between streptococcal infection and obsessive-compulsive disorder, Tourette's syndrome, and tic disorder. Pediatrics. 2005;116:56–60.

Meyer KC, Arend RA, Kalayoglu MV, Rosenthal NS, Byrne GI, Brown RR. Tryptophan metabolism in chronic inflammatory lung disease. J Lab Clin Med. 1995;126:530–40.

Mittleman BB, Castellanos FX, Jacobsen LK, Rapoport JL, Swedo SE, Shearer GM. Cerebrospinal fluid cytokines in pediatric neuro-psychiatric disease. J Immunol. 1997;159:2994–9.

Mol Debes NM. Co-morbid disorders in Tourette syndrome. Behav Neurol. 2013;27:7–14.

Morer A, Vinas O, Lazaro L, Bosch J, Toro J, Castro J. D8/17 monoclonal antibody: an unclear neuro-psychiatric marker. Behav Neurol. 2005;16:1–8.

Morer A, Vinas O, Lazaro L, Calvo R, Andres S, Bosch J, et al. Subtyping obsessive-compulsive disorder: clinical and immunological findings in child and adult onset. J Psychiatr Res. 2006;40:207–13.

Morer A, Chae W, Henegariu O, Bothwell AL, Leckman JF, Kawikova I. Elevated expression of MCP-1, IL-2 and PTPR-N in basal ganglia of Tourette syndrome cases. Brain Behav Immun. 2010;24:1069–73.

Morris CM, Pardo-Villamizar C, Gause CD, Singer HS. Serum autoantibodies measured by immunofluorescence confirm a failure to differentiate PANDAS and Tourette syndrome from controls. J Neurol Sci. 2009;276:45–8.

Morris-Berry CM, Pollard M, Gao S, Thompson C, Tourette Syndrome Study Group, Singer HS. Anti-streptococcal, tubulin, and dopamine receptor 2 antibodies in children with PANDAS and Tourette syndrome: single-point and longitudinal assessments. J Neuroimmunol. 2013;264:106–13.

Morshed SA, Parveen S, Leckman JF, Mercadante MT, Bittencourt Kiss MH, Miguel EC, et al. Antibodies against neural, nuclear, cytoskeletal and streptococcal epitopes in children and adults with Tourette's syndrome, Sydenham's chorea and autoimmune disorders. Biol Psychiatry. 2002;50:566–77.

Muller N, Riedel M, Erfurth A, Moller HJ. Immunoglobulin therapy in Gilles de la Tourette syndrome. Nervenarzt. 1997;68:914–6.

Muller N, Riedel M, Straube A, Gunther W, Wilske B, et al. Increased anti-streptococcal antibodies in patients with Tourette's syndrome. Psychiatry Res. 2000;94:43–9.

Muller N, Kroll B, Schwarz MJ, Riedel M, Straube A, Lutticken R, et al. Increased titers of antibodies against streptococcal M12 and M19 proteins in patients with Tourette's syndrome. Psychiatry Res. 2001;25:187–93.

Muller N, Riedel M, Blendinger C, Oberle K, Jacobs E, Abele-Horn M. Mycoplasma pneumoniae infection and Tourette's syndrome. Psychiatry Res. 2004;129:119–25.

Murphy TK, Goodman WK, Fudge MW, Williams Jr RC, Ayoub EM, Dalal M, et al. B lymphocyte antigen D8/17: a peripheral marker for childhood-onset obsessive-compulsive disorder and Tourette's syndrome? Am J Psychiatry. 1997;154:402–7.

Murphy TK, Sajid M, Soto O, Shapira N, Edge P, Yang M, et al. Detecting pediatric autoimmune neuropsychiatric disorders associated with Streptococcus in children with obsessive-compulsive disorder and tics. Biol Psychiatry. 2004;55:61–8.

Murphy TK, Snider LA, Mutch PJ, Harden E, Zaytoun A, Edge PJ, et al. Relationship of movements and behaviors to Group A Streptococcus infections in elementary school children. Biol Psychiatry. 2007;61:279–84.

Murr C, Widner B, Gerlach D, Werner-Felmayer G, Dierich MP, Wachter H, et al. Streptococcal erythrogenic toxins induce tryptophan degradation in human peripheral blood mononuclear cells. Int Arch Allergy Immunol. 1997;114:224–8.

Olfson M, Crystal S, Gerhard T, Huang C, Walkup JT, Scahill L, et al. Patterns and correlates of tic disorder diagnoses in privately and publicly insured youth. J Am Acad Child Adolesc Psychiatry. 2011;50:119–31.

Pavone P, Bianchini R, Parano E, Incorpora G, Rizzo R, Mazzone L, et al. Anti-brain antibodies in PANDAS versus uncomplicated streptococcal infection. Pediatr Neurol. 2004;30:107–10.

Perlmutter SJ, Garvey MA, Castellanos X, Mittleman BB, Giedd J, Rapoport JL, et al. A case of pediatric autoimmune neuropsychiatric disorders associated with streptococcal infections. Am J Psychiatry. 1998;155:1592–8.

Perlmutter SJ, Leitman SF, Garvey MA, Hamburger S, Feldman E, Leonard HL, et al. Therapeutic plasma exchange and intravenous immunoglobulin for obsessive-compulsive disorder and tic dis-orders in childhood. Lancet. 1999;354:1153–8.

Perrin EM, Murphy ML, Casey JR, Pichichero ME, Runyan DK, Miller WC, et al. Does group A beta-hemolytic streptococcal infection increase risk for behavioral and neuropsychiatric symptoms in children? Arch Pediatr Adolesc Med. 2004;158:848–56.

Peterson BS, Leckman JF. The temporal dynamics of tics in Gilles de la Tourette syndrome. Biol Psychiatry. 1998;44:1337–48.

Peterson BS, Leckman JF, Tucker D, Scahill L, Staib L, Zhang H, et al. Preliminary findings of antistreptococcal antibody titers and basal ganglia volumes in tic, obsessive-compulsive, and attention deficit/hyperactivity disorders. Arch Gen Psychiatry. 2000;57:364–72.

Rickards H, Dursun SM, Farrar G, Betts T, Corbett JA, Handley SL. Increased plasma kynurenine and its relationship to neopterin and tryptophan in Tourette's syndrome. Psychol Med. 1996;26:857–62.

Rizzo R, Gulisano M, Pavone P, Fogliani F, Robertson MM. Increased antistreptococcal antibody titers and anti-basal ganglia antibodies in patients with Tourette syndrome: controlled cross-sectional study. J Child Neurol. 2006;21:747–53.

Robertson MM. The prevalence and epidemiology of Gilles de la Tourette syndrome. Part 1: The epidemiological and prevalence studies. J Psychosom Res. 2008;65:461–72.

Schrag A, Gilbert R, Giovannoni G, Robertson MM, Metcalfe C, Ben-Shlomo Y, et al. Streptococcal infection, Tourette syndrome, and OCD: is there a connection? Neurology. 2009;73:1256–63.

Shulman ST. Pediatric autoimmune neuropsychiatric disorders associated with streptococci (PANDAS). Pediatr Infect Dis J. 1999;18:281–2.

Singer HS, Loiselle C. PANDAS. A commentary. J Psychosom Res. 2003;55:31–9.

Singer HS, Giuliano JD, Hansen BH, Hallett JJ, Laurino JP, Benson M, et al. Antibodies against human putamen in children with Tourette syndrome. Neurology. 1998;50:1618–24.

Singer HS, Giuliano JD, Hansen BH, Hallett JJ, Laurino JP, Benson M, et al. Antibodies against a neuron-like (HTB-10 neuroblastoma) cell in children with Tourette syndrome. Biol Psychiatry. 1999;46:775–80.

Singer HS, Hong JJ, Yoon DY, Williams PN. Serum autoantibodies do not differentiate PANDAS and Tourette syndrome from controls. Neurology. 2005a;13:1701–7.

Singer HS, Mink JW, Loiselle CR, Burke KA, Ruchkina I, Morshed S, et al. Micro-infusion of anti-neuronal antibodies into rodent striatum: failure to differentiate between elevated and low titers. J Neuroimmunol. 2005b;163:8–14.

Singer HS, Gause C, Morris C, Lopez P, Tourette Syndrome Study Group. Serial immune markers do not correlate with clinical exacerbations in pediatric autoimmune neuropsychiatric disorders associated with streptococcal infections. Pediatrics. 2008;121:1198–205.

Singer HS, Gilbert DL, Wolf DS, Mink JW, Kurlan R. Moving from PANDAS to CANS. J Pediatr. 2012;160:725–31.

Snider LA, Lougee L, Slattery M, Grant P, Swedo SE. Antibiotic prophylaxis with azithromycin or penicillin for childhood-onset neuropsychiatric disorders. Biol Psychiatry. 2005;57:788–92.

Steingard R, Dillon-Stout D. Tourette's syndrome and obsessive compulsive disorder. Clinical aspects. Psychiatr Clin North Am. 1992;15:849–60.

Storch EA, Gerdes AC, Adkins JW, Geffken GR, Star J, Murphy T. Behavioral treatment of a child with PANDAS. J Am Acad Child Adolesc Psychiatry. 2004;43:510–1.

Swedo SE. Sydenham's chorea. A model for childhood auto-immune neuropsychiatric disorders. JAMA. 1994;272:1788–91.
Swedo SE, Leonard HL, Schapiro MB, Casey BJ, Mannheim GB, Lenane MC, et al. Sydenham's chorea: physical and psychological symptoms of St Vitus dance. Pediatrics. 1993;91:706–13.
Swedo SE, Leonard HL, Garvey M, Mittleman B, Allen AJ, Perlmutter S, et al. Pediatric autoimmune neuropsychiatric disorders associated with streptococcal infections: clinical description of the first 50 cases. Am J Psychiatry. 1998;199:264–71.
Tang Y, Gilbert DL, Glauser TA, Hershey AD, Sharp FR. Blood gene expression profiling of neurologic diseases: a pilot micro-array study. Arch Neurol. 2005;62:210–5.
Tanner CM, Goldman SM. Epidemiology of Tourette syndrome. Neurol Clin. 1997;15:395–402.
Taylor JR, Morshed SA, Parveen S, Mercadante MT, Scahill L, Peterson BS, et al. An animal model of Tourette's syndrome. Am J Psychiatry. 2002;159:657–60.
Tucker DM, Leckman JF, Scahill L, Wilf GE, LaCamera R, Cardona L, et al. A putative poststreptococcal case of OCD with chronic tic disorder, not otherwise specified. J Am Acad Child Adolesc Psychiatry. 1996;35:1684–91.
van Toorn R, Weyers HH, Schoeman JF. Distinguishing PANDAS from Sydenham's chorea: case report and review of the literature. Eur J Paediatr Neurol. 2004;8:211–6.
Wanderer S, Roessner V, Freeman R, Bock N, Rothenberger A, Becker A. Relationship of obsessive-compulsive disorder to age-related comorbidity in children and adolescents with Tourette syndrome. J Dev Behav Pediatr. 2012;33:124–33.
Weidinger E, Krause D, Wildenauer A, Meyer S, Gruber R, Schwarz MJ, et al. Impaired activation of the innate immune response to bacterial challenge in Tourette syndrome. World J Biol Psychiatry. 2014;15:453–8.
Weisz JL, McMahon WM, Moore JC, Augustine NH, Bohnsack JF, Bale JF, et al. D8/17 and CD19 expression on lymphocytes of patients with acute rheumatic fever and Tourette's disorder. Clin Diagn Lab Immunol. 2004;11:330–6.
Wenzel C, Wurster U, Müller-Vahl KR. Oligoclonal bands in cerebrospinal fluid in patients with Tourette's syndrome. Mov Disord. 2011;26:343–6.
Yaddanapudi K, Hornig M, Serge R, De Miranda J, Baghban A, Villar G, et al. Passive transfer of Streptococcus-induced antibodies reproduces behavioral disturbances in a mouse model of pediatric autoimmune neuropsychiatric disorders associated with streptococcal infection. Mol Psychiatry. 2010;15:712–26.
Yeh CB, Wu CH, Tsung HC, Chen CW, Shyu JF, Leckman JF. Antineural antibody in patients with Tourette's syndrome and their family members. J Biomed Sci. 2006;13(1):101–12.
Yuce M, Guner SN, Karabekiroglu K, Baykal S, Kilic M, Sancak R, et al. Association of Tourette syndrome and obsessive-compulsive disorder with allergic diseases in children and adolescents: a preliminary study. Eur Rev Med Pharmacol Sci. 2014;18:303–10.
Zykov VP, Shcherbina AY, Novikova EB, Shvabrina TV. Neuroimmune aspects of the pathogenesis of Tourette's syndrome and experience in the use of immunoglobulins in children. Neurosci Behav Physiol. 2009;39:635–8.

Chapter 14
The Role of Inflammation in Autism Spectrum Disorder

Casara Jean Ferretti and Eric Hollander

Abstract Autism spectrum disorders (ASD) are a group of heterogeneous neurodevelopmental disabilities whose chief manifestations are qualitative impairment in social interaction and social communication and restricted, repetitive, and stereotyped patterns of behavior, interests, and activities. The clinical heterogeneity of ASD makes it difficult to identify biomarkers, etiologies, and treatments; and identification of subgroups within the diagnosis is necessary. There are two distinct research communities with findings relevant to the genetics and molecular biology of ASD; those that posit etiology related to immune dysfunction and those suggesting disorder of synaptic function and neuronal connectivity. The role of immune aberrations in neural connectivity suggests a link between these two hypotheses, which will be explored in this chapter. Knowledge of immune dysregulation and inflammation is relevant to the prevention, diagnosis, and treatment of ASD. This chapter will review the literature related to inflammation and immune dysfunction in ASD and summarize the development of treatments based on this research.

Keywords Autism spectrum disorder • Inflammation • Immune System • Autoimmune • Cytokines • Microglia • Microbiome • Immune dysfunction • Treatments • Etiology • Biomarkers • Neuronal connectivity

C.J. Ferretti (✉)
Autism and Obsessive Compulsive Spectrum Program, Montefiore Medical Center, Albert Einstein College of Medicine, 111 East 210th Street, Bronx, NY 10467, USA
e-mail: cferrett@montefiore.org

E. Hollander
Autism and Obsessive Compulsive Spectrum Program, Montefiore Medical Center, Albert Einstein College of Medicine, 111 East 210th Street, Bronx, NY 10467, USA

Department of Psychiatry and Behavioral Sciences, Montefiore Medical Center, Albert Einstein College of Medicine, 111 East 210th Street, Bronx, NY 10467, USA
e-mail: eholland@montefiore.org

N. Müller et al. (eds.), *Immunology and Psychiatry*, Current Topics in Neurotoxicity 8, DOI 10.1007/978-3-319-13602-8_14

Introduction

Autism spectrum disorders (ASD) are a group of heterogeneous neurodevelopmental disabilities whose chief manifestations are qualitative impairment in social interaction and social communication and restricted, repetitive, and stereotyped patterns of behavior, interests, and activities (American Psychiatric Association 2013). The Centers for Disease Control's (CDC) Autism and Developmental Disabilities Monitoring Network (ADDM) has shown increasing prevalence rates, with the current estimate of ASD affecting 1 in 68 children in the USA (Baio 2010). ASD is more prevalent in males, with 1 in 42 boys identified as having ASD compared to 1 in 189 girls. It is currently estimated that approximately 1 % of the human population has an ASD (Muhle et al. 2004; Fombonne 2009). According to the fifth edition of the Diagnostic and Statistical Manual of Mental Disorders (American Psychiatric Association 2013) ASD is diagnosed in individuals with three social communication or interaction deficits and at least two symptoms of repetitive and restricted behaviors. The variance in presentation of these symptoms makes ASD an extremely heterogeneous disorder with individuals on a continuum of severity. The DSM5 also allows for descriptions of severity and specifiers that characterize those with accompanying language or intellectual impairment, co-morbid medical or genetic conditions or environmental exposure, catatonia, or associations with other neurodevelopmental, mental, or behavioral disorders (Ousley and Cermak 2014). In addition to core diagnostic features, there are a variety of co-morbid associated symptoms of ASD, including aggression, hyperactivity, epilepsy, anxiety, sleep disturbances, gastrointestinal disturbances, and immune dysfunction (Hsiao 2013; Ousley and Cermak 2014; Kohane et al. 2012).

The clinical heterogeneity of ASD makes it difficult to identify biomarkers, etiologies, and treatments. The identification of subgroups within the diagnosis is necessary both to understand the etiology and to establish a personalized treatment approach for both core and associated symptoms. A review of biomedical literature highlights two distinct research communities with findings relevant to the genetics and molecular biology of ASD; studies that posit etiology related to immune dysfunction and those related to disorders of synaptic function and neuronal connectivity (Kohane 2015; Ashwood et al. 2006; Noriega and Savelkoul 2014; Onore et al. 2012). This chapter will explore the research related to inflammation and immune dysfunction in ASD and provide a potential link between the two prominent hypotheses.

Genetic Factors vs. Environment Factors in ASD

Genetic variation is recognized as a major etiological factor of ASD. Concordance between identical twins is estimated to be between 70 and 90 %, and there is a tenfold higher risk if a previous sibling has a diagnosis (Ronemus et al. 2014). Recent years have seen increased research in the relationship between de novo mutations, single-nucleotide polymorphisms, and copy number variants (CNV), and increased

ASD risk (Gaugler et al. 2014). De novo CNVs are observed in 5–10 % of those with ASD who have been screened and affect high risk genes, including *NRXN1* (Béna et al. 2013); *CNTNAP2* (Peñagarikano and Geschwind 2012) and *SHANK3* (Boccuto et al. 2013; Guilmatre et al. 2014), in addition to 15q11.2-q13 duplications and deletions; 16p11.2 and duplications and deletions (Pinto et al. 2014). Despite these findings it is estimated that only 20 % of ASD cases can be explained by gene mutations and CNVs (Abrahams and Geshwind 2010), and a recent twin study concluded that the influence of genetic factors on ASD risk had been overestimated, based on increased rates of concordance in dizygotic twins (Hallmayer et al. 2011). A two-hit gene x environment model involving interactions between genetic predisposition and environmental insults during early prenatal or postnatal development may contribute to the etiology of ASD, and increase the severity of symptoms. As such, there has been further exploration into the influence of environmental factors on ASD, particularly those that occur prenatally or early on postnatally, including, but not limited to, maternal infection and immune system insults and abnormalities (Goyal and Miyan 2014; Hallmayer et al. 2011). The enriched gene expression of immune and microglial genes in ASD appears to have a non-genetic etiology and may indicate internal or external environmental influences (Tetreault et al. 2012), as seen in the detailed review below (Fig. 14.1).

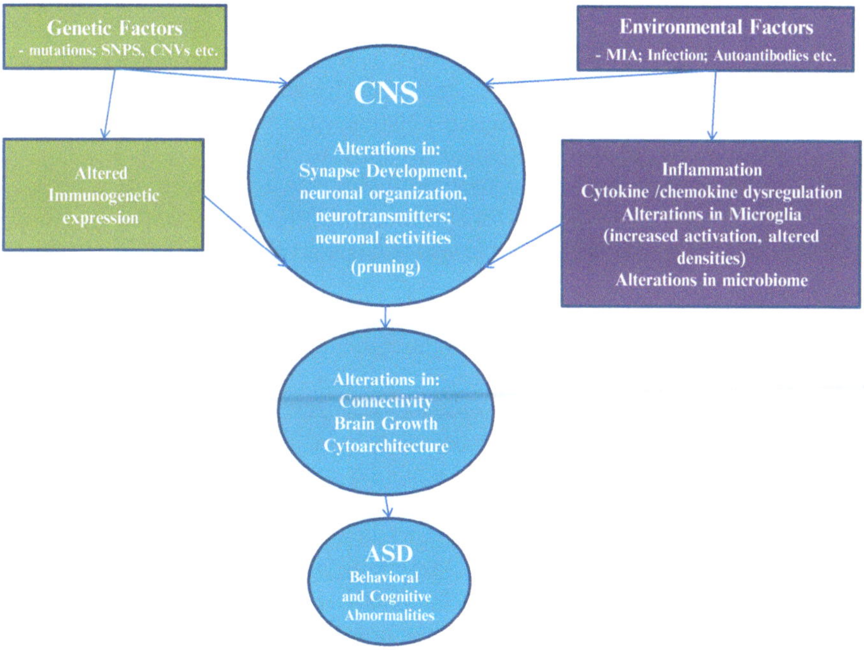

Fig. 14.1 Contributions of genetic and environmental factors to alterations in the CNS

Animal Models of ASD

BTBR Mouse Model

Animal models of human conditions allow researchers to directly test hypotheses generated from clinical observations. Development of effective animal models requires extensive research regarding the biological and behavioral pathologies of the condition. The BTBR mouse is recognized to be a valid genetic model of ASD. It is characterized by reductions in sociability, impaired social communication, and increased repetitive/compulsive behaviors that are the core features of ASD. Impaired sociability is observed through reduced social approach and social motivation during the three-chamber social approach task. BTBR mice also spend significantly less time engaging in social interactions, such as sniffing. Social communication deficits are observed through altered ultrasonic vocalizations (USV), both in frequency and quality. As olfactory cues are important in mouse communication, it is noted that the BTBR mouse also has fewer scent marking behaviors than the more typically developed C57BL/6J mouse model. The third core feature of ASD, repetitive behavior, restricted interests, and insistence on sameness, is also observed in the BTBR mouse model. BTBR mice spend excessive time engaging in self-grooming behaviors, and the behaviors themselves are altered and do not occur in the typical behavioral sequences. Perseverative behaviors are represented by an increased frequency of compulsive marble burying and persistent errors in the modified T-maze. These behaviors result in cognitive deficits, including impairments in reversal learning and reduced behavioral and cognitive flexibility. In addition to representing the core impairments of ASD the BTBR genetic mouse model also has a number of immune abnormalities, similar to the profile seen in children with ASD. This includes elevated expression of proinflammatory cytokines IL-1β, IL-6, and IL-12 in brain tissue, in addition to increased microglia activation, and increased numbers of microglia (Careaga et al. 2015). The expression of IL-33 in BTBR brain tissue has also been correlated with increased impairment in behavior (Heo et al. 2011). BTBR mice show skewing in macrophage and microglia subtype, with M1 skewing and increased production of IL-12 and less production of anti-inflammatory IL-10 (Onore et al. 2013).

The immune system abnormalities in this genetic model of ASD demonstrate a potential role for epigenetic changes as causative factors of ASD behaviors. Interestingly, when BTBR mice are exposed to a maternal immune activation (MIA) insult, there are more persistent immune alterations and more severe behavioral impairments. This is potentially due to a two-hit gene x environment interaction. Crosses between different strains of mice also demonstrate the role that maternal environmental factors and immune insults play in ASD etiology. For example, when the BTBR mouse is crossed with the healthy C57BL/6J mouse, both its immune and behavioral phenotypes are improved. Additionally, if BTBR embryos are placed in C57BL/6J dams, there are also significant improvements in both social and repetitive behaviors. On the other hand, when the C57BL/6J mouse is given an MIA insult, such as exposure to poly(I:C) its offspring resemble BTBR mice in both their immunological alterations and behavioral impairments (Careaga et al. 2015).

MIA Mouse Model

The MIA mouse is the only environmental model of ASD that has all three core impairments. The MIA insult varies, and can be exposure to IL-6, poly(I:C), LPS, or influenza, all with similar results in the immunological and behavioral effects on the offspring. Offspring of dams immune-activated with poly(I:C) have both the behavioral and neuropathological symptoms of human ASD in addition to altered immune profiles. Poly(I:C) MIA offspring have deficits in prepulse inhibition, a feature common in individuals with ASD, in addition to increased stereotypic behavior and decreased social preference. MIA offspring do not appear to have global changes in the levels of activated or total brain microglia/macrophages. Rather, dysregulated immune function in MIA offspring is characterized by systemic deficits in CD4+ TCRβ+ Foxp3+ CD25+ T regulatory cells, increased production of IL-6 and IL-17 by CD4+ T cells, and elevated levels of peripheral Gr-1+ cells. Regulatory T cells are suppressors of the innate and adaptive immune response, and deficits in these cells could result in the proinflammatory phenotype observed in both animal models and humans with ASD. If poly(I:C) MIA offspring are irradiated and transplanted with donor bone marrow from healthy saline exposed mice, their repetitive behavior deficits are normalized, and they have significantly less anxiety-like behavior. Social preference is not improved by the healthy bone marrow transplant. Interestingly, if irradiated poly(I:C) MIA offspring are transplanted with poly(I:C) affected bone marrow their repetitive and anxiety-like behaviors also normalize. Both types of transplants appear to immunologically normalize MIA offspring, suggesting that the peripheral environment of the MIA offspring is critical. These findings imply that some ASD-related behavioral abnormalities may be corrected by repairing immune function. However, the authors caution that transplant procedures may have their own ameliorative effects outside of the type of bone marrow used. In particular, irradiation could alter metabolic function, gastrointestinal microbial composition, and oxidative stress, all of which could directly influence behavior (Hsiao et al. 2012). Further detail about the MIA mouse model is provided below, in the Maternal Immune Activation section of this chapter.

MIA Non-human Primate Model

In addition to mouse models, non-human primate models are also used to study the longitudinal immunological and behavioral effects of MIA. Non-human primate models help to bridge the gap between the clinical population and rodent models due to their similarities with humans both genetically and behaviorally. Rodent models also have limitations in understanding the effects of MIA on fetal brain development, as the human third trimester occurs during the early postnatal period in rodents. Additionally, there are challenges in relating rodent brains and behavior to human brains and behavior. Recent MIA models have focused on challenges during the first and second trimester, which are known gestational windows of vulnerability for fetal development. Pregnant rhesus monkeys injected with poly(I:C) during the end of the first or second trimester had a strong inflammatory response, including sickness behavior, increased temperature, and altered cytokine profiles. MIA offspring were subsequently observed for 24 months. Initial observations of these primates showed increased distress and self-soothing behaviors and increased maternal attachment. When compared to control animals, the MIA offspring had more motor stereotypic and self-directed behaviors in addition to aberrant vocal communication and social interactions that mirror the core symptoms of ASD. Just as with rodent models, non-human primate models also have limitations, including small sample sizes and delayed brain pathology studies that may not reflect the mechanistic neural basis for behavioral abnormalities (Bauman et al. 2014).

Hygiene Hypothesis and Its Relationship to ASD

The hygiene hypothesis suggests that stimulation of the immune system by healthy gut microbes is protective against the development of inflammatory diseases, and that a rise in hygiene in urban settings has been associated with less protective microbes in humans and an increase in autoimmune inflammatory disorders such as multiple sclerosis, inflammatory bowel disease (IBD), asthma, allergic rhinitis, and potentially ASD. One of the major differences between infectious diseases in the developing world versus industrialized nations is the prevalence of parasitic infections (Rook 2008). Studies from the developing world have demonstrated that parasitic infection is statistically significantly associated with reduced risk of skin sensitization to allergens (Feary et al. 2011). Studies examining allergy status before and after anti-helminthic treatment showed an overall increase in skin allergen sensitivity after de-worming (Jouvin and Kinet 2012). Epidemiological studies in humans vs. the industrialized world have shown that infection with helminth parasites is associated with a lower incidence of allergy/asthma and autoimmunity. The primary factor associated with allergic and autoimmune disease is apparently loss of species diversity, both parasitic and symbiotic, from the human biome. In those with a genetic predisposition and adequate environmental triggers, the absence of species from the human biome, many of which co-evolved with the human immune system, leads to hypersensitive states that could foster allergic and autoimmune disease. The high prevalence of autoimmunity and allergy in the industrialized world and the well-established effects of inflammation on cognitive development supports the hypothesis that gene by environment (both external environment and gut microbiome) effects could lie at the center of the pathophysiology of ASD (Rook et al. 2014; Parker and Ollerton 2013).

The Microbiome and ASD

Definition and Roles of the Microbiome

It is estimated that up to 100 trillion microbial cells are found in the human body and that they outnumber human body cells by an order of magnitude. These cells as a whole are known as the human microbiome and recent research has demonstrated that these microbial species and their interactions with the host (human body) play key roles in homeostasis and the growth and continued development and function of the nervous system. Advances in technology have led to rapid DNA sequencing techniques, known as metagenomics, and have allowed for classifications of new microbial species and definitions of phylogenetic relationships between species (Mulle et al. 2013). This advancement in the understanding of the microbiome's role in brain health and disease has contributed to our knowledge of the microbiome–brain–gut axis (Wang and Kasper 2014). Communication between the nervous system and the gut occurs through various mechanisms: (1) neural, including parasympathetic innervation, through the vagus nerve, and sympathetic innervation (2) endocrine, including the hypothalamic–pituitary–adrenal (HPA) axis (3) hormonal (4) metabolic, via nutrients and (5) immune-mediated mechanisms. Dysfunction of this axis is implicated in stress-related disorders such as depression, anxiety, and irritable bowel syndrome in addition to neurodevelopmental disorders such as ASD. Thus strategies that modulate the microbiome may be a novel therapeutic approach to developing treatments for neurodevelopmental, CNS and immune-related disorders (Mulle et al. 2013; Borre et al. 2014).

The gut microbiome is composed of bacteria belonging to two major phyla, *Firmicutes* and *Bacteriodetes*, and three enterotypes comprised of one of three genera, *Bacteroides, Prevotella*, and *Ruminococcus*. These enterotypes are largely related to dietary intake, in addition to host genetics, age, and other environmental factors. The core microbiome in healthy adults is typically developed by age 3, and functions in nutrient intake, immunological development, and disease pathology. Microbial species can break down plant polysaccharides and carbohydrates, and are believed to have evolved to help the human body extract the maximal energy supply from food sources. Certain species have also been shown to interact with the host immune system to produce intestinal T-cells and correct the balance of Th1/Th2 cytokine response (Mulle et al. 2013).

The symbiotic relationship between the immune system and the gut microbiome begins early in life, and is influenced by many factors, including breastfeeding, diet, antibiotics, birth order, and method of birth (cesarean or natural). There is also evidence that some microbial species may be passed from mother to child while in utero, and that prenatal stressors may further affect the development of a healthy microbiome (Borre et al. 2014). For instance, rhesus monkeys exposed to perinatal stressors have lower amounts of *bifidobacteria* and *lactobacilli* in adulthood (Bailey et al. 2004). Additionally, germ-free mice are shown to have increased behavioral abnormalities, altered gene expression, and altered HPA axis function that can only be corrected if they are reconstituted with normal microbiota early in life (Sudo et al. 2004). Thus, early life events modulate the microbiota and impact the ability of the immune system to create a healthy microbiome which aids in CNS development.

Alterations in the ASD Microbiome

Individuals with ASD have a reduced richness and diversity and altered composition of gut microflora (Kang et al. 2013). The first hypothesis of an impaired microbiome being linked with ASD was in 1998 (Bolte 1998). It was initially hypothesized that clostridia toxin could adversely affect neurotransmitter function and result in the neurobehavioral changes seen in ASD. Since then studies have repeatedly found abnormal levels of *Clostridia* species in those with ASD. Those with ASD and bowel problems have been found to have significantly increased levels of *clostridium histolytica* when compared to healthy controls (Goyal and Miyan 2014; Finegold et al. 2002; Song et al. 2004; Martirosian et al. 2011). Although another study in regressive ASD showed the reverse (lower levels of *Clostridia* species in ASD compared to controls), they did find that these children had significantly higher levels of *Disulfovibrio* species, and that these levels positively correlated with severity of autism symptoms (Finegold et al. 2010; Finegold 2011a, b). There is also a noted lack of the *prevotella* enterotype, which is key in digesting plant polysaccharides and carbohydrate-rich foods. This altered composition of the microbiome has been observed in multiple studies of children with ASD that compare them to both healthy controls and sibling controls. The lack of this enterotype may be linked to the deficiencies in disaccharide metabolism that are observed in some individuals with ASD (Kang et al. 2013). Insufficiencies in the digestion and absorption of disaccharides and monosaccharides could alter the balance of growth substrates by creating an environment more suited to competitive bacteria phylotypes that grow on undigested and unabsorbed carbohydrates (Williams et al. 2011). These changes in the intestinal environment have also been linked to the abnormal *Firmicutes* to *Bacteriodetes* ratios found in biopsies of children with ASD when compared to those with IBD (Goyal and Miyan 2014). In contrast, *Sutterella* spp. has been found to be increased relative to controls, both in fecal samples and biopsies of the gastrointestinal (GI) tract (De Angelis 2013). Although *Sutterella* was believed to be more prominent in those with GI disturbances, including those with ASD and GI disturbances and those with IBD, these results were not replicated in more recent studies, and it appears that *Sutterella* species may be more common to the human gut than previously thought (Williams et al. 2012; Hansen et al. 2013; Mukhopadhya et al. 2011; Wang et al. 2013, 2014; Parracho et al. 2005).

An attempted meta-analysis of the above studies demonstrates that despite small sample sizes and varied methodologies there are significant differences in the prevalence of different GI microbial species in children with ASD compared to controls, but further work is needed to better understand these differences and how they relate to ASD symptomatology and etiology (Cao et al. 2013). Overall, most studies have demonstrated an altered microbiome in patients with ASD, with the largest issue being a reduction in diversity and richness of microbial species when compared to controls. Increased diversity of the microbiome may allow for better microbial integrity, and the ability to protect the host from environmental stresses and reduce incidence of GI disorders and infections. There is a high prevalence of GI distress, between 23 and 70 %, in those with ASD, including abdominal pain, bloating, diarrhea, constipation, and food selectivity (Cao et al. 2013). Further study is needed to determine if the prevalence of GI symptoms in children with ASD is higher than typically developing children, however, data shows that those with ASD have higher prevalence rates of constipation (33.9 vs. 17.6 %) and food selectivity (24.5 vs. 16.1 %) than controls (Ibrahim et al. 2009). There may be a significant correlation between an altered microbiome and the GI symptoms present in ASD, but further study is needed (Jyonouchi et al. 2005a, b). However, GI symptoms have been linked to ASD symptom severity. For example, constipation, the most common GI symptom in ASD, is significantly correlated with increased irritability and insomnia (Kang et al. 2013; Adams et al. 2011; Buie et al. 2010; Coury et al. 2012; Hsiao et al. 2013). Approximately 90 % of children with ASD experience food selectivity, a form of behavioral rigidity that results in eating a narrow variety of foods by type, texture, or presentation. Issues with feeding may begin early in development, as breastfeeding is less prevalent and occurs for a shorter duration in those with ASD. This could be due to issues in metabolizing lactate due to an altered microbiome, or to difficulties in the child's social engagement with the feeding process (Mulle et al. 2013; Williams et al. 2011). These deviations could further perpetuate an altered microbial ecosystem in ASD that is not remedied as the child develops. As noted above, a healthy and balanced microbiome is needed in order to properly digest foods, of note, di- and poly-saccharides. It is therefore possible that deviations in the establishment and maintenance of a healthy microbiome could result in difficulties in digesting certain foods, such as carbohydrates, resulting in GI distress and pain and leading to rigidity in food selections that maintain the altered microbiome. The irritability, social withdrawal, and anxiety symptoms observed more frequently in those with ASD and GI distress could thus be remedied by interventions that restore microbial balance.

Evidence of Altered Cytokine and Chemokine Levels in ASD

Brief Review of Cytokines

Cytokines are important modulators in both the initiation (innate immune system) and maintenance (adaptive immune system) of immune responses that also help to facilitate the exchange of information between cells of the immune system and other tissues. Involved in both proinflammatory and anti-inflammatory pathways, the effect of these proteins can cross the blood–brain barrier via active transport mechanisms. These proteins that are important in cell signaling include chemokines, interferons, interleukins, lymphokines, and tumor necrosis factor. They carry out the effector functions of T cells by altering the behavior of the target cells. All effector T-cells produce cytokines of different types in different combinations. These small secreted proteins and related membrane-bound proteins act through cell-surface receptors and induce changes in gene expression within their target cell (Masi et al. 2014; Parham 2009). Antigen-presenting dendritic cells take up pathogens and antigens within infected tissue and carry them in the lymph to the secondary lymphoid tissues where they stimulate T-cell activation by prompting naïve T cells to differentiate into effector T cells: CD4 Th1 cells and Th2 cells. Proinflammatory CD4 Th1 cells leave the lymphoid tissues and enter the blood to seek out the site of infection and stimulate the production of more effector T cells. When Th1 cells reach the site of infection they activate tissue macrophages for phagocytosis of microorganisms. Anti-inflammatory CD4 Th2 cells remain in the secondary lymphoid tissues where they activate naïve B cells specific to the antigen of the T cell which differentiate into antibody-secreting plasma cells. Th1 and Th2 CD4 cells are distinguished by the sets of cytokines they make and the effects that these have on the immune response. Proinflammatory Th1 cells work mainly with macrophages in developing a cell-mediated immune response, and secrete cytokines IFN-γ, TNF-α, IL-2. The activated macrophages will then secrete cytokines IL-1, IL-6, IL-8 (CXCL8), IL-12, and more TNF-α. Anti-inflammatory Th2 cells work only with B-cells to develop an antibody-mediated immune response, and secrete IL-4, IL-5, IL-6, IL-10, IL-13, and TGF-β. Once an infection has been brought under control by Th1 cells, Th2 regulatory cells act as immunosuppressors by preventing the production of new effector T cells and suppressing the functions of existing effector T cells. Of note, IL-6 can act as both a proinflammatory and anti-inflammatory cytokine depending on its route of production. An additional subpopulation of CD4 T cells is Th17 cells, which produce IL-17 (Parham 2009). These cells appear to have a proinflammatory role and have been linked to the inflammation seen in MS and IBD (Fleming et al. 2011). The relationship between proinflammatory and anti-inflammatory cytokines keeps the immune system at a balance, which, as described below, is shown to be disrupted in ASD (Fig. 14.2).

Cytokine Profiles of Individuals with ASD

Individuals with ASD have been observed to have altered cytokine profiles compared to healthy controls, and to be in a chronic state of cytokine induction. Many studies have shown those with ASD have a dampened Th2 anti-inflammatory cytokine response and an increased Th1 pro-inflammatory cytokine response, including an increased innate and adaptive immune response through the Th1 pathway, suggesting that localized brain inflammation and autoimmune dysfunction may be involved. More specifically, increases in IL-6, IL-8, and IL-23 have been observed, with an increased Th1 to Th2 ratio when compared to controls. These aberrations have been shown in amniotic fluid (elevated IL-4, IL-10, TNF-α,

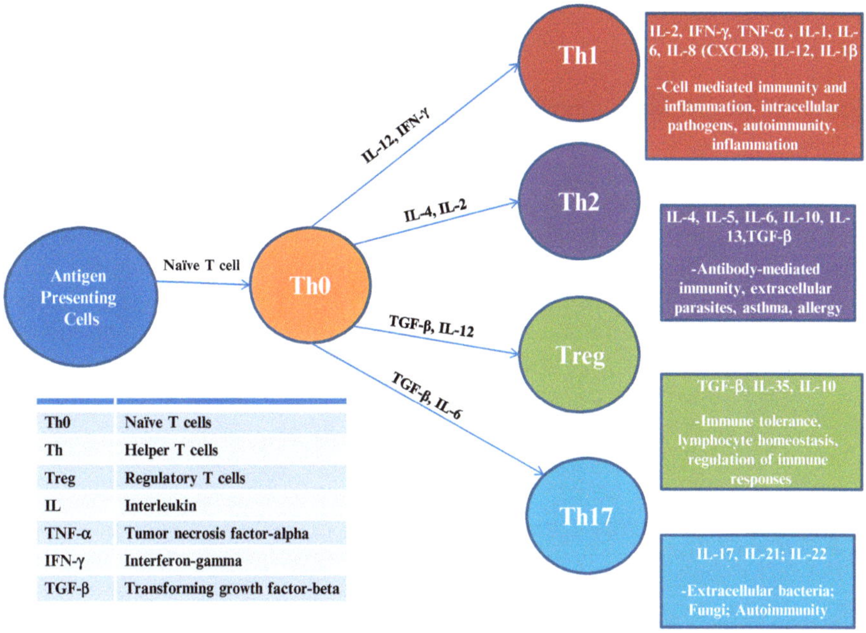

Fig. 14.2 Illustration of cytokine response. *Note*: Depending on context certain cytokines can act in Th1, Th2, or Treg roles

TGF-β), plasma, and peripheral blood mononuclear cells. Further studies have shown increased proinflammatory cytokines in the brain and cerebrospinal fluid (CSF) of ASD patients, including higher levels of proinflammatory IL-6, IL-8, IL-12 and interferon-gamma (IFN-γ) and tumor necrosis factor-α (TNF-α); in addition to elevated proinflammatory Th17 cytokines (Pardo et al. 2005; Abdallah et al. 2013a; Al-Ayadhi and Mostafa 2012; Masi et al. 2014; Croonenberghs et al. 2002a, b; Li et al. 2009; Ricci et al. 2013; Chez and Guido-Estrada 2010; Chez et al. 2007; Garbett et al. 2008; Goines and Ashwood 2013). This upregulation of cytokines has been linked to the activation of microglia and astroglia in ASD patients. Meta-analysis on studies that compare the plasma and serum concentrations of cytokines in unmedicated ASD and healthy participants found significantly altered concentrations of cytokines, further strengthening the hypothesis of an abnormal cytokine profile and immune system dysregulation in ASD that results in increased proinflammatory signals and CNS impairment. The largest effect sizes showed an elevation of proinflammatory IFN-γ and a reduction in anti-inflammatory TGF-β1, while increased levels of proinflammatory IL-1β, IL-6, eotaxin, IL-8, and MCP-1 were also significant. These cytokines have diverse and significant roles in the immune system. TNF-α is a proinflammatory cytokine involved in synapse maturation and stabilization, which at high levels has been shown to impair the development of the visual system by prematurely stabilizing developing synapses leading to increased local connectivity, and potentially epilepsy (Lee et al. 2010; Masi et al. 2014). IFN-γ is responsible for coordinating the transition from innate to adaptive immunity by amplifying immune system sensitivity and promoting macrophage activation. By skewing the immune response toward the Th1 phenotype IFN-γ inhibits the proliferation of Th2 cells and subsequently anti-inflammatory cytokines. TGF-β1 is an immunosuppressant required for maintaining homeostasis of the immune system in addition to promoting cell growth, proliferation, and differentiation (Masi et al. 2014). TGF-β is also a key regulator of neuronal C1q expression required to initiate microglia-mediated synaptic pruning and additionally modulates programmed cell death (Bialas and Stvens 2013). This is relevant to the hypotheses involving aberrant neuroconnectivity in ASD due to either excessive or insufficient synaptic pruning. IL-1β is known to affect the HPA axis and is involved in both acute and chronic inflammation. The proposed impaired chemokines in ASD, eotaxin, IL-8, and MCP-1 are involved in the recruiting neutrophils, eosinophils, monocytes, and T-cells to areas of tissue inflammation (Masi et al. 2014).

A Review of IL-6

IL-6 is an important neuroimmune factor that is shown to be involved in brain development and several neurological disorders. It can act to both promote neural growth and cause neuronal death, depending on its concentration, brain region, and the cell type affected. IL-6 facilitates the communication between the CNS and immune system, and has been linked to the impairment of neural cell adhesion and migration, synapse formation, and synaptic plasticity. Elevated levels have been linked to detrimental effects on synaptic plasticity in addition to physiological and pathological effects on learning and memory. Its role in promoting neuronal differentiation in the adult hippocampus may be compromised if it is chronically elevated, as seen with altered levels of synaptic proteins in the hippocampus during chronic IL-6 administration. Alterations in neural cell adhesion and migration may lead to an imbalance of inhibitory and excitatory synaptic transmissions, including elevated excitatory synapse formation and impaired inhibitory synapses, resulting in reduced paired-pulse inhibition and short-term synaptic plasticity. The impaired excitatory/inhibitory balance theory is one of the leading hypotheses in ASD etiology, and the role of IL-6 demonstrates that it may be linked to immune-inflammatory hypotheses that have spurred recent research. In fact, IL-6 is also proposed as a key factor in the MIA mechanistic pathway (Wei et al. 2013).

Neonatal Cytokine Levels in ASD

The abnormal levels of cytokines and chemokines in patients with ASD may indicate a pathological inflammatory disease process that could result in increased susceptibility to pathogens and allergens resulting in abnormalities in the immune system and CNS impairments. Additional research demonstrates that these abnormalities may begin early in life, with neonatal blood samples showing depressed immune cell activity, including decreased levels of IFN-γ, IL-2, IL-4, IL-6 and Regulated upon Activation Normal T-cell Expressed and Secreted (RANTES, or CCL5) and increased levels of IL-8 and soluble IL-6 alpha. Increased levels of these chemokines, MCP-1 and RANTES, were also observed in the plasma of ASD children. However, as seen with other etiological factors of ASD, results across studies are inconsistent. An additional study of neonates only found increased levels of MCP-1 and decreased levels of RANTES (CCL5) compared to healthy controls, and did not report changes in other cytokines or a hypoactive immune system. Nevertheless, the altered levels of these chemokines that are key to healthy neuronal migration could alter neurodevelopment. Studies that demonstrate links between MIA and developmental delays and ASD features in offspring support this hypothesis (Abdallah et al. 2012, 2013b; Zerbo et al. 2014).

Associations Between Cytokine Levels and Behaviors

CNS impairment may be indicated in ASD through the severity of behaviors and diagnostic features. Similar to evidence of the altered microbiome, an altered inflammatory response, observed as elevated cytokines, has also been linked to ASD symptoms. Worsening GI symptoms have been shown to aggravate behavioral symptoms. Those who have fluctuating behavioral symptoms following immune insults have additional innate immune abnormalities and significantly altered transcript profiles of peripheral blood monocytes (Jyonouchi et al. 2011). Constipation, the most prominent GI symptom in those with ASD, has also been associated with increased levels of blood myeloid dendritic cells (Breece et al. 2013). These cells act as messengers between the innate and adaptive immune system, and present antigen material to the T-cells of the immune system. Higher levels of these cells is also associated with increased repetitive behaviors in those with ASD, larger amygdala volumes, developmental regression, and a later onset of ASD symptoms. Typically developing children with altered dendritic cell frequencies are shown to have poorer adaptive behavior skills. Associations with the amygdala are relevant to ASD etiology, as it is implicated in alterations of core social abilities in ASD and the modulation of anxiety, in addition to having a role in modulating GI activity (Breece et al. 2013). Reduced adaptive behavior skills were also associated with lower plasma levels of TGF-β1, a crucial regulator of the immune system and mediator in the development of autoimmune and systemic inflammation. These children with ASD and low plasma levels of TGF-β1 were found to have worse behavioral scores on the Aberrant Behavior Checklist (ABC) including higher levels of irritability, social withdrawal, hyperactivity, and stereotypies. These results were seen in both children with early onset ASD and those who had regressed, but not for typically developing children or those with developmental disabilities (Ashwood et al. 2008). Other studies have also shown no differences in immune abnormalities between those with early onset ASD and those with developmental regression (Ashwood et al. 2011b). Both groups have a skewed Th1/Th2 response to stimulation with phytohemagglutin (PHA), including increased levels of Granulocyte-macrophage colony-stimulating factor (GM-CSF), TNF-α and Th2 cytokine IL-13 and reduced Th1 cytokine IL-12p40. Increases in Th2 cytokine production are associated with better behavior scores on the ABC (IL-13), improved adaptive behaviors on the Vineland Adaptive Behavior Scale (VABS), and better developmental and cognition scores on the Mullen Scales of Early Learning (MSEL) (IL-10 and IL-5). In contrast, increases in Th1 proinflammatory cytokines were associated with severity of ASD behaviors on gold standard measures of ASD, the Autism Diagnostic Interview-Revised (ADI-R), the Autism Diagnostic Observation Schedule (ADOS), and aberrant behaviors on the ABC. The improvement in behaviors and cognition observed to be associated with increased GM-CSF is significant, as it can cross the blood brain barrier and act as a neuronal growth factor and promote axonal regeneration. However, these receptors are also found on microglia, astrocytes, neurons, and oligodendrocytes, and blocking GM-CSF suppresses microglia activity and the neuroinflammation seen in disorders such as Alzheimer's disease and multiple sclerosis. In contrast, other studies have showed higher cytokine profiles in those with ASD and regression compared to those with ASD without regression (Ashwood et al. 2011c). Both populations had higher levels than typically developing children, and higher levels of Th1 proinflammatory cytokines IL-1β, IL-6, IL-8, and IL-12p40 were associated with severity of core behaviors on the ADI-R and more severe aberrant behaviors on the ABC, in particular increased stereotypical behaviors. Levels of chemokines MCP-1, eotaxin, and RANTES observed to be increased in the plasma of ASD children compared to typically developing and developmentally disabled controls were also associated with severity of behaviors, cognition, and adaptive function observed on the MSEL, ABC, and VABS (Ashwood et al. 2011a).

Cytokine Levels in Siblings of Individuals with ASD

Of note, siblings of children with ASD have been shown to have immune profiles that are significantly similar to ASD profiles. This autism endophenotype in unaffected siblings was proven in a recent study that compared multiple immune parameters between the two groups. Both children with ASD and their siblings had increased levels of proinflammatory cytokine IL-6, and anti-inflammatory cytokine IL-10 (Saresella et al. 2009). This study and others (Manzardo et al. 2012; Napolioni et al. 2013) highlight that the immune dysregulation observed in ASD is also visible in siblings, indicating the need of further research into genetic and environmental insults that may affect both populations. It also further emphasizes that siblings should not be used as healthy comparators in studies of ASD immune dysfunction.

Summary

The altered cytokine profile, in addition to observations of altered T-cell activation, increased circulation of autoantibodies to brain tissue, increased monocyte cell activation, natural killer (NK) cell abnormalities, skewed immunoglobulin profiles, and alterations in complement components, suggest a definite underlying immune dysfunction in ASD (Hsiao 2013; Goyal and Miyan 2014). There has been significant variation in the results seen in these studies, which could be due to varied experimental designs, diagnostic criteria, ages of probands and control populations, use of non-clinically assessed siblings as controls, and small sample sizes. Nevertheless, a definite immune dysfunction is present, and ongoing inflammatory responses may be linked to disturbances in behavior and cognition in ASD.

The Role of Microglia in Inflammation and ASD

Definition of Microglia

The genesis, development, and functional remodeling of our highly complex central nervous system requires strict homeostatic control. This homeostasis is maintained by neuroglial cells, which include macroglia (astroglia, oligodendoglia, and NG2 cells) and microglia. Microglial cells originate from myeloid progenitors that enter the CNS during early embryonic development. They are the first and only glial cells populating the early embryonic brain and have multiple physiological functions that are important for the development and shaping of synaptic connectivity. Their functions include (1) early synaptogenesis during which microglia can provide growth factors and thrombospondins, (2) the elimination of redundant synapses, (3) the direct modulation of synaptic transmission through secretion of factors such as BDNF or TNF-α, (4) provision of trophic support, and (5) the regulation of neurogenesis. Microglial cell bodies are relatively stationary, but their processes continually probe tissue, so that complete coverage of the brain is completed every few hours as they search for signs of damaged tissue (Zeidan-Chulia et al. 2014). Additionally, expression of IL-6, IL-1β, and TNF-α, cytokines observed to be increased in ASD, is correlated with microglia activation (Tetreault et al. 2012; Frick et al. 2013). In sum, they are resident immune cells in the CNS that serve to detect damage, secrete cytokines, and control neuroinflammation, in addition to having a significant role in synaptic pruning during critical periods of development (Zhan et al. 2014; Pardo et al. 2005).

Microglia Pathology in ASD

Microglial pathology has been suggested to have a role in neurodevelopmental disorders such as ASD. Functional connectivity appears to be decreased across cortical regions and is linked to decreased expression of genes associated with synaptic transmission and abnormalities in synaptic scaffolding proteins. Multiple hypotheses about the etiology of ASD revolve around circuit-level deficits in the disorder, including the underconnectivity model, weak central coherence model, and the enhanced excitatory–inhibitory model. It is possible that microglial abnormalities in ASD lead to deficits in synaptic pruning during critical periods of development resulting in connectivity issues that lead to behavioral and cognitive deficits. Microglia's role in synaptic elimination also provides a link between the ASD hypotheses involving the immune system's role in etiology, and those revolving around synaptic connectivity (Zhan et al. 2014). Questions remain about whether activated microglia serves to protect or harm within patients with ASD. Typical neurons prevent phagocytosis by microglia by emitting the cytokine protein fractalkine (CX3CR1) (Tetreault et al. 2012; Zhan et al. 2014; Derecki et al. 2014). Mouse models that delete the microglial fractalkine receptor have more dendritic spines in CA1 neurons and reduced microglia-mediated synaptic pruning which results in abnormal brain development and immature connectivity. Additionally, disorders where microglial pruning is impaired, such as Nasu-Hakola disease, are characterized by deterioration of social behavior and dementia (Tetreault et al. 2012). This suggests that microglia typically have a neuroprotective role, and disruption can cause widespread neuronal damage that can result in neurobehavioral deficits.

Post-mortem studies of the brains of individuals with ASD demonstrate a marked increase in neuroglial responses, characterized by activation of the microglia and astroglia and increased production of both proinflammatory and anti-inflammatory cytokines. This occurs in the absence of lymphocyte infiltration and immunoglobulin deposition. Although microglial responses were diffusely distributed across the cortex and subcortical white matter, there is significant neuroglial activation in the cerebellum. This is of particular interest in ASD due to neuropathological studies that indicate abnormalities in both cortical organization and neuronal packing, and reduced cerebellar Purkinje cell numbers (Vargas et al. 2005). Microglial cells are also described to be of increased density in the dorsolateral prefrontal cortex of post-mortem ASD brain tissue (Morgan et al. 2010, 2012). Additionally, there are significant increases in average microglial somal volume in white matter, microglial density in gray matter, and trends of increased microglial somal volume in gray matter. Microglia also appear to be closer to neurons in the dorsolateral prefrontal cortex in ASD. These changes were present across age groups, including children under the age of 6, during periods of known brain overgrowth and macrocephaly in ASD. Soma enlargement characterizes a shift toward the amoeboid morphology of microglia, a sign of activation and differentiation towards a cytotoxic phenotype. ASD brain tissue also has increased microglial densities in the visual cortex and the fronto-insular cortex. Visual cortices play a role in the deficits in social cognition and social communication observed in the early years of ASD as reduced eye contact, deficits in visual tracking, and prolonged and perseverative fixation. The fronto-insular cortex is involved in social behaviors and is shown to have reduced activity in those with ASD (Tetreault et al. 2012).

In Vivo Microglia Activation Study in ASD

In the first in vivo study of microglia activation in ASD, both men with ASD and age and IQ matched controls received positron emission tomography (PET) using the microglia radiotracer [^{11}C] (R)-PK11195. As observed in the post-mortem brain studies, there was an increased microglia activation across a wide range of brain areas, including the brainstem, anterior cingulate cortex, orbitalfrontal cortex, mid-frontal cortex, superior temporal cortex, fusiform, parietal cortex, and corpus collosum. The most prominent increase in microglial activation was observed in the cerebellum. The pattern of distribution of [^{11}C] (R)-PK11195 binding potential values was similar between the ASD and control subjects, which could indicate augmented but not altered microglial activation in ASD. Correlation studies between levels of microglial activation and severity of behaviors initially were not significant. However, subtyping the subjects by level of binding potential (high BP versus low BP) revealed more severe behaviors on the ADI-R and ADOS in the high BP group. The limitations of this study may have contributed to the minimal links between microglial activation levels and clinical features. As described by the authors, the group was higher functioning and did not include subjects with the most severe ASD behaviors, or those that had a history of regression or immunologic abnormalities. Additionally, the radiotracer used has a high incidence of non-specific binding (Suzuki et al. 2013). Previous studies also suggest that microglial alteration in ASD may be heterogeneous to the disorder, just as many of the clinical features are. Nevertheless, this initial in vivo study provides further evidence of immune alterations and increased microglial activation in those with ASD.

Microglia in Rett's Syndrome, an X-Linked ASD

Interestingly, ground-breaking research in Rett's syndrome, an X-linked ASD, further demonstrates the importance of microglial phagocytic activity and posits bone marrow transplantation as a possible therapeutic approach. Rett's syndrome is characterized by the mutation of the *MeCP2* gene, which is attributed to neuronal dysfunction, including altered glia. When the Rett's syndrome mouse model (MeCP2-null) is irradiated and receives a bone marrow transplant from healthy wild-type mice, bone-marrow derived myeloid cells of microglial phenotype engraft brain parenchyma, resulting in the stoppage of disease development. However, when the MeCP2-null mouse does not receive cranial irradiation, the engraftment by healthy microglia does not occur, and the disease development continues. Additionally, when MeCP2 mice with newly engrafted microglia receive annexin V, a pharmacological agent known to block the phagocytic activity of microglia, the benefits of arresting disease development are diminished. Although future studies need to be conducted to better understand the roles of microglia and their phagocytic activities play in Rett's syndrome, the amelioration of symptoms in this mouse model and the link between this disorder and ASD are significant to therapeutics developments for both disorders (Derecki et al. 2012).

Summary

In sum, microglia are necessary to modulate synaptic activities and plasticity and to aid in the structured remodeling of neuronal circuits. They may play different roles depending on the spatial and temporal context of the afflicting neurodevelopmental or neurodegenerative disease and they may be both a response to the underlying condition and a cause of further abnormalities and insult (Wu et al. 2013). It is clear that they are important in the pathology of ASD, but further research is needed to clarify their role at each developmental stage, and how this may result in aberrant neuronal connectivity and resulting behavioral abnormalities.

The MIA Hypothesis and ASD

Animal Models of MIA

MIA during pregnancy is an environmental risk factor for psychiatric illnesses, such as ASD, in offspring. The effect of MIA on offspring neurodevelopment and behavior has been demonstrated in multiple animal models using a variety of immune system activators, including the influenza virus, the viral mimetic polyinosinic-polycytidylic acid (polyI:C), IL-6, and the bacterial endotoxin lipo-polysaccharide (LPS). Early prenatal stress (EPS) can also act as an immune system activator, and has been shown to produce changes in gene expression patterns, including the up-regulation of immune-related genes of proinflammatory cytokines and chemokines such as IL-6, IL-1β, MCP-1, MIP-1α, and CCL5 (RANTES) (Bronson and Bale 2014). The timing of the MIA is critical to the subsequent effect on the fetus, with early first trimester and second trimester insults causing the most significant insult (Garay et al. 2013). Alterations in brain cytokine levels are found throughout the lifespan of MIA offspring. In both the frontal cortex and cingulate cortex, proinflammatory and anti-inflammatory cytokines are elevated in the early postnatal period, lower than controls during early adolescence, and elevated again in the adult brain. The hippocampus also has alterations of cytokine levels at each age level, but these are distinct compared to other brain areas. There also appears to be a widespread decrease in cytokine levels during critical periods of synaptogenesis and plasticity in MIA offspring compared to controls. This is in contrast to the expected pro-inflammatory phenotype and may indicate an alternative pathway for cytokines to impact brain connectivity and subsequently behavior in ASD. Of note, brain cytokine levels are different than serum cytokine levels, indicating that the levels in brain tissue are not contaminated by serum. Additionally, MIA offspring have no changes in BBB permeability, immune cell infiltration, or microglial density (Garay et al. 2013; Patterson 2012; Ito et al. 2010; Shi et al. 2009; Arrode-Brusés and Brusés 2012).

Offspring of MIA mice exhibit the three core features of ASD, including deficits in communication, sociability, and the presence of repetitive and stereotyped behavior. These ASD features are demonstrated through alterations in their USV, including fewer harmonic and more complex and short syllables, fewer USVs in response to social encounters; reduced scent marking; and deficits in marble burying and self-grooming (Malkova et al. 2012). MIA offspring also have alterations in fear regulation, with marked deficits in the recall and extinction of context. This is an indication of behavioral inflexibility, or rigidity, and is a common feature of the ASD phenotype (Sangha et al. 2014). The MIA model is the only model of an environmental risk factor for ASD, where all three of the core features are present. MIA with either influenza, poly(I:C), or IL-6 causes gene expression changes in the fetal brain transcriptome, most notably in the upregulation of α, β, and γ crystalline gene families (Garbett et al. 2012). This upregulation is correlated with the weight of the placenta, which is an indicator of the severity of MIA, and may be an attempt of the developing brain to counteract the environmental stress of the insult. Crystallin is a protein and neurite promoting factor involved in neuronal differentiation and axonal growth that may also be involved in neuroprotection in response to immune activation. Brain tissue of ASD patients has been shown to have altered αβ-crystallin protein in the frontal cortex. Additionally, altered crystallin levels are implicated in neurodegenerative disorders, and αβ-crystallin knockout mice used to study Alzheimer's disease or autoimmune encephalitis are shown to have worsening symptoms, further indicating the role of crystallin as a neuroprotector. Upregulated crystallin appears to be transient, and characteristic of a response to the MIA exposure, as adolescent mice do not have these higher levels. However, this neuroprotective response appears to have detrimental consequences, and could alter neurogenesis and differentiation in the embryonic brain, causing delays in development that result in changes to connectivity and neurochemistry (Garbett et al. 2012).

Gene expression patterns across MIA offspring also demonstrate the importance of IL-6 as a critical mediator of the behavioral and transcriptional changes in the offspring. As mentioned above, IL-6 is significant in ASD pathology as it is elevated in both the CNS and peripherally. Additionally, when injected with IL-6 prenatally, adult mice exhibit behavioral and cognitive deficits. IL-6 may act at different locations in the MIA model, most likely at one or more of three major sites of signaling: the maternal immune system; the maternal/fetal interface (the placenta); and the fetal brain (Hsiao and Patterson 2011). IL-6 genes are upregulated in response to MIA or EPS in the placenta and in the fetal brain, although this increase may occur acutely (Pratt et al. 2013; Bronson and Bale 2014). After MIA the placenta shows increases in both IL-6 mRNA and the maternally derived IL-6 protein (Hsiao and Patterson 2011). IL-6 could alter the vascular permeability of the placenta, and hence the transfer of nutrients, hormones, or other key molecules to the fetus. Additionally, IL-6 could degrade the maternal tolerance of the fetus by activating lymphocyte migration and cytotoxicity, and enhancing the production of maternal antibodies that cross into the fetal brain. Interestingly, IL-6 could result in changes to fetal programming. Typically there is an increase in TGF-β in pregnant women that is responsible for developing lymphocytes into T regulatory cells that inhibit maternal responses which could be harmful to the developing fetus. TGF-β in combination with IL-6 may result in these cells instead being differentiated into TH17 cells in the offspring. TH17 cells produce pro-inflammatory cytokines, such as IL-1β, IL-6, IL-12, and TNF-α, which could further contribute to the neuroinflammation observed in those with ASD (Mandal et al. 2011). High levels of maternal IL-6 are required to see the behavioral changes observed in MIA offspring, as when IL-6 is blocked and eliminated from the maternal immune response to infection or insult, both gene expression and behavioral changes are almost completely normalized (Smith et al. 2007). However, certain types of MIA insults may require IL-6 for the survival of the mother. If mice mothers injected with influenza are given an anti-IL6 injection, they become severely ill and often die or have a miscarriage. Additionally, IL-6 KO mice that are infected with influenza have the same results, with changes in body temperature, weight loss, and anorexia. Poly(I:C) MIA mice may one of the few that can tolerate the removal of IL-6 from the immune response (Smith et al. 2007). Contrary to the above, an additional study of IL-6 KO mice showed that offspring have increased levels of inflammatory cytokines and chemokines compared to offspring of wild-type saline injected mice, even without an inflammatory insult. This heightened level of inflammation in IL-6 KO offspring does not increase when mothers are injected with poly(I:C), suggesting that the already inflamed brain tissue is not further insulted as a result of MIA (Pratt et al. 2013). The heightened inflammatory state of offspring of IL-6 KO mice is accompanied by higher levels of choline acetyltransferase (ChAT) and decreased expression of glutamic acid decarboxylase (GAD)65, even without MIA. These results are also observed in MIA offspring and are accompanied by an increase in the number of cholinergic neurons in the basal forebrain, an area that is known to have enlarged neurons in those with ASD. Associations between these changes and heightened cytokine production have been made, but it is unclear if this is a result of microglial activation. Decreases in GAD65 expression with heightened ChAT activity could be representative of skewed differentiation of GABAergic interneurons and hence alterations in ASD connectivity and excitatory/inhibitory balance (Pratt et al. 2013). The role of IL-6 in these alterations could indicate it interferes with the normal function of microglia, or, alternatively, it should act as a shut-off switch to regulate the inflammatory environment. IL-6 is also involved in the pathophysiology of epilepsy and may increase seizure susceptibility in offspring following MIA. This is relevant to ASD due to the comorbidity of epilepsy and epileptogenic activity. Inflammatory factors are known to be involved in epilepsy. Brain inflammation facilitates seizures through both enhanced excitation and compromised inhibition. MIA is shown to increase hippocampal excitability, prolong the persistence of kindling states, and increase the speed of the progression of kindled seizures, in addition to impairing social interaction. Both IL-6 and IL1-β are required for this increased seizure propensity, and their activity creates lasting changes in the offspring neocortex and hippocampus. Blockades of either of these cytokines during pregnancy remove the increased propensity for seizures in MIA offspring (Pineda et al. 2013). The ability of IL-6 to cause neuronal survival or neuronal degeneration, balance neurogenesis and gliogenesis, and affect brain development, learning, and memory and the CNS response to injury, clearly indicate it has a complex role in fetal development that requires further investigation.

Although all MIA models do exhibit commonalities in that they each cause levels of increased inflammation exhibited through cytokine and chemokine expression, the type of MIA insult may result in differences in the immunological, behavioral, and cognitive abnormalities observed in offspring. Microglia activation has not been observed in all MIA model studies, although some speculate that its role can be observed through changes in cytokine and chemokine levels. In addition to changes in cytokines, such as IL1-β, dams infected with LPS had higher levels of corticosterone, increased HPA activity, and an increased risk of placental injury. Their offspring did not have any changes in the HPA axis, but did have higher levels of IL1-β than controls when exposed to an immune insult (LPS infection) as an adult. This supports the two-hit model of ASD, in that changes due to an immune insult prenatally can negatively alter the immune response when re-challenged in adulthood (Kirsten et al. 2013). Poly(I:C) infected dams have been shown to have offspring with lasting changes in macrophage function (Onore et al. 2014). Even a non-infectious insult, such as EPS could negatively alter fetal development. Mouse models exposed to EPS had increased gene expression of proinflammatory cytokines IL-6 and IL1-β, particularly in male placentas. These sex-specific effects of inflammation resulted in increased locomotor hyperactivity in adult males. Interestingly, there was partial amelioration of both cytokine response and behavioral abnormalities when EPS mothers were treated with non-steroidal anti-inflammatory drugs (NSAIDs) (Bronson and Bale 2014).

MIA in combination with other insults may also result in greater abnormalities in the offspring. The combination of prenatal immune activation with iron deficiency in mothers resulted in an additive effect on offspring, including a greater induction of proinflammatory cytokines IL-6 and TNF-α and worsened behavior abnormalities (Harvey and Boksa 2014). In addition to variations in MIA insult, the maternal response to the insult, such as anhedonia, could result in alterations to offspring phenotype. Pregnant dams that lost weight following MIA exhibited higher levels of TNF-α. Weight loss is a symptom of anhedonia in the mouse model, and offspring of these dams were more impaired than those of dams that had gained weight following MIA insult (Missault et al. 2014).

The choice of the strain of mouse that is used to study MIA is also relevant to the immunological and behavioral changes observed in offspring. Mouse models have successfully demonstrated that either genetic factors or environmental factors can contribute to the ASD phenotype. However, few studies have considered the two-hit model of a gene x environment interaction, where a genetic predisposition is exacerbated by an immunological insult. The BTBR mouse is the best genetic model of ASD, while the MIA mouse is the best environmental model of ASD. Both models independently demonstrate aberrant immune responses that contribute to the ASD phenotype. In combining the two models, by exposing the genetically predisposed BTBR mouse to the environmental MIA insult poly(I:C) a synergistic result is observed, with increased behavioral impairment indicating greater ASD severity. These observations are made in comparison with the effects of the environmental insult on the healthy mouse strain (C57), which incidentally results in the C57 offspring resembling the BTBR genetically altered strain. In summary, ASD symptomatology may result from either a genetic predisposition or an environmental insult, but in combination this two-hit model would result in greater impairment and severity of symptoms (Schwartzer et al. 2013).

Human Epidemiological Studies

Mirroring the results observed in animal models, large epidemiological studies have found significant associations between maternal viral infections during the first trimester of pregnancy and increased autism risk in their children (Atladóttir et al. 2012). Maternal infections from a variety of microorganisms, including influenza, varicella, and rubella, have been associated with an increased risk for developing ASD, demonstrating that it is not the particular pathogen that is responsible, but rather the generalized activation of the maternal immune system during pregnancy (Atladóttir et al. 2012; Zerbo et al. 2014). Additionally, prolonged febrile episodes and antibiotic use have been linked to an increased risk for infantile ASD (Atladóttir et al. 2012). A study of children that had ASD, developmental delays (DD), or were typically developing completed by the Childhood Autism Risks from Genetics and Environment (CHARGE) group evaluated both maternal influenza and maternal febrile episodes during pregnancy and its relationship with ASD risk. The development of ASD or DD was significantly associated with maternal febrile episodes but not with maternal influenza during pregnancy. Additionally, the ASD risk due to fever was attenuated when mothers reported taking antipyretic medications (Zerbo et al. 2013). These results are comparable to a study completed in Denmark that showed an association between admission to the hospital in the first trimester for maternal viral infection and the second trimester for maternal bacterial infection, with a diagnosis of ASD in offspring (Atladóttir et al. 2010). However, not all infections have the same result, as some viral infections, i.e. influenza, appear to be heavily correlated with increased ASD risk, while urinary tract infections, upper respiratory infections, and cystitis do not appear to be risk factors. Although there is significant data from mouse models on the detriment of MIA, further work is needed in humans to expand on the epidemiological research mentioned above.

The Link Between Maternal Autoantibodies and ASD

Human Studies of the Effect of Maternal Autoantibodies on ASD

The MIA model is extremely important and relevant to learning about the pathophysiology of ASD. However, it can be expanded upon by further mirroring the exposure that would be seen in human mothers of children with ASD. For example, most mothers are not immunologically naïve and have been exposed to infections, antigens, and vaccines that create antibodies. This initial exposure may alter how their immune system responds to an MIA insult during pregnancy, and subsequently how the offspring is affected. Pregnant dams with an immunological memory exposed to an MIA insult have offspring with a preferential differentiation of Th cells to become Th17 cells, compared with offspring of immunologically naïve dams. The observed preferential Th17 cell differentiation occurs in an antigen non-specific manner. As described above, Th17 cells are known to produce proinflammatory cytokines that can contribute to neuroinflammation (Mandal et al. 2011).

Gestational exposure to maternal antibodies to proteins in the fetal brain may also play a role in the pathophysiology of ASD. Typically, maternal antibodies have a protective role and cross the placenta to prepare the immunologically naïve fetus for postnatal exposures. They are detectable in fetal circulation by 13 weeks gestation and often exceed maternal levels at delivery in full term infants and are still detectable in the infant's circulation beyond 6 months of age. Their protective role may thus continue until the infant's immune system is more fully developed (Nordahl et al. 2013). However, these gestationally transferred maternal antibodies could also be potential etiologic factors in a subset of ASD. Large cohort studies of mothers of children with ASD have been completed yielding significant associations between the presence of IgG reactivity to fetal brain proteins by maternal antibodies and a diagnosis of ASD. In particular, maternal IgG reactivity to proteins at 37 and 73 kDa was strongly associated with an autism diagnosis when compared to mothers of typically developing or developmentally delayed children. Paired reactivity to bands at 39 and 73 kDa was also significantly higher in the ASD group compared to controls, as was the presence of reactivity to the 39 kDa band alone. In a sample of 131 children with ASD, 12 % had mothers with reactivity to fetal brain proteins at 37 and 73 kDa. This group was more likely to have abnormal brain enlargement, particularly in the frontal lobe, when compared to both the typical developing group and the ASD group with mothers without fetal brain protein reactivity (Nordahl et al. 2013). Children with ASD and these antibodies in maternal plasma had significantly impaired expressive language on the MSEL, and higher scores on the irritability subscale of the ABC. Maternal antibody reactivity may represent a biomarker of increased severity of ASD features. At minimum these antibodies may characterize an ASD-related maternal immune phenomenon (Braunschweig et al. 2012).

Identification of the target antigens for maternal autoantibody-related autism (MAR) diagnosis is necessary to identify potentially clinically significant biomarkers and to further the understanding of both this mechanism and subset of ASD. In a sample of 246 children with ASD and autism 23 % of their mothers had reactivity to specific antigen combinations, compared to 1 % of controls. Mothers with autoantibodies targeting lactate dehydrogenase (LDH), stress induced phosphoprotein 1 (STIP1), and collapsin response mediator proteins 1 (CRMP1) had children with increased levels of stereotypic behavior, observed on the stereotyped behavior scale of the ABC. These behavioral effects of maternal autoantibody exposure are consistent with those observed in non-human primate (Braunschweig et al. 2013), and mouse model studies (Singer et al. 2009).

Animal Models of the Effect of Maternal Autoantibodies on ASD

As a follow-up to these observations, animal models have been created using IgG from mothers of children with ASD and IgG from mothers of typically developing children. Mice injected with ASD IgG have offspring with changes in both anxiety and sociability. These antibodies, in particular those that target 37 and 73 kDa bands, have access to the fetal mouse brain and can alter both the physical and social development of offspring. These studies were also completed in a non-human primate, the rhesus monkey. Monkeys prenatally exposed to ASD IgG antibodies had increased motor stereotypies and hyperactivity when compared to controls exposed to IgG from mothers of typically developing children. Further study was completed to observe the effects of antibodies that target the 37 and 73 kDa bands in particular. These monkeys were observed for 2 years postnatally to monitor the longitudinal effects of prenatal exposure. Mothers of IgG-ASD$^{37/73\ \mathrm{kDa}}$ offspring had heightened protectiveness during early development and perceived a greater risk to their infants. Consistent with observations in MRIs of children with ASD who may have been exposed to 37/73 kDa antibodies prenatally, monkeys in this group also had increased brain volumes. This was primarily seen in male IgG-ASD$^{37/73\ \mathrm{kDa}}$ primates, congruous with neuroimaging data that shows increased brain size in boys with ASD. Exposure to these antibodies during critical periods of neuronal differentiation, synaptogenesis, dendritic and axonal arborization, myelination, and apoptosis could disrupt the trajectory of brain development and result in abnormal brain size. IgG-ASD$^{37/73\ \mathrm{kDa}}$ offspring also had abnormal social approaches and aberrant social reciprocity to other primates representing clear deviations in social development (Bauman et al. 2013).

The Relationship of Maternal and Family History of Autoimmune Diseases to ASD

Maternal or familial autoimmune disease may also be significant to the etiology and pathophysiology of ASD. Family surveys have found a greater mean number of autoimmune disorders in families of individuals with ASD compared to healthy controls. Additionally, the risk of ASD appears to correlate with the number of family members that have an autoimmune disease. The most common autoimmune disorders observed in these families are type 1 diabetes, adult rheumatoid arthritis, hypothyroidism, and systemic lupus erythematosus (Gesundheit et al. 2013; Comi et al. 1999). Larger epidemiological studies both support and refine these initial results. A study of 689,196 children born in Denmark from 1993 to 2004 indicated that 3,325 had an ASD diagnosis and that an increased risk for ASD was significantly correlated with maternal history of rheumatoid arthritis and celiac disease. Additionally, family history of type 1 diabetes was significantly associated with an increased risk of infantile autism (Atladóttir et al. 2009).

Maternal autoimmune disease does appear to impact ASD risk more severely than family history or paternal history (Keil et al. 2010), and may be of relevance in that maternal onset of autoimmune diseases is typically during the childbearing years. There may be limitations in these studies in the methodology of history data collection, the sample sizes, and how ASD is diagnosed in the sampled populations. A more recent study using data collected from the CHARGE study found that maternal autoimmune disease was significantly associated with a modest increase of developmental disorders and ASD, but not of ASD alone, compared to mothers with typically developing children. Thus, maternal autoimmune disease may be significant to developmental disorders as a whole, and not only to ASD. This study did not examine most autoimmune diseases at the individual level due to low prevalence rates of each condition, but did find trends in the presence of maternal eczema and ASD, and maternal psoriasis and ASD. Additionally, there was a significant association between maternal diary allergy and ASD, which is of note due to the common dietary and food issues in children with ASD. Interestingly, maternal autoantibodies to 37/73 kDa, observed to be linked to ASD behaviors in other studies, were not significantly associated with maternal immune-mediated conditions. In fact, mothers with immune-mediated conditions, especially asthma, were less likely to have these maternal autoantibodies. The authors posit that it is possible maternal antibodies specifically associated with ASD do not share a target with other autoimmune diseases. However, these immune-mediated conditions could have an impact on the fetal brain development through other mechanisms, such as cytokine cascades and oxidative stress (Lyall et al. 2014).

Treatments for ASD Based on Inflammation and Autoimmune Dysfunction

There are currently no FDA approved treatments that target the core features of ASD. The approved treatments risperidone and aripiprazole target irritability and disruptive behaviors, and further investigation is needed to develop treatments that ameliorate core symptoms, with increased focus on using the immune system as a mechanism. Interestingly, the atypical antipsychotic risperidone has been shown to affect cytokines in children and adolescents with ASD. An 8-week treatment of risperidone resulted in statistically significant decreases in chemokines eotaxin and MCP-1, which are known to be abnormally elevated in children with ASD (Choi et al. 2014). A prior study did not see cytokine effects as a result of risperidone treatment (Tobiasova et al. 2011) and further studies are needed to determine if the immune system is balanced by risperidone treatment, and what role this may play in ASD therapeutics.

Helminth-Based Treatments: Trichuris suis *Ova*

Helminth treatment is another possible immune based avenue for developing ASD treatments. The porcine whipworm *Trichuris suis* is genetically related to the human whipworm *Trichuris trichiura*, though in contrast, it only colonizes humans briefly (Beer 1976). *Trichuris suis* Ova (TSO) has been proposed to work through multiple mechanisms, including interference with antigen presentation, cell proliferation, activation, and antibody production. Down-regulation of the host's immune system facilitates the helminth's survival and until recent was thought to be beneficial only to the parasite (Wang et al. 2008). More recently some have suggested that the interaction of the developing immune system with microorganisms, including helminthes, may be an important component of the normal immune system maturation. Helminthes, including TSO, are well known to induce tolerance in their hosts via differential modulation of anti-inflammatory Th2 cytokine (IL-4, IL-5, IL-10, and IL-13) and pro-inflammatory Th1 cytokine (IL-1, IL-12, IFN-γ, TNF-α, and IL-6) response (Jackson et al. 2009a, b). Additional studies have also demonstrated the positive effect of helminthes on the microbiome. A study in macaques with GI distress has shown that the helminth *T. trichiura* favorably impacts the gut microbiome, while improving the gut immune response, mimicking the results seen with TSO in human IBD (Broadhurst et al. 2012). Mouse studies have also demonstrated that regulatory immune responses induced by helminthes can suppress Th2 and Th1/Th17 responses that mediate allergy and autoimmunity respectively. Additionally, the MIA mouse model demonstrates GI barrier defects and alterations in the microbiome and immune system, which correlate with severity of ASD symptoms, and are both corrected with the probiotic *B. fragilis* (Hsiao et al. 2012). These observations have provided a rational explanation for the hygiene hypothesis and the utilization of helminthes as immunomodulatory agents for inflammatory disease in humans. Currently TSO is being studied in multiple clinical trials of immune-inflammatory disorders including Crohn's, ulcerative colitis, relapsing-remitting multiple sclerosis, celiac disease, psoriasis, and ASD (Sandborn et al. 2012; Summers et al. 2003, 2005a, b, 2010). High functioning adults with ASD treated with TSO had improvements in behavioral rigidity and repetitive behaviors compared to treatment with placebo. TSO also has a low side effect profile (Hollander et al. 2013). Further studies are being completed with this novel treatment in children with ASD and immune-inflammatory risk factors, such as family/maternal history of immune illness, presence of co-morbid GI distress, or macrocephaly.

Febrile Hypothesis: Hyperthermia-Based Treatments

The observation that some ASD patients manifest transient clinical improvement in response to fever further supports the idea that symptoms may be modulated by immune-inflammatory factors. The febrile hypothesis of ASD stems from this observation and could be due to the (1) direct effect of temperature, (2) a resulting change in the immune-inflammatory system function associated with the infection of fever, or (3) an increase in the functionality of a previously dysfunctional locus coeruleus-noradrenergic (LC-NA) system. An initial prospective study of the effects of fever on ASD symptoms found fewer aberrant behaviors for febrile patients on the ABC subscales of irritability, hyperactivity, stereotypy, and inappropriate speech compared to control subjects (Curran et al. 2007). To collect further information on the febrile hypothesis a double blind crossover study was completed in children with ASD and a history of febrile response using two treatment conditions, a hyperthermia condition (102 °F) and a control condition (98 °F) in a Hydroworx aquatic therapy pool. Children were immersed in the pool at each temperature condition for 30 min in order to raise their temperatures to the febrile or control levels. Improvement in social cognition and repetitive/restrictive behaviors were observed at the hyperthermia condition (102 °F) on parent (Social Responsiveness Scale; Repetitive Behavior Scale-Revised) and clinician assessments, compared to the control condition. This study demonstrates the feasibility of observing the direct effect of temperature in children with ASD and provides preliminary data on the relationship between body temperature and changes in social/behavioral measures (Ferretti et al. 2013). Further investigation is needed to determine how these results could be beneficial in elucidating the mechanisms behind the observed symptom amelioration and in developing related therapeutics.

Minocycline

Minocycline, a derivative of tetracycline and antibacterial agent, readily crosses the blood brain barrier and is shown to have direct neuroprotective effects and anti-inflammatory properties. Therapeutic trials have been completed in a number of neuroinflammatory conditions, including multiple sclerosis, fragile X, and other neurodegenerative and cerebrovascular diseases. Contradictory results have been obtained in these clinical trials, with beneficial results observed in fragile X, but adverse side effects in amyotrophic lateral sclerosis (ALS). Although its mechanism is not completely understood, it is believed to target multiple inflammatory pathways via complex interactions with proteins in addition to decreasing microglial activation. As both microglial activation and neuroinflammation are observed in ASD, a trial of minocycline in children with ASD and developmental regression was recently completed. Although no clinical improvements were observed, changes were seen in some inflammatory markers, including brain derived neurotrophic factor (BDNF) isoforms in the CSF and blood, hepatic growth factor (HGF) in the CSF and CXCL8 (IL-8) in the serum. This suggests that minocycline may have effects in the CNS despite the lack of clinical changes in this subset of the ASD population (Pardo et al. 2013).

Probiotics

Inflammation, immune imbalance, and the high prevalence of GI symptoms in ASD have led many to look at probiotics as a potential treatment. Interestingly, according to a recent survey, one in five physicians recommends the use of probiotics for treatment of GI symptoms in ASD. The large interplay between the brain, gut, and immune system suggests that if one system improves from treatment others will also benefit. Probiotics are living nonpathogenic microorganisms that can benefit the host's health as either a food ingredient or supplement. They consist mainly of lactic acid-producing organisms, such as lactobacilli, lactococci, and bifidobacteria or yeasts. Research in disorders with known gastrointestinal difficulties, such as IBD and irritable bowel syndrome, has shown there is a beneficial response to probiotic treatment, which may include alterations to the host immune system. It is pertinent to note that the species of probiotic used is relevant to the benefits received. For example, *Bifidobacterium infantis* has been shown to significantly reduce GI symptoms in irritable bowel disease, compared to other probiotics.

The evidence of improved GI symptoms in other disorders alone is enough to warrant further study in ASD, due to the large prevalence of these issues within the population. Additionally, alterations in microbiome, mucosal immunity, and immune system dysregulation may also benefit from treatment with probiotics. Probiotics can restore the balance of intestinal microbiota and inhibit the growth of more toxic species, such as *Clostridium difficile*, which is shown to be elevated in those with ASD. Although research regarding the integrity of the mucosal barrier in ASD is controversial, if there are alterations in ASD intestinal permeability probiotics are able to stabilize the mucosal barrier, stimulate mucosal immunity, and potentially prevent further inflammation. Modulation of the immune system, including cytokine responses, is also possible with probiotics. This suggests that they could be beneficial to those with autoimmune diseases, such as rheumatoid arthritis, where a recent pilot study of probiotics was successful. Probiotic bacteria can suppress MCP-1 activation, a chemokine known to be elevated in ASD, in addition to stimulating production of anti-inflammatory cytokine IL-10. As mentioned above, these changes are highly strain specific and mechanisms of action still need further research (Critchfield et al. 2011). It is clear that probiotics may be beneficial to a subset of those with ASD and GI disorders. With increased investigation into their mechanisms and role in the immune system probiotics could also be beneficial for other subsets of ASD patients as well.

Stem Cell Based Therapeutics

The immune and neural dysregulations observed in ASD also provide targets for stem cell therapies. Stem cells are promising for the treatment of neurodegenerative and neurodevelopmental disorders due to their ability to self-renew, to differentiate, and to release complex biomolecules capable of repairing tissues and organs, regulating cell differentiation and acting as anti-inflammatory agents. Their immunomodulatory properties are of interest in ASD therapeutics, as stem cell therapies could counterbalance the immune system aberrations while restoring damaged tissues that contribute to functional and behavioral deficits.

There are four major types of stem cells that have the potential to be developed as ASD therapies, (1) embryonic stem cells (ESCs), (2) fetal stem cells (FSCs), (3) neural stem cells (NSCs), and (4) adult stem cells. ESCs are pluripotent cells capable of differentiating into all three potential germ layers, ectoderm, mesoderm, and endoderm. Their ability to differentiate into multiple hematopoietic cell lineages, including ASD relevant immune cells (T cells, NK cells, and dendritic cells) is of interest, as they could be used to restore balance to the aberrant immune system. Unfortunately, clinical trial literature on the use of ESCs is sparse, and due to their derivation from early stages of embryonic development, they can cause teratoma formation and uncontrollable cell proliferation in the recipient. Due to safety and efficacy issues, ESCs are not ready for clinical therapy, but may be more useful as a part of a restorative secretome-derived pharmacological cocktail, rather than for transplantation. FSCs are found within fetal tissues and are already differentiated into three subtypes: ectodermal, mesodermal, and endodermal. FSCs have enhanced plasticity and are better suited to clinical use as they possess immune-regulatory functions and are more efficiently reprogrammed to pluripotency. Importantly, FSCs have little to no expression of MHC-I and MHC-II, and confer increased viability post-transplantation. Fetal mesenchymal stem cells (MSCs) exert strong immunomodulatory effects, have a stable phenotype, less senescence and do not form teratomas post-transplantation. Rather than replacing cells, FSCs act on host tissues using paracrine trophic actions. They are of interest in ASD therapeutics, as they both produce and release several neurotropic and growth factors, in addition to working to suppress proinflammatory cytokines known to be elevated in ASD. NSCs can be isolated from both fetal and adult stem cells and are able to generate multiple cell types of the CNS. Their ability to integrate themselves into the neural tissue and reconstruct neural circuitry makes them of interest in ASD due to hypotheses regarding excitatory and inhibitory imbalance that could be remedied with the integration of these stem cells.

Adult stem cells can be further characterized into MSCs and their subtypes, (1) adipose derived and (2) umbilical cord derived. MSCs are self-renewing precursor cells found in the bone marrow and adipose tissue that show immune-modulating capabilities. They are both easily obtained and clinically safe for transplantation. Important to ASD research, MSCs have strong anti-inflammatory and immunosuppressive activity, and once implanted will migrate to the sites of tissue injury. It is thought that when implanted in ASD patients MSCs could integrate themselves into the existing neural and synaptic network in order to restore synaptic transmitter release and plasticity and secrete needed growth factors. These cells have long lasting immunosuppressive activity and are capable of inhibiting proliferation of CD8+ and CD4+ T lymphocytes and Natural Killer cells and downregulating production of T lymphocyte proinflammatory cytokine production. Their ability to restore immune balance in ASD patients is seen in their ability to inhibit proinflammatory TNF-α and INF-γ while increasing immunomodulatory IL-10. With regard to the different subtypes of mesenchymal cells, adipose derived MSCs, while having anti-inflammatory characteristics, have not been shown to be useful in clinical therapies. Umbilical cord derived mesenchymal cells, on the other hand, show good promise for therapeutic development in ASD with multiple clinical trials currently ongoing. The umbilical cord and placenta are abundant sources of MSCs that have all of the beneficial characteristics previously described. They are also a source of primitive stem cells that reside in Wharton's jelly, which have low immunogenicity, produce paracrine trophic factors, and have immunomodulation properties (Siniscalco et al. 2013). There is currently a lot of interest in using umbilical cord derived stem cells for ASD therapeutics. A search of clinicaltrials.gov shows multiple ongoing clinical trials, and a recent publication from the Simons Foundation for Autism Research (SFARI) (Subbaraman 2014) shows that there are large amounts of funding dedicated to developing these therapies. There is potential for this treatment to succeed and the scientific community will be eagerly awaiting the results of these trials.

Conclusion

Immune dysregulation and inflammation are important components of ASD and are relevant to prevention, diagnosis, and treatment of the disorder. There may be multiple mechanisms that result in the aberrations in gut microbiome, cytokine immune response, and neural connectivity and inflammation observed in ASD. Each of these mechanisms could be unique to individual subgroups of ASD with different resulting behavioral abnormalities, or they all could be a part of the disorder as a whole. It will be important in future research to understand if immune dysfunction and inflammation is a final common pathway for ASD or if they only identify particular subgroups of the disorder. The use of each of these abnormalities as biomarkers will help further our goals of early intervention, and may allow for preventative treatments prenatally. Biomarkers will also be important as we better understand the links between immune dysregulation and resulting changes in neural connectivity and behavior. Recent years have seen amazing strides in autism research and the convergence of many hypotheses. The role of immune aberrations in neural connectivity links inflammatory hypotheses with those of excitatory/inhibitory balance, while further research into the genetics of immune markers has led to work in the two-hit models of gene x environment interactions. It is the hope that with continued work in each of these areas we will be able to develop both better diagnostic approaches and better therapeutics for the growing population of individuals and families affected by ASD.

References

Abdallah MW, Larsen N, Mortensen EL, Atladóttir HO, Nørgaard-Pedersen B, Bonefeld-Jørgensen EC, Grove J, Hougaard DM. Neonatal levels of cytokines and risk of autism spectrum disorders: an exploratory register-based historic birth cohort study utilizing the Danish Newborn Screening Biobank. J Neuroimmunol. 2012;252:75–82.

Abdallah MW, Larsen N, Grove J, Nørgaard-Pedersen B, Thorsen P, Mortensen EL, Hougaard DM. Amniotic fluid inflammatory cytokines: potential markers of immunologic dysfunction in autism spectrum disorders. World J Biol Psychiatry. 2013a;14:528–38.

Abdallah MW, Larsen N, Grove J, Bonefeld-Jørgensen EC, Nørgaard-Pedersen B, Hougaard DM, Mortensen EL. Neonatal chemokine levels and risk of autism spectrum disorders: findings from a Danish historic birth cohort follow-up study. Cytokine. 2013b;61:370–6.

Abrahams BS, Geshwind DH. Connecting genes to brain in the autism spectrum disorders. Arch Neurol. 2010;67(4):395–9.

Adams JB, Johansen LJ, Powell LD, Quig D, Rubin RA. Gastrointestinal flora and gastrointestinal status in children with autism – comparisons to typical children and correlation with autism severity. BMC Gastroenterol. 2011;11:22. doi:10.1186/1471-230X-11-22.

Al-Ayadhi LY, Mostafa GA. Elevated serum levels of interleukin-17A in children with autism. J Neuroinflammation. 2012;9(158):1–6.

American Psychiatric Association. Diagnostic and statistical manual of mental disorders. 5th ed. Arlington: American Psychiatric Publishing; 2013.

Arrode-Brusés G, Brusés JL. Maternal immune activation by poly(I:C) induces expression of cytokines IL-1β and IL-13, chemokine MCP-1 and colony stimulating factor VEGF in fetal mouse brain. J Neuroinflammation. 2012;9(83):1–16.

Ashwood P, Wills S, Van de Water J. The immune response in autism: a new frontier for autism research. J Leukoc Biol. 2006;80(1):1–15. Epub 12 May 2006. Review. PMID: 16698940.

Ashwood P, Enstrom A, Krakowiak P, Hertz-Picciotto I, Hansen R, Croen LA, Ozonoff S, Pessah I, Van de Water J. Decreased transforming growth factor beta1 in autism: a potential link between immune dysregulation and impairment in clinical behavioral outcomes. J Neuroimmunol. 2008;204(1–2):149–53.

Ashwood P, Krakowiak P, Hertz-Picciotto I, Hansen R, Pessah IN, Van de Water J. Associations of impaired behaviors with elevated plasma chemokines in autism spectrum disorders. J Neuroimmunol. 2011a;232(1–2):196–9.

Ashwood P, Krakowiak P, Hertz-Picciotto I, Hansen R, Pessah IN, Van de Water J. Altered T cell responses in children with autism. Brain Behav Immun. 2011b;25(5):840–9.

Ashwood P, Krakowiak P, Hertz-Picciotto I, Hansen R, Pessah I, Van de Water J. Elevated plasma cytokines in autism spectrum disorders provide evidence of immune dysfunction and are associated with impaired behavioral outcome. Brain Behav Immun. 2011c;25(1):40–5.

Atladóttir HO, Pedersen MG, Thorsen P, Mortensen PB, Deleuran B, Eaton WW, Parner ET. Association of family history of autoimmune diseases and autism spectrum disorders. Pediatrics. 2009;124:687–94.

Atladóttir HO, Thorsen P, Østergaard L, Schendel DE, Lemcke S, Abdallah M, Parner ET. Maternal infection requiring hospitalization during pregnancy and autism spectrum disorders. J Autism Dev Disord. 2010;40:1423–30.

Atladóttir HO, Henriksen TB, Schendel DE, Parner ET. Autism after infection, febrile episodes, and antibiotic use during pregnancy: an exploratory study. Pediatrics. 2012;130:e1447–54.

Bailey MT, Lubach GR, Coe CL. Prenatal stress alters bacterial colonization of the gut in infant monkeys. J Pediatr Gastroenterol Nutr. 2004;38:414–21.

Baio J. Prevalence of autism spectrum disorder among children aged 8 years – Autism and Developmental Disabilities Monitoring Network, 11 sites, United States, 2010. Centers for Disease Control and Prevention, Morbidity and Mortality Weekly Report. 2014 March 28; 63(SS02):1–21.

Bauman MD, Iosif A-M, Ashwood P, Braunschweig D, Schumann CM, Ven de Water J J. Amaral DG maternal antibodies from mothers of children with autism alter brain growth and social behavior development in the rhesus monkey. Transl Psychiatry. 2013;3(e278):1–12.

Bauman MD, Iosif A, Smith SEP, Bregere C, Amaral DG, Patterson PH. Activation of the maternal immune system during pregnancy alters behavioral development of rhesus monkey offspring. Biol Psychiatry. 2014;75:332–41. http://dx.doi.org/10.1016/j.biopsych.2013.06.025.

Beer RJ. The relationship between *Trichuris trichiura* (Linnaeus 1758) of man and *Trichuris suis* (Schrank 1788) of the pig. Res Vet Sci. 1976;20(1):47–54. PMID: 1257627.

Béna F, Bruno DL, Eriksson M, van Ravenswaaij-Arts C, Stark Z, Dijkhuizen T, Gerkes E, et al. Molecular and clinical characterization of 25 individuals with exonic deletions of NRXN1 and comprehensive review of the literature. Am J Med Genet B Neuropsychiatr Genet. 2013; 162B(4):388–403.

Bialas AR, Stvens B. TGF-β signaling regulates neuronal C1q expression and developmental synaptic refinement. Nat Neurosci. 2013;16(12):1773–82.

Boccuto L, Lauri M, Sarasua SM, Skinner CD, Buccella D, Dwivedi A, Orteschi D, et al. Prevalence of SHANK3 variants in patients with different subtypes of autism spectrum disorders. Eur J Hum Genet. 2013;21:310–6.

Bolte ER. Autism and *Clostridium tetani*. Med Hypotheses. 1998;51(2):133–44.

Borre YE, Moloney RD, Clarke G, Dinan TG, Cryan JF. The impact of microbiota on brain and behavior: mechanisms & therapeutic potential. Adv Exp Med Biol. 2014;817:373–403. doi:10.1007/978-1-4939-0897-4_17.

Braunschweig D, Duncanson P, Boyce R, Hansen R, Ashwood P, Pessah IN, Hertz-Picciotto I, Van de Water J. Behavioral correlates of maternal antibody status among children with autism. J Autism Dev Disord. 2012;42:1435–45.

Braunschweig D, Krakowiak P, Duncanson P, Boyce R, Hansen RL, Ashwood P, Hertz-Picciotto I, Pessah IN, Van de Water J. Autism-specific maternal autoantibodies recognize critical proteins in developing brain. Transl Psychiatry. 2013;3(e277):1–9.

Breece E, Paciotti B, Nordahl CW, Ozonoff S, Van de Water JA, Rogers SJ, Amaral D, Ashwood P. Myeloid dendritic cells frequencies are increased in children with autism spectrum disorder and associated with amygdala volume and repetitive behaviors. Brain Behav Immun. 2013;31: 69–75.

Broadhurst MJ, Ardeshir A, Kanwar B, Mirpuri J, Gundra UM, Leung JM, Wiens KE, Vujkovic-Cvijin I, Kim CC, Yarovinsky F, Lerche NW, McCune JM, Loke P. Therapeutic helminth infection of macaques with idiopathic chronic diarrhea alters the inflammatory signature and mucosal microbiota of the colon. PLoS Pathogen. 2012;8(11):e1003000. doi:10.1371/journal.ppat.1003000. Epub 15 November 2012.

Bronson SL, Bale TL. Prenatal stress-induced increases in placental inflammation and offspring hyperactivity are male-specific and ameliorated by maternal anti-inflammatory treatment. Neuroendocrinology. 2014;155(7):2635–46.

Buie T, Campbell DB, Fuchs 3rd GJ, Furuta GT, Levy J, Vandewater J, Whitaker AH, Atkins D, Bauman ML, Beaudet AL, Carr EG, Gershon MD, Hyman SL, Jirapinyo P, Jyonouchi H, Kooros K, Kushak R, Levitt P, Levy SE, Lewis JD, Murray KF, Natowicz MR, Sabra A, Wershil BK, Weston SC, Zeltzer L, Winter H. Evaluation, diagnosis, and treatment of gastrointestinal disorders in individuals with ASDs: a consensus report. Pediatrics. 2010;125 Suppl 1:S1–18. doi:10.1542/peds.2009-1878C.

Cao X, Lin P, Jiang P, Li C. Characteristics of the gastrointestinal microbiome in children with autism spectrum disorder: a systemic review. Shanghai Arch Psychiatry. 2013;25(6):342–52.

Careaga M, Scwartzer J, Ashwood P. Inflammatory profiles in the BTBR mouse: how relevant are they to autism spectrum disorders. Brain Behav Immun. 2015;43C:11–6. http://dx.doi.org/10.1016/j.bbi.2014.06.006.

Chez MG, Guido-Estrada N. Immune therapy in autism: historical experience and future directions with immunomodulatory therapy. Neurotherapeutics. 2010;7(3):293–301. doi:10.1016/j.nurt.2010.05.008. Review. PMID: 20643382.

Chez MG, Dowling T, Patel PB, Khanna P, Kominsky M. Elevation of tumor necrosis factor-alpha in cerebrospinal fluid of autistic children. Pediatr Neurol. 2007;36(6):361–5. PMID: 17560496.

Choi JE, Widjaja F, Careaga M, Bent S, Ashwood P, Hendren RL. Change in plasma cytokine levels during risperidone treatment in children with autism. J Child Adolesc Psychopharmacol. 2014;24:1–4. doi:10.1089/cap.2013.0108.

Comi AM, Zimmerman AW, Frye VH, Law PA, Peeden JN. Familial clustering of autoimmune disorders and evaluation of medical risk factors in autism. J Child Neurol. 1999;14:388–94.

Coury DL, Ashwood P, Fasano A, Fuchs G, Geraghty M, Kaul A, Mawe G, Patterson P, Jones NE. Gastrointestinal conditions in children with autism spectrum disorder: developing a research agenda. Pediatrics. 2012;130 Suppl 2:S160–8. doi:10.1542/peds.2012-0900N.

Critchfield JW, van Hemert S, Ash M, Mulder L, Ashwood P. The potential role of probiotics in the management of childhood autism spectrum disorders. Gastroenterol Res Pract. 2011; 2011:161358. doi:10.1155/2011/161358.

Croonenberghs J, Wauters A, Devreese K, Verkerk R, Scharpe S, Bosmans E, Egyed B, Deboutte D, Maes M. Increased serum albumin, gamma globulin, immunoglobulin IgG, and IgG2 and IgG4 in autism. Psychol Med. 2002a;32(8):1457–63. PMID: 12455944.

Croonenberghs J, Bosmans E, Deboutte D, Kenis G, Maes M. Activation of the inflammatory response system in autism. Neuropsychobiology. 2002b;45(1):1–6. PMID: 11803234.

Curran LK, Newschaffer CJ, Lee LC, Crawford SO, Johnston MV, Zimmerman AW. Behaviors associated with fever in children with autism spectrum disorders. Pediatrics. 2007;120:e1386–92. doi:10.1542/peds. 2007-0360.

De Angelis M, Piccolo M, Vannini L, Siragusa S, De Giacomo A, Serrazzanetti I, Cristofori F, Guerzoni ME, Gobbetti M, Francavilla R. Fecal microbiota and metabolome of children with autism and pervasive developmental disorder not otherwise specified. PLoS One. 2013;8(10):e76993. doi:10.1371/journal.pone.0076993. eCollection 2013.

Derecki NC, Cronk JC, Lu Z, Xu E, Abbott SBG, Guyenet PG, Kipnis J. Wild type microglia arrest pathology in a mouse model of Rett syndrome. Nature. 2012;484(7392):105–9. doi:10.1038/nature10907.

Derecki NC, Katzmarski N, Kipnis J, Meyer-Luehmann M. Microglia as a critical player in both developmental and late-life CNS pathologies. Acta Neuropathol. 2014;128:333–45. doi:10.1007/s00404-014-1324-z. Epub 24 July 2014.

Feary J, Britton J, Leonardi-Bee J. Atopy and current intestinal parasite infection: a systematic review and meta-analysis. Allergy. 2011;66(4):569–78. doi:10.1111/j.1398-9995.2010.02512.x. Epub 18 November 2010. Review. PMID: 21087217.

Ferretti CJ, Taylor BP, Noone R, Racine E, Kirsch J, Hollander E. Hyperthermia and the improvement of ASD symptoms. ACNP Conference 2013. Poster 231.

Finegold SM. State of the art; microbiology in health and disease. Intestinal bacterial flora in autism. Anaerobe. 2011a;17:367–8.

Finegold SM. *Desulfovibrio* species are potentially important in regressive autism. Med Hypotheses. 2011b;77(2):270–4. doi:10.1016/j.mehy.2011.04.032. Epub 17 May 2011.

Finegold SM, Molitoris D, Song Y, Liu C, Vaisanen ML, Bolte E, et al. Gastrointestinal microflora studies in late-onset autism. Clin Infect Dis. 2002;35 Suppl 1:S6–16.

Finegold SM, Dowd SE, Gontcharovav V, Lio C, Henley KE, Wolcott RD, et al. Pyrosequencing study of fecal microflora of autistic and control children. Anaerobe. 2010;16:444–53.

Fleming JO, Isaak A, Lee JE, Luzzio CC, Carrithers MD, Cook TD, Field AS, Boland J, Fabry Z. Probiotic helminth administration in relapsing-remitting multiple sclerosis: a phase 1 study. Mult Scler. 2011;17(6):743–54. doi:10.1177/1352458511398054. Epub 3 March 2011.

Fombonne E. Epidemiology of pervasive developmental disorders. Pediatr Res. 2009;65:591–8.

Frick LR, Williams K, Pittenger C. Microglial dysregulation in psychiatric disease. Clin Dev Immunol. 2013;2013:1–10. doi:10.1155/2013/608654.

Garay PA, Hsiao EY, Patterson PH, McAllister AK. Maternal immune activation causes age- and region-specific changes in brain cytokines in offspring throughout development. Brain Behav Immun. 2013;31:54–68.

Garbett K, Ebert PJ, Mitchell A, Lintas C, Manzi B, Mirnics K, Persico AM. Immune transcriptome alterations in the temporal cortex of subjects with autism. Neurobiol Dis. 2008;30(3): 303–11. doi:10.1016/j.nbd.2008.01.012. Epub 10 March 2008. PMID: 18378158.

Garbett KA, Hsiao EY, Káimán S, Patterson PH, Mirnics K. Effects of maternal immune activation on gene expression patterns in the fetal brain. Transl Psychiatry. 2012;2(e98):1–8.

Gaugler T, Klei L, Sanders SJ, Bodea CA, Goldberg AP, Lee AB, Mahajan M, et al. Most genetic risk for autism resides with common variation. Nat Genet. 2014;46(8):881–5.

Gesundheit B, Rosenzweig JP, Naor D, Lerer B, Zachor DA, Prochazka V, Melmad M, Kristt DA, Steinberg A, Shulman C, Hwang P, Koren G, Walfisch A, Passweg JR, Snowden JA, Tamouza R, Leboyer M, Farge-Bancel D, Ashwood P. Immunological and autoimmune considerations of autism spectrum disorders. J Autoimmun. 2013;44:1–7.

Goines PE, Ashwood P. Cytokine dysregulation in autism spectrum disorders (ASD): possible role of the environment. Neurotoxicol Teratol. 2013;36:67–81. doi:10.1016/j.ntt.2012.07.006.

Goyal DK, Miyan JA. Neuro-immune abnormalities in autism and their relationship with the environment: a variable insult model for autism. Front Endocrinol. 2014;5(29):1–10. doi:10.3389/fendo.2014.00029.

Guilmatre A, Huguet G, Delorme R, Bourgeron T. The emerging role of SHANK genes in neuropsychiatric disorders. Dev Neurobiol. 2014;74:113–1220.

Hallmayer J, Cleveland S, Torres A, Phillips J, Cohen B, Torigoe T, Miller J, Fedele A, Colling J, Smith K, Lotspeich L, Croen L, Ozonoff S, Lajonchere C, Grether JK, Risch N. Genetic heritability and shared environmental factors among twin pairs with autism. Arch Gen Psychiatry. 2011;68(11):1095–102.

Hansen R, Berry SH, Mukhopadhya I, Thomson JM, Saunders KA, Nicholl CE, Bisset WM, et al. The microaerophilic microbiota of de-novo paediatric inflammatory bowel disease: the BISCUIT Study. PLoS One. 2013;8(3):e58825. doi:10.1371/journal.pone.0058825.

Harvey L, Boksa P. Additive effects of maternal iron deficiency and prenatal immune activation on adult behaviors in rat offspring. Brain Behav Immun. 2014;40:27–37.

Heo Y, Zhang Y, Gao D, Miller VM, Lawrence DA. Aberrant immune responses in a mouse with behavioral disorders. PLoS One. 2011;6:e20912.

Hollander E, Ferretti CJ, Taylor BP, Noone R, Kirsch J, Racine E. *Trichuris suis* ova (TSO) as an immune-inflammatory treatment for repetitive behaviors in autism spectrum disorder (ASD). ACNP Conference 2013. Poster 177.

Hsiao EY. Immune dysregulation in autism spectrum disorder (chapter 9). Int Rev Neurobiol. 2013;113:269–302.

Hsiao EY, Patterson PH. Activation of the maternal immune system induces endocrine changes in the placenta via IL-6. Brain Behav Immun. 2011;25(4):604–15.

Hsiao EY, McBride SW, Chow J, Mazmanian SK, Patterson PH. Modeling an autism risk factor in mice leads to permanent immune dysregulation. Proc Natl Acad Sci U S A. 2012;109(31): 12776–81.

Hsiao EY, McBride SW, Hsien S, Sharon G, Hyde ER, McCue T, Codelli JA, Chow J, Reisman SE, Petrosino JF, Patterson PH, Mazmanian SK. Microbiota modulate behavioral and physiological abnormalities associated with neurodevelopmental disorders. Cell. 2013;155(7):1451–63. doi:10.1016/j.cell.2013.11.024. Epub 5 December 2013.

Ibrahim SH, Voigt RG, Katusic SK, Weaver AL, Barbaresi WJ. Incidence of gastrointestinal symptoms in children with autism: a population-based study. Pediatrics. 2009;124(2):680–6. doi:10.1542/peds.2008-2933. Epub 27 July 2009.

Ito HT, Smith SEP, Hsiao E, Patterson PH. Maternal immune activation alters nonspatial information processing in the hippocampus of the adult offspring. Brain Behav Immun. 2010;24(6): 930–41.

Jackson JA, Friberg IM, Bolch L, Lowe A, Ralli C, Harris PD, Behnke JM, Bradley JE. Immunomodulatory parasites and toll-like receptor-mediated tumour necrosis factor alpha responsiveness in wild mammals. BMC Biol. 2009a;7:16. doi:10.1186/1741-7007-7-16. PMID: 19386086.

Jackson JA, Friberg IM, Little S, Bradley JE. Review series on helminths, immune modulation and the hygiene hypothesis: immunity against helminths and immunological phenomena in modern human populations: coevolutionary legacies? Immunology. 2009;126(1):18–27. doi:10.1111/j.1365-2567.2008.03010.x. Review. Erratum in: Immunology. 2009;126(3):446. PMID: 19120495.

Jouvin MH, Kinet JP. *Trichuris suis* ova: testing a helminth-based therapy as an extension of the hygiene hypothesis. J Allergy Clin Immunol. 2012;130(1):3–10; quiz 11–2. doi:10.1016/j.jaci.2012.05.028. Review. PMID: 22742834.

Jyonouchi H, Geng L, Ruby A, Reddy C, Zimmerman-Bier B. Evaluation of an association between gastrointestinal symptoms and cytokine production against common dietary proteins in children with autism spectrum disorders. J Pediatr. 2005a;146(5):605–10. PMID:15870662.

Jyonouchi H, Geng L, Ruby A, Zimmerman-Bier B. Dysregulated innate immune responses in young children with autism spectrum disorders: their relationship to gastrointestinal symptoms and dietary intervention. Neuropsychobiology. 2005b;51(2):77–85. PMID: 15741748.

Jyonouchi H, Geng L, Streck DL, Toruner GA. Children with autism spectrum disorders (ASD) who exhibit chronic gastrointestinal (GI) symptoms and marked fluctuation of behavioral symptoms exhibit distinct innate immune abnormalities and transcriptional profiles of peripheral blood (PB) monocytes. J Neuroimmunol. 2011;235:73–80.

Kang DW, Park JG, Ilhan ZE, Wallstrom G, LaBaer J, Adams JB, Krajmalnik-Brown R. Reduced incidence of prevotella and other fermenters in intestinal microflora of autistic children. PLoS One. 2013;8(7):e68322. doi:10.1371/journal.pone.0068322.

Keil A, Daniels JL, Forssen U, Hultman C, Cnattinguis S, Soderberg KC, Feychting M, Sparen P. Parental autoimmune diseases associated with autism spectrum disorders in offspring. Epidemiology. 2010;21(6):805–8. doi:10.1097/EDE.0b013e3181f26e3f.

Kirsten TB, Lippi LL, Bevilacqua E, Bernardi MM. LPS exposure increases maternal corticosterone levels, causes placental injury and increases IL-1β levels in adult rat offspring: relevance to autism. PLoS One. 2013;8(12):e82244, 1–10.

Kohane IS. An autism case history to review the systemic analysis of large-scale data to refine the diagnosis and treatment of neuropsychiatric disorders. Biol Psychiatry. 2015;77:59–65. http://dx.doi.org/10.1016/j.biopsych.2014.05.024.

Kohane IS, McMurry A, Weber G, MacFadden D, Rappaport L, Kunkel L, Bickel J, Wattanasin N, Spence S, Murphy S, Churchill S. The co-morbidity burden of children and young adults with autism spectrum disorders. PLoS One. 2012;7(4): e33224, 1–7.

Lee RH, Mills EA, Schwartz N, Bell MR, Deeg KE, Ruthazer ES, Marsh-Armstrong N, Aizenman CD. Neurodevelopmental effects of chronic exposure to elevated levels of pro-inflammatory cytokines in a developing visual system. Neural Dev. 2010;5(2):1–16.

Li X, Chauhan A, Sheikh AM, Patil S, Chauhan V, Li XM, Ji L, Brown T, Malik M. Elevated immune response in the brain of autistic patients. J Neuroimmunol. 2009;207(1–2):111–6. doi:10.1016/j.jneuroim.2008.12.002. Epub 20 January 2009. PMID: 19157572.

Lyall K, Ashwood P, Van de Water J, Hertz-Picciotto I. Maternal immune-mediated conditions, autism spectrum disorders and developmental delay. J Autism Dev Disord. 2014;44:1546–55.

Malkova NV, Yu CZ, Hsiao EY, Moore MJ, Patterson PH. Maternal immune activation yields offspring displaying mouse versions of the three core symptoms of autism. Brain Behav Immun. 2012;26(4):607–16.

Mandal M, Marzouk AC, Donnelly R, Ponzio NM. Maternal immune stimulation during pregnancy affects adaptive immunity in offspring to promote development of TH17 cells. Brain Behav Immun. 2011;25:863–71.

Manzardo AM, Henkhaus R, Dhillon S, Butler MG. Plasma cytokine levels in children with autistic disorder and unrelated siblings. Int J Dev Neurosci. 2012;30:121–7.

Martirosian G, Ekiel A, Aptekorz M, Wiechula B, Kazek B, Jankowska-Stiefer E, et al. Fecal lactoferrin and *Clostridium* spp. in stools of autistic children. Anaerobe. 2011;17(1):43–5.

Masi A, Quintana DS, Glozier N, Lloyd AR, Hickie IB, Guastella AJ. Cytokine aberrations in autism spectrum disorder: a systematic review and meta-analysis. Mol Psychiatry. 2014. doi:10.1038/mp.2014.59.

Missault S, Van den Eynde K, Vanden Berghe W, Fransen E, Weeren A, Timmermans JP, Kumar-Singh S, Dedeurwaerdere S. The risk for behavioural deficits is determined by the maternal immune response to prenatal immune challenge in a neurodevelopmental mode. Brain Behav Immun. 2014;42:138–46. http://dx.doi.org/10.1016/j.bbi.2014.06.013.

Morgan JT, Chana G, Pardo CA, Achim C, Semendeferi K, Buckwalter J, Courchesne E, Everall IP. Microglial activation and increased microglial density observed in the dorsolateral prefrontal cortex in autism. Biol Psychiatry. 2010;68:368–76. doi:10.1016/j.biopsych.2010.05.024.

Morgan JT, Chana G, Abramson I, Semendeferi K, Courchesne E, Everall IP. Abnormal microglial-neuronal spatial organization in the dorsolateral cortex in autism. Brain Res. 2012;1456:72–81. doi:10.1016/j.brainres.2012.03.036.

Muhle R, Trentacoste SV, Rapin I. The genetics of autism. Pediatrics. 2004;113(5):e472–86.

Mukhopadhya I, Hansen R, Nicholl CE, Alhaiden YA, Thomson JM, Berry SH, et al. A comprehensive evaluation of colonic mucosal isolates of *Sutterella wadsworthensis* from inflammatory bowel disease. PLoS One. 2011;6(10):e27076. doi:10.1371/journal.pone.0027076.

Mulle JG, Sharp WG, Cubells JF. The gut microbiome: a new frontier in autism research. Curr Psychiatry Rep. 2013;15(2):337. doi:10.1007/s11920-012-0337-0.

Napolioni V, Ober-Reynolds B, Szelinger S, Corneveaux JJ, Pawlowski T, Ober-Reynolds S, Kirwan J, Persico AM, Melmed RD, Craig DW, Smith CJ, Huentelman MJ. Plasma cytokine profiling in sibling pairs discordant for autism spectrum disorder. J Neuroinflammation. 2013;10(38):1–11.

Nordahl CW, Braunschweig D, Iosif AM, Lee A, Rogers S, Ashwood P, Amaral DG, Van de Water J. Maternal autoantibodies are associated with abnormal brain enlargement in a subgroup of children with autism spectrum disorder. Brain Behav Immun. 2013;30:61–5.

Noriega DB, Savelkoul HFJ. Immune dysregulation in autism spectrum disorder. Eur J Pediatr. 2014;173:33–43. doi:10.1007/s00431-013-2183-4.

Onore C, Careaga M, Ashwood P. The role of immune dysfunction in the pathophysiology of autism. Brain Behav Immun. 2012;26(3):383–92. doi:10.1016/j.bbi.2011.08.007.

Onore CE, Careaga M, Baibineau BA, Schwartzer JJ, Berman RF, Ashwood P. Inflammatory macrophage phenotype in BTBR T+tf/J mice. Front Neurosci. 2013;7:158.

Onore CE, Schwartzer JJ, Careaga M, Berman RF, Ashwood P. Maternal immune activation leads to activated inflammatory macrophages in offspring. Brain Behav Immun. 2014;38:220–6.

Ousley O, Cermak T. Autism spectrum disorder: defining dimensions and subgroups. Curr Dev Disord Rep. 2014;1:20–8.

Pardo CA, Vargas DL, Zimmerman AW. Immunity, neuroglia and neuroinflammation in autism. Int Rev Psychiatry. 2005;17(6):485–95. doi:10.1080/02646830500381930.

Pardo CA, Buckley A, Thurm A, Lee L, Azhagiri A, Neville DM, Swedo SE. A pilot open-label trial of minocycline in patients with autism and regressive features. J Neurodev Disord. 2013; 5(9):1–9.

Parham P. The immune system. New York: Garland Science; 2009.

Parker W, Ollerton J. Evolutionary biology and anthropology suggest biome reconstitution as a necessary approach toward dealing with immune disorders. Evol Med Public health. 2013;2013(1):89–103. doi:10.1093/emph/eot008. Epub 19 April 2013.

Parracho HM, Bingham MO, Gibson GR, McCartney AL. Differences between the gut microflora of children with autistic spectrum disorders and that of healthy children. J Med Microbiol. 2005;54(10):987–92.

Patterson PH. Maternal infection and autism. Brain Behav Immun. 2012;26:393.

Peñagarikano O, Geschwind DH. What does CNTNAP2 reveal about autism spectrum disorder? Trends Mol Med. 2012;18(3):156–63.

Pineda E, Shin D, You SJ, Auvin S, Sankar R, Mazarati A. Maternal immune activation promotes hippocampal kindling epileptogenesis in mice. Ann Neurol. 2013;74:11–9.

Pinto D, Delaby E, Merico D, Barbosa M, Merikangas A, Klei L, et al. Convergence of genes and cellular pathways dysregulated in autism spectrum disorder. Am J Hum Genet. 2014;94: 677–94.

Pratt L, Ni L, Ponzio NM, Jonakait GM. Maternal inflammation promotes fetal microglia activation and increased cholinergic expression in the fetal basal forebrain: role of interleukin-6. Pediatr Res. 2013;74(4):393–401.

Ricci S, Businaro R, Ippoliti F, Lo Vasco VR, Massoni F, Onofri E, Troili GM, Pontecorvi V, Morelli M, Rapp Ricciardi M, Archer T. Altered cytokine and BDNF levels in autism spectrum disorder. Neurotox Res. 2013;24:491–501. Epub 19 April 2013. PMID: 23604965.

Ronemus M, Iossifov I, Levy D, Wigler M. The role of de novo mutations in the genetics of autism spectrum disorders. Nat Rev Genet. 2014;15:133–41.

Rook GA. The changing microbial environment and chronic inflammatory disorders. Allergy Asthma Clin Immunol. 2008;15;4(3):117–24. doi:10.1186/1710-1492-4-3-117. Epub 15 September 2008. PMID: 20525133.

Rook GAW, Lowry CA, Raison CL. Hygiene and other early childhood influences on the subsequent function of the immune system. Brain Res. Epub 13 April 2014. http://dx.doi.org/10.1016/j.brainres.2014.01.004.

Sandborn WJ, Elliott DE, Weinstock J, Summers RW, Landry-Wheeler A, Silver N, Harnett MD, Hanauer SB. Randomised clinical trial: the safety and tolerability of *Trichuris suis* ova in patients with Crohn's disease. Aliment Pharmacol Ther. 2013;38:255–63. doi:10.1111/apt.12366. Epub 3 June 2013. PMID: 23730956.

Sangha S, Greba Q, Robinson PD, Ballendine SA, Howland JG. Heightened fear in response to a safety cue and extinguished fear cue in a rat model of maternal immune activation. Front Behav Neurosci. 2014;8(168):1–11.

Saresella M, Marventano I, Guerini FR, Mancuso R, Ceresa L, Zanzottera M, Rusconi B, Maggioni E, Tinelli C, Clerici M. An autistic endophenotype results in complex immune dysfunction in healthy siblings of autistic children. Biol Psychiatry. 2009;66:978–84.

Schwartzer JJ, Careaga M, Onore CE, Rushakoff JA, Berman RF, Ashwood P. Maternal immune activation and strain specific interactions in the development of autism-like behaviors in mice. Transl Psychiatry. 2013;3(e240):1–9.

Shi L, Smith SEP, Malkova N, Tse D, Su Y, Patterson PH. Activation of the maternal immune system alters cerebellar development in the offspring. Brain Behav Immun. 2009;23(1):116–23.

Singer HS, Morris C, Gause C, Pollard M, Zimmerman AW, Pletnikov M. Prenatal exposure to antibodies from mothers of children with autism produces neurobehavioral alterations: a pregnant dam mouse model. J Neuroimmunol. 2009;211:39–48.

Siniscalco D, Bradstreet JJ, Sych N, Antonucci N. Perspectives on the use of stem cells for autism treatment. Stem Cells Int. 2013;2013:262438. http://dx.doi.org/10.1155/2013/262438.

Smith SEP, Li J, Garbett K, Mirnic K, Patterson PH. Maternal immune activation alters fetal brain development through IL-6. J Neurosci. 2007;27(40):10695–702.

Song Y, Liu C, Finegold SM. Real-time PCR quantification of *Clostridia* in feces of autistic children. Appl Environ Microbiol. 2004;70(11):6459–65.

Subbaraman N. Experts balk at large trial of stem cells for autism. Simons Foundation Autism Research Initiative (SFARI) News and Opinion. 2014 July 14. http://sfari.org/news-and-opinion/news/2014/experts-balk-at-large-trial-of-stem-cells-for-autism.

Sudo N, Chida Y, Aiba Y, Sonoda J, Oyama N, Yu XN, Kubo C, Koga Y. Postnatal microbial colonization programs the hypothalamic–pituitary–adrenal system for stress response in mice. J Physiol. 2004;558(1):263–75.

Summers RW, Elliott DE, Qadir K, Urban Jr JF, Thompson R, Weinstock JV. *Trichuris suis* seems to be safe and possibly effective in the treatment of inflammatory bowel disease. Am J Gastroenterol. 2003;98(9):2034–41. PMID: 14499784.

Summers RW, Elliott DE, Urban Jr JF, Thompson R, Weinstock JV. *Trichuris suis* therapy in Crohn's disease. Gut. 2005a;54(1):87–90. PMID: 15591509.

Summers RW, Elliott DE, Urban JF Jr., Thompson RA, Weinstock JV. *Trichuris suis* therapy for active ulcerative colitis: a randomized controlled trial. Gastroenterology. 2005b;128(4):825–32. PMID: 15825065.

Summers RW, Elliott DE, Weinstock JV. *Trichuris suis* might be effective in treating allergic rhinitis. J Allergy Clin Immunol. 2010;125(3):766–7. doi:10.1016/j.jaci.2009.12.937. Epub 11 February 2010. No abstract available. PMID: 20153033.

Suzuki K, Sugihara G, Ouchi Y, Nakamura K, Futatsubashi M, Takebayashi K, Yoshihara Y, Omata K, Matsumoto K, Tsuchiya KJ, Iwata Y, Tsujii M, Sugiyama T, Mori N. Microglia activation in young adults with autism spectrum disorder. JAMA Psychiatry. 2013;70(1): 49–58.

Tetreault NA, Hakeem AY, Jiang S, Williams BA, Allman E, Wold BJ, Allman JM. Microglia in the cerebral cortex in autism. J Autism Dev Disord. 2012;42:2569–84. doi:10.1007/s10803-012-1513-0.

Tobiasova Z, van der Lingen KHB, Scahill L, Leckman J, Zhang Y, Chae W, McCracken JT, McDougle CJ, Vitiello B, Tierney E, Aman MG, Arnold LE, Katsovich L, Hoekstra PJ, Volkmar F, Bothwell ALM, Kawikova I. Risperidone-related improvement of irritability in children with autism is not associated with changes in serum of epidermal growth factor and interleukin-13. J Child Adolesc Psychopharmacol. 2011;21(6):555–64. doi:10.1089/cap.2010.0134.

Vargas DL, Nascimbene C, Krishnan C, Zimmerman AW, Pardo CA. Neuroglial activation and neuroinflammation in the brain of patients with autism. Ann Neurol. 2005;57:67–81.

Wang Y, Kasper LH. The role of microbiome in central nervous system disorders. Brain Behav Immun. 2014;38:1–12.

Wang LJ, Cao Y, Shi HN. Helminth infections and intestinal inflammation. World J Gastroenterol. 2008;14(33):5125–32. Review. PMID: 18777588.

Wang L, Christophersen CT, Sorich MJ, Gerber JP, Angley MT, Conlon MA. Increased abundance of *Sutterella* spp. and *Ruminococcus torques* in feces of children with autism spectrum disorder. Mol Autism. 2013;4(42):1–4.

Wang L, Conlon MA, Christophersen CT, Sorich MJ, Angley MT. Gastrointestinal microbiota and metabolite biomarkers in children with autism spectrum disorders. Biomark Med. 2014;8(3): 331–44. doi:10.2217/bmm.14.12.

Wei H, Alberts I, Li X. Brain IL-6 and autism. Neuroscience. 2013;252:320–5.

Williams BL, Hornig M, Buie T, Bauman ML, Cho Paik M, Wick I, Bennett A, Jabado O, Hirschberg DL, Lipkin WI. Impaired carbohydrate digestion and transport and mucosal dysbiosis in the intestines of children with autism and gastrointestinal disturbances. PLoS One. 2011;6(9):e24585. doi:10.1371/journal.pone.0024585.

Williams BL, Hornig M, Parekh T, Lipkin WI. Application of novel PCR-based methods for detection, quantification, and phylogenetic characterization of *Sutterella* species in intestinal biopsy samples from children with autism and gastrointestinal disturbances. MBio. 2012;3(1). pii: e00261-11. doi:10.1128/mbio.00261-11.

Wu L, Stevens B, Duan S, MacVicar BA. Microglia in neuronal circuits. Neural Plast. 2013;2013:586426. http://dx.doi.org/10.1155/2013/586426.

Zeidan-Chulia F, Salmina AB, Malinovskaya NA, Noda M, Verkhratsky A, Moreira JCF. The glial perspective of autism spectrum disorders. Neurosci Biobehav Rev. 2014;38:160–72. http://dx.doi.org/10.1016/j.neubiorev.2013.11.008.

Zerbo O, Iosif AM, Walker C, Ozonoff S, Hansen RL, Hertz-Picciotto I. Is maternal influenza or fever during pregnancy associated with autism or developmental delays? Results from the CHARGE (Childhood Autism Risks from Genetics and Environment) Study. J Autism Dev Disord. 2013;43(1):25–33.

Zerbo O, Yoshida C, Grether JK, Van de Water J, Ashwood P, Delorenze GN, Hansen RL, Kharrazi M, Croen LA. Neonatal cytokines and chemokines and risk of autism spectrum disorder: the early markers for autism (EMA): a case–control study. J Neuroinflammation. 2014;11(113): 1–9.

Zhan Y, Paolicelli RC, Sforazzini F, Weinhard L, Bolasco G, Pagani F, Vyssotski AL, Bifone A, Gozzi A, Ragozzino D, Gross CT. Deficient neuron-microglia signaling results in impaired functional brain connectivity and social behavior. Nat Neurosci. 2014;17(3):400–6. doi:10.1038/nn.3641.

Chapter 15
The Role of Inflammation in Alzheimer's Disease

Norbert Müller, Daniela L. Krause, Markus J. Schwarz, Elif Weidinger, and Veronika M. Reinisch

Abstract The pathogenetic mechanisms of Alzheimer's disease (AD) are only partly understood. There is no doubt that 'immunosenescence', the aging of the (healthy) immune system, leads to impaired immune function and that aging is the main risk factor for AD. Also beyond doubt is that neuroinflammation plays a key role in the pathophysiology of the disease. However, whether inflammation is an underlying cause or a resulting condition in AD remains unresolved. At higher ages, communication in the peripheral and central nervous system (CNS) immune systems, including both the initiation of the immune process and the downregulation of inflammation, is impaired; this impaired communication might be one of the main factors contributing to the immune pathology of AD. The innate and adaptive immune systems (T and B cells) have been shown to be upregulated in aging and AD. Mounting evidence indicates that microglia activation contributes to neuronal damage in neurodegenerative diseases, but beneficial aspects of microglia activation have also been identified. The purpose of this chapter is to highlight new insights into the detrimental and beneficial role of neuroinflammation in AD. In this regard, we discuss the limitations and advantages of the protective effects of non-steroidal anti-inflammatory drugs (NSAIDs) and anti-inflammatory treatment options and identify possible future implications for AD therapy that might result from this underlying neuroinflammation. A further focus is put on treatment with cyclo-oxygenase-1 and -2 (COX-1, COX-2) inhibitors and anti-Aβ antibodies.

Keywords Inflammation • Alzheimer • Microglia • COX inhibitors • Aβ • DNA vaccination

N. Müller (✉) • D.L. Krause • E. Weidinger • V.M. Reinisch
Department of Psychiatry and Psychotherapy, Ludwig-Maximilians University Munich, Nussbaumstr. 7, 80336 Munich, Germany
e-mail: Norbert.Mueller@med.uni-muenchen.de; daniela.krause@med.uni-muenchen.de; elif.weidinger@med.uni-muenchen.de; veronika.reinisch@med.uni-muenchen.de

M.J. Schwarz
Institut fuer Laboratoriumsmedizin, Klinikum der Universitaet Muenchen, Muenchen, Germany
e-mail: markus.schwarz@med.uni-muenchen.de

N. Müller et al. (eds.), *Immunology and Psychiatry*, Current Topics in Neurotoxicity 8, DOI 10.1007/978-3-319-13602-8_15

Introduction

Alzheimer's disease (AD) is the most common form of dementia, for one reason because people are living longer. The most important long-term aim is to diagnose AD at an earlier stage, so that treatment can start before most clinical symptoms are present (Sperling et al. 2011). Earlier diagnosis is possible with new diagnostic imaging, e.g. amyloid PET imaging, which can show the amyloid burden in the brain (Rosenberg et al. 2013). The pathophysiology of AD and a lot of factors that play a role in AD are still unclear, limiting the identification of effective disease-modifying therapies. The main neuropathological hallmarks of AD are peptide deposition (senile plaques), extracellular β-amyloid (Aβ), and intracellular neurofibrillary tangles containing hyperphosphorylated tau protein (Panza et al. 2012). Senile plaques are composed of a 4-kDa Aβ peptide derived from the cleavage of the amyloid precursor protein (APP) by two proteases, β- and γ-secretase. The Aβ peptide is thought to play a crucial role in the pathogenesis of AD. It adopts several neurotoxic forms, including soluble and highly insoluble oligomers and proteolysis-resistant fibrils. The deposition of Aβ is the result of an imbalance between Aβ production and clearance; this imbalance leads to chronic inflammation in the central nervous system (CNS) (Serpente et al. 2014). Apart from the disease's distinct pathological markers, its neurodegenerative conditions are characterized by chronic neuroinflammatory processes. However, those inflammatory markers are not exclusively associated with AD and aging is also associated with 'immunosenescence', the aging of the immune system. Accordingly, higher concentrations of serum markers related to inflammation have been described in elderly people. Increased pro-inflammatory immune markers, homocysteine, and cholesterol homeostasis are associated with cognitive functioning in the non-demented healthy aging population (Teunissen et al. 2003a). In AD pathology, these aging-related inflammatory processes are increased. The suggestion that inflammation may participate in AD was first put forward more than two decades ago.

After several clinical trials showed a beneficial effect of non-steroidal anti-inflammatory drugs (NSAIDs) on the occurrence and course of AD, the inflammatory hypothesis in AD gained a lot of attention. In regard to treatment and prevention of AD, several classes of medications have emerged on the market, which improve the cognitive symptoms of this disorder (e.g. the cholinesterase inhibitors). The relief that these drugs provide remains symptomatic, however, so the development of effective disease-modifying therapy remains a major goal for the future. Various substantial efforts have been made to identify potential strategies to ameliorate or prevent AD pathology, with data stemming from both basic research, animal studies and epidemiological studies. Because many investigators have concluded that neuroinflammation contributes to neuronal damage in the brain during AD (McGeer and McGeer 2010; Akiyama et al. 2000), the use of anti-inflammatory drugs as a possible treatment option has been widely investigated (in t' Veld et al. 2001; Szekely et al. 2008; Anthony et al. 2000). Anti-inflammatory therapy has therefore been credited as a strategy for reducing the risk or slowing the progression of AD.

However, the results of these studies remain inconsistent (Reines et al. 2004). Until now, many questions regarding the inflammatory response are still unresolved. Discussion continues whether neuroinflammation is an underlying cause or a consequence of AD (Krause and Müller 2010).

Impaired Cognition Is Associated with a Pro-inflammatory Immune State

Several studies have shown that an intact immune response, including intact T cell immunity, is a prerequisite for cognitive function. An elegant animal study found that T cell-deficient mice have impaired learning abilities; substitution of the T cells reversed the cognitive deficits in the animals (Kipnis et al. 2004; Ziv et al. 2006).

Aging of the T cellular immune response starts from about the age of 55, so that immunosenescence may also explain cognitive decline during aging. The 'Maastricht Aging Study' followed this approach. Nearly 100 healthy people, mean age 57 years, were followed up in a longitudinal analysis that included inflammatory markers and cognitive tests. The study found a significant, negative correlation between high levels of haptoglobin (an acute phase protein) and cognitive abilities, measured by the Stroop test and the auditory verbal learning test. High levels of C-reactive protein (CRP) also correlated negatively with the auditory verbal learning test after 3 and 6 years' follow-up. Lower cognitive abilities were associated with higher concentrations of CRP and haptoglobin (Teunissen et al. 2003b). A similarly designed occupational prospective cohort study in 4,200 healthy persons, the so-called Whitehall II study, combined measurements of CRP and interleukin-6 (IL-6) as inflammatory markers with cognitive tests at the 7- and 12-year follow-up. CRP and IL-6 were significantly associated with cognitive performance, in particular in men. Higher levels of pro-inflammatory markers during midlife correlated significantly with lower cognitive abilities and weakly with the decline of cognitive abilities (Gimeno et al. 2008).

Pro-inflammatory Cytokines and Memory

Although IL-6 enhancement shows no short-term effect on learning and memory in rats (Bianchi et al. 1997), over a longer period increased levels of pro-inflammatory cytokines have a negative effect on cognition in animal models. Interestingly, in animal experiments an increased secretion of IL-6 was shown to lead to deficits in learning and memory (Heyser et al. 1997), while IL-6 KO mice were less prone to forget learned skills and exhibited better cognitive performance than wild-type mice (Braida et al. 2004). Accordingly, the intraventricular administration of anti-IL-6 antibodies improved memory function (Balschun et al. 2004).

Similar effects were observed for the pro-inflammatory cytokine tumour necrosis factor α (TNF-α): adult mice with TNF-α over-expression within the CNS showed impaired learning (Fiore et al. 1996). An intraventricular injection of TNF-α before a training period impaired learning and memory in mice (Bjugstad et al. 1998). TNF-α injection into the hippocampus resulted in an impairment of working memory in mice (Matsumoto et al. 2002). Partly inconsistent results might be age dependent: TNF-α, which increases with age, is involved in age-dependent memory loss (Fiore et al. 2000).

On the other hand, the anti-inflammatory cytokine IL-4 may have pro-cognitive effects in the mouse model. For example, IL-4-deficient mice exhibited cognitive deficits, but the adoptive transfer of (IL-4–expressing) T cells into IL-4-deficient mice reversed the cognitive impairment (Derecki et al. 2010). T cells that during learning accumulate in the meninges of the animal expressed high levels of IL-4 (Derecki et al. 2010).

Is (Chronic) Infection the Cause of Inflammation-Related Cognitive Decline?

In the peripheral organs of the body, infections are the main cause for immune activation and a pro-inflammatory immune state. Interestingly, the influence of infectious pathogens on cognitive decline has rarely been studied. Only during the last few years, associated with the growing knowledge about the interaction of pathogens with the immune system, has research focussed more on the possible role of pathogens in CNS disorders, including psychiatric disorders and AD. Understanding the role of microbes and their interaction with the immune system—microbiomics—is one of the challenges of the near future. An interesting study, performed between 1993 and 2001, enrolled 3298 healthy multi-ethnic participants aged ≥40 years living in Manhattan, NY. The participants' cognitive state was assessed at baseline with the Mini Mental State Examination (MMSE) and afterwards by an annual Telephone Interview for Cognitive Status (TICS-M). An 'infections burden', which included the estimation of antibody titres against *Chlamydia pneumoniae* (IgA), *Helicobacter pylori* (IgG), cytomegalovirus (IgG) and herpes virus-1 (HSV-1) and -2 (HSV-2; IgG) was measured in 1,625 of these healthy persons. Interestingly, a higher infections burden index was associated with lower MMSE scores ($p \leq 0.0001$), even after adjusting for vascular risk factors and demographics at baseline ($p \leq 0.06$). During the course of the study, infections burden was significantly associated with a low cognitive status (telephone interviews), also after adjustment for interfering factors at baseline ($p \leq 0.0001$). The viral burden contributed most to the result, whereas the burden with *C. pneumoniae* or *H. pylori* (Katan et al. 2013; Swardfager and Black 2013) did not contribute.

Another study investigated the association between HSV-1 exposure and HSV-1 titres and cognition. It examined the role of HSV-1 in cognitive deficits in schizophrenia patients but found no specific role of HSV-1. However, it did find an association between HSV-1 exposure and reduced function in specific cognitive domains in both the schizophrenia patients and healthy controls (Thomas et al. 2013).

These significant relationships between infectious titres and cognition pinpoint the possible association between cognitive decline, a pro-inflammatory immune state and infections.

Microglia Cells as the Cellular Key of CNS Immunity

Inflammation in the brain is characterized by activation of glial cells (mainly microglia and astrocytes) and expression of key inflammatory mediators and neurotoxic free radicals. It has been suggested that neuroinflammation is associated with neurodegenerative disorders—both acute (e.g. stroke, injury) and chronic (e.g. multiple sclerosis, AD). Microglia cells play a crucial role in this context and therefore microglia and cytokines have been extensively studied in these conditions. In the CNS, microglia cells are the resident phagocytes of the innate immune system and microglia are found in a highly activated state in close anatomical proximity to senile plaques within the AD brain. In the activated state, they produce various pro-inflammatory cytokines and other immune mediators that create a neurotoxic milieu leading to disease progression (Akiyama et al. 2000; Wyss-Coray 2006).

As a first step in considering this issue, we focus on newer controversies in the field of microglia activation and its role in AD pathology. Consequently, we ask two questions: Are neuroinflammatory alterations neuroprotective or rather an underlying cause of AD? What strategies for future treatment options result from this underlying neuroinflammation?

Characteristics of Neuroinflammation in AD

The relevance of neuroinflammation for AD pathology has been established by multiple lines of direct and indirect evidence. One argument is that microglia activation is increased in regions associated with Aβ deposition (Edison et al. 2008). Upregulated inflammatory mechanisms co-localize in the AD brain with those regions that exhibit high levels of AD pathology (e.g. frontal and limbic cortex) and are minimal in brain regions with low AD pathologic susceptibility (e.g. cerebellum) (Rogers and Shen 2000).

As a second point, many of the inflammatory mechanisms that have been uncovered in the AD brain are established as being cytotoxic in the periphery of the body. Therefore it seems likely that they are also cytotoxic in the brain, an organ that is sensitive to inflammation (e.g. meningitis, oedema). However, inflammation in the brain is different from inflammation in the peripheral body. AD brains lack the classical hallmarks of inflammation such as neutrophil infiltration and perivascular mononuclear cuffing. As for other neurodegenerative diseases, a local inflammatory reaction is sustained by activated microglia and reactive astrocytes (Krause and Müller 2010), as indicated by the presence of antigens associated with microglia activation and inflammatory mediators, such as factors of the complement system, cytokines and free radicals (Perry et al. 2003).

Only modest elevations of inflammatory markers are found post-mortem in persons who had no clinical presentation of dementia but whose brains exhibit sufficient Aβ and neurofibrillary tangles to otherwise qualify for a diagnosis of AD. The level of inflammatory markers in these patients is significantly greater than in non-demented patients, but dramatically less than in AD patients (Lue et al. 1996). These findings further strengthen the hypothesis that inflammation is necessary for clinical symptoms of AD.

A huge variety of pro-inflammatory markers have been identified for AD but not for other forms of dementia. A relevant reduction of monocyte chemotactic protein-1 levels in the grey matter in dementia patients has been shown. Decreases of IL-6 and related markers of this pro-inflammatory cytokine system were observed in the brain and cerebrospinal fluid (CSF) of demented patients (Mulugeta et al. 2008; Stübner et al. 1999). It is discussed, however, whether this decrease is related to further psychopathological symptoms such as depression (Stübner et al. 1999). On the other hand, IL-6 also has neuroprotective properties, and decreased IL-6 might be associated with decreased neuroprotection (Wang et al. 2009).

There also is direct evidence of inflammatory toxicity in the AD brain. For instance, complement fixation and lysis of neurites could be demonstrated ultrastructurally in the cortex of AD patients, but only very weakly in the cortex of non-demented elderly (Webster et al. 1997).

Finally, many clinical and animal studies have strongly suggested that especially NSAIDs could be used as preventive or treatment strategies in AD. This aspect is further discussed in a later section of this chapter, where we focus on anti-inflammatory treatment.

Even though there are many indicators that neuroinflammation plays a key role in AD pathology, we still do not know which of these inflammatory activities causes disease progression. The question remains whether some of these processes help to fight the disease. The role of microglia seems important when addressing this question, because these cells are known to have neuroprotective and degenerative functions.

Are Activated Microglia Neuroprotective or Neurodegenerative in the Brain of AD Patients?

Microglia cells are one of three glial cell types in the CNS. Since the 1970s it has been widely recognized that microglia are immune effectors in the CNS which respond to pathological conditions and participate in the initiation and progression of neurological disorders (including AD) by releasing potentially cytotoxic molecules such as pro-inflammatory cytokines, reactive oxygen intermediates, proteinases and complement proteins (Block et al. 2007). This means that their phagocytic function can be beneficial while their inflammation-related functions might be detrimental.

Several studies give evidence for an increased number of morphologically reactive microglia in AD brains compared to non-demented individuals (Cras et al. 1990; Styren et al. 1990). The location of these reactive microglia has been identified directly around plaques (Perlmutter et al. 1999). This finding was verified in a recent imaging study that used a specific ligand for positron emission tomography (PET) and showed increased microglia activation in regions associated with amyloid deposition (Edison et al. 2008). The exact timing of this association could not yet be identified. Microgliosis might be an early component of the disease process and not necessarily dependent upon Aβ plaque interaction as a stimulus. What is known so far is that activation of microglia by Aβ fibrils is associated with a chemotactic response and extensive clustering of microglia around Aβ plaques in the AD brain (Lue et al. 2001). These findings indicate the prominent role of microglia cells in AD. Nonetheless it remains unclear whether their functions are beneficial or detrimental.

The following section explains the chequered role of activated microglia in AD pathology.

Neuroprotective Properties of Microglia in AD

Activated microglia cells may perhaps be beneficial in neurodegenerative diseases. Evidence for such a beneficial role is the fact that glutamate removal by microglia results in neuroprotection. Glutamate has been identified as a relevant neurotoxic substance that acts through *N*-methyl-D-aspartic acid (NMDA) receptors on neurons and can lead to increased neuronal cell death. Microglia cells can increase their capacity to take up glutamate upon stimulation with lipopolysaccaride (LPS) via a mechanism that is TNFα dependent (Krause and Müller 2010; Persson et al. 2005). This microglia function could be relevant because the NMDA receptor antagonist memantine has been shown to improve cognition, function (activities of daily living), agitation and delusions in AD patients (Francis 2009). Taken together, findings show that microglia cells are important for the control of glutamate levels and might therefore contribute to neuronal survival. There is also evidence that microglia are capable of secreting neurotrophic or neuron survival factors (e.g. nerve growth factor and neurotrophin 3) upon activation via inflammation or injury (Kim and de Vellis 2005).

Furthermore it has been suggested that newly in the CNS invaded activated macrophages show morphological and functional similarities to the intrinsic, yolk-sac derived microglia, which is important for Aβ elimination. Lysosomes from the macrophage cell line are more acidic than microglia lysosomes (Majumdar et al. 2007), which indicate that microglia derived from the periphery might be more efficient in eliminating Aβ than brain microglia. Furthermore, phagocytic activity of microglia is dampened by pro-inflammatory cytokines like TNF-α (Koenigsknecht-Talboo and Landreth 2005). These findings show that microglia cells that are committed to an inflammatory response may have a lower phagocytic capacity than newly recruited macrophage derived cells. It could be demonstrated in mouse models of AD that anti-inflammatory drugs like minocycline improve cognitive functions and reduce the activation of microglia cells but do not alter the deposition and distribution of Aβ plaques (Fan et al. 2007). Seabrook et al. (2006) showed an age-dependent effect of minocycline in APP transgenic mice: the drug increased the amyloid load in young animals, indicating a beneficial effect of microglia in clearing amyloid. Minocycline was investigated as a potential treatment not only for AD but also for schizophrenia; add-on therapy with minocycline appeared to be effective in improving cognitive performance in schizophrenia patients by reducing a broad range of psychotic symptoms (Chaves et al. 2009). An additional mechanism might help microglia cells with the elimination process: transforming growth factor-β 1 has been shown to promote microglia Aβ clearance and reduce plaque burden (Wyss-Coray et al. 2001), which could support the idea that microglia activation is useful in the clearance of Aβ.

A recent review explains that, when challenged, microglia may adapt to different stimulatory contexts and pass through a sequence of reactive profiles. This is in line with the finding that microglia are not just 'resting' but have active sensor and versatile functions (Krause and Müller 2010; Hanisch and Kettenmann 2007).

Are most microglia cells functions beneficial in AD? Several studies suggest an overbalance of the detrimental properties of microglia. This issue is discussed in the next section.

Neurodegenerative Aspects of Microglia

In order to address this question, it is important to focus on timing, i.e. one must investigate when microglia activity begins during the disease. An increase in microglia activation was observed in very early stages of AD that surprisingly disappeared over time, strengthening the assumption that microglia activation begins early in disease progression (Vehmas et al. 2003). This could be a hint that microglia initially try to eliminate Aβ but fail to do so over time, as the disease progresses, and consequently decrease their activity. Alternatively, the role of microglia in AD could be detrimental and they could initiate the underlying AD pathology. To further evaluate this issue, we need to take a closer look at the causes of microglia activation in AD. It thereby seems important to distinguish between acute and chronic stimulation of microglia cells. While an acute insult may trigger oxidative and nitrosative stress, it is typically short-lived and unlikely to be harmful to long-term neuronal survival. Therefore an acute neuroinflammatory response is believed to be generally beneficial to the CNS, since it tends to minimize further injury and contributes to repair of damaged tissue. The opposite is the case for chronic stimulation: chronic neuroinflammation is most often detrimental and damaging to nervous tissue. Thus, whether neuroinflammation has beneficial or harmful outcomes in the brain may depend critically on the duration of the inflammatory response. The progressive deposition of Aβ in AD might provide a chronic stimulus to microglia cells. Also, the chemotactic functions of Aβ to attract microglia might contribute further to the ongoing inflammatory process (Lue et al. 2001). The ratio of the pro-inflammatory cytokine IL-1β to the anti-inflammatory cytokine IL-10 is drastically elevated in the serum of AD patients, giving these patients a definite long-term pro-inflammatory profile (Remarque et al. 2001) and indicating a chronic neuroinflammatory state of the CNS. In addition, the accumulating loss of neurons that characterize AD further contributes to the generation of debris and keeps microglia activated indefinitely. These data indicate that in AD the inflammation might be rather chronic and therefore contribute to disease progression (Krause and Müller 2010).

The idea is also emerging that an inflamed CNS environment may influence the ability of microglia to contribute to plaque deposition rather than plaque removal (Koenigsknecht-Talboo and Landreth 2005). This idea strongly suggests that the microenvironment of the brain can influence whether microglia perform beneficial or deleterious functions in pathophysiological states, which means that microglia cells functionally adapt to their environment (Hanisch and Kettenmann 2007). Recent studies show that in response to certain environmental toxins and endogenous proteins, microglia can enter an over-activated state and release reactive oxygen species (ROS) that cause neurotoxicity (Innamorato et al. 2009). Over-activated microglia can be detected with imaging techniques, so that this knowledge offers an opportunity not only for early diagnosis but perhaps also for the development of targeted anti-inflammatory therapies that might diminish the progression of the disease (Block et al. 2007).

In addition, activated microglia release the excitotoxin quinolinic acid (QUIN) (Espey et al. 1997), and microglia activated by AD plaques produce an apparently novel amine that evokes fulminant excitotoxicity (Giulian et al. 1995). One interesting implication of an excitotoxic contribution to inflammatory mechanisms is the potential for limited damage to functional cellular compartments. Because excitatory amino acid receptors are restricted to synapses and dendrites, these subcellular compartments are preferentially vulnerable. As a result, microglia-produced excitotoxins may lead to cognitive impairment that is not necessarily correlated with neuronal cell loss (McGeer and McGeer 2010). However, activated microglia not only produce neurotoxic metabolites. Some of their products, like 3-hydroxyanthranilic acid (which, like QUIN, is one of the downstream products of tryptophan x exert antioxidant and anti-inflammatory functions (Leipnitz et al. 2007; Thomas et al. 1996).

Microglia-derived factors capable of inducing neuronal death are numerous and range from ROS to pro-inflammatory cytokines. TNF-α is one such factor that has received much attention because of its ability to promote progression in Parkinson's disease (PD) (McCoy et al. 2011), while TNF-receptor 1 knockout protects against AD- and PD-like diseases in mice (He et al. 2007; Sriram et al. 2002; Lucin and Wyss-Coray 2009).

Aging is the strongest risk factor for neurodegenerative diseases and, while aging is not considered a disease, it results in a significant increase in glial activation, complement factors, inflammatory mediators and brain atrophy (Lu et al. 2004; Streit et al. 2008; West et al. 1994). Microarrays of aged human and mouse brains extend these findings by showing that genes related to cellular stress and inflammation increase with age, while genes related to synaptic function/transport, growth factors and trophic support decrease (Lu et al. 2004; Lee et al. 2000).

Monocytes and Innate Immunity in Alzheimer's Disease

For a long time, monocytes/macrophages have been discussed to be one of the origins of microglia because of their morphologic, phenotypic and functional similarities. In the meantime it is clear that the source of microglia is the 'yolk sac', but monocytes/macrophages invading the brain have similar morphology and functions.

The effective amyloid clearance observed in models has recently been attributed to infiltrating monocytes or perivascular macrophages which seem to readily degrade amyloid (Hawkes and McLaurin 2009; Majumdar et al. 2008). This disparity may be explained by the fact that microglial lysosomes are less acidic than macrophage lysosomes, resulting in reduced activity of lysosomal enzymes (Majumdar et al. 2008). Although CNS-infiltrating monocytes appear beneficial for amyloid clearance, their role in human neurodegenerative diseases remains unclear (Lucin and Wyss-Coray 2009). A microarray study showed a widespread upregulation of genes reflecting activation of microglia and perivascular macrophages in the aging brain, coupled with downregulation of select factors (TOLLIP, fractalkine) that, when present, curtail microglial/macrophage activation. Notably, essentially all pathways of the innate immune system are upregulated in aging, including numerous complement components, genes involved in toll-like receptor signalling and inflammasome signalling, as well as genes coding for immunoglobulin (Fc) receptors and human leucocyte antigens I and II (Cribbs et al. 2012).

Another important factor for immune/CNS communication and the role of innate immunity in the CNS is cell migration from peripheral blood into the CNS. An overview of the components of the innate and adaptive immune systems is shown in Table 15.1. With the example of neutrophil leucocytes, the migration has been shown to possibly be impaired during aging—a defect that was associated with lowered expression of the intercellular adhesion molecule-1 (ICAM-1) during aging (Brubaker et al. 2013; Shaw et al. 2013). Lower levels of soluble ICAM-1 were previously described to be associated with another CNS disorder, schizophrenia (Schwarz et al. 2000). Lower expression of ICAM-1 might not only be associated with diminished chemotaxis, resulting in reduced migration to the sites of inflammation, but also with defective cellular egress from inflamed tissues (Shaw et al. 2013). Migration can also be facilitated or inhibited in vitro, depending on the presence of pro-inflammatory environmental stimuli, including IL-8 and TNF-α, and such age-associated and tissue-related factors probably contribute to the dysregulation of (neutrophil) activation and migration. Taken together, these results suggest that certain components of innate immunity, including neutrophils, monocytes and dendritic cells, have an impaired ability to traffic into and out of sites of infection.

Table 15.1 Components of the innate and adaptive immune systems

Component	Innate Immune System	Adaptive IS
Cellular	Monocyte, macrophage, granulocyte, NK-cell, γ/δ-cell	T- & B-cell
Humoral	Complement, APP, mannose binding lectin (MBL), neopterin	Antibody

Notably, defects in the migration of mouse dendritic cells and human monocyte-derived dendritic cells have also been reported (Agrawal et al. 2007; Zhao et al. 2011). Such alterations in cell trafficking with aging could affect the initiation and downregulation of innate immune responses at sites of infection.

The view that a decline of the immune response might be a pathogenetic factor in AD was discussed earlier when lower levels of pro-inflammatory cytokines, including the mainly monocyte-derived cytokine TNF-α, were described in the CSF of AD patients (Richartz et al. 2005).

The Possible Role of Immune-Driven Tryptophan/Kynurenine Metabolism in AD

The tryptophan/kynurenine metabolism—i.e. the degradation of tryptophan to its partly neuroprotective, partly neurotoxic metabolites such as kynurenic acid (KYNA), 3-hydroxykynurenine (3-HK) and QUIN—is driven by the enzyme indoleamine 2,3 dioxygenase (IDO). Immune mechanisms are key players in this system: IDO is activated by pro-inflammatory cytokines such as interferon-γ or IL-2. An immune activation is associated with an increased degradation of tryptophan and kynurenine. An interesting study investigated whether an imbalance between neurotoxic and neuroprotective kynurenine metabolites could be detected in patients with AD. The study measured serum levels of tryptophan, KYNA, 3-HK, picolinic acid and QUIN in patients with AD and found that the serum levels of 3-HK were markedly higher in AD patients than in the comparison groups ($p < 0.0001$), while serum levels of the other metabolites of the kynurenine pathway did not significantly differ between groups. 3-HK generates toxic free radicals, leading to neurodegeneration and apoptosis of neurons (Chiarugi et al. 2001; Okuda et al. 1998). In contrast to its downstream metabolites QUIN and picolinic acid, 3-HK can cross the blood–brain barrier via an active transport process. These data therefore indicate an enhanced availability of 3-HK in the brain of AD patients, which may be related to the previously reported higher production of QUIN in AD brains (Schwarz et al. 2013). Accordingly, an immunocytochemical study showed that both 3-HK and IDO—the first, rate-limiting enzyme in tryptophan/kynurenine metabolism—were increased in AD brains compared to controls (Bonda et al. 2010). The increased 3-HK concentration in the serum of AD patients encourages further studies in the search for a peripheral diagnostic biomarker for AD. In this study, the peripheral concentration of QUIN was not elevated, although studies suggest that QUIN—downstream from 3-HK and the product of 3-HK degradation—might also be involved in AD. QUIN facilitates the formation of amyloid plaques and tau phosphorylation, one of the major mechanisms in the generation of neurofibrillary tangles (Rahman et al. 2009), a key factor in AD. Moreover, QUIN is an NMDA receptor antagonist that affects glutamatergic neurotransmission. One study showed increased levels of QUIN in the plasma of AD (Gulaj et al. 2010), but did not find differences in 3-HK concentrations. An explanation could be the stage-dependent increase of QUIN in AD. Increased brain levels of QUIN were also found in AD (Guillemin et al. 2001, 2003), but primarily in early to moderate stages (Guillemin et al. 2007), while no elevated levels of 3-HK or QUIN were found in post-mortem brains of AD, which represent late stages of the disease (Baran et al. 1999).

The balance of the products that result from activated microglia is important for the inflammation process.

To sum up, the results from microglia studies provide clear indications for the important role of neuroinflammation in disease progression in AD. However, some parts of microglia activation might also be beneficial during the course of AD. As explained above, neuroinflammation is a critical event in AD. Consequently, anti-inflammatory therapy has been suggested to be beneficial in delaying the onset or slowing the progression of AD. Cyclooxygenase (COX) is a unique enzyme. First, it exhibits two catalytic activities, bis-oxygenase activity, which catalyses prostaglandin (PG) G_2 formation from arachidonic acid, and peroxidase activity, which reduces PG G_2 to PG H_2. The peroxidase activity also results in the production of free radicals, which are in part utilized by COX itself (Smith et al. 1996). Although NSAIDs may have other effects as well, their primary mechanism of action is generally assumed to be competitive inhibition of COX activity, which reduces the production of inflammatory prostaglandins from membrane-derived arachidonate. COX not only helps mediate production of prostaglandins and other inflammatory factors, but is itself upregulated by pro-inflammatory mediators (Krause and Müller 2010; Aisen and Davis 1997).

In AD, Aβ neurotoxicity may result from several mechanisms, most likely in combination. These mechanisms include oxidative damage, direct cytotoxicity, and induction of destructive inflammatory mechanisms; efforts have been directed at the control of each of these processes (Aisen and Davis 1997), as described below.

Possible Mechanisms of Action of NSAIDs in Alzheimer's Disease

The treatment of AD with NSAIDs is one of the most promising approaches. If NSAIDs are beneficial in AD, their presumed mechanism of action would be inhibition of COX expressed in the brain, where both COX-1 and COX-2 are expressed. COX-2 plays a unique role in the brain compared to the periphery: only in the brain it is expressed constitutively, whereas elsewhere its expression is activation-dependent. Although in vivo the majority of COX-2 appears to be made in neurons, COX-2 was also seen in rat astrocytes and microglia (Hirst et al. 1999). COX-inhibiting NSAIDs have been shown to reduce microglia activation after infusion of Aβ in rats (Hauss-Wegrzyniak et al. 1999). Neuronal stress, such as ischaemia and excitotoxicity, is associated with strong upregulation of neuronal COX-2 expression, suggesting that COX-2 is involved in neurotoxic mechanisms and may therefore represent a target for drug therapy in the treatment of AD (Planas et al. 1995; Tocco et al. 1997).

Several epidemiological studies provide the background for possible mechanisms of action of NSAIDs in AD. Most studies use COX-2 inhibitors, because neuronal COX-2 is upregulated in response to exposure to Aβ (Pasinetti and Aisen 1998) and focal increases in COX-2 have been shown in the region of amyloid plaques in double transgenic mice carrying genes that encode both mutant APP and mutant presenilin 1 (Matsuoka et al. 2001). Many studies seem to show that COX-2 inhibition confers neuroprotection (Hewett et al. 2000; Willard et al. 2000; Kunz and Oliw 2001; Araki et al. 2001). Some studies have revealed an upregulation of neuronal COX-2 in the brains of patients with AD (Yasojima et al. 1999; Ho et al. 2001), although this has not been a universal finding (Lukiw and Bazan 1997; Chang et al. 1996). Explanations for the variation in COX expression are the short half-life of COX-2 transcripts and individual variability of inflammatory-related processes.

COX-1 is also localized in microglia and is actively involved in brain injury induced by pro-inflammatory stimuli, including Aβ, LPS and interleukins. A study in 20-month-old triple transgenic AD mice showed that their memory function increased after treatment with COX-1 inhibitors and amyloid deposits and tau hyperphosphorylation in the hippocampus decreased (Choi et al. 2013). Trifusal, a platelet anti-aggregant and irreversible COX-1 inhibitor, was able to rescue cognitive deficits by reducing the dense-core amyloid plaque load, associated glial cell activation and pro-inflammatory cytokine levels in a transgenic mouse model (Coma et al. 2010). Unfortunately this process could be clearly demonstrated only in animal experiments.

Another possible principle for the action of NSAIDs comes from the finding that prostaglandin E_2 levels are elevated in patients with AD, especially in early stages of the disease (Montine et al. 1999). Therefore NSAIDs that block prostaglandin E_2 synthesis might be beneficial. This idea is further strengthened by glial culture studies indicating that prostaglandins, particularly prostaglandin E, alter the production of several inflammation-related molecules, including IL-6, chemokines and APP (Lee et al. 1999; Blom et al. 1997; Fiebich et al. 1997).

In addition to the more traditional inflammatory mechanisms associated with COX, unique functions of COX-mediated damage may also occur in the AD brain. For example, several of the prostanoid products of arachidonate metabolism potentiate glutamate excitotoxicity, and COX-2—overexpressing transgenic mice exhibit increased neuronal susceptibility to excitotoxic insult (Kelley et al. 1999).

Some of the previously mentioned studies of COX in ischaemia also suggest that intraneuronal COX-2 levels may contribute to neuronal death by production of free radicals (Pasinetti 1998). In addition, increased COX-2 levels in AD neurons may directly damage neurons or increase their vulnerability to other detrimental processes occurring in the AD brain (Pasinetti 1998). Thus, the actions of NSAIDs to inhibit COX-mediated production of apoptotic factors by neurons could be one of the mechanisms by which these drugs seem to exert beneficial effects in AD.

Another non-COX-dependent mechanism of NSAIDs is to attenuate inflammatory processes by directly activating the peroxisome proliferator-activated receptor gamma (PPARγ), a receptor and nuclear transcription factor (Jiang et al. 1998; Lehmann et al. 1997; Ricote et al. 1998). PPARγ is a member of the orphan nuclear receptor family. In cells of the monocytic lineage, including microglia, PPARγ acts to suppress the expression of a broad range of pro-inflammatory genes (Jiang et al. 1998; Ricote et al. 1998). Some NSAIDs act as PPARγ agonists, directly binding to it and initiating its transcriptional activity. Activation of PPARγ inhibits the Aβ-stimulated activation of microglia and monocytes and their secretion of pro-inflammatory and neurotoxic products. For example, PPARγ agonists act to inhibit the Aβ-stimulated expression of IL-6 and TNF-alpha (Combs et al. 1999), by microglia and monocytes, and to prevent Aβ-mediated conversion of microglia into an activated phenotype (Krause and Müller 2010; Combs et al. 2000).

A further underlying mechanism of AD pathology is oxidative stress (Ansari and Scheff 2010; Smith et al. 2010), possibly caused by ROS, which are known to be released from activated microglia cells. On the other hand, glia cells can also exhibit antioxidative functions by releasing hemeoxygenase-1 (HO-1) triggered by accumulation of 3-hydroxyanthranilic acid (3-HAA), a downstream product of tryptophan metabolism. The association of neuronal injury in AD and oxidative stress has been demonstrated by over-expression of immunoreactive HO-1 protein in neurons and astrocytes of the cerebral cortex and hippocampus. HO-1 was found to be co-localized to senile plaques, neurofibrillary tangles and corpora amylacea (Schipper et al. 2009). A moderate activation of heme catabolism is widely accepted to be neuroprotective and to contribute to degradation of neurotoxic protein aggregates. Regulatory interactions between HO-1 and COX pathways have also been reported (Alcaraz et al. 2003). However, experimental observations indicate that the extent of HO-1 induction may be critical, because excessive heme degradation may result in toxic levels of carbon monoxide, bilirubin and iron. Pharmacological modulation of HO-1 levels in the brain shows promising results in models of AD and PD (Cuadrado and Rojo 2008).

A further aspect of ROS includes activation of COX-1 and -2, which are blocked by NSAIDs. Daily doses of NSAIDs have been shown to increase circulating levels of antioxidants (Kimura 1997). A study in a rat model of AD suggested that treatment with a COX-2 inhibitor reduces oxidative stress and might therefore be beneficial for the course of AD (Nivsarkar et al. 2008).

Another possible mechanism of effect of NSAIDs has been suggested to be a direct effect on amyloid pathology in the brain by reducing Aβ-42 peptide levels via gamma-secretase activity independently of COX activity (Guardia-Laguarta et al. 2010). Weggen et al. (2001) reported that the NSAIDs ibuprofen, indomethacin and sulindac sulphide preferentially decrease the highly amyloidogenic Aβ-42 peptide produced from a variety of cultured cells by as much as 80 %. However, the lowering effect on Aβ-42 could not be shown for some NSAIDs; instead, an increase in Aβ-42 levels was observed (Kukar et al. 2005). The underlying mechanism of reduction of Aβ-42 by NSAIDs was clarified by Lleo et al. (2004), who demonstrated that Aβ-42—lowering NSAIDs specifically affect the proximity between APP and presenilin 1 and alter the presenelin 1 confirmation in vivo and in vitro, suggesting a novel allosteric mechanism of action.

Anti-inflammatory Treatment Studies in AD

In recent years it has become widely accepted that inflammatory processes are an underlying condition of AD. As a result, a number of clinical trials have investigated different anti-inflammatory treatment regimens. In the following paragraph, we summarize the most import findings in regard first to mainly COX-2 dominant inhibitors and second to COX-1 inhibitors.

COX-1 and COX-2 Inhibitors

A prospective cohort study with 6,989 participants showed that long-term use of NSAIDs protects against AD but not against vascular dementia (in t' Veld et al. 2001). More recently, Szekely et al. provided very similar findings and concluded that NSAID use reduced the risk of AD more than the risk of vascular dementia but mainly in those individuals with an apolipoprotein E (APO) epsilon 4 allele. This study was performed in over 3,000 subjects aged 65 years and older (Szekely et al. 2008). Not only were selective COX-2 inhibitors shown to be associated with a decreased risk of AD, but a reduced occurrence of AD could also be demonstrated with the mixed COX-1/COX-2 inhibitor aspirin (Anthony et al. 2000). A meta-analysis of 17 epidemiological studies yielded strong, generally consistent statistical evidence that NSAID and steroid use is associated with a reduced risk of AD (McGeer et al. 1996). Vlad et al. investigated 49,349 patients with AD and 196,850 controls and showed that long-term (>5 years) use of NSAIDs is protective against AD. These findings were clearest for ibuprofen, but did not appear for other NSAIDs (Vlad et al. 2008).

Naproxen, which is slightly more selective for COX 1 than COX 2, reduced the risk of developing AD in 117 patients with mild cognitive impairment (MCI) from whom CSF was collected 21–41 months after treatment was terminated. The tau-to-Aβ42 ratio was reduced by more than 40 % in the group treated with naproxen (Breitner et al. 2011). Another NSAID with preferential COX 1 selectivity, indomethacin, also reduced amyloid burden in transgenic mice (Jantzen et al. 2002).

To conclude, at least ten studies show effects of NSAIDs on amyloid burden and inflammation in mice.

In humans, not all studies showed a positive outcome for COX inhibitors. For example, a one-year randomized controlled study failed to find an advantage of the selective COX-2 inhibitor rofecoxib over placebo. The authors argued that their results could indicate that the disease process was too advanced to be modified, because the goal of the study was to slow the progression of dementia in patients with established AD (Reines et al. 2004). No beneficial effect on the occurrence of AD could be demonstrated for another COX-2 inhibitor, celecoxib, in a group aged over 70 years (Martin et al. 2008). Wolfson et al. (2002) looked retrospectively at a case–control population and found no support for a beneficial effect of NSAIDs in AD. However, this negative result may have been caused by an insufficient period of data collection before disease onset.

Passive and Active Anti-Aβ Immunotherapies

In the last few years, most of the pharmaceutical industry's efforts have been directed towards preventing the production and accumulation of Aβ. The most revolutionary development is the removal of brain β-amyloid via anti-Aβ antibodies. Both passive and active anti-Aβ immunotherapies can clear Aβ deposits from the brain of AD patients. AN1792 was used in AD patients with some indications of clinical efficacy but caused aseptic meningoencephalitis in about 6 % of patients and was therefore abandoned. The next generation of active and passive vaccines has been developed in the past few years and is currently under clinical investigation. The aim of these substances is to clear Aβ deposits from the brain and to stop the progression of AD.

Bapineuzumab, composed of humanized anti-Aβ monoclonal antibodies, is the most advanced substance. It has been tested in two phase II trials and reduced Aβ burden in the brain of AD patients. Some patients, especially apolipoprotein E4 carriers, developed a vasogenic oedema, so that the clinical use of bapineuzumab is limited, especially in higher doses (2 mg/kg). The proposed remedy is to treat AD patients, particularly APOE4 carriers, with lower doses. A large phase III trial is currently being conducted with bapineuzumab which will tell us if passive anti-Aβ immunization is able to reduce progression of the disease (Panza et al. 2011). Of course improvements in vaccine design are needed to improve the safety and efficacy of anti-Aβ immunotherapy. Unfortunately at this point we cannot definitely identify individuals in the preclinical stages of AD, therefore passive immunotherapy is indicated only in patients who have been diagnosed with AD, i.e. who have clinical symptoms. At that point, such individuals have accumulated substantial neuropathology in affected regions of the brain (Cribbs 2010).

Perhaps amyloid PET imaging combined with genetic markers and last but not least clinical symptoms such as MCI will potentially be able to detect individuals with a higher risk for developing AD (Fleisher et al. 2012). The development of valid biomarkers for AD should be the aim of research in the next few years.

Conclusion

Neuroinflammation plays a key role in the pathophysiology of AD. Mechanisms that parallel those encountered in localized peripheral inflammatory responses are identified, along with detailed pathways for how the mechanisms interact. On balance, it is likely that AD neuroinflammation exacerbates AD pathogenesis.

A general treatment principle in neurology and psychiatry, i.e. that as early an intervention as possible leads to the best outcome, seems to be especially true for AD, but appropriate biomarkers such as genetic risk factors or imaging techniques are currently lacking. This problem needs to be solved in the next few years. Until then, we will be unable to start treatment early enough.

Many lines of evidence show that Aβ-induced neuroinflammation is an early event in the neurodegeneration of AD (Craft et al. 2006), because increases in microglia activation were observed in very early stages of AD that disappeared over the course of the disease (Vehmas et al. 2003). The fact that neuroinflammation occurs very early in AD could explain why anti-inflammatory treatment seems to be most efficient as a preventive or early treatment. One reason why an early use of NSAIDs is superior to a late one is that COX expression in the brain decreases over time in AD brains (Yermakova and O'Banion 2001). Another reason is that CSF PG E_2 levels in patients with AD are high when their short-term memory scores are just below those of controls, but low in later stages of the disease. These findings further support the hypothesis that inflammatory processes predominate early in Alzheimer's disease (Combrinck et al. 2006) and therefore require early intervention with anti-inflammatory treatment.

The change in inflammation processes over the course of AD could explain the failure of some prospective clinical trials of selective COX-2 inhibitors, i.e. treatment onset was too late. However, the failure could also be due to drug selection (COX-1 and COX-2 have different effects) and dose and duration of treatment. Drug selection in particular seems essential because some NSAIDs have recently been shown to increase Aβ-42 levels. It also has to be noted that NSAIDs may exert their protective effects via non-COX-inhibitory mechanisms, such as lowering of Aβ levels and activation of the peroxisome proliferator-activated receptor-[gamma] (Aisen 2002). These non-COX-dependent mechanisms might be differentially distributed among different COX inhibitors.

In summary, harmful inflammatory processes seem to dominate AD pathology, but inflammatory subsets also have some beneficial functions. If AD neuroinflammation is approached with realistic expectations and rational drug design, AD patients could significantly benefit from anti-inflammatory treatment, especially with NSAIDs. Another important aspect of treatment could be to utilize not only the efficient treatment properties of NSAIDs in early AD, but also to make use of the neuroprotective aspects of neuroinflammation by applying a combination therapy that maximizes the potential of glial activation. This would include treatment with NSAIDs and drugs that enforce anti-inflammatory and antioxidative properties (e.g. with 3-HAA and HO-1 enhancement).

The most promising therapeutic pathway seems to be the passive and active immunization against Aβ. In the next few years we should be able to detect AD much earlier so that ideally we could start treatment, e.g. by immunization, before clinical symptoms appear.

Acknowledgement Part of this review has been published before (Reinisch et al. 2014).

References

Agrawal A, Agrawal S, Cao JN, Su H, Osann K, Gupta S. Altered innate immune functioning of dendritic cells in elderly humans: a role of phosphoinositide 3-kinase-signaling pathway. J Immunol. 2007;178(11):6912–22.

Aisen PS. The potential of anti-inflammatory drugs for the treatment of Alzheimer's disease. Lancet Neurol. 2002;1(5):279–84.

Aisen PS, Davis KL. The search for disease-modifying treatment for Alzheimer's disease. Neurology. 1997;48(5 Suppl 6):S35–41.
Akiyama H, Barger S, Barnum S, Bradt B, Bauer J, Cole GM, et al. Inflammation and Alzheimer's disease. Neurobiol Aging. 2000;21(3):383–421.
Alcaraz MJ, Fernandez P, Guillen MI. Anti-inflammatory actions of the heme oxygenase-1 pathway. Curr Pharm Des. 2003;9(30):2541–51.
Ansari MA, Scheff SW. Oxidative stress in the progression of Alzheimer disease in the frontal cortex. J Neuropathol Exp Neurol. 2010;69(2):155–67.
Anthony JC, Breitner JC, Zandi PP, Meyer MR, Jurasova I, Norton MC, et al. Reduced prevalence of AD in users of NSAIDs and H2 receptor antagonists: the Cache County study. Neurology. 2000;54(11):2066–71.
Araki E, Forster C, Dubinsky JM, Ross ME, Iadecola C. Cyclooxygenase-2 inhibitor ns-398 protects neuronal cultures from lipopolysaccharide-induced neurotoxicity. Stroke. 2001; 32(10):2370–5.
Balschun D, Wetzel W, Del Ray A, Pitossi F, Schneider H, Zuschratter W, et al. Interleukin-6: a cytokine to forget. FASEB J. 2004;18(14):1788–90.
Baran H, Jellinger K, Deecke L. Kynurenine metabolism in Alzheimer's disease. J Neural Transm. 1999;106(2):165–81.
Bianchi M, Ferrario P, Clavenna A, Panerai AE. Interleukin-6 affects scopolamine-induced amnesia, but not brain amino acid levels in mice. Neuroreport. 1997;8(7):1775–8.
Bjugstad KB, Flitter WD, Garland WA, Su GC, Arendash GW. Preventive actions of a synthetic antioxidant in a novel animal model of AIDS dementia. Brain Res. 1998;795(1–2):349–57.
Block ML, Zecca L, Hong JS. Microglia-mediated neurotoxicity: uncovering the molecular mechanisms. Nat Rev Neurosci. 2007;8(1):57–69.
Blom MA, van Twillert MG, de Vries SC, Engels F, Finch CE, Veerhuis R, et al. NSAIDS inhibit the IL-1 beta-induced IL-6 release from human post-mortem astrocytes: the involvement of prostaglandin E2. Brain Res. 1997;777(1–2):210–8.
Bonda DJ, Mailankot M, Stone JG, Garrett MR, Staniszewska M, Castellani RJ, et al. Indoleamine 2,3-dioxygenase and 3-hydroxykynurenine modifications are found in the neuropathology of Alzheimer's disease. Redox Rep. 2010;15(4):161–8.
Braida D, Sacerdote P, Panerai AE, Bianchi M, Aloisi AM, Iosue S, et al. Cognitive function in young and adult IL (interleukin)-6 deficient mice. Behav Brain Res. 2004;153(2):423–9.
Breitner JC, Baker LD, Montine TJ, Meinert CL, Lyketsos CG, Ashe KH, et al. Extended results of the Alzheimer's disease anti-inflammatory prevention trial. Alzheimers Dement. 2011;7(4): 402–11.
Brubaker AL, Rendon JL, Ramirez L, Choudhry MA, Kovacs EJ. Reduced neutrophil chemotaxis and infiltration contributes to delayed resolution of cutaneous wound infection with advanced age. J Immunol. 2013;190(4):1746–57.
Chang JW, Coleman PD, O'Banion MK. Prostaglandin G/H synthase-2 (cyclooxygenase-2) mRNA expression is decreased in Alzheimer's disease. Neurobiol Aging. 1996;17(5):801–8.
Chaves C, Marque CR, Trzesniak C, Machado de Sousa JP, Zuardi AW, Crippa JA, et al. Glutamate-N-methyl-D-aspartate receptor modulation and minocycline for the treatment of patients with schizophrenia: an update. Braz J Med Biol Res. 2009;42(11):1002–14.
Chiarugi A, Meli E, Moroni F. Similarities and differences in the neuronal death processes activated by 3OH-kynurenine and quinolinic acid. J Neurochem. 2001;77(5):1310–8.
Choi SH, Aid S, Caracciolo L, Minami SS, Niikura T, Matsuoka Y, et al. Cyclooxygenase-1 inhibition reduces amyloid pathology and improves memory deficits in a mouse model of Alzheimer's disease. J Neurochem. 2013;124(1):59–68.
Coma M, Sereno L, Da Rocha-Souto B, Scotton TC, Espana J, Sanchez MB, et al. Triflusal reduces dense-core plaque load, associated axonal alterations and inflammatory changes, and rescues cognition in a transgenic mouse model of Alzheimer's disease. Neurobiol Dis. 2010;38(3): 482–91.
Combrinck M, Williams J, De Berardinis MA, Warden D, Puopolo M, Smith AD, et al. Levels of CSF prostaglandin E2, cognitive decline, and survival in Alzheimer's disease. J Neurol Neurosurg Psychiatry. 2006;77(1):85–8.

Combs CK, Johnson DE, Cannady SB, Lehman TM, Landreth GE. Identification of microglial signal transduction pathways mediating a neurotoxic response to amyloidogenic fragments of beta-amyloid and prion proteins. J Neurosci. 1999;19(3):928–39.

Combs CK, Johnson DE, Karlo JC, Cannady SB, Landreth GE. Inflammatory mechanisms in Alzheimer's disease: inhibition of beta-amyloid-stimulated proinflammatory responses and neurotoxicity by PPARgamma agonists. J Neurosci. 2000;20(2):558–67.

Craft JM, Watterson DM, Van Eldik LJ. Human amyloid beta-induced neuroinflammation is an early event in neurodegeneration. Glia. 2006;53(5):484–90.

Cras P, Kawai M, Siedlak S, Mulvihill P, Gambetti P, Lowery D, et al. Neuronal and microglial involvement in beta-amyloid protein deposition in Alzheimer's disease. Am J Pathol. 1990;137(2):241–6.

Cribbs DH. Abeta DNA, vaccination for Alzheimer's disease: focus on disease prevention. CNS Neurol Disord Drug Targets. 2010;9(2):207–16.

Cribbs DH, Berchtold NC, Perreau V, Coleman PD, Rogers J, Tenner AJ, et al. Extensive innate immune gene activation accompanies brain aging, increasing vulnerability to cognitive decline and neurodegeneration: a microarray study. J Neuroinflammation. 2012;9:179. doi:10.1186/1742-2094-9-179.

Cuadrado A, Rojo AI. Heme oxygenase-1 as a therapeutic target in neurodegenerative diseases and brain infections. Curr Pharm Des. 2008;14(5):429–42.

Derecki NC, Cardani AN, Yang CH, Quinnies KM, Crihfield A, Lynch KR, et al. Regulation of learning and memory by meningeal immunity: a key role for IL-4. J Exp Med. 2010; 207(5):1067–80.

Edison P, Archer HA, Gerhard A, Hinz R, Pavese N, Turkheimer FE, et al. Microglia, amyloid, and cognition in Alzheimer's disease: An [11C](R)PK11195-PET and [11C]PIB-PET study. Neurobiol Dis. 2008;32(3):412–9.

Espey MG, Chernyshev ON, Reinhard Jr JF, Namboodiri MA, Colton CA. Activated human microglia produce the excitotoxin quinolinic acid. Neuroreport. 1997;8(2):431–4.

Fan R, Xu F, Previti ML, Davis J, Grande AM, Robinson JK, et al. Minocycline reduces microglial activation and improves behavioral deficits in a transgenic model of cerebral microvascular amyloid. J Neurosci. 2007;27(12):3057–63.

Fiebich BL, Hull M, Lieb K, Gyufko K, Berger M, Bauer J. Prostaglandin E2 induces interleukin-6 synthesis in human astrocytoma cells. J Neurochem. 1997;68(2):704–9.

Fiore M, Probert L, Kollias G, Akassoglou K, Alleva E, Aloe L. Neurobehavioral alterations in developing transgenic mice expressing TNF-alpha in the brain. Brain Behav Immun. 1996; 10(2):126–38.

Fiore M, Angelucci F, Alleva E, Branchi I, Probert L, Aloe L. Learning performances, brain NGF distribution and NPY levels in transgenic mice expressing TNF-alpha. Behav Brain Res. 2000;112(1–2):165–75.

Fleisher AS, Chen K, Quiroz YT, Jakimovich LJ, Gomez MG, Langois CM, et al. Florbetapir PET analysis of amyloid-beta deposition in the presenilin 1 E280A autosomal dominant Alzheimer's disease kindred: a cross-sectional study. Lancet Neurol. 2012;11(12):1057–65.

Francis PT. Altered glutamate neurotransmission and behaviour in dementia: evidence from studies of memantine. Curr Mol Pharmacol. 2009;2(1):77–82.

Gimeno D, Marmot MG, Singh-Manoux A. Inflammatory markers and cognitive function in middle-aged adults: the Whitehall II study. Psychoneuroendocrinology. 2008;33(10):1322–34.

Giulian D, Haverkamp LJ, Li J, Karshin WL, Yu J, Tom D, et al. Senile plaques stimulate microglia to release a neurotoxin found in Alzheimer brain. Neurochem Int. 1995;27(1):119–37.

Guardia-Laguarta C, Pera M, Lleo A. gamma-Secretase as a therapeutic target in Alzheimer's disease. Curr Drug Targets. 2010;11(4):506–17.

Guillemin GJ, Kerr SJ, Smythe GA, Smith DG, Kapoor V, Armati PJ, et al. Kynurenine pathway metabolism in human astrocytes: a paradox for neuronal protection. J Neurochem. 2001;78(4):842–53.

Guillemin GJ, Smythe GA, Veas LA, Takikawa O, Brew BJ. A beta 1-42 induces production of quinolinic acid by human macrophages and microglia. Neuroreport. 2003;14(18):2311–5.

Guillemin GJ, Brew BJ, Noonan CE, Knight TG, Smythe GA, Cullen KM. Mass spectrometric detection of quinolinic acid in microdissected Alzheimer's disease plaques. In: Takai K, editor. 2007. p. 404–8.

Gulaj E, Pawlak K, Bien B, Pawlak D. Kynurenine and its metabolites in Alzheimer's disease patients. Adv Med Sci. 2010;55(2):204–11.
Hanisch UK, Kettenmann H. Microglia: active sensor and versatile effector cells in the normal and pathologic brain. Nat Neurosci. 2007;10(11):1387–94.
Hauss-Wegrzyniak B, Vraniak P, Wenk GL. The effects of a novel NSAID on chronic neuroinflammation are age dependent. Neurobiol Aging. 1999;20(3):305–13.
Hawkes CA, McLaurin J. Selective targeting of perivascular macrophages for clearance of beta-amyloid in cerebral amyloid angiopathy. Proc Natl Acad Sci U S A. 2009;106(4):1261–6.
He P, Zhong Z, Lindholm K, Berning L, Lee W, Lemere C, et al. Deletion of tumor necrosis factor death receptor inhibits amyloid beta generation and prevents learning and memory deficits in Alzheimer's mice. J Cell Biol. 2007;178(5):829–41.
Hewett SJ, Uliasz TF, Vidwans AS, Hewett JA. Cyclooxygenase-2 contributes to *N*-methyl-D-ASPARTATE-mediated neuronal cell death in primary cortical cell culture. J Pharmacol Exp Ther. 2000;293(2):417–25.
Heyser CJ, Masliah E, Samimi A, Campbell IL, Gold LH. Progressive decline in avoidance learning paralleled by inflammatory neurodegeneration in transgenic mice expressing interleukin 6 in the brain. Proc Natl Acad Sci U S A. 1997;94(4):1500–5.
Hirst WD, Young KA, Newton R, Allport VC, Marriott DR, Wilkin GP. Expression of COX-2 by normal and reactive astrocytes in the adult rat central nervous system. Mol Cell Neurosci. 1999;13(1):57–68.
Ho L, Purohit D, Haroutunian V, Luterman JD, Willis F, Naslund J, et al. Neuronal cyclooxygenase 2 expression in the hippocampal formation as a function of the clinical progression of Alzheimer disease. Arch Neurol. 2001;58(3):487–92.
in t' Veld BA, Ruitenberg A, Hofman A, Launer LJ, van Duijn CM, Stijnen T, et al. Nonsteroidal antiinflammatory drugs and the risk of Alzheimer's disease. N Engl J Med. 2001;345(21):1515–21.
Innamorato NG, Lastres-Becker I, Cuadrado A. Role of microglial redox balance in modulation of neuroinflammation. Curr Opin Neurol. 2009;22(3):308–14.
Jantzen PT, Connor KE, DiCarlo G, Wenk GL, Wallace JL, Rojiani AM, et al. Microglial activation and beta -amyloid deposit reduction caused by a nitric oxide-releasing nonsteroidal anti-inflammatory drug in amyloid precursor protein plus presenilin-1 transgenic mice. J Neurosci. 2002;22(6):2246–54.
Jiang C, Ting AT, Seed B. PPAR-gamma agonists inhibit production of monocyte inflammatory cytokines. Nature. 1998;391(6662):82–6.
Katan M, Moon YP, Paik MC, Sacco RL, Wright CB, Elkind MS. Infectious burden and cognitive function: the Northern Manhattan Study. Neurology. 2013;80(13):1209–15.
Kelley KA, Ho L, Winger D, Freire-Moar J, Borelli CB, Aisen PS, et al. Potentiation of excitotoxicity in transgenic mice overexpressing neuronal cyclooxygenase-2. Am J Pathol. 1999;155(3):995–1004.
Kim SU, de Vellis J. Microglia in health and disease. J Neurosci Res. 2005;81(3):302–13.
Kimura K. Mechanisms of active oxygen species reduction by non-steroidal anti-inflammatory drugs. Int J Biochem Cell Biol. 1997;29(3):437–46.
Kipnis J, Cohen H, Cardon M, Ziv Y, Schwartz M. T cell deficiency leads to cognitive dysfunction: implications for therapeutic vaccination for schizophrenia and other psychiatric conditions. Proc Natl Acad Sci U S A. 2004;101(21):8180–5.
Koenigsknecht-Talboo J, Landreth GE. Microglial phagocytosis induced by fibrillar beta-amyloid and IgGs are differentially regulated by proinflammatory cytokines. J Neurosci. 2005;25(36):8240–9.
Krause DL, Muller N. Neuroinflammation, microglia and implications for anti-inflammatory treatment in Alzheimer's disease. Int J Alzheimer Dis. 2010;2010. pii: 732806. doi:10.4061/2010/732806.
Kukar T, Murphy MP, Eriksen JL, Sagi SA, Weggen S, Smith TE, et al. Diverse compounds mimic Alzheimer disease-causing mutations by augmenting Abeta42 production. Nat Med. 2005; 11(5):545–50.
Kunz T, Oliw EH. The selective cyclooxygenase-2 inhibitor rofecoxib reduces kainate-induced cell death in the rat hippocampus. Eur J Neurosci. 2001;13(3):569–75.
Lee RK, Knapp S, Wurtman RJ. Prostaglandin E2 stimulates amyloid precursor protein gene expression: inhibition by immunosuppressants. J Neurosci. 1999;19(3):940–7.

Lee CK, Weindruch R, Prolla TA. Gene-expression profile of the ageing brain in mice. Nat Genet. 2000;25(3):294–7.

Lehmann JM, Lenhard JM, Oliver BB, Ringold GM, Kliewer SA. Peroxisome proliferator-activated receptors alpha and gamma are activated by indomethacin and other non-steroidal anti-inflammatory drugs. J Biol Chem. 1997;272(6):3406–10.

Leipnitz G, Schumacher C, Dalcin KB, Scussiato K, Solano A, Funchal C, et al. In vitro evidence for an antioxidant role of 3-hydroxykynurenine and 3-hydroxyanthranilic acid in the brain. Neurochem Int. 2007;50(1):83–94.

Lleo A, Berezovska O, Herl L, Raju S, Deng A, Bacskai BJ, et al. Nonsteroidal anti-inflammatory drugs lower Abeta42 and change presenilin 1 conformation. Nat Med. 2004;10(10):1065–6.

Lu T, Pan Y, Kao SY, Li C, Kohane I, Chan J, et al. Gene regulation and DNA damage in the ageing human brain. Nature. 2004;429(6994):883–91.

Lucin KM, Wyss-Coray T. Immune activation in brain aging and neurodegeneration: too much or too little? Neuron. 2009;64(1):110–22.

Lue LF, Brachova L, Civin WH, Rogers J. Inflammation, a beta deposition, and neurofibrillary tangle formation as correlates of Alzheimer's disease neurodegeneration. J Neuropathol Exp Neurol. 1996;55(10):1083–8.

Lue LF, Rydel R, Brigham EF, Yang LB, Hampel H, Murphy Jr GM, et al. Inflammatory repertoire of Alzheimer's disease and nondemented elderly microglia in vitro. Glia. 2001;35(1):72–9.

Lukiw WJ, Bazan NG. Cyclooxygenase 2 RNA message abundance, stability, and hypervariability in sporadic Alzheimer neocortex. J Neurosci Res. 1997;50(6):937–45.

Majumdar A, Cruz D, Asamoah N, Buxbaum A, Sohar I, Lobel P, et al. Activation of microglia acidifies lysosomes and leads to degradation of Alzheimer amyloid fibrils. Mol Biol Cell. 2007;18(4):1490–6.

Majumdar A, Chung H, Dolios G, Wang R, Asamoah N, Lobel P, et al. Degradation of fibrillar forms of Alzheimer's amyloid beta-peptide by macrophages. Neurobiol Aging. 2008;29(5): 707–15.

Martin BK, Szekely C, Brandt J, Piantadosi S, Breitner JC, Craft S, et al. Cognitive function over time in the Alzheimer's disease anti-inflammatory prevention trial (ADAPT): results of a randomized, controlled trial of naproxen and celecoxib. Arch Neurol. 2008;65(7): 896–905.

Matsumoto Y, Watanabe S, Suh YH, Yamamoto T. Effects of intrahippocampal CT105, a carboxyl terminal fragment of beta-amyloid precursor protein, alone/with inflammatory cytokines on working memory in rats. J Neurochem. 2002;82(2):234–9.

Matsuoka Y, Picciano M, Malester B, LaFrancois J, Zehr C, Daeschner JM, et al. Inflammatory responses to amyloidosis in a transgenic mouse model of Alzheimer's disease. Am J Pathol. 2001;158(4):1345–54.

McCoy MK, Ruhn KA, Blesch A, Tansey MG. TNF: a key neuroinflammatory mediator of neurotoxicity and neurodegeneration in models of Parkinson's disease. Adv Exp Med Biol. 2011;691:539–40. doi:10.1007/978-1-4419-6612-4_56.

McGeer EG, McGeer PL. Neuroinflammation in Alzheimer's disease and mild cognitive impairment: a field in its infancy. J Alzheimers Dis. 2010;19(1):355–61.

McGeer PL, Schulzer M, McGeer EG. Arthritis and anti-inflammatory agents as possible protective factors for Alzheimer's disease: a review of 17 epidemiologic studies. Neurology. 1996;47(2):425–32.

Montine TJ, Sidell KR, Crews BC, Markesbery WR, Marnett LJ, Roberts LJ, et al. Elevated CSF prostaglandin E2 levels in patients with probable AD. Neurology. 1999;53(7):1495–8.

Mulugeta E, Molina-Holgado F, Elliott MS, Hortobagyi T, Perry R, Kalaria RN, et al. Inflammatory mediators in the frontal lobe of patients with mixed and vascular dementia. Dement Geriatr Cogn Disord. 2008;25(3):278–86.

Nivsarkar M, Banerjee A, Padh H. Cyclooxygenase inhibitors: a novel direction for Alzheimer's management. Pharmacol Rep. 2008;60(5):692–8.

Okuda S, Nishiyama N, Saito H, Katsuki H. 3-Hydroxykynurenine, an endogenous oxidative stress generator, causes neuronal cell death with apoptotic features and region selectivity. J Neurochem. 1998;70(1):299–307.

Panza F, Frisardi V, Imbimbo BP, Seripa D, Paris F, Santamato A, et al. Anti-beta-amyloid immunotherapy for Alzheimer's disease: focus on bapineuzumab. Curr Alzheimer Res. 2011;8(8):808–17.

Panza F, Frisardi V, Solfrizzi V, Imbimbo BP, Logroscino G, Santamato A, et al. Immunotherapy for Alzheimer's disease: from anti-beta-amyloid to tau-based immunization strategies. Immunotherapy. 2012;4(2):213–38.

Pasinetti GM. Cyclooxygenase and inflammation in Alzheimer's disease: experimental approaches and clinical interventions. J Neurosci Res. 1998;54(1):1–6.

Pasinetti GM, Aisen PS. Cyclooxygenase-2 expression is increased in frontal cortex of Alzheimer's disease brain. Neuroscience. 1998;87(2):319–24.

Perlmutter SJ, Leitman SF, Garvey MA, Hamburger S, Feldman E, Leonard HL, et al. Therapeutic plasma exchange and intravenous immunoglobulin for obsessive-compulsive disorder and tic disorders in childhood. Lancet. 1999;354(9185):1153–8.

Perry VH, Newman TA, Cunningham C. The impact of systemic infection on the progression of neurodegenerative disease. Nat Rev Neurosci. 2003;4(2):103–12.

Persson M, Brantefjord M, Hansson E, Ronnback L. Lipopolysaccharide increases microglial GLT-1 expression and glutamate uptake capacity in vitro by a mechanism dependent on TNF-alpha. Glia. 2005;51(2):111–20.

Planas AM, Soriano MA, Rodriguez-Farre E, Ferrer I. Induction of cyclooxygenase-2 mRNA and protein following transient focal ischemia in the rat brain. Neurosci Lett. 1995;200(3):187–90.

Rahman A, Ting K, Cullen KM, Braidy N, Brew BJ, Guillemin GJ. The excitotoxin quinolinic acid induces tau phosphorylation in human neurons. PLoS One. 2009;4(7):e6344.

Reines SA, Block GA, Morris JC, Liu G, Nessly ML, Lines CR, et al. Rofecoxib: no effect on Alzheimer's disease in a 1-year, randomized, blinded, controlled study. Neurology. 2004;62(1):66–71.

Reinisch VM, Krause DL, Müller N. Neuroinflammation in Alzheimer's disease. In: Peterson, editor. Neuroinflammation and Neurodegeneration. Springer; 2014. in press.

Remarque EJ, Bollen EL, Weverling-Rijnsburger AW, Laterveer JC, Blauw GJ, Westendorp RG. Patients with Alzheimer's disease display a pro-inflammatory phenotype. Exp Gerontol. 2001;36(1):171–6.

Richartz E, Stransky E, Batra A, Simon P, Lewczuk P, Buchkremer G, et al. Decline of immune responsiveness: a pathogenetic factor in Alzheimer's disease? J Psychiatr Res. 2005;39(5):535–43.

Ricote M, Li AC, Willson TM, Kelly CJ, Glass CK. The peroxisome proliferator-activated receptor-gamma is a negative regulator of macrophage activation. Nature. 1998;391(6662):79–82.

Rogers J, Shen Y. A perspective on inflammation in Alzheimer's disease. Ann N Y Acad Sci. 2000;924:132–5.

Rosenberg PB, Wong DF, Edell SL, Ross JS, Joshi AD, Brasic JR, et al. Cognition and amyloid load in Alzheimer disease imaged with florbetapir F 18(AV-45) positron emission tomography. Am J Geriatr Psychiatry. 2013;21(3):272–8.

Schipper HM, Song W, Zukor H, Hascalovici JR, Zeligman D. Heme oxygenase-1 and neurodegeneration: expanding frontiers of engagement. J Neurochem. 2009;110(2):469–85.

Schwarz MJ, Riedel M, Ackenheil M, Müller N. Decreased levels of soluble intercellular adhesion molecule-1 (sICAM-1) in unmedicated and medicated schizophrenic patients. Biol Psychiatry. 2000;47(1):29–33.

Schwarz MJ, Guillemin GJ, Teipel SJ, Buerger K, Hampel H. Increased 3-hydroxykynurenine serum concentrations differentiate Alzheimer's disease patients from controls. Eur Arch Psychiatry Clin Neurosci. 2013;263(4):345–52.

Seabrook TJ, Jiang L, Maier M, Lemere CA. Minocycline affects microglia activation, Abeta deposition, and behavior in APP-tg mice. Glia. 2006;53(7):776–82.

Serpente M, Bonsi R, Scarpini E, Galimberti D. Innate immune system and inflammation in Alzheimer's disease: from pathogenesis to treatment. Neuroimmunomodulation. 2014;21(2–3):79–87.

Shaw AC, Goldstein DR, Montgomery RR. Age-dependent dysregulation of innate immunity. Nat Rev Immunol. 2013;13(12):875–87.

Smith WL, Garavito RM, DeWitt DL. Prostaglandin endoperoxide H synthases (cyclooxygenases)-1 and -2. J Biol Chem. 1996;271(52):33157–60.

Smith MA, Zhu X, Tabaton M, Liu G, McKeel Jr DW, Cohen ML, et al. Increased iron and free radical generation in preclinical Alzheimer disease and mild cognitive impairment. J Alzheimers Dis. 2010;19(1):363–72.

Sperling RA, Aisen PS, Beckett LA, Bennett DA, Craft S, Fagan AM, et al. Toward defining the preclinical stages of Alzheimer's disease: recommendations from the National Institute on Aging-Alzheimer's Association workgroups on diagnostic guidelines for Alzheimer's disease. Alzheimers Dement. 2011;7(3):280–92.

Sriram K, Matheson JM, Benkovic SA, Miller DB, Luster MI, O'Callaghan JP. Mice deficient in TNF receptors are protected against dopaminergic neurotoxicity: implications for Parkinson's disease. FASEB J. 2002;16(11):1474–6.

Streit WJ, Miller KR, Lopes KO, Njie E. Microglial degeneration in the aging brain—bad news for neurons? Front Biosci. 2008;13:3423–38.

Stübner S, Schön T, Padberg F, Teipel SJ, Schwarz MJ, Haslinger A, et al. Interleukin-6 and the soluble IL-6 receptor are decreased in cerebrospinal fluid of geriatric patients with major depression: no alteration of soluble gp130. Neurosci Lett. 1999;259(3):145–8.

Styren SD, Civin WH, Rogers J. Molecular, cellular, and pathologic characterization of HLA-DR immunoreactivity in normal elderly and Alzheimer's disease brain. Exp Neurol. 1990;110(1):93–104.

Swardfager W, Black SE. Dementia – a link between microbial infection and cognition? Nature. 2013;9:301–2.

Szekely CA, Breitner JC, Fitzpatrick AL, Rea TD, Psaty BM, Kuller LH, et al. NSAID use and dementia risk in the Cardiovascular Health Study: role of APOE and NSAID type. Neurology. 2008;70(1):17–24.

Teunissen CE, van Boxtel MP, Bosma H, Jolles J, Lutjohann D, von Bergmann K, et al. Serum markers in relation to cognitive functioning in an aging population: results of the Maastricht Aging Study (MAAS). Tijdschr Gerontol Geriatr. 2003a;34(1):6–12.

Teunissen CE, van Boxtel MP, Bosma H, Bosmans E, Delanghe J, De BC, et al. Inflammation markers in relation to cognition in a healthy aging population. J Neuroimmunol. 2003b; 134(1–2):142–50.

Thomas SR, Witting PK, Stocker R. 3-Hydroxyanthranilic acid is an efficient, cell-derived co-antioxidant for alpha-tocopherol, inhibiting human low density lipoprotein and plasma lipid peroxidation. J Biol Chem. 1996;271(51):32714–21.

Thomas P, Bhatia T, Gauba D, Wood J, Long C, Prasad K, et al. Exposure to herpes simplex virus, type 1 and reduced cognitive function. J Psychiatr Res. 2013;47(11):1680–5.

Tocco G, Freire-Moar J, Schreiber SS, Sakhi SH, Aisen PS, Pasinetti GM. Maturational regulation and regional induction of cyclooxygenase-2 in rat brain: implications for Alzheimer's disease. Exp Neurol. 1997;144(2):339–49.

Vehmas AK, Kawas CH, Stewart WF, Troncoso JC. Immune reactive cells in senile plaques and cognitive decline in Alzheimer's disease. Neurobiol Aging. 2003;24(2):321–31.

Vlad SC, Miller DR, Kowall NW, Felson DT. Protective effects of NSAIDs on the development of Alzheimer disease. Neurology. 2008;70(19):1672–7.

Wang XQ, Peng YP, Lu JH, Cao BB, Qiu YH. Neuroprotection of interleukin-6 against NMDA attack and its signal transduction by JAK and MAPK. Neurosci Lett. 2009;450(2): 122–6.

Webster S, Lue LF, Brachova L, Tenner AJ, McGeer PL, Terai K, et al. Molecular and cellular characterization of the membrane attack complex, C5b-9, in Alzheimer's disease. Neurobiol Aging. 1997;18(4):415–21.

Weggen S, Eriksen JL, Das P, Sagi SA, Wang R, Pietrzik CU, et al. A subset of NSAIDs lower amyloidogenic Abeta42 independently of cyclooxygenase activity. Nature. 2001;414(6860):212–6.

West MJ, Coleman PD, Flood DG, Troncoso JC. Differences in the pattern of hippocampal neuronal loss in normal ageing and Alzheimer's disease. Lancet. 1994;344(8925):769–72.

Willard LB, Hauss-Wegrzyniak B, Danysz W, Wenk GL. The cytotoxicity of chronic neuroinflammation upon basal forebrain cholinergic neurons of rats can be attenuated by glutamatergic antagonism or cyclooxygenase-2 inhibition. Exp Brain Res. 2000;134(1):58–65.

Wolfson C, Perrault A, Moride Y, Esdaile JM, Abenhaim L, Momoli F. A case-control analysis of nonsteroidal anti-inflammatory drugs and Alzheimer's disease: are they protective? Neuroepidemiology. 2002;21(2):81–6.

Wyss-Coray T. Inflammation in Alzheimer disease: driving force, bystander or beneficial response? Nat Med. 2006;12(9):1005–15.

Wyss-Coray T, Lin C, Yan F, Yu GQ, Rohde M, McConlogue L, et al. TGF-beta1 promotes microglial amyloid-beta clearance and reduces plaque burden in transgenic mice. Nat Med. 2001;7(5):612–8.

Yasojima K, Schwab C, McGeer EG, McGeer PL. Distribution of cyclooxygenase-1 and cyclooxygenase-2 mRNAs and proteins in human brain and peripheral organs. Brain Res. 1999; 830(2):226–36.

Yermakova AV, O'Banion MK. Downregulation of neuronal cyclooxygenase-2 expression in end stage Alzheimer's disease. Neurobiol Aging. 2001;22(6):823–36.

Zhao J, Zhao J, Legge K, Perlman S. Age-related increases in PGD(2) expression impair respiratory DC migration, resulting in diminished T cell responses upon respiratory virus infection in mice. J Clin Invest. 2011;121(12):4921–30.

Ziv Y, Ron N, Butovsky O, Landa G, Sudai E, Greenberg N, et al. Immune cells contribute to the maintenance of neurogenesis and spatial learning abilities in adulthood. Nat Neurosci. 2006;9(2):268–75.

Part III
Therapeutic Application

Chapter 16
Do Antidepressants Exert Effects on the Immune System?

Angelos Halaris

Abstract The role of the immune system and interactions between the endocrine, neurotransmitter, and immune systems in affective disorders have opened up new avenues to pursue in psychiatric research. Exploration of a potential causal relationship between inflammation and depression has lent support to the contention that a bidirectional relationship is at play. Exogenously administered cytokines can induce a syndrome mimicking many symptoms of the spontaneously occurring syndrome of depression. The majority of clinical studies of depression, in which pro-inflammatory biomarkers were measured, confirmed that a chronic inflammatory status is present, as measured mostly in blood but also in cerebrospinal fluid and postmortem brain of suicide victims with a history of depression. However, it is unclear as to what happens to the pro-inflammatory status following symptom resolution and remission of the depressive episode. In this context the precise role antidepressant agents play in the resolution of inflammation and return of the immune system to homeostasis has been the object of numerous studies. This chapter seeks to provide a critical overview of the available evidence in the literature and address key factors that contribute to the discrepant findings. Specifically, select basic and clinical studies are discussed and recommendations are made how future studies can lead to greater consistency amongst reported findings.

Keywords Depression • Antidepressants • Inflammation • Cytokines • CRP

Introduction

Major depressive disorder (MDD) is a disease of high prevalence worldwide. Investigators have attempted to unravel the pathophysiology of depression for decades but, in spite of significant breakthroughs, definitive elucidation of the complex mechanisms underlying this disorder (or disorders) still eludes us. The monoaminergic theory of depression, spurned in part by the discovery of the

A. Halaris (✉)
Department of Psychiatry, Loyola University Chicago Stritch School of Medicine, 2160 South First Avenue, Maywood, IL 60153, USA
e-mail: ahalaris@lumc.edu

N. Müller et al. (eds.), *Immunology and Psychiatry*, Current Topics in Neurotoxicity 8, DOI 10.1007/978-3-319-13602-8_16

first antidepressant agents and serendipitous observations of pharmacologic agents used to treat other disease entities, guided research for several decades. Enhancing neurotransmitter availability via neurotransmitter reuptake inhibition thus remained the cornerstone of antidepressant drug development. More recently, however, the role of the immune system and interactions between the endocrine, neurotransmitter, and immune systems have opened up new avenues to pursue. Based in part on observations that administration of pro-inflammatory cytokines, therapeutically or experimentally, induces a syndrome, commonly referred to as sickness behavior, in many ways akin to the human depressive syndrome, a new era of biomedical research was ushered in. The Macrophage Theory of Depression was postulated by Smith (1991) and paved the way for a proliferation of studies aimed at elucidating the connection between inflammation and depression. Exploration of a potential causal relationship between inflammation and depression lent support to the contention that there probably is a bidirectional relationship at play. Exogenously administered cytokines can induce a syndrome mimicking many symptoms of the spontaneously occurring syndrome of depression. The majority of clinical studies of depression, in which pro-inflammatory biomarkers were measured, confirmed that a chronic inflammatory status is present, as measured mostly in blood but also in cerebrospinal fluid and postmortem brain of suicide victims with a history of depression. So, if immune system activation and inflammatory responses can lead to depression, and, if depression is accompanied by chronic inflammation, what should be expected to happen when the depressed patient recovers—emotionally, cognitively, and behaviorally—from an episode of depression as a result of successful pharmacologic intervention, or, for that matter, non-somatic interventions? In other words, once a euthymic state is achieved, should inflammation be expected to return to control levels concomitantly? And, if this should be the case, what instigates or mediates normalization of the prior inflammatory status? A direct anti-inflammatory effect of the antidepressant agent used is a theoretical possibility. However, other factors could be involved singly or interactively, notably restoration of homeostatic balance in the autonomic nervous system, normalization of the hypothalamo–pituitary–adrenal (HPA) axis, gradual down-regulation of up-regulated enzymatic activity in critical metabolic pathways, such as the tryptophan/kynurenine pathway, receptors and second messenger systems, or some other hitherto unknown mechanism.

In seeking answers to these questions, the available literature is controversial. Some studies claim that inflammation returns to normal levels as the patient responds to antidepressant therapy and becomes euthymic. However, a larger number of studies do not support this claim. It appears there may also be a difference amongst the measured cytokines with some authors reporting a return to normalcy of one cytokine but not another. No doubt methodological and design differences amongst reported studies play a key role in explaining the discrepancies. The heterogeneity of depressive illness unquestionably plays a key role and renders comparisons across studies often impossible. However, there is substantially greater agreement amongst studies with respect to baseline measurements of inflammation indices than there is about post-treatment effects. This raises the issue whether the length of exposure to the antidepressant agent might not be a crucial variable indicating that the time frame within which euthymia is achieved may not have to be paralleled by immune and endocrine system responses and normalization. In this chapter I will attempt to shed some light on these issues. This review article cannot be an exhaustive review of the literature due to space limitations. Rather I have selected representative articles to support my interpretation of the relevant literature.

Basic Studies

Under basic studies we refer to either animal studies or human studies conducted in vitro, or ex vivo. While such studies are of significant scientific value and can stimulate translational research, caution should be exercised in extrapolating from such findings to the human condition.

Animal Studies

Earlier studies had provided animal based and/or in vitro evidence that antidepressant agents exert suppressive effects on cytokine production. For example, Sommer et al. (1995) reported that rolipram, a selective type IV phosphodiesterase inhibitor, stereospecifically suppressed the production of tumor necrosis factor (TNF)-α and less strongly also interferon (IFN)-γ in human and rat auto-reactive T cells. Although never marketed as an antidepressant in the USA, rolipram had been shown to possess antidepressant activity. Yirmiya et al. (2001) studied effects of antidepressant drugs on behavioral and physiological responses to lipopolysaccharide (LPS) stimulation in rodents. While imipramine and fluoxetine attenuated anorexia, weight loss, CRH depletion and adrenocortical activation, the authors concluded that these effects are probably not mediated by drug effects on peripheral pro-inflammatory cytokine production. They based their conclusion on their finding that LPS-induced expression of TNF-α and interleukin (IL)-1β mRNA in the spleen (assessed by semiquantitative in situ hybridization) was not altered following chronic treatment with either fluoxetine or imipramine.

However, they did allow for the possibility that antidepressants produce some of their effects by modulating the cytokine response for the following reasons. First, the effects of antidepressants were examined only with respect to splenic cytokine production. It is still possible that production in other peripheral tissues, such as liver cells, endothelial cells, peritoneal macrophages, and circulating macrophages, may have been affected. Second, mediators other than TNF-α and IL-1β could be affected by antidepressants. The interaction of desipramine with TNF-α in animal brain produced intriguing results as reported by Reynolds et al. (2005). Their innovative study illustrated the complex interactions between cytokines, neurotransmitters, and antidepressants inasmuch as they demonstrated the efficacy of desipramine is due to decreased levels of TNF-α in the brain induced by this drug, ultimately modifying noradrenergic transmission.

Roumestan et al. (2007) compared the antidepressants desipramine and fluoxetine in vivo to the glucocorticoid prednisolone, as an anti-inflammatory drug of reference. In a murine model of LPS-induced septic shock, animals received the drugs either before or after injection of LPS. Desipramine and fluoxetine reduced the inflammatory reaction in this animal model. The authors concluded that these antidepressants act directly on relevant peripheral cell types to decrease expression of inflammatory mediators probably by affecting their gene transcription.

Whether antidepressants inhibit the release of cytokines from activated microglia is also a matter of some controversy. Horikawa et al. (2010) examined the effects of paroxetine and sertraline on IFN-γ-induced microglial activation in vitro. They showed significant inhibition of TNF-α release from IFN-γ-activated 6-3 microglia and suggested that the inhibition is mediated via IFN-γ-induced elevation of intracellular Ca^{2+}. They commented, however, that the inhibitory effect of paroxetine and sertraline on microglial activation may not be a prerequisite for antidepressant function, but an additional beneficial effect. In an earlier publication these investigators had examined the effects of various types of antidepressants and lithium chloride on IFN-γ-induced microglial production of IL-6 and nitric oxide (NO) (Hashioka et al. 2007). Treatment of the murine microglial 6-3 cells with 100 U/ml of IFN-γ resulted in an eightfold increase in IL-6 and a tenfold increase in NO into the culture medium. Pretreatment with fluvoxamine, reboxetine, or imipramine significantly inhibited IL-6 and NO production in a dose-dependent manner. Since these three agents differentially affect monoaminergic systems, it appears no specific monoamine is responsible for the observed effects. Indeed these inhibitions were reversed significantly by inhibition of cyclic adenosine monophosphate (cAMP), and, except for reboxetine, by protein kinase A (PKA) inhibition. Interestingly, lithium chloride enhanced IFN-γ-stimulated IL-6 production and inhibited NO production. The authors concluded that antidepressants have inhibitory effects on IFN-γ-activated microglia and these effects are, at least partially, mediated by the cAMP-dependent PKA pathway. In his recent critical review of the pertinent literature, Leonard (2014) aptly concluded: "Thus it would appear that the inhibition of activated microglia may be important in the mode of action of some types of antidepressants but would not appear to be a unifying mechanism of action of all antidepressants!"

In Vitro Studies

Typically such studies are conducted using either human whole blood samples or peripheral blood mononuclear cells (PBMC) which are incubated with the investigative agent and then stimulated to induce the production of cytokines. The first such study was reported by Xia et al. (1996). These authors investigated how tricyclic antidepressants (TCAs) and the SSRI, citalopram, influence cytokine release by T lymphocytes and monocytes from human blood, with and without LPS exposure. They used PBMC from healthy volunteers. All tested agents exhibited inhibitory action albeit in varying degrees. Subsequently, Maes et al. (1999) reported similar findings with clomipramine, sertraline, and trazodone. A number of studies followed most of which demonstrated that antidepressant agents produce a strong immunosuppressive effect on whole blood of both depressed and healthy control subjects (Janssen et al. 2010). Of note is the report by Diamond et al. (2006) who demonstrated that individual antidepressants may exert differential effects on cytokine release.

Ex Vivo Studies

The ex vivo studies reported on cytokines, depression and antidepressant drug effects have been typically conducted in whole blood samples or PBMCs obtained from depressed subjects before and after treatment. The studies are fraught with markedly conflicting results. Cytokine production in these studies has been reported to be increased, decreased, or unchanged in depression and the effects of antidepressants can be facilitatory or inhibitory. The possible factors contributing to the contradictory findings are critically reviewed by Janssen et al. (2010).

Clinical (In Vivo) Studies

Numerous studies suggest that MDD is accompanied by immune dysregulation and activation of the inflammatory response system (IRS). It is beyond the scope of this article to discuss all of these studies and the reader is referred to two relevant meta-analytic studies. The analysis by Dowlati et al. (2010) included 24 studies involving unstimulated measurements of cytokines in patients meeting DSM criteria for MDD. This meta-analysis confirmed significantly higher concentrations of the pro-inflammatory cytokines, TNFα and IL-6, in depressed subjects compared with control subjects. While both positive and negative results have been reported in individual studies, this meta-analytic result strengthens the evidence that depression is accompanied by activation of the IRS. Another meta-analytic study published by Howren et al. (2009) assessed the magnitude and direction of associations of depression with C-reactive protein (CRP), IL-1, and IL-6 in community and clinical samples. Each inflammatory marker was positively associated with depression; CRP ($p<0.001$); IL-6 ($p<0.001$); IL-1 ($p=0.03$); IL-1ra ($p=0.02$). Associations were strongest in clinically depressed patient samples—but were also significant in community-based samples.

The possible effects of antidepressant agents on blood (plasma, serum) levels of inflammatory cytokines remains a topic of investigation inasmuch as the literature on the topic is far from unanimous. The reader is referred again to the comprehensive review of the literature and summary of findings published by Janssen et al. (2010). A summary statement worth repeating here stipulates that "there are cytokine changes in depression, which often resolve with treatment but the nature of the association between cytokines and depression remains to be fully understood…the immunosuppressive effect of antidepressant medication is found independently of treatment outcome." Of note also is the meta-analytic study of Hannestad et al. (2011) which critically examined 22 studies fulfilling basic criteria of including MDD patients, providing pre- and post-treatment data and utilizing approved pharmacologic treatments. The authors concluded that resolution of a depressive episode is *not* associated with normalization of levels of circulating inflammatory cytokines. The reader is referred to this excellent meta-analytic study for further details. In the following sections of this chapter I will attempt to provide the reader with a selective review of published data on individual cytokines as it appears that not all cytokines are "equal" in their response to diverse antidepressant agents and treatments.

C-Reactive Protein

CRP is an annular, pentameric protein synthesized by the liver in response to factors released by macrophages and adipocytes. It is an acute-phase protein and its plasma levels rise in response to inflammation. Patients with elevated basal levels of CRP are at increased risk of diabetes, hypertension, and cardiovascular disease but other disease entities have been associated with abnormal CRP levels. When studying low-grade inflammation, a high-sensitivity assay must be used. Such low-grade inflammation has been reported in many studies of psychological distress and depression but not all studies agree. The controversial findings could be due to a host of confounding factors as we will discuss later. Of relevance to this review is the question whether antidepressant drug therapies exert anti-inflammatory activity and hence should be expected to normalize elevated basal CRP levels in depressed patients.

In their retrospective study, Hornig et al. (1998) examined positive and negative acute-phase proteins in affective subtypes. They focused in particular on CRP and the effects of lithium as compared to subjects not maintained on lithium. They established that patients on lithium monotherapy were less likely to demonstrate elevated CRP and concluded that lithium may reverse immune system activation (as reflected in CRP elevations) and thereby normalize rather than enhance immune parameters. Unfortunately, the retrospective nature of this study, the heterogeneous groups studied and the diversity in affective state of their subjects and medications used renders any definitive conclusions difficult.

O'Brien et al. (2006) studied depressed patients in two separate component studies. In study 1 they compared three groups, a euthymic group, a currently depressed group, and a healthy control group and found no significant differences in CRP levels amongst them. In component study 2 they treated 20 depressed women with fluoxetine or sertraline for 3 weeks and measured plasma CRP before and after treatment. A significant reduction in CRP was noted after treatment when compared to baseline. Combining the observations from these two studies, they concluded that antidepressant treatment leads to a significant reduction in CRP levels independent of resolution of symptoms. One limitation of that study was that their assay did not measure high-sensitivity CRP.

Entirely different results were reported by Dawood et al. (2007). Among other parameters, they measured hsCRP in 25 patients with MDD, with no history of coronary heart disease, and in 15 healthy subjects. Treatment consisted of SSRIs for approximately 12 weeks. While no significant differences were observed between untreated MDD patients and healthy subjects in other parameters, pulse pressure and hsCRP were significantly *elevated* in the MDD subjects prior to treatment ($p = 0.025$). Moreover, while pharmacotherapy was effective in alleviating depression, pulse pressure and plasma hsCRP levels *increased* raising concern about potential cardiovascular risk in association with extended SSRI use.

Chang et al. (2012) explored the association between CRP and cognitive impairment in major depression in a cohort of 112 MDD patients none of whom had ever received antidepressant treatment. They were randomly assigned to receive either fluoxetine or venlafaxine for 6 weeks. CRP levels were measured at baseline and 2, 4, and 6 weeks. Although the baseline HAMD score did not show an association with CRP, baseline CRP was significantly correlated with treatment response. Patients with a higher baseline CRP had a poorer early treatment response at 2 weeks. Additionally, CRP levels *increased* significantly after 6 weeks of treatment with either antidepressant. They interpreted their findings to be in line with previous studies showing that patients with a pro-inflammatory status are more likely to show treatment resistance (Miller et al. 2009; O'Brien et al. 2007).

Results from two population studies reported by Hamer et al. (2011) using a representative sample of community dwelling adults suggested that the use of antidepressants was associated with elevated levels of systemic inflammation. The authors observed an association between antidepressant medication use and CRP and this phenomenon was robust with TCAs but less so with SSRIs. A longitudinal association between any antidepressant use and subsequent CRP was replicated in the Whitehall study and "these associations were independent of psychological distress." In an effort to explain the possible mechanism(s) by which antidepressants can cause low-grade inflammation they discussed weight gain, and effects on glucocorticoid receptors and the HPA. Whether these findings explain why antidepressant medication has the potential to increase CVD risk by raising the level of systemic inflammation remains an open question.

A large study reported by Vogelzangs et al. (2012) selected MDD and control subjects from the Netherlands Study of Depression and Anxiety. In addition to confirming elevated inflammation in depressed men, especially those with late-onset depression, differential effects of the antidepressants used were observed. Specifically, inflammation was *increased* in men using SNRIs (CRP, IL-6) and in men and women using tri- or tetracyclic compounds, but *decreased* among men using SSRIs (IL-6). These findings obtained from a large cohort of controls and MDD subjects underscore the need to avoid generalizations, namely, that antidepressant treatments invariably lead to reduction in inflammation. It appears that we should be looking specifically at individual cytokines and individual pharmacologic agents rather than at entire classes of agents. Additionally, the length of exposure to the antidepressant agent and the specific time point(s) of cytokine measurement are critical factors to be considered before drawing definitive conclusions. Time series measurements rather than single time point snapshots should go a long way toward resolving the existing discrepancies.

Cytokines

The majority of studies have focused on one to three cytokines. The most commonly assessed cytokines are the predominantly pro-inflammatory cytokines: TNF-α, IL-6, and IL-1. Far fewer studies have assessed anti-inflammatory cytokines or cytokines that may exert dual action. In my view, the question whether antidepressant treatments enhance the release of anti-inflammatory compounds is of equal scientific importance.

We have reported on MMD patients treated for 8 weeks with venlafaxine and who were carefully selected to be free of any source of inflammation, current or recent, and had no evidence of an autoimmune disorder (Piletz et al. 2009). While we confirmed that concentrations of TNF-α and IL-1β were significantly elevated at pretreatment compared to matched healthy controls, these inflammation biomarkers did not normalize after 4 or 8 weeks of treatment. MMD patients pre-venlafaxine had a significantly higher mean TNF-α concentration compared to controls which *increased* significantly after 4 weeks and then slightly declined after 8 weeks on venlafaxine but still remained significantly higher than baseline and was still statistically higher than healthy controls. Results with IL-1β were somewhat different. Patients had a significantly higher baseline concentration compared to controls which did not change after 4 weeks but showed a trend toward decline after 8 weeks of treatment but without reaching statistical significance. The lack of cytokine changes occurred in the presence of treatment response by the majority of the patient cohort. The observed "spike" in TNF-α levels at week 4 is noteworthy as we have observed it in a subsequent study with a different antidepressant (in preparation) and therefore we believe it may not be purely coincidental. While it may relate to the rising concentrations of venlafaxine and the associated interplay between noradrenergic and serotonergic reuptake inhibition and the presumptive cAMP and PKA interplay, it does point to the need to obtain a time course of blood measurements at least at monthly intervals for perhaps as long as 6 months in order to avoid reaching misleading conclusions about whether or not antidepressants exert specific anti-inflammatory activity, or for that matter, increase the overall level of inflammation with potentially adverse health consequences.

Somewhat similar results were reported later by Li et al. (2013) who assessed plasma levels of TNF-α in MDD subjects at baseline and after 8 weeks of treatment with venlafaxine. MDD subjects had significantly higher levels compared to healthy controls and responders did not differ from non-responders at baseline. At the end of treatment all patients had lower levels, but the responder group had shown a greater reduction which was associated with a greater reduction in the HAM-D17 score. Suicidality did not appear to correlate with the elevation in TNF-α plasma levels in their patients.

In the study by Chen et al. (2010) 43 young depressed males were treated with one of four antidepressants as inpatients for 4 weeks. Among other parameters, hsCRP, TNF-α, and IL-6 were measured before and after treatment. Since no comparison group was included, comparisons could only be made between pre- and post-treatment effects. Neither hsCRP nor TNF-α showed any changes during the course of treatment. However, IL-6 levels *increased* significantly. Due to the lack of statistical power no determination could be made whether any of the observed cytokine effects were related to one or more of the antidepressants utilized. Opposite results with IL-6 had been reported previously by Basterzi et al. (2005). These investigators had focused on IL-6 exclusively. Their patients received an SSRI for 6 weeks but no details were provided about the specific agent and dose used. Baseline IL-6 levels did not differ between patients and controls. At post-treatment IL-6 levels had *decreased* and were significantly lower than baseline or those of the control group. These two studies exemplify the existing discrepancies in reported effects in spite of the fact that perhaps the most consistent finding to date is that antidepressant treatment significantly *reduces* IL-6 plasma levels and generally within the time frame of treatment response.

Discussion

It should be clear from this brief review that untreated depression is more robustly associated with immune dysregulation and activation of the IRS than the status of the immune system during the course of antidepressant drug therapy and the potentially immunosuppressive effects of antidepressant agents. At the very least the discrepant findings in the world literature point out that there is no compelling concordance between treatment outcome, both in time frame and magnitude, and changes in blood levels of inflammatory biomarkers (e.g., CRP, cytokines). The putative factors contributing to the discrepant findings are ostensibly multi-faceted and complex. Nevertheless we stand to learn useful lessons from the contradictory findings that should allow us to design future studies with a greater likelihood of producing definitive and consistent findings. Here I present a list, probably not complete, of confounding factors:

1. *Heterogeneity*: Undoubtedly the heterogeneity of depressive illness is at the core of the problem. Including in future study designs a "typology" of depression that transcends the old dichotomies, such as psychotic vs. non-psychotic, typical vs. atypical, melancholic vs. non-melancholic, neurotic vs. non-neurotic, to name a few, and assesses instead specific symptoms and symptom clusters should predictably allow more clinically relevant correlations with inflammation (and other) biomarkers. For example, appetite changes, weight loss or gain, sleep-wakefulness changes, anhedonia, energy level, motivation, irritability, among others, would be critical in correlating psychopathological and biomarker profiles ultimately leading to "signature" based typologies for each patient. It is fully understood that associations do not necessarily imply causation but can provide useful clues for further research. Indeed I fully anticipate that we will ultimately confirm the existence of an "immune" type of depression as a distinct subclass of depression, in accord with the suggestion

by Janssen et al. (2010). Such a subclass might even cut across the traditional typologies and even the dichotomy between unipolar and bipolar depression since there is strong evidence that bipolar disorder is also associated with immune dysregulation.

2. *Co-morbidity*: Careful assessment of co-morbid conditions in the study cohort and exclusion of such subjects from enrollment is of utmost importance. This is particularly critical when cardiovascular disease, diabetes and other endocrinopathies, autoimmune disorders, indeed any inflammatory locus anywhere in the body are present. Such patients should not be allowed into the study when the immune system is being investigated with respect to depressive illness. Community-based studies, as informative as they are in their own right, are difficult to interpret due, among other reasons, to the confounding co-morbidity amongst portions of the study populations.
3. *Addictions and addictive behaviors*: The immune system is known to respond to substances of abuse, the most common of which is smoking. Allowing current or very recent smokers is bound to skew the results. Of equal importance is to exclude alcohol abuse and to instruct the subject to refrain from consuming even a small amount of alcohol preceding the blood test. Assessment of addictive behaviors beyond substances of abuse should also be ascertained and recorded as a potential confound.
4. *Lifestyle issues*: Level of physical activity is important to assess since it is established that a sedentary lifestyle and lack of exercise, especially in conjunction with overweight and obesity is a source of inflammation.
5. *Age* must be assessed as a covariate since advancing age has been associated with immune activation.
6. *Sex* and menopausal status must be included as female hormonal status is potentially a confound.
7. *Cohort size*: Many of the published studies have included relatively small patient cohorts and have used at times more than one antidepressant agent. From a statistical viewpoint it is impossible to conduct meaningful analyses, especially when numerous variables must be analyzed to determine the influence of confounds and still achieve statistical power.
8. *Genetic polymorphisms*: Most of all the commonly assessed biomarkers of inflammation have established genetic polymorphisms which could at least theoretically explain some of the variance in the reported findings across in vivo studies. It is highly desirable in future studies to include polymorphism of the biomarker being tested although such a proposition necessitates much larger cohorts.
9. *Indices of autonomic function*: It is recommended to obtain indices of autonomic function, such as determination of heart rate variability, at the same time that immune system activity is assessed. The role of vagal input in inflammation control is well established.
10. *Study design issues*: Every effort should be made not to enroll subjects who are currently receiving antidepressant medication or have gone through a short washout of just a few weeks. The ideal design should include subjects who either have never been treated or who are in depressive relapse following discontinuation of

medication of several weeks or months' duration. This is because biological changes take weeks to re-equilibrate following the removal of a stimulus, notably, the antidepressant agent in this instance. Of paramount importance also is to build into the study repeated measures of both symptom assessment and biomarker measurements. A single time point design aimed to correspond to the anticipated behavioral outcome has proven to be woefully inadequate and likely to lead to misinterpretations of the obtained data.

Concluding Remarks

The introduction of psychoneuroimmunology into biological psychiatry has marked the dawn of a new era in psychiatric research. Cytokine research has produced a wealth of new data in spite of controversies and outright contradictions in the reported findings. Assessment of genetic variants in cytokine expression will predictably help at least some of the current inconsistencies in the data. To that end, studies on a far larger scale need to be designed and executed involving a multi-site approach with highly standardized procedures in psychiatric assessments as well as in laboratory procedures as the latter is also a major source of variance in the reported findings. Such large scale multi-center studies will be costly and difficult for a single team of researchers and institution to bear thereby requiring the involvement of governmental agencies. The investment is expected to pay off, however, with the ultimate development of more effective therapeutic agents and associated reduction in the mental and financial cost to the individual and society associated with one of the most widespread and debilitating diseases of mankind.

References

Basterzi AD, Aydemir C, Kisa C, et al. IL-6 levels decrease with SSRI treatment in patients with major depression. Hum Psychopharmacol. 2005;20:473–6.

Chang HH, Lee HI, Gean PW, et al. Treatment response and cognitive impairment in major depression: association with C-reactive protein. Brain Behav Immun. 2012;26:90–5.

Chen Y-C, Lin W-W, Chen Y-J, et al. Antidepressant effects on insulin sensitivity and proinflammatory cytokines in the depressed males mediators of inflammation. Mediat Inflamm. 2010, Article ID 573594, 7 pages. doi:10.1155/2010/573594.

Dawood T, Lambert EA, Barton DA, et al. Specific serotonin reuptake inhibition in major depressive disorder adversely affects novel markers of cardiac risk. Hypertens Res. 2007; 30:285–93.

Diamond M, Kelly JP, Connor TJ. Antidepressants suppress production of the Th1 cytokine interferon-gamma, independent of monoamine transporter blockade. Eur Neuropsychopharmacol. 2006;16:481–90.

Dowlati Y, Herrmann N, Swardfager W, et al. A Meta-analysis of cytokines in major depression. Biol Psychiatry. 2010;67:446–57.

Hamer M, Batty GD, Marmot MG, et al. Anti-depressant medication use and C-reactive protein: results from two population-based studies. Brain Behav Immun. 2011;25(1):168–73.

Hannestad J, DellaGioia N, Bloch M. The effect of antidepressant medication treatment on serum levels of inflammatory cytokines: a meta-analysis. Neuropsychopharmacology. 2011;36:2452–9.

Hashioka S, Klegeris A, Monji A, et al. Antidepressants inhibit interferon-γ-induced microglial production of IL-6 and nitric oxide. Exp Neurol. 2007;206(1):33–42.

Horikawa H, Kato TA, Yoshito Mizoguchi Y, et al. Inhibitory effects of SSRIs on IFN-γ induced microglial activation through the regulation of intracellular calcium. Prog Neuropsychopharmacol Biol Psychiatry. 2010;34(7):1306–16.

Hornig M, Goodman DBP, Kamoun M, et al. Positive and negative acute phase proteins in affective disorders. J Affect Disord. 1998;49:9–18.

Howren MB, Lamkin DM, Suls J. Associations of depression with C-reactive protein, IL-a, and IL-6: a meta-analysis. Psychosom Med. 2009;71:171–86.

Janssen DG, Caniato RN, Verster JC. A psychoneuroimmunological review on cytokines involved in antidepressant treatment response. Hum Psychopharmacol. 2010;25:201–15.

Leonard BE. Impact of inflammation on neurotransmitter changes in major depression: an insight into the action of antidepressants. Prog Neuropsychopharmacol Biol Psychiatry. 2014;48:261–7.

Li Z, Qi D, Chen J, et al. Venlafaxine inhibits the upregulation of plasma tumor necrosis factor-alpha (TNF-α) in the Chinese patients with major depressive disorder: a prospective longitudinal study. Neuroendocrinology. 2013;38(1):107–14.

Maes M, Song C, Lin A, et al. Negative immunoregulatory effects of antidepressants: inhibition of interferon-gamma and stimulation of interleukin-10 secretion. Neuropsychopharmacology. 1999;20:370–9.

Miller AH, Maletic V, Raison CL. Inflammation and its discontents: the role of cytokines in the pathophysiology of major depression. Biol Psychiatry. 2009;65:732–41.

O'Brien SM, Scott LV, Dinan TC. Antidepressant therapy and C-reactive protein levels. Br J Psychiatry. 2006;188:449–52.

O'Brien SM, Scully P, et al. Plasma cytokine profiles in depressed patients who fail to respond to selective serotonin reuptake inhibitor therapy. J Psychiatr Res. 2007;41:326–31.

Piletz JE, Halaris A, Iqbal O, et al. Pro-inflammatory biomarkers in depression: treatment with venlafaxine. World J Biol Psychiatry. 2009;10:313–23.

Reynolds JL, Ignatowski TA, Sud R, et al. An antidepressant mechanism of desipramine is to decrease tumor necrosis factor-alpha production culminating in increases in noradrenergic neurotransmission. Neuroscience. 2005;133(2):519–31.

Roumestan C, Michel A, Bichon F, et al. Anti-inflammatory properties of desipramine and fluoxetine. Respir Res. 2007;8:35.

Smith RS. The macrophage theory of depression. Med Hypotheses. 1991;35(4):298–306.

Sommer N, Löschmann P-A, Northoff GH, et al. The antidepressant rolipram suppresses cytokine production and prevents autoimmune encephalomyelitis. Nat Med. 1995;1:244–8.

Vogelzangs N, Dulvis HE, Beekman ATF, et al. Association of depressive disorders, depression characteristics and antidepressant medication with inflammation. Transl Psychiatry. 2012;2:e79. doi:10.1038/tp.2012.8.

Xia Z, DePierre JW, Nässberger L. Tricyclic antidepressants inhibit IL-6, IL-1 beta and TNF-alpha release in human blood monocytes and IL-2 and interferon-gamma in T cells. Immunopharmacology. 1996;34(1):27–37.

Yirmiya R, Pollak Y, Barak O, et al. Effects of antidepressant drugs on the behavioral and physiological responses to lipopolysaccharide (LPS) in rodents. Neuropsychopharmacology. 2001;24:531–44.

Chapter 17
Immunomodulation as Therapeutic Approach in Schizophrenia and Depression: State of the Art

Norbert Müller

Abstract Inflammation has been discussed for decades as an underlying cause of psychiatric disorders such as major depression (MD) and schizophrenia. Almost a hundred years ago, an anti-inflammatory therapeutic approach, so-called vaccination therapy, was proposed by Wagner von Jauregg. In schizophrenia and MD, opposite patterns of the type-1 and type-2 immune response seem to be associated with differences in the activation of the enzyme indoleamine 2,3-dioxygenase (IDO) and in tryptophan-kynurenine metabolism. These differences are associated with an imbalance in glutamatergic neurotransmission, which may contribute to an excess of *N*-methyl-D-aspartate (NMDA) agonism in depression and NMDA antagonism in schizophrenia. In both schizophrenia and depression the immunological imbalance results in increased prostaglandin E_2 (PE_2) production and probably also in increased cyclooxygenase-2 (COX-2) expression. Although there is strong evidence for the view that the interactions of the immune system, IDO, serotonergic system and glutamatergic neurotransmission play a key role in schizophrenia and depression, several gaps remain that need to be closed by intense research. Accordingly, anti-inflammatory or immune-modulating substances might have beneficial effects in schizophrenia and MD. COX-2 inhibitors have shown encouraging results in animal models. Moreover, during the last decade many clinical studies have been performed with COX-2 inhibitors, mostly celecoxib, in schizophrenia and MD. For ethical reasons, all these studies used an add-on design, i.e. the COX-2 inhibitor was given adjunctive to either antipsychotics (in schizophrenia) or antidepressants (in MD). Although an add-on design is a methodological challenge, favourable effects of COX-2 inhibitors were observed in placebo-controlled double-blind studies in both indications. Meta-analytic studies proved a significant therapeutic effect of COX-2 inhibitors in MD and in early stages of schizophrenia. Other pharmacological approaches based on

N. Müller (✉)
Department of Psychiatry and Psychotherapy, Ludwig-Maximilians University Munich, Nussbaumstr. 7, 80336 Munich, Germany
e-mail: Norbert.Mueller@med.uni-muenchen.de

N. Müller et al. (eds.), *Immunology and Psychiatry*, Current Topics in Neurotoxicity 8, DOI 10.1007/978-3-319-13602-8_17

immunological effects are discussed. Anti-inflammatory and immune-modulating compounds are promising but still need careful further scientific evaluation, including clinical studies in larger samples of patients.

Keywords Schizophrenia • Major Depression • Inflammation • Immunomodulation • COX-2 inhibition • Cytokines • Therapy

Introduction

Inflammation has been discussed for decades as an underlying cause of psychiatric disorders such as major depression (MD) and schizophrenia. As early as 1890, one of the founders of modern psychiatry, Emil Kraepelin, described during an influenza epidemic 11 cases of psychiatric disorders that presented with different symptoms such as depressed mood, a paranoid and hallucinatory syndrome, involuntary movements, cognitive deterioration and a delirious state (Kraepelin 1890). Later, Kraepelin proposed in his programmatic essay 'Objectives and methods of psychiatric research' to make the immunological defence and adaption system a focus of psychiatric research (Kraepelin 1918; Steinberg and Himmerich 2013). An immune-based therapeutic approach for psychiatric disorders was first proposed decades ago, when the Nobel Laureate Julius Ritter Wagner von Jauregg developed a vaccination therapy for psychoses (Wagner von Jauregg 1926). He treated patients successfully with vaccines derived from attenuated *Mycobacterium tuberculosis*, *Plasmodium malariae* or *Salmonella typhi*, all of which stimulate a type 1 immune response (Müller et al. 2005a). This therapy showed the best results in syphilis infection, but Wagner von Jauregg also administered it in other psychiatric disorders. Earlier, Wagner von Jauregg had observed that around half of the patients from asylums who had recovered from typhus infections showed improvement of their psychiatric disorder or recovered from it, which led to the development of 'artificial fever' and fever therapy in psychiatric disorders.

After the introduction of penicillin for the treatment of syphilis and later as a consequence of the overwhelming success of the catecholaminergic neurotransmitter approach in psychiatry, scientific interest in psychiatry shifted away from the immune system and the therapeutic vaccination approach was not pursued further. However, in the past decades the emerging limitations of the neurotransmitter approach for both pathogenetic research and therapeutic progress have resulted in renewed and increased interest in other therapeutic approaches, including the possible use of modern anti-inflammatory agents in schizophrenia (Fond et al. 2013) and MD.

In order to understand the rationale for the use of anti-inflammatory or immune-modulating compounds, disease-associated immune alterations in MD and schizophrenia are first discussed. Cognitive impairment as a major consequence of and a risk factor for inflammation is an aspect which might play a role in MD and schizophrenia, because both disorders are associated with impaired cognition. The discussion of the possible role of inflammation in impaired cognition, however, is part of the chapter on Alzheimer's disease.

Major Depression and Inflammation

One inflammatory model of MD is 'sickness behaviour', an organism's reaction to infection and inflammation. Sickness behaviour is characterised by weakness, malaise, listlessness, inability to concentrate, lethargy, decreased interest in the surroundings and reduced food intake—all of which are depression-like symptoms. Sickness-related psychopathological symptomatology during infection and inflammation is mediated by pro-inflammatory cytokines such as interleukin-1 (IL-1), interleukin-6 (IL-6), tumour necrosis factor alpha (TNF-α) and interferon gamma (IFN-γ) (Dantzer 2001).

In humans, the involvement of cytokines in the regulation of the behavioural symptoms of sickness behaviour has been studied by administering the bacterial endotoxin lipopolysaccharide (LPS) to human volunteers (Reichenberg et al. 2001). LPS, a potent activator of pro-inflammatory cytokines, was found to induce mild fever, anorexia, anxiety, depressed mood and cognitive impairment. The levels of anxiety, depression and cognitive impairment were found to be related to the levels of circulating cytokines (Reichenberg et al. 2001, 2002).

Microglial cells and astrocytes are mediators of inflammation in the brain. Moreover, disturbances to the blood–brain barrier facilitate the invasion of inflammatory molecules and immune cells including monocytes and T and B cells. The active pathway of these cytokines from the peripheral immune system to the brain is via afferent neurons and through direct targeting of the amygdala and other brain regions after diffusion at the circumventricular organs and choroid plexus (Dantzer 2001). Undoubtedly, there is a strong relationship between the cytokine and neurotransmitter systems, but a more differentiated analysis may be required to understand the specific mechanisms underlying the heterogeneous disease.

The activation of the inflammatory response system in MD is well documented (Maes 1994; Maes et al. 1992; Müller et al. 1993; Myint et al. 2005; Rothermundt et al. 2001). Two recent meta-analyses clearly showed elevated IL-6 levels in patients with MD (Dowlati et al. 2010; Howren et al. 2009). However, the findings of the two meta-analyses differed regarding levels of the inflammatory markers C-reactive protein (CRP), IL-1, IL-1RA and TNF-α. In general, the inflammatory response system appears to be activated, but the levels of the different markers vary across studies. MD is a disorder often triggered by stress. It has been shown that—often based on genetic disposition—early-life stress or separation stress is associated with an increase of pro-inflammatory cytokines, leading to an activation of the immune system and pro-inflammatory prostaglandins. Prostaglandin E2 (PGE2) is an important mediator of inflammation (Song et al. 1998). Increased PGE2 in the saliva, serum and cerebrospinal fluid of depressed patients has been described previously (Calabrese et al. 1986; Linnoila et al. 1983; Nishino et al. 1989; Ohishi et al. 1988). The enzyme cyclooxygenase-2 (COX-2) is involved in the function of PGE2 in the inflammatory pathway. In the brain, the activation of microglia cells and astrocytes is crucial, because pro-inflammatory molecules are produced and released in the brain. The interactions between the immune system and neurotransmitters, the tryptophan-kynurenine system and glutamatergic neurotransmission are further links between stress, depression and the immune system.

Results of an interesting population-based Danish register study support the view that an infection or autoimmune disease significantly increases the risk to later develop a depressive disorder. This population-based prospective cohort study of 78 million person-years included 3.6 million people (born between 1945 and 1996). The follow-up was documented from 1977 to 2010. All people with a diagnosis of an affective disorder according to ICD-8, ICD-9 or ICD-10 were included if they had had at least one hospital contact as an in- or outpatient because of an affective disorder (including bipolar disorder). Every hospital contact because of an autoimmune disorder or infection (excluding HIV/AIDS) prior to psychiatric diagnosis was recorded. More than 91,000 cases of affective disorder were identified, approximately 30,000 cases of infection and >4,000 diagnoses of autoimmune disease. Hospitalization for infection significantly increased the risk for later a mood disorder by 62 % (incidence rate ratio [IRR] 1.62), while hospitalization for an autoimmune disease significantly increased the risk for a later mood disorder by 45 % (IRR 1.45). Both risk factors interacted and increased the risk to IRR 2.35. Interestingly, the risk for mood disorder increased with the proximity to the infection, with the highest risk within the first year (IRR 2.70) (Benros et al. 2013).

Schizophrenia and Inflammation

Infection during pregnancy, in particular in the second trimester, has been repeatedly described in mothers of offspring who later develop schizophrenia (Brown et al. 2004; Buka et al. 2001; Westergaard et al. 1999). As opposed to any single pathogen, the mother's immune response itself may be related to the offspring's increased risk for schizophrenia (Zuckerman and Weiner 2005). A fivefold increased risk for developing psychoses later on, however, was also observed after infection of the central nervous system (CNS) in early childhood (Gattaz et al. 2004; Koponen et al. 2004). Additional recent data also provide evidence for an increased risk to develop schizophrenia after a childhood CNS infection (Brown 2008; Dalman et al. 2008).

Prenatal immune activation—infection triggered or not—is an important risk factor for schizophrenia (Meyer et al. 2011). In humans, increased maternal levels of the pro-inflammatory cytokine IL-8 during pregnancy were shown to be associated with an increased risk for schizophrenia in the offspring—whatever the reason for the increase in IL-8 (Brown 2006). Moreover, increased maternal IL-8 levels in pregnancy were also significantly related to decreased brain volume in the schizophrenic offspring, i.e. lower volumes of the right posterior cingulum and left enterorhinal cortex and higher volumes of the cerebrospinal fluid ventricles (Ellman et al. 2010).

A recent, large-scale study from the same Danish group mentioned above showed, however, that severe infections and autoimmune disorders during a person's lifetime additively increase the risk for schizophrenia and schizophrenia spectrum disorders (Benros et al. 2011), whereas it did not confirm infections of the parents—including intrauterine infections—as definite risk factors (Benros et al. 2011). The sensitivity of the study was not very high, despite its large scale, so it may have clearly identified only the tip of an iceberg of risk factors.

One of our own studies on the relationship between infectious agents (measured by antibody titres) and schizophrenia revealed higher titres of several different infectious agents ('susceptibility index') in schizophrenia, indicating that not a specific virus, bacterium, protozoon, etc. might be responsible for the schizophrenia risk, but the inflammatory process (Krause et al. 2010).

Signs of inflammation have been found in schizophrenic brains (Körschenhausen et al. 1996) and the term 'mild localized chronic encephalitis' was proposed to describe a slight but chronic inflammatory process in schizophrenia (Bechter et al. 2003).

Effects of Cyclooxygenase-2 Inhibition

One class of modern drugs is well known to induce a shift from a type-2- to a type-1-dominated immune response: the selective COX-2 inhibitors (Padol and Hunt 2010; Aid and Bosetti 2011). Several studies have demonstrated the type-2-inducing effect of PGE_2—the major product of COX-2—and that inhibition of COX-2 is accompanied by inhibition of type 2 cytokines and induction of type 1 cytokines (Pyeon et al. 2000; Stolina et al. 2000). Recently, PGE_2 was shown to enhance the production of type 2 cytokines such as IL-4, IL-5, IL-6 and IL-10; PGE_2 also drastically inhibits the production of the type 1 cytokines IFN-γ, IL-2 and IL-12 (Stolina et al. 2000). Therefore, inhibition of PGE_2 synthesis is hypothesized to be beneficial in the treatment of disorders with dysregulated T helper cell responses (Harris et al. 2002). COX-2 inhibition seems to re-balance the type 1/type 2 immune response by inhibiting IL-6 and PGE_2 and stimulating the type 1 immune response (Litherland et al. 1999). Therefore COX-2 inhibition seems to be a promising approach in the therapy of schizophrenia, in particular since increased COX-2 expression was found in this disorder (Das and Khan 1998).

COX inhibition has differential effects on kynurenine metabolism: while COX-1 inhibition increases the levels of kynurenic acid (KYNA), a metabolite of kynurenine, COX-2 inhibition decreases them (Schwieler et al. 2005). Therefore, psychotic symptoms and cognitive dysfunctions, observed during therapy with COX-1 inhibitors, were assigned to the COX-1-mediated increase of KYNA. The reduction of KYNA levels—by a prostaglandin-mediated mechanism—might be an additional mechanism to the above described immunological mechanism for therapeutic effects of selective COX-2 inhibitors in schizophrenia (Schwieler et al. 2005).

Cyclooxygenase Inhibitors as Adjunctive Therapy in Schizophrenia

Indeed, a prospective, randomized, double-blind study of the COX-2 inhibitor celecoxib as an add-on to risperidone in acute exacerbation of schizophrenia found a therapeutic effect of celecoxib (Müller and Schwarz 2002). Immunologically, an increase of the type 1 immune response was found in the celecoxib treatment group (Müller et al. 2004a). The clinical effect of COX-2 inhibition was especially pronounced regarding cognition in schizophrenia (Müller et al. 2005b). The finding of a clinical advantage of COX-2 inhibition, however, could not be replicated in a second study. Further analysis of the data revealed that the outcome depends on the duration of the disease (Müller et al. 2004b). The efficacy of therapy with a COX-2 inhibitor seems most pronounced in the first years of the schizophrenic disease process. This observation is in accordance with results from animal studies showing that the effects of COX-2 inhibition on cytokines, hormones and particularly behavioural symptoms are dependent on the duration of the preceding changes and the time of application of the COX-2 inhibitor (Casolini et al. 2002). Subsequent clinical studies with a similar randomized double-blind placebo-controlled design of 400 mg celecoxib as an add-on to risperidone (in one study risperidone or olanzapine) in partly different patient populations obtained similar positive results of cyclooxygenase inhibition: in a Chinese population of first manifestation schizophrenics (Zhang et al. 2006) and in an Iranian sample of chronic schizophrenics (Akhondzadeh et al. 2007). No advantage of celecoxib could be found in continuously ill chronic schizophrenia patients (Rapaport et al. 2005). An open-label study from a group in India that used 400 mg celecoxib in addition to olanzapine in schizophrenia patients with an acute exacerbation revealed a significant superiority of the celecoxib group (Baheti et al. 2013).

On the basis of the findings discussed above, a point of no return for therapeutic effects regarding the pathological changes during the long-term course of an inflammatory process has to be postulated.

Since therapy with COX-2 inhibitors is suggested to be most beneficial in the early stages of the disease, we conducted a 6-week, double-blind, placebo-controlled, randomized study of celecoxib added on to amisulpride in 49 patients with a first manifestation of schizophrenia. Patients were treated with either amisulpride (200–1,000 mg) plus celecoxib (400 mg) or amisulpride (200–1,000 mg) plus placebo. Inclusion criteria were a diagnosis of schizophrenia according to DSM-IV within the past 2 years. Psychopathology was assessed at weekly intervals with the positive and negative syndrome scale (PANSS). The outcome in the PANSS negative, global and total scales was significantly better in the amisulpride plus celecoxib group than in the amisulpride plus placebo group. In addition, the clinical global impression (CGI) scale also improved significantly more in the amisulpride and celecoxib group. This was the first study to show a superior effect of celecoxib on negative symptoms. Future studies of the effect of COX-2 inhibitors on prodromal and negative symptoms of schizophrenia are needed (Müller 2010; Müller et al. 2010). A recent study found a beneficial effect also of acetylsalicylic acid in schizophrenia spectrum disorders (Laan et al. 2010).

In the meantime, meta-analytic studies have been performed on the use of nonsteroidal anti-inflammatory drugs (NSAIDs) in schizophrenia. The first study found a significant benefit of add-on treatment with NSAIDs on positive and negative symptoms and total symptomatology over all studies (Sommer et al. 2012). A second meta-analysis on the same topic but based on a broader database found a benefit only in early stages of the disease, in particular in first manifestations of schizophrenia (Nitta et al. 2013) (Table 17.1).

Table 17.1 Studies of COX-2 inhibitors in the therapy of schizophrenia

Authors	Diagnosis	Course and duration	Duration of trial	*N*	Study design	Concomitant drug	COX-2 inhibitor	Outcome
Zhang et al. (2006)	Schizophrenia	First manifestation	12 weeks	40		Risperidone (flexible dose)	Celecoxib 400 mg/day	Significant advantage of the COX-2 inhibitor
Müller et al. (2002)	Schizophrenia	Not specified, mean 5.9 years	5 weeks	50		Risperidone (flexible dose)	Celecoxib 400 mg/day	Significant advantage of the COX-2 inhibitor
Rappard and Müller (2004)	Schizophrenia	≤10 years	11 weeks	270		Risperidone (flexible dose)	Celecoxib 400 mg/day	No advantage of the COX-2 inhibitor
Rapaport et al. (2005)	Schizophrenia	Continuously ill mean 20 years	8 weeks	38		Risperidone or olanzapine (fixed dose)	Celecoxib 400 mg/day	No advantage on the COX-2 inhibitor
Akhondzadeh et al. (2007)	Schizophrenia	Chronic type (active phase)	8 weeks	60		Risperidone (fixed dose)	Celecoxib 400 mg/day	Significant advantage of the COX-2 inhibitor
Müller et al. (2010)	Schizophrenia	First manifestation (<2 years)	6 weeks	50		Amisulpride (flexible dose)	Celecoxib 400 mg/day	Significant advantage of the COX-2 inhibitor
Baheti et al. (2013)	Schizophrenia	Acute exacerbation	6 weeks	62	Double-blind, randomized, placebo-controlled, add-on	Olanzapine	Celecoxib 400 mg/day	Significant advantage of the COX-2 inhibitor

Further Immunomodulatory Approaches to Treat Schizophrenia

Several studies have been performed with omega-3 fatty acids in schizophrenia. However, the results are inconsistent and the effect size is small in both first manifestation schizophrenia and chronic schizophrenia (Ross et al. 2007). Of great interest are the findings of a 1-year study by Amminger et al. (2010) in a sample of people at high risk for schizophrenia who were already showing prodromal symptoms. This study found a significantly lower rate of progression to psychoses in those people who received omega-3 fatty acids compared to the placebo group. The use of omega-3 fatty acids is discussed in detail in Chap. 18 of this book.

In addition to its other effects, erythropoietin has immunomodulatory effects. In a 12-week placebo-controlled study in chronic schizophrenia patients, cognition improved significantly more with rh-erythropoietin than with placebo. However, rh-erythropoietin was not superior to placebo in overall psychopathology, measured with the PANSS scale, or social functioning (Ehrenreich et al. 2007). Another interesting effect of rh-erythropoietin is that it is able to slow the loss of CNS volume in schizophrenia (Wüstenberg et al. 2011).

Recent studies have shown that the modern tetracycline antibiotic minocycline can also be effective as an add-on treatment in schizophrenia (Chaves et al. 2009). Although minocycline has numerous effects in the brain, the most relevant one is the anti-inflammatory effect via modulation of the oxidative system. Minocycline also inhibits microglia activation. PET studies show that microglia are more strongly activated in schizophrenia patients than in healthy controls (van Berckel et al. 2008). In addition to case reports of clinical effects of minocycline in schizophrenia (Ahuja and Carroll 2007) and reports from animal experiments (Mizoguchi et al. 2008), double-blind, controlled studies have also shown positive effects of minocycline on cognitive functions and negative symptoms (Levkovitz et al. 2010; Chaudhry et al. 2012).

A double-blind, prospective, randomized study of the virostatic agent valaciclovir found no advantage with regard to schizophrenic symptoms in patients seropositive for cytomegalovirus (Dickerson et al. 2003, 2009a). Another study with azithromycine also found no superiority over place in schizophrenia patients seropositive for toxoplasmosis (Dickerson et al. 2009b).

Type-1 Immune Activation in Schizophrenia: A Possible Therapeutic Mechanism?

A blunted type 1 (acute) immune response and shift to a type 2 (chronic) response have been described previously in schizophrenia: serum levels of the proinflammatory type 1 cytokine IFN-γ and after in vitro stimulation, the IFN-γ production were lower in (unmedicated) schizophrenia patients than in healthy controls (Rothermundt et al. 2000; Schwarz et al. 2001), although the findings are in part controversial (Miller et al. 2011). Stimulation of the blunted part of the immune response and downregulation of the (upregulated) immune response by anti-inflammatory medication might be two sides of the same coin. Therefore the type 1 response stimulant IFN-γ was hypothesized to have a therapeutic effect in schizophrenia. The effects of adjunctive IFN-γ in two treatment-resistant schizophrenia inpatients were evaluated and both patients showed an impressive therapeutic benefit from the IFN-γ therapy (Grüber et al. 2014). However, these results have to be considered cautiously because of the very limited experiences in two patients, possible side effects of IFN-γ and the fact that the data are preliminary.

Possible unwanted immune effects of IFN-γ, a strong type-1 immune activator, have to be taken into account. On the other hand, the type 1 stimulation might be a further option to re-balance the type 1/type 2 imbalance, in particular in patients showing a blunted type 1 immune response.

Celecoxib as Adjunctive Therapy in Major Depression

Anti-inflammatory treatment would be expected to show antidepressant effects also in depressed patients, because of the increase of pro-inflammatory cytokines and PGE_2 in depression. COX-2 inhibitors in particular seem to have beneficial effects: in animal studies COX-2 inhibition can lower the increase of the pro-inflammatory cytokines IL-1α, TNF-α and of PGE_2 and it can also prevent clinical symptoms such as anxiety and cognitive decline that are associated with this increase of pro-inflammatory cytokines (Casolini et al. 2002).

Additionally, COX-2 inhibitors influence the CNS serotonergic system, either directly or via CNS immune mechanisms. In a rat model, treatment with rofecoxib was followed by an increase of serotonin in the frontal and temporo-parietal cortex (Sandrini et al. 2002). Since a lack of serotonin is one of the key features in the pathophysiology of depression, a clinical antidepressant effect of COX-2 inhibitors would therefore be expected. A possible mechanism of the antidepressant action of COX-2 inhibitors is the inhibition of IL-1 and IL-6 release. Moreover, COX-2 inhibitors also protect the CNS from the effects of quinolinic acid (Salzberg-Brenhouse et al. 2003). In the depression model of the bulbectomized rat, a decrease of cytokine levels in the hypothalamus and a change in behaviour have been observed after chronic celecoxib treatment (Myint et al. 2007). In another animal model of depression, however, the mixed COX-1/COX-2 inhibitor acetylsalicylic acid showed an additional antidepressant effect by accelerating the antidepressant effect of fluoxetine (Brunello et al. 2006).

Accordingly, a clinical antidepressant effect of rofecoxib was found in 2,228 patients with osteoarthritis, 15 % of whom had a co-morbid depressive syndrome, evaluated by a specific depression self-report. Co-morbid depression was a significant predictor for worse outcome (assessed by osteoarthritis-related pain) after rofecoxib therapy. Surprisingly, during therapy with 25 mg rofecoxib the rate of substantive depression decreased significantly from 15 to 3 % of the patients (Collantes-Esteves and Fernandez-Perrez 2003).

Moreover, we were able to demonstrate a significant therapeutic effect of the selective COX-2 inhibitor celecoxib on depressive symptoms in a randomized double-blind pilot add-on study in MD (Müller et al. 2006). In another clinical study, the mixed COX-1/COX-2 inhibitor acetylsalicylic acid accelerated the antidepressant effect of fluoxetine and increased the response rate in depressed non-responders to a monotherapy with fluoxetine in an open-label pilot study (Mendlewicz et al. 2006).

COX-2 inhibitors have also showed interesting effects in animal models of depression. Treatment with the COX-2 inhibitor celecoxib, but not with a COX-1 inhibitor, prevented the dysregulation of the HPA axis, in particular the increase of cortisol, one of the key biological features associated with depression (Casolini et al. 2002; Hu et al. 2005). This effect was expected because PGE_2, which stimulates the HPA axis in the CNS (Song and Leonard 2000), is inhibited by COX-2 inhibition. The functional effects of IL-1 in the CNS, which include sickness behaviour, were also shown to be antagonized by treatment with a selective COX-2 inhibitor (Cao et al. 1999).

Another randomized, double-blind study in 50 patients with MD also showed a significantly better outcome with the COX-2 inhibitor celecoxib plus fluoxetine than with fluoxetine alone (Akhondzadeh et al. 2009). A similar result was obtained with a celecoxib add-on approach to sertraline in MD (Abbasi et al. 2012). A meta-analysis on the use of COX-2 inhibitors in MD found an overall benefit of celecoxib add-on therapy (Na et al. 2013).

Although those preliminary data have to be interpreted cautiously and intense research is required in order to further evaluate the therapeutic effects of COX-2 inhibitors in MD, those results are encouraging for further studies on the inflammatory hypothesis of depression with regard to pathogenesis, course and therapy (Table 17.2).

Table 17.2 Studies of COX-2 inhibitors in the therapy of major depression

Authors	Diagnosis	Duration of trial	*N*	Study design	Concomitant drug	COX-2 inhibitor	Outcome
Collantes-Esteves and Fernandez-Perrez (2003)	Depressive syndrome, co-morbid to osteoarthritis	Mean 33 days	343	Open	Not specified	Rofecoxib 12.5 or 25 mg/day	Significant reduction of self-reported depression
Müller et al. (2006)	Major depression	6 weeks		Randomized double-blind, placebo-controlled add-on	Reboxetine (flexible dose)	Celecoxib 400 mg/day	Significant superiority of the COX-2 inhibitor
Akhondzadeh et al. (2007)	Major depression	6 weeks	50	Randomized, double-blind, placebo-controlled add-on	Fluoxetine (flexible dose)	Celecoxib 400 mg/day	Significant superiority of celecoxib
Abbasi et al. (2012)	Major depression	6 weeks	40	Randomized, double-blind, placebo-controlled	Sertralin 200 mg/day	Celecoxib 400 mg/day	Significant superiority and more responders in celecoxib group; IL-6 levels predict response and remission
Nery et al. (2008)	Bipolar disorder, depressive or mixed episode	6 weeks	28	Randomized, double-blind, placebo-controlled	Mood stabilizer or atypical antipsychotics	Celecoxib 400 mg/day	Significant superiority after 1 week, no difference at endpoint
Begemann et al. (2008)	Bipolar depression, rapid cycling	>5 months	1	Open	Not specified	Celecoxib 400 mg/day	Significant improvement of depressed and manic symptoms
Müller et al. in preparation	Major depression	6 weeks	66 (30)	Randomized, double-blind, placebo-controlled	Sertraline (100–150 mg)	Cimicoxib 50 mg	No difference in total group, significant superiority in severely depressed (HamD $\geq$ 25)

Further Immunomodulatory Approaches to Treat Major Depression

Interestingly, in a study of patients with rheumatoid arthritis, eternacept, which blocks the interaction of TNF-α with the TNF-α cell surface receptors, showed a highly significant antidepressant effect on the beck depression inventory (BDI), a self-rating scale (Raison et al. 2012). Depression, however, was not the primary outcome criterion in this study (Tyring et al. 2006).

Another interesting study with the TNF-α receptor blocker infliximab in treatment-resistant patients with MD found no overall benefit, but did find a benefit in those with higher levels of inflammatory markers, such as CRP, TNF-α or soluble TNF-receptors (Raison et al. 2012), i.e. those patients who showed pronounced signs of inflammation responded better to infliximab than depressed patients without signs of inflammation.

Although those preliminary data have to be interpreted cautiously and further research is needed to evaluate the therapeutic effects of COX-2 inhibitors in MD, these results are encouraging for further studies on the inflammatory hypothesis of depression and the pathogenesis, course and therapy of the disease.

Risks Associated with COX-2 Inhibitors

Several COX-2-inhibiting substances have been withdrawn from the market because of side effects, especially during long-term use. Compared to mixed COX-1/COX-2 inhibitors such as aspirin, the selective COX-2 inhibitors show fewer gastrointestinal side effects such as bleeding but a higher rate of cardiovascular side effects (Katz 2013). In a placebo-controlled study of rofecoxib—withdrawn from the market because of cardiovascular side effects—the increased relative risk for a cardiovascular event became apparent after 18 months of treatment; during the first 18 months, the event rates were similar in the rofecoxib and placebo groups. The results primarily reflect a greater number of myocardial infarctions and ischaemic cerebrovascular events in the rofecoxib group, while the overall and cardiovascular mortality was similar in the rofecoxib and placebo groups (Bresalier et al. 2005). These data for rofecoxib (at a dose of 25 mg/day) show that short-term treatment with COX-2 inhibitors is safe with regard to cardiovascular side effects. The cardiovascular effects, however, vary between the individual COX-2-inhibiting drugs. The strongest evidence for an increased risk of serious cardiovascular events is with rofecoxib therapy. Celecoxib therapy may be associated with an increased risk of cardiovascular events, but only when used at doses substantially higher than those recommended for the treatment of arthritis. There is a greater body of evidence supporting the relative cardiovascular safety of celecoxib when used at the doses recommended for the treatment of arthritis than for any of the other selective COX-2 inhibitors or NSAIDs (Howes 2007).

Increased risk for cardiovascular events was an exclusion criterion in our studies with celecoxib, during which we did not observe any cardiovascular events. In our recent studies, we estimated N-terminal probrain natriuretic peptide (NT-proBNP) as a marker for an increased risk for cardiovascular events; only patients showing normal NT-proBNP values were included into the studies. Plasma NT-proBNP is the best diagnostic marker for increased risks for cardiovascular events and has high sensitivity and specificity (Toufan et al. 2014).

In addition to the above-mentioned side effects of COX-2 inhibitors, some further concerns have been raised for the use of COX-2 inhibitors in depression (Maes 2012), for example the rebalancing effect of the type 1/type 2 immune response (Aid and Bosetti 2011). However, this effect seems to drive the therapeutic effect in schizophrenia. Many studies—including meta-analyses—have proven the beneficial effect of celecoxib and its tolerability in MD is well established. Therefore concerns about the use of COX-2 inhibitors, raised particularly on the basis of in vitro studies, have been disproved by clinical studies, although long-term studies are lacking.

Methodological Aspects of Clinical Studies That Use the Immunomodulatory and Anti-inflammatory Therapeutic Approach

Limited therapeutic progress has been made in schizophrenia and depression in the past few years. One of the limitations is the situation in the pharmaceutical market for psychiatric disorders. Although antipsychotics, including second generation antipsychotics, are known to have good effects on schizophrenic positive symptoms, their effects on negative symptoms and cognitive performance are limited. It is even under discussion whether antipsychotics influence the natural course of schizophrenia at all. However, new therapeutic approaches have to be compared with state-of-the-art therapy. For ethical reasons, antipsychotics cannot be denied to schizophrenia patients. Therefore, studies of new therapeutic approaches have to be tested in an add-on design. Showing an add-on effect to an antipsychotic in randomized controlled trials, especially in short-term studies over a few weeks, represents a great methodological challenge for the investigator.

One possible solution to overcome this ethical dilemma is to choose a subgroup of treatment-resistant patients, for example, but this requires that the drug being tested has a special effect regarding response to therapy in this negative selection of patients. In MD, however, the challenge is even greater: responses of up to 40 % have been observed in the placebo group in randomized controlled trials of an antidepressant. The lack of a statistically significant difference between the antidepressant and placebo is one reason why a lot of antidepressant studies fail. Showing a therapeutic effect of a drug given as an add-on to an effective antidepressant is an even bigger challenge and may not be possible, even in large studies.

The placebo response becomes smaller as the severity of depression increases. Therefore the therapeutic effect can be shown more easily in more severely depressed patients. However, for ethical reasons the more severely depressed the patients are, the more difficult it is to include them in placebo-controlled trials.

Inflammation may play a role in subgroups of schizophrenia and depressed patients, but this role cannot yet be clearly defined. Therefore it is difficult to show superiority in an add-on study in an unselected group of patients with schizophrenia or MD and the add-on compound would need to have a pronounced effect. A study in a selected subgroup of patients, e.g. patients showing signs of inflammation, might reveal better therapeutic effects. The selection of patients, however, requires selection criteria, especially for the therapeutic response, which are unknown before studies have been performed with a certain compound.

These facts and considerations have to be taken into account when interpreting the results of studies and planning future studies based on anti-inflammatory or immune-modulating modes of action in psychiatric disorders.

Acknowledgements A part of tis overview has been published before (Müller 2014a, b). The author thanks J. Klesing for help with the manuscript.

References

Abbasi SH, Hosseini F, Modabbernia A, Ashrafi M, Akhondzadeh S. Effect of celecoxib add-on treatment on symptoms and serum IL-6 concentrations in patients with major depressive disorder: randomized double-blind placebo-controlled study. J Affect Disord. 2012;18.

Ahuja N, Carroll BT. Possible anti-catatonic effects of minocycline in patients with schizophrenia. Prog Neuropsychopharmacol Biol Psychiatry. 2007;31(4):968–9.

Aid S, Bosetti F. Targeting cyclooxygenases-1 and -2 in neuroinflammation: therapeutic implications. Biochimie. 2011;93(1):46–51.

Akhondzadeh S, Tabatabaee M, Amini H, Ahmadi Abhari SA, Abbasi SH, Behnam B. Celecoxib as adjunctive therapy in schizophrenia: a double-blind, randomized and placebo-controlled trial. Schizophr Res. 2007;90(1–3):179–85.

Akhondzadeh S, Jafari S, Raisi F, Nasehi AA, Ghoreishi A, Salehi B, et al. Clinical trial of adjunctive celecoxib treatment in patients with major depression: a double blind and placebo controlled trial. Depress Anxiety. 2009;26(7):607–11.

Amminger GP, Schafer MR, Papageorgiou K, Klier CM, Cotton SM, Harrigan SM, et al. Long-chain omega-3 fatty acids for indicated prevention of psychotic disorders: a randomized, placebo-controlled trial. Arch Gen Psychiatry. 2010;67(2):146–54.

Baheti T, Nischal A, Nischal A, Khattri S, Arya A, Tripathi A, et al. A study to evaluate the effect of celecoxib as add-on to olanzapine therapy in schizophrenia. Schizophr Res. 2013;147(1):201–2.

Bechter K, Schreiner V, Herzog S, Breitinger N, Wollinsky KH, Brinkmeier H, et al. Cerebrospinal fluid filtration as experimental therapy in therapy refractory psychoses in Borna disease virus seropositive patients. Therapeutic effects, findings. Psychiatr Prax. 2003;30(Suppl 2):216–20.

Begemann M, Sargin D, Rossner MJ, Bartels C, Theis F, Wichert SP, et al. Episode-specific differential gene expression of peripheral blood mononuclear cells in rapid cycling supports novel treatment approaches. Mol Med. 2008;14(9–10):546–52. doi: 10.2119/2008-00053.

Benros ME, Nielsen PR, Nordentoft M, Eaton WW, Dalton SO, Mortensen PB. Autoimmune diseases and severe infections as risk factors for schizophrenia: a 30-year population-based register study. Am J Psychiatry. 2011;168(12):1303–10.

Benros ME, Waltoft BL, Nordentoft M, Ostergaard SD, Eaton WW, Krogh J, et al. Autoimmune diseases and severe infections as risk factors for mood disorders: a nationwide study. JAMA Psychiatry. 2013;70(8):812–20.

Bresalier RS, Sandler RS, Quan H, Bolognese JA, Oxenius B, Horgan K, et al. Cardiovascular events associated with rofecoxib in a colorectal adenoma chemoprevention trial. N Engl J Med. 2005;352(11):1092–102.

Brown AS. Prenatal infection as a risk factor for schizophrenia. Schizophr Bull. 2006;32(2): 200–2.

Brown AS. The risk for schizophrenia from childhood and adult infections. Am J Psychiatry. 2008;165(1):7–10.

Brown AS, Begg MD, Gravenstein S, Schaefer CA, Wyatt RJ, Bresnahan M, et al. Serologic evidence of prenatal influenza in the etiology of schizophrenia. Arch Gen Psychiatry. 2004;61(8): 774–80.

Brunello N, Alboni S, Capone G, Benatti C, Blom JM, Tascedda F, et al. Acetylsalicylic acid accelerates the antidepressant effect of fluoxetine in the chronic escape deficit model of depression. Int Clin Psychopharmacol. 2006;21(4):219–25.

Buka SL, Tsuang MT, Torrey EF, Klebanoff MA, Bernstein D, Yolken RH. Maternal infections and subsequent psychosis among offspring. Arch Gen Psychiatry. 2001;58(11):1032–7.

Calabrese JR, Skwerer RG, Barna B, Gulledge AD, Valenzuela R, Butkus A, et al. Depression, immunocompetence, and prostaglandins of the E series. Psychiatry Res. 1986;17(1):41–7.

Cao C, Matsumura K, Ozaki M, Watanabe Y. Lipopolysaccharide injected into the cerebral ventricle evokes fever through induction of cyclooxygenase-2 in brain endothelial cells. J Neurosci. 1999;19(2):716–25.

Casolini P, Catalani A, Zuena AR, Angelucci L. Inhibition of COX-2 reduces the age-dependent increase of hippocampal inflammatory markers, corticosterone secretion, and behavioral impairments in the rat. J Neurosci Res. 2002;68(3):337–43.

Chaudhry IB, Hallak J, Husain N, Minhas F, Stirling J, Richardson P, et al. Minocycline benefits negative symptoms in early schizophrenia: a randomised double-blind placebo-controlled clinical trial in patients on standard treatment. J Psychopharmacol. 2012;26(9):1185–93.

Chaves C, Marque CR, Chaudhry IB, Hussain FM, Oliveira JP, Wichert-Ana L. Short-term improvement by minocycline added to olanzapine antipsychotic treatment in paranoid schizophrenia. Schizophr Bull. 2009;35 Suppl 1:354.

Collantes-Esteves E, Fernandez-Perrez C. Improved self-control of ostheoarthritis pain and self-reported health status in non-responders to celecoxib switched to rofecoxib: results of PAVIA, an open-label post-marketing survey in Spain. Curr Med Res Opin. 2003;19:402–10.

Dalman C, Allebeck P, Gunnell D, Harrison G, Kristensson K, Lewis G, et al. Infections in the CNS during childhood and the risk of subsequent psychotic illness: a cohort study of more than one million Swedish subjects. Am J Psychiatry. 2008;165(1):59–65.

Dantzer R. Cytokine-induced sickness behavior: where do we stand? Brain Behav Immun. 2001; 15(1):7–24.

Das I, Khan NS. Increased arachidonic acid induced platelet chemiluminescence indicates cyclooxygenase overactivity in schizophrenic subjects. Prostaglandins Leukot Essent Fatty Acids. 1998;58(3):165–8.

Dickerson FB, Boronow JJ, Stallings CR, Origoni AE, Yolken RH. Reduction of symptoms by valacyclovir in cytomegalovirus-seropositive individuals with schizophrenia. Am J Psychiatry. 2003;160(12):2234–6.

Dickerson FB, Stallings CR, Boronow JJ, Origoni AE, Sullens A, Yolken RH. Double blind trial of adjunctive valacyclovir in individuals with schizophrenia who are seropositive for cytomegalovirus. Schizophr Res. 2009a;107(2–3):147–9.

Dickerson FB, Stallings CR, Boronow JJ, Origoni AE, Yolken RH. A double-blind trial of adjunctive azithromycin in individuals with schizophrenia who are seropositive for Toxoplasma gondii. Schizophr Res. 2009b;112(1–3):198–9.

Dowlati Y, Herrmann N, Swardfager W, Liu H, Sham L, Reim EK, et al. A meta-analysis of cytokines in major depression. Biol Psychiatry. 2010;67(5):446–57.

Ehrenreich H, Hinze-Selch D, Stawicki S, Aust C, Knolle-Veentjer S, Wilms S, et al. Improvement of cognitive functions in chronic schizophrenic patients by recombinant human erythropoietin. Mol Psychiatry. 2007;12(2):206–20.

Ellman LM, Deicken RF, Vinogradov S, Kremen WS, Poole JH, Kern DM, et al. Structural brain alterations in schizophrenia following fetal exposure to the inflammatory cytokine interleukin-8. Schizophr Res. 2010;121(1–3):46–54.

Fond G, Hamdani N, Kapczinski F, Boukouaci W, Drancourt N, Dargel A, et al. Effectiveness and tolerance of anti-inflammatory drugs' add-on therapy in major mental disorders: a systematic qualitative review. Acta Psychiatr Scand. 2013;11:10.

Gattaz WF, Abrahao AL, Foccacia R. Childhood meningitis, brain maturation and the risk of psychosis. Eur Arch Psychiatry Clin Neurosci. 2004;254(1):23–6.

Grüber L, Bunse T, Weidinger E, Reichard H, Müller N. Adjunctive recombinant human interferon gamma-1b for treatment-resistant schizophrenia in 2 patients. J Clin Psychiatry. 2014;75:1266–7.

Harris SG, Padilla J, Koumas L, Ray D, Phipps RP. Prostaglandins as modulators of immunity. Trends Immunol. 2002;23(3):144–50.

Howes LG. Selective COX-2 inhibitors, NSAIDs and cardiovascular events – is celecoxib the safest choice? Ther Clin Risk Manag. 2007;3(5):831–45.

Howren MB, Lamkin DM, Suls J. Associations of depression with C-reactive protein, IL-1, and IL-6: a meta-analysis. Psychosom Med. 2009;71(2):171–86.

Hu F, Wang X, Pace TW, Wu H, Miller AH. Inhibition of COX-2 by celecoxib enhances glucocorticoid receptor function. Mol Psychiatry. 2005;10(5):426–8.

Katz JA. COX-2 inhibition: what we learned—a controversial update on safety data. Pain Med. 2013;14 Suppl 1:S29–34. doi:10.1111/pme.12252.:S29-S34.

Koponen H, Rantakallio P, Veijola J, Jones P, Jokelainen J, Isohanni M. Childhood central nervous system infections and risk for schizophrenia. Eur Arch Psychiatry Clin Neurosci. 2004;254(1):9–13.

Körschenhausen DA, Hampel HJ, Ackenheil M, Penning R, Müller N. Fibrin degradation products in post mortem brain tissue of schizophrenics: a possible marker for underlying inflammatory processes. Schizophr Res. 1996;19(2–3):103–9.

Kraepelin E. Über Psychosen nach Influenza. Deutsche Medicinische Wochenschrift. 1890;11:209–12.

Kraepelin E. Ziele und Wege der Psychiatrischen Forschung. Z Ges Neurol Psychiatrie. 1918;42:169–205.

Krause D, Matz J, Weidinger E, Wagner J, Wildenauer A, Obermeier M, et al. The association of infectious agents and schizophrenia. World J Biol Psychiatry. 2010;11(5):739–43.

Laan W, Grobbee DE, Selten JP, Heijnen CJ, Kahn RS, Burger H. Adjuvant aspirin therapy reduces symptoms of schizophrenia spectrum disorders: results from a randomized, double-blind, placebo-controlled trial. J Clin Psychiatry. 2010;71(5):520–7.

Levkovitz Y, Mendlovich S, Riwkes S, Braw Y, Levkovitch-Verbin H, Gal G, et al. A double-blind, randomized study of minocycline for the treatment of negative and cognitive symptoms in early-phase schizophrenia. J Clin Psychiatry. 2010;71(2):138–49.

Linnoila M, Whorton AR, Rubinow DR, Cowdry RW, Ninan PT, Waters RN. CSF prostaglandin levels in depressed and schizophrenic patients. Arch Gen Psychiatry. 1983;40(4):405–6.

Litherland SA, Xie XT, Hutson AD, Wasserfall C, Whittaker DS, She JX, et al. Aberrant prostaglandin synthase 2 expression defines an antigen-presenting cell defect for insulin-dependent diabetes mellitus. J Clin Invest. 1999;104(4):515–23.

Maes M. Cytokines in major depression [letter; comment]. Biol Psychiatry. 1994;36(7):498–9.

Maes M. Targeting cyclooxygenase-2 in depression is not a viable therapeutic approach and may even aggravate the pathophysiology underpinning depression. Metab Brain Dis. 2012;27(4):405–13.

Maes M, Stevens W, DeClerck L, Bridts C, Peeters D, Schotte C, et al. Immune disorders in depression: higher T helper/T suppressor-cytotoxic cell ratio. Acta Psychiatr Scand. 1992;86(6):423–31.

Mendlewicz J, Kriwin P, Oswald P, Souery D, Alboni S, Brunello N. Shortened onset of action of antidepressants in major depression using acetylsalicylic acid augmentation: a pilot open-label study. Int Clin Psychopharmacol. 2006;21(4):227–31.

Meyer U, Schwarz MJ, Müller N. Inflammatory processes in schizophrenia: a promising neuroimmunological target for the treatment of negative/cognitive symptoms and beyond. Pharmacol Ther. 2011;132(1):96–110.

Miller BJ, Buckley P, Seabolt W, Mellor A, Kirkpatrick B. Meta-analysis of cytokine alterations in schizophrenia: clinical status and antipsychotic effects. Biol Psychiatry. 2011;70(7):663–71.

Mizoguchi H, Takuma K, Fukakusa A, Ito Y, Nakatani A, Ibi D, et al. Improvement by minocycline of methamphetamine-induced impairment of recognition memory in mice. Psychopharmacology (Berl). 2008;196(2):233–41.

Müller N. COX-2 inhibitors as antidepressants and antipsychotics: clinical evidence. Curr Opin Investig Drugs. 2010;11(1):31–42.

Müller N. Immunology of major depression. Neuroimmunomodulation. 2014a;21:123–30.

Müller N. Immunology of schizophrenia. Neuroimmunomodulation. 2014b;(21):109–16.

Müller N, Schwarz MJ. Immunology in anxiety and depression. In: Kasper S, den Boer JA, Sitsen JMA, editors. Handbook of depression and anxiety. 2nd ed. New York: Marcel Dekker; 2002. p. 267–88.

Müller N, Hofschuster E, Ackenheil M, Mempel W, Eckstein R. Investigations of the cellular immunity during depression and the free interval: evidence for an immune activation in affective psychosis. Prog Neuropsychopharmacol Biol Psychiatry. 1993;17(5):713–30.

Müller N, Riedel M, Scheppach C, Brandstätter B, Sokullu S, Krampe K, et al. Beneficial antipsychotic effects of celecoxib add-on therapy compared to risperidone alone in schizophrenia. Am J Psychiatry. 2002;159(6):1029–34.

Müller N, Riedel M, Dehning S, Spellmann I, Müller-Arends A, Cerovecki A, et al. Is the therapeutic effect of celecoxib in schizophrenia depending from duration of disease? Neuropsychopharmacology. 2004a;29(1):176.

Müller N, Ulmschneider M, Scheppach C, Schwarz MJ, Ackenheil M, Möller HJ, et al. COX-2 inhibition as a treatment approach in schizophrenia: immunological considerations and clinical effects of celecoxib add-on therapy. Eur Arch Psychiatry Clin Neurosci. 2004b;254(1):14–22.

Müller N, Schwarz MJ, Riedel M. COX-2 inhibition in schizophrenia: focus on clinical effects of celecoxib therapy and the role of TNF-alpha. In: Eaton WW, editor. Medical and psychiatric comorbidity over the course of life. Washington DC: American Psychiatric Publishing; 2005a. p. 265–76.

Müller N, Riedel M, Schwarz MJ, Engel RR. Clinical effects of COX-2 inhibitors on cognition in schizophrenia. Eur Arch Psychiatry Clin Neurosci. 2005b;255(2):149–51.

Müller N, Schwarz MJ, Dehning S, Douhet A, Cerovecki A, Goldstein-Müller B, et al. The cyclooxygenase-2 inhibitor celecoxib has therapeutic effects in major depression: results of a double-blind, randomized, placebo controlled, add-on pilot study to reboxetine. Mol Psychiatry. 2006;11(7):680–4.

Müller N, Krause D, Dehning S, Musil R, Schennach-Wolff R, Obermeier M, et al. Celecoxib treatment in an early stage of schizophrenia: results of a randomized, double-blind, placebo-controlled trial of celecoxib augmentation of amisulpride treatment. Schizophr Res. 2010; 121:119–24.

Myint AM, Leonard BE, Steinbusch HW, Kim YK. Th1, Th2, and Th3 cytokine alterations in major depression. J Affect Disord. 2005;88(2):167–73.

Myint AM, Steinbusch HW, Goeghegan L, Luchtman D, Kim YK, Leonard BE. Effect of the COX-2 inhibitor celecoxib on behavioural and immune changes in an olfactory bulbectomised rat model of depression. Neuroimmunomodulation. 2007;14(2):65–71.

Na KS, Lee KJ, Lee JS, Cho YS, Jung HY. Efficacy of adjunctive celecoxib treatment for patients with major depressive disorder: a meta-analysis. Prog Neuropsychopharmacol Biol Psychiatry. 2013;48C:79–85. doi:10.1016/j.pnpbp.2013.09.006.

Nery FG, Monkul ES, Hatch JP, Fonseca M, Zunta-Soares GB, Frey BN, et al. Celecoxib as an adjunct in the treatment of depressive or mixed episodes of bipolar disorder: a double-blind, randomized, placebo-controlled study. Hum Psychopharmacol. 2008;23(2):87–94.

Nishino S, Ueno R, Ohishi K, Sakai T, Hayaishi O. Salivary prostaglandin concentrations: possible state indicators for major depression. Am J Psychiatry. 1989;146(3):365–8.

Nitta M, Kishimoto T, Müller N, Weiser M, Davidson M, Kane JM, et al. Adjunctive use of nonsteroidal anti-inflammatory drugs for schizophrenia: a meta-analytic investigation of randomized controlled trials. Schizophr Bull. 2013;39:1230–41.

Ohishi K, Ueno R, Nishino S, Sakai T, Hayaishi O. Increased level of salivary prostaglandins in patients with major depression. Biol Psychiatry. 1988;23(4):326–34.

Padol IT, Hunt RH. Association of myocardial infarctions with COX-2 inhibition may be related to immunomodulation towards a Th1 response resulting in atheromatous plaque instability: an evidence-based interpretation. Rheumatology (Oxford). 2010;49(5):837–43.

Pyeon D, Diaz FJ, Splitter GA. Prostaglandin E(2) increases bovine leukemia virus tax and pol mRNA levels via cyclooxygenase 2: regulation by interleukin-2, interleukin-10, and bovine leukemia virus. J Virol. 2000;74(12):5740–5.

Raison CL, Rutherford RE, Woolwine BJ, Shuo C, Schettler P, Drake DF, et al. A randomized controlled trial of the tumor necrosis factor antagonist infliximab for treatment-resistant depression: the role of baseline inflammatory biomarkers. Arch Gen Psychiatry. 2012;3:1–11.

Rapaport MH, Delrahim KK, Bresee CJ, Maddux RE, Ahmadpour O, Dolnak D. Celecoxib augmentation of continuously ill patients with schizophrenia. Biol Psychiatry. 2005;57(12): 1594–6.

Rappard F, Müller N. Celecoxib add-on therapy does not have beneficial antipsychotic effects over risperidone alone in schizophrenia. Neuropsychopharmacology. 2004;29(Supp 1):222.

Reichenberg A, Yirmiya R, Schuld A, Kraus T, Haack M, Morag A, et al. Cytokine-associated emotional and cognitive disturbances in humans. Arch Gen Psychiatry. 2001;58(5):445–52.

Reichenberg A, Kraus T, Haack M, Schuld A, Pollmacher T, Yirmiya R. Endotoxin-induced changes in food consumption in healthy volunteers are associated with TNF-alpha and IL-6 secretion. Psychoneuroendocrinology. 2002;27(8):945–56.

Ross BM, Seguin J, Sieswerda LE. Omega-3 fatty acids as treatments for mental illness: which disorder and which fatty acid? Lipids Health Dis. 2007;6:21.

Rothermundt M, Arolt V, Leadbeater J, Peters M, Rudolf S, Kirchner H. Cytokine production in unmedicated and treated schizophrenic patients. Neuroreport. 2000;11(15):3385–8.

Rothermundt M, Arolt V, Peters M, Gutbrodt H, Fenker J, Kersting A, et al. Inflammatory markers in major depression and melancholia. J Affect Disord. 2001;63(1–3):93–102.

Salzberg-Brenhouse HC, Chen EY, Emerich DF, Baldwin S, Hogeland K, Ranelli S, et al. Inhibitors of cyclooxygenase-2, but not cyclooxygenase-1 provide structural and functional protection against quinolinic acid-induced neurodegeneration. J Pharmacol Exp Ther. 2003;306(1): 218–28.

Sandrini M, Vitale G, Pini LA. Effect of rofecoxib on nociception and the serotonin system in the rat brain. Inflamm Res. 2002;51(3):154–9.

Schwarz MJ, Chiang S, Müller N, Ackenheil M. T-helper-1 and T-helper-2 responses in psychiatric disorders. Brain Behav Immun. 2001;15(4):340–70.

Schwieler L, Erhardt S, Erhardt C, Engberg G. Prostaglandin-mediated control of rat brain kynurenic acid synthesis–opposite actions by COX-1 and COX-2 isoforms. J Neural Transm. 2005;112(7):863–72.

Sommer IE, de Witte L, Begemann M, Kahn RS. Nonsteroidal anti-inflammatory drugs in schizophrenia: ready for practice or a good start? A meta-analysis. J Clin Psychiatry. 2012;73(4): 414–9.

Song C, Leonard BE. Fundamentals of psychoneuroimmunology. Chichester, New York: Wiley; 2000.

Song C, Lin A, Bonaccorso S, Heide C, Verkerk R, Kenis G, et al. The inflammatory response system and the availability of plasma tryptophan in patients with primary sleep disorders and major depression. J Affect Disord. 1998;49(3):211–9.

Steinberg H, Himmerich H. Emil Kraepelin's habilitation and his thesis: a pioneer work for modern systematic reviews, psychoimmunological research and categories of psychiatric diseases. World J Biol Psychiatry. 2013;14(4):248–57.

Stolina M, Sharma S, Lin Y, Dohadwala M, Gardner B, Luo J, et al. Specific inhibition of cyclooxygenase 2 restores antitumor reactivity by altering the balance of IL-10 and IL-12 synthesis. J Immunol. 2000;164(1):361–70.

Toufan M, Namdar H, Abbasnezhad M, Habibzadeh A, Esmaeili H, Yaraghi S, et al. Diagnostic values of plasma, fresh and frozen urine NT-proBNP in heart failure patients. J Cardiovasc Thorac Res. 2014;6(2):111–5.

Tyring S, Gottlieb A, Papp K, Gordon K, Leonardi C, Wang A, et al. Etanercept and clinical outcomes, fatigue, and depression in psoriasis: double-blind placebo-controlled randomised phase III trial. Lancet. 2006;367(9504):29–35.

van Berckel BN, Bossong MG, Boellaard R, Kloet R, Schuitemaker A, Caspers E, et al. Microglia activation in recent-onset schizophrenia: a quantitative (R)-[11C]PK11195 positron emission tomography study. Biol Psychiatry. 2008;64(9):820–2.

Wagner von Jauregg J. Fieberbehandlung bei Psychosen. Wien Med Wochenschr. 1926;76: 79–82.

Westergaard T, Mortensen PB, Pedersen CB, Wohlfahrt J, Melbye M. Exposure to prenatal and childhood infections and the risk of schizophrenia: suggestions from a study of sibship characteristics and influenza prevalence. Arch Gen Psychiatry. 1999;56(11):993–8.

Wüstenberg T, Begemann M, Bartels C, Gefeller O, Stawicki S, Hinze-Selch D, et al. Recombinant human erythropoietin delays loss of gray matter in chronic schizophrenia. Mol Psychiatry. 2011;16:26–36.

Zhang Y, Chun Chen D, Long Tan Y, Zhou DF. A double-blind, placebo-controlled trial of celecoxib addes to risperidone in first-episode and drug-naive patients with schizophrenia. Eur Arch Psychiatry Clin Neurosci. 2006;256(Suppl 2):50 (Ref Type: Abstract).

Zuckerman L, Weiner I. Maternal immune activation leads to behavioral and pharmacological changes in the adult offspring. J Psychiatr Res. 2005;39(3):311–23.

Chapter 18
Polyunsaturated Fatty Acids in Adult Psychiatric Disorders: A Comprehensive Overview

Tammy Saah, Steven J. Garlow, and Mark Hyman Rapaport

Abstract Essential fatty acids are critical for normal development and function of the brain and central nervous system (CNS). Long-chain poly-unsaturated fatty acids (PUFA), such as omega-3 polyunsaturated fatty acids (ω-3), are crucial to membrane synthesis, structure and fluidity, proper gene expression, and neuronal development. In examining the connection between inflammatory states, physical health, and mental health, it has become apparent that there is an intimate link between chronic inflammation and psychiatric disorders. Because of the moderate benefits found with ω-3 supplementation in other pro-inflammatory states, ω-3 supplementation has since been investigated as a possible alternative or complimentary therapy in psychiatric disorders. We present an overview of the biochemistry and physiology of PUFA followed by a review of the existing clinical literature on the efficacy of ω-3 supplementation in major mental illnesses. The literature on PUFA in mental illnesses is highly heterogeneous and wide-ranging, encompassing many different experimental designs and methods. We have focused on the randomized clinical trials designed to demonstrate efficacy of ω-3 supplementation in recognized psychiatric disorders.

Keywords Omega-3 • Fatty acids • Psychiatric disorders inflammation • MDD • Bipolar • Schizophrenia

Abbreviations

AA	Arachidonic acid
AD	Alzheimer's disease
ADAS-Cog	Alzheimer's Disease Assessment Scale-Cognition
ALA	α-Linolenic acid
ARCD	Age-related cognitive decline

T. Saah (✉) • S.J. Garlow • M.H. Rapaport
Department of Psychiatry and Behavioral Science, Emory University School of Medicine, Emory University, Atlanta, GA 30322, USA
e-mail: tammy.saah@emory.edu

N. Müller et al. (eds.), *Immunology and Psychiatry*, Current Topics in Neurotoxicity 8, DOI 10.1007/978-3-319-13602-8_18

BAI	Beck anxiety inventory
BDI	Beck Depression Inventory
BPAD	Bipolar affective disorder
CANTAB PAL	Cambridge neuropsychological test automated battery paired associates learning
CARDIA	Coronary artery risk development in young adults
CDR	Clinician dementia rating
CDRS	Children's Depression Rating Scale
CES-D	Center for Epidemiologic Studies Depression Scale
CIBIC-plus	Clinician interview-based impression of change plus caregiver input
CNS	Central nervous system
COX	Cyclooxygenase
CVD	Cardiovascular disease
DASS	Depression Anxiety Stress Scale
DHA	Docosahexaenoic acid
DM	Diabetes mellitus
EET	Epoxyeicosantrienoic acid
EPA	Eicosapentaenoic acid
EPDS	Edinburgh Postnatal Depression Scale
ESRS	Extrapyramidal Symptom Rating Scale
GAF	Global assessment of functioning
GDS	Geriatric Depression Scale
HDRS	Hamilton Depression Rating Scale
HETE	Hydroxyeicosatetraenoic acid
IDS-SR30	Inventory of depressive symptoms-self report 30-item
LA	Linoleic acid
LEIDS-R	Leiden Index of Depression Sensitivity-Revised
LOX	Lipooxygenase
MADRS	Montgomery-Asperg Depression Rating Scale
MCI	Mild cognitive impairment
MDD	Major depressive disorder
MDE	Major depressive episode
MI	myocardial infarction
MMSE	Mini Mental Status Exam
NOS	Not otherwise specified
NTB	Neuropsychological test battery
OBI	Organic brain injury
OCD	Obsessive compulsive disorder
PANSS	Positive and Negative Syndrome Scale
PD	Perinatal depression
PLB	Placebo
POMS	Profile of mood states
PUFA	Polyunsaturated fatty acids
RBANS	Repeatable battery for the assessment of neuropsychological status
RCT	Randomized controlled trial
SPM	Specialized pro-resolving mediator

TD	Tardive dyskinesia
YBOCS	Yale–Brown Obsessive Compulsive Scale
YMRS	Young Mania Rating Scale

Introduction

Essential fatty acids are critical for normal development and function of the brain and central nervous system (CNS). Long-chain poly-unsaturated fatty acids (PUFA), such as omega-3 polyunsaturated fatty acids (ω-3), are crucial to membrane synthesis, structure and fluidity, proper gene expression, and neuronal development (Perica and Delas 2011). These molecules have been implicated in contributing to various pathological states in the CNS. Besides their role in membrane structure and cellular physiology, ω-3 fatty acids have anti-inflammatory physiological effects when present and a pro-inflammatory state develops when deficient. Low levels of ω-3 have long been implicated in chronic inflammatory diseases, including cardiovascular disease (CVD) and autoimmune disorders, which are often associated with comorbid psychiatric disorders.

Whereas ω-3 are thought to be anti-inflammatory, ω-6 have been implicated in pro-inflammatory states. This has become important when scrutinizing associations between cultural diet trends and chronic inflammatory diseases. Epidemiological studies in Japan, Greenland, and Alaska documented diets that are rich in ω-3 acquired from marine animal sources (Fenton et al. 2013; Freeman et al. 2006a; Freeman and Rapaport 2011; Giordano and Visioli 2014; Lopresti et al. 2014; Rosenblat et al. 2014; Sanchez-Villegas et al. 2007; Murakami et al. 2010; Tanskanen et al. 2001; Timonen et al. 2004). Conversely, Western society has seen a significant decrease in dietary ω-3 with an increase in corn products and soy-based protein intake, both rich in ω-6 fatty acids (Freeman and Rapaport 2011). Initial epidemiological studies on populations with high dietary ω-3 intake suggested that diets rich in ω-3 correlated with lower levels of chronic inflammatory disease, most consistently CVD but also including cognitive decline (Yates et al. 2014). As the decreased ratio of dietary ω-3:ω-6 in Western society has been thought to be a contributor to high rates of cardiovascular and other inflammatory diseases, ω-3 dietary supplementation has become a target for investigational therapeutic interventions in pro-inflammatory states. However when examining other inflammatory diseases, such as asthma, rheumatoid arthritis, and inflammatory bowel disease, evidence is inconclusive to suggest therapeutic benefit from increased dietary ω-3 consumption. Reviews of current data suggest that dietary supplementation of healthy individuals with ω-3 may prevent development of some inflammatory diseases, but only limited evidence supports therapeutic benefit in those with established inflammatory disease.

Many authors have described decreased rates of cognitive decline and decreased onset of dementia including decreased rates of Alzheimer's disease (AD) in populations consuming at least one serving of fatty fish one or more times per week. These same studies have concluded that increased fish intake slows overall cognitive decline (Mazza et al. 2007). Dietary ω-3 and plasma ω-3 levels have been linked, although not directly correlated. As the dietary consumption of the ω-3 fatty acid docosahexaenoic acid (DHA) decreases below 1 g/day, plasma DHA drops below 4 % of total plasma lipids, however never becomes fully depleted even in strict vegan diets that contain no DHA (Cunnane et al. 2013). This eventually led to investigations into ω-3 supplementation and the possible protective cognitive benefit of increasing ω-3 intake.

In examining the connection between inflammatory states, physical health, and mental health, it has become apparent that there is an intimate link between chronic inflammation and psychiatric disorders. Pro-inflammatory states are increasingly recognized as having a significant impact on the development and progression of a number of major psychiatric disorders. Mood disorders and chronic inflammation are thought of as bidirectional in nature. Pathologic mood states may aggravate inflammatory conditions and in turn may be exacerbated by the inflammatory disease itself (Rosenblat et al. 2014). Also, low levels of serum and neuronal membrane ω-3 have been found in a subset of people with mood disorders. Similarly, investigations into the pathogenesis of schizophrenia have shown altered neuronal membranes and impaired PUFA metabolism contribute to psychosis, suggesting that low ω-3 levels may be related to the progression of psychotic symptoms in schizophrenia (Goren and Tewksbury 2011). Because of the moderate benefits found with ω-3 supplementation in other pro-inflammatory states, ω-3 supplementation has since been investigated as a possible alternative or complimentary therapy in psychiatric disorders.

We present an overview of the biochemistry and physiology of PUFA followed by a review of the existing clinical literature on the efficacy of ω-3 supplementation in major mental illnesses. The literature on PUFA in mental illnesses is highly heterogeneous and wide-ranging, encompassing many different experimental designs and methods. We have focused on the randomized clinical trials designed to demonstrate efficacy of ω-3 supplementation in recognized psychiatric disorders.

Essential Fatty Acids: Composition, Biochemistry, and Physiological Importance

Fatty acids are hydrocarbon chains of varying length with terminal methyl or carboxyl groups (Perica and Delas 2011; Fenton et al. 2013; Freeman et al. 2006a; Freeman and Rapaport 2011; Giordano and Visioli 2014; Rosenblat et al. 2014; Yates et al. 2014; Goren and Tewksbury 2011; Ferguson et al. 2014; Giles et al. 2013; Mischoulon and Freeman 2013; Persons et al. 2014). They are classified as either (1) saturated, not containing double bonds, (2) monounsaturated, indicating they contain one double bond between adjacent carbons, or (3) PUFA, having more than one double bond within the chain. PUFA are further divided into groups based on the position of first double bond (ω-3 or ω-6). There are long-chain and short-chain variants of both ω-3 and ω-6 fatty acids. Within the ω-6 class is the short-chain linoleic acid (LA) and long-chain arachidonic acid (AA) (Fig. 18.1). Similarly, ω-3 examples include the short-chain α-linolenic acid (ALA) and long-chains eicosapentaenoic acid (EPA) and DHA (Fig. 18.2). Short-chain PUFA cannot be

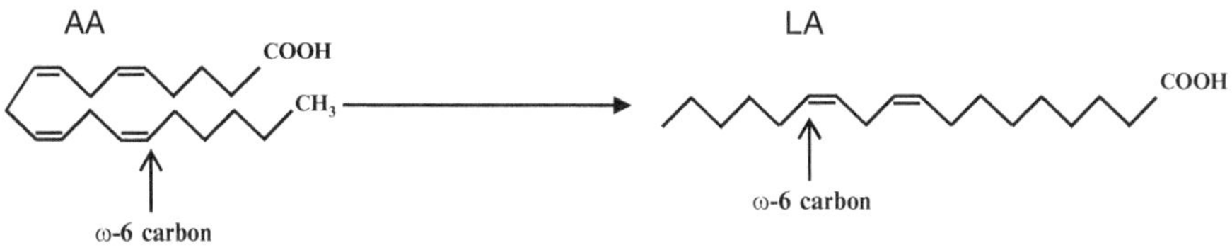

Fig. 18.1 Chemical structure of short-chain ω-6 arachidonic acid (*AA*) and long-chain ω-6 linolenic acid (*LA*)

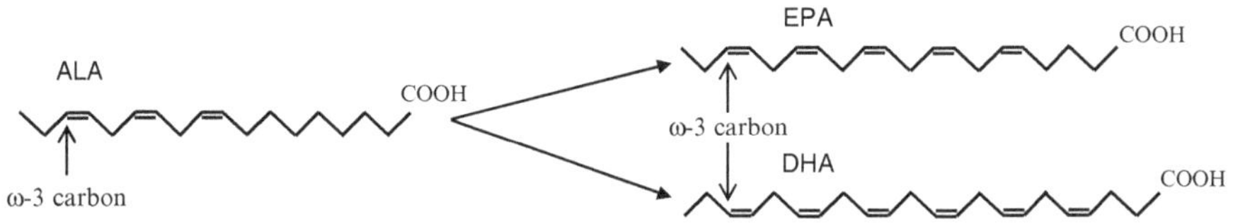

Fig. 18.2 Chemical structure of short-chain ω-3 α-linolenic acid (*ALA*) and long-chain ω-3s eicosapentaenoic acid (*EPA*) and docosahexaenoic acid (*DHA*)

endogenously produced in humans, and are therefore referred to as "essential fatty acids," however long-chain PUFA may be synthesized from 10 to 15 % of short-chain PUFA through endogenous processes.

DHA and EPA contribute to phospholipid membrane fluidity, which allows for normal lipid and protein, receptor, channel, and transporter function within a cell membrane. When depleted, EPA/DHA are replaced by ω-6 resulting in stiffer membranes. DHA, found in large concentration in the cone and rod cells of the retina as well as the brain, is vital to neuronal health, growth, maturation, and function. Depletion of DHA in the CNS negatively impacts membrane structures, as well as the activity of membrane-bound proteins, receptors, and enzymes (Perica and Delas 2011). This may subsequently impair neurotransmitter binding, metabolism, and reuptake. Both EPA and DHA decrease pro-inflammatory effects in endothelial cells (increase nitric oxide production and improve arterial compliance), adhesion molecules (inhibit plaque formation within vessels), and sodium channels (reduce arrhythmias). Some early evidence suggested ω-3 may inhibit platelet aggregation by reducing thromboxane and inhibiting cyclooxygenase (COX) enzymes, however it was later deemed that supplementation with ω-3 did not adversely affect coagulation when taken at proper doses.

Both EPA and DHA can be synthesized endogenously from the essential fatty acid ALA, which must be derived entirely from dietary sources. The synthetic reaction converting ALA to EPA and DHA is very inefficient in humans (Perica and Delas 2011; Freeman and Rapaport 2011; Sanchez-Villegas et al. 2007; Yates et al. 2014; Goren and Tewksbury 2011; Shen et al. 2014). Diets rich in APA tend to increase concentrations of EPA but not DHA. Exogenously, ALA is found in plant sources, while EPA and DHA are found in marine animals and algae sources. Human bioavailability of PUFA is variable from marine algae yet high from fish. Therefore, human dietary intake of EPA and DHA from oily fish (herring, salmon, mackerel) is more effective at maintaining PUFA levels than relying on the inherently inefficient endogenous synthesis from dietary short-chain precursors. The long-chain ω-6 AA, like the long-chain ω-3, can be synthesized from ALA or acquired from dietary intake. These dietary fatty acids are then stored in membrane phospholipids and assist in modulation of the inflammatory response. It has been found that high levels of EPA and DHA in cellular phospholipids can decrease levels of AA, in turn decreasing the inflammatory response as both EPA and AA utilize the same metabolic enzymes. This may result in decreased production of eicosanoids, cytokines, and adhesion molecules by ω-6.

Inflammation and Fatty Acids

Acute inflammatory processes are adaptive and protective physiologic responses to infection and injury (Yates et al. 2014). At the site of local injury or infection, inflammatory cytokines (chemokines, interleukins, lymphokines, tumor necrosis factor) and histamine attach to vascular endothelial cell receptors, which in turn triggers the leukocyte adhesion cascade. Leukocytes within tissues trigger remodeling and cytotoxicity reactions, that in the setting of acute injury or infection are protective. Acute systemic inflammation stimulates the expression of acute phase proteins as well as the development of sickness behaviors. Systemic inflammation causes a decrease in appetite and concentration, as well as increased fatigue and irritability. In chronic inflammatory states however, leukocytes are continually captured from blood, infiltrating across affected endothelial cells into underlying tissue. This may result in tissue remodeling and subsequent damage. On a systemic level, chronic inflammation is associated with a host of potential pathological consequences including: fatigue and depressive symptoms, anorexia, muscle-coasting, obesity, insulin resistance, dyslipidemia, decreased androgen production, increased cortisol, decreased para-sympathetic tone, increased sympathetic tone, hypertension, inflammatory anemia, and osteopenia (Straub 2012).

Eicosanoids and docosanoids are molecules that modulate the immune system, inflammation, and platelet activity (Goren and Tewksbury 2011). Eicosanoids are derived from AA and EPA and docosanoids are derived from DHA. While AA-produced eicosanoids are critical for acute inflammation, ongoing or excessive production will lead to chronic inflammatory and subsequent tissue damage. This AA-mediated inflammatory cascade is modulated by the effects of EPA-produced anti-inflammatory or less potently inflammatory eicosanoids as well as anti-inflammatory DHA-derived docosanoids. Consequently the level of activity of inflammatory reactions is dependent upon the ω-3:ω-6 ratio.

Pro-inflammatory cytokines trigger the production and release of eicosanoids by stimulating PUFA release from membrane phospholipids (Shen et al. 2014). When released, the ω-6 AA is metabolized into pro-inflammatory eicosanoids in one of three ways: by COX to produce prostaglandins and thromboxanes; by lipo-oxygenases (LOX) to produce lipoxins, hydroxyeicosatetraenoic acids (HETE), and leukotrienes (LT); or by p450 epoxygenases to produce epoxyeicosantrienoic acids (EET) and HETEs. These eicosanoids subsequently generate the inflammatory symptoms of erythema, swelling, pain, and heat.

COX and LOX metabolize EPA and DHA into alternative eicosanoids, including specialized pro-resolving mediators (SPMs), called resolvins and protectins, which generate a less intense inflammatory response or have an outright anti-inflammatory action. SPMs block production of prostaglandins, leukotrienes and cytokines, inhibit neutrophil recruitment, and induce phagocytosis of apoptotic cells by macrophages. It was therefore believed that increasing dietary EPA/DHA would tilt the balance towards resolution of inflammation and improved tissue homeostasis by decreasing production of inflammatory eicosanoids and cytokines as well as increasing SPM synthesis.

ω-3 PUFA incorporated into cell phospholipid membranes increases membrane resistance and integrity of epithelial cells and block inflammatory cytokine action on those cells. They also inhibit synthesis of the ω-6 AA, further decreasing levels of pro-inflammatory eicosanoids. Despite the impacts of ω-3 on the inflammatory reactions described above, studies in humans have failed to demonstrate direct benefit from ω-3 dietary supplementation in inflammatory diseases. There may be indirect benefit of ω-3 supplementation from alterations in intestinal flora and downstream effects on microbiota-induced inflammatory reactions (Rosenblat et al. 2014; Yates et al. 2014; Shen et al. 2014).

Major Depressive Disorder

Initial epidemiologic studies suggested a correlation between increased fish consumption and lower rates of depression and suicide related behaviors (Murakami et al. 2010; Tanskanen et al. 2001; Timonen et al. 2004; Sublette et al. 2006). This led to a large number of investigations into the direct antidepressant action of ω-3 PUFA in patients with mood disorders. Antidepressant activity of ω-3 PUFA supplementation in Major Depressive Disorder (MDD) has been scrutinized in a wide variety of heterogeneous studies (Table 18.1). Only randomized controlled trials specifically in subjects with unipolar depression are included here. The largest trial of ω-3 PUFA in subjects with MDD is the SU.FOL.OM3 study by Adreeva et al. (2012). This study showed no significant benefit of ω-3 supplementation in a sample of 2,501 geriatric patients. There are five other large studies that include a total of 898 patients (Antypa et al. 2012; Carney et al. 2009; Grenyer et al. 2007; Lesperance et al. 2011; Rogers et al. 2008) and two smaller studies (Marangell et al. 2003; Mischoulon et al. 2009) with another 70 patients that found no benefit of ω-3 supplementation in MDD. These findings were contradicted by four mid-size trials (Mozaffari-Khosravi et al. 2013; Peet and Horrobin 2002; Sinn et al. 2012; Rizzo et al. 2012) and five small studies (da Silva et al. 2008; Gertsik et al. 2012; Jazayeri et al. 2008; Nemets et al. 2002, 2006), altogether including 406 patients that found significant decreases in depression severity with ω-3 supplementation. One study by Antypa et al. found conflicting results between the symptom measurement scales utilized; reporting no difference on the Beck Depression Inventory (BDI-II) or the Leiden Index of Depression Sensitivity-Revised (LEIDS-R) but significant decreases in depression severity scores on the profile of mood states (POMS) in response to ω-3 consumption. Most of the positive studies were augmentation studies, with

Table 18.1 ω-3 efficacy in major depressive disorder

Study	Population	*N*	Intervention	Duration (weeks)	Primary outcome measure	Results	Comments
Unipolar depression RCTs							
Andreeva et al. (2012)	MDD, CHD, geriatric	2,501	Combination B-vit+EPA/DHA 0.6 g; EPA/DHA 0.6 g; B-vit; PLB	156 and 260	GDS	No difference; increase in males (GDS>10)	Predominantly males
Antypa et al. (2012)	MDD	71	EPA/DHA 2.3 g; PLB	4	POMS, BDI-II, LEIDS-R	No difference in BDI-II & LEIDS-R; decreased depression, tension, fatigue in POMS	Cognition & mood study
Carney et al. (2009)	MDD, CHD	122	EPA/DHA 2 g+antidepressant; PLB+antidepressant	10	BDI-II, HDRS	No difference	Augmentation study
da Silva et al. (2008)	MDD, PD	29	EPA/DHA 0.3 g; EPA/DHA+antidepressant; PLB; PLB+antidepressant	12	MADRS	Decreased depression in all EPA/DHA groups	Compares monotherapy vs. augmentation
Gertsik et al. (2012)	MDD	42	EPA/DHA+citalopram; PLB+citalopram	9	HDRS	Decreased depression	Augmentation study; groups separated at week 4
Grenyer et al. (2007)	MDD	83	EPA/DHA 3 g; PLB	16	HDRS, BDI	No difference	Augmentation study; all subjects underwent weekly psychotherapy
Jazayeri et al. (2008)	MDD	48	EPA 1 g; fluoxetine 20 mg; EPA 1 g+fluoxetine 20 mg	8	HDRS	Decreased depression with combination EPA/fluoxetine	Monotherapy and augmentation study; Monotherapies equivalent

Lesperance et al. (2011)	MDD in MDE ≥4 weeks, ±anxiety	432	EPA/DHA 1.2 g; PLB	4	IDS-SR_{30}	No difference overall	40 % maintained on antidepressants, 15 % maintained in psychotherapy; decreased depression in group without anxiety
Marangell et al. (2003)	MDD	35	DHA 2 g; PLB	6	HDRS	No difference	
Mischoulon et al. (2009)	MDD	35	EPA 1 g; PLB	8	HDRS	No difference	
Mozaffari-Khosravi et al. (2013)	MDD	81	EPA 1 g; DHA 1 g; PLB	12	HDRS	Decreased depression with EPA only	Mild-moderate baseline MDD
Nemets et al. (2002)	MDD	20	EPA + antidepressant; PLB + antidepressant	4	HDRS	Decreased depression	Augmentation study
Nemets et al. (2006)	MDD, children	20	EPA/DHA 0.28–0.6 mg; PLB	16	CDRS	Decreased depression	Low placebo response rates
Peet and Horrobin (2002)	MDD	70	EPA 1 g; EPA 2 g; EPA 4 g; PLB	12	HDRS	Decreased depression in 1 g	Augmentation study
Rizzo et al. (2012)	MDD, geriatric females	46	EPA/DHA 2.5 g; PLB	8	GDS	Decreased depression	No change in immune markers
Rogers et al. (2008)	MDD	190	EPA/DHA 1.5 g; PLB	12	DASS	No difference	
Sinn et al. (2012)	MDD, geriatric	50	EPA 1.67 g; DHA 1.55 g; ω-6PUFA 2.2 g	24	GDS	Decreased depression with both EPA and DHA	Improvement in cognition also noted
Cardiac patient studies on depression outcomes							
Bot et al. (2010)	DM type II	25	EPA 1 g + antidepressant; PLB + antidepressant	12	MADRS	No difference	Augmentation study
Colangelo et al. (2009)	Young adult, CARDIA study	3,317	EPA; DHA; EPA/DHA	20 year	CES-D	Decreased depression in females	No psych history

(continued)

Table 18.1 (continued)

Study	Population	*N*	Intervention	Duration (weeks)	Primary outcome measure	Results	Comments
Haberka et al. (2013)	Acute MI	52	EPA/DHA 1 g; PLB	4	BDI	Decreased depression	No psych history
Zimmer et al. (2013)	Post-MI, OMEGA trial	2,081	EPA/DHA 0.84 g; PLB	52	BDI-II	No difference	Both monotherapy and adjunctive therapy; post hoc positive for efficacy
Perinatal female studies on depression outcomes							
Doornbos et al. (2009)	Healthy, perinatal females	182	DHA 0.22 g; DHA/AA	14–20	EPDS	No difference	No psych history
Freeman et al. (2006b)	Postpartum females	16	EPA/DHA 0.5 g, 1.5 g or 2.8 g; PLB	8	EPDS	Decreased depression	No psych history; No difference between doses of EPA/DHA
Freeman et al. (2008)	MDD, perinatal females	51	EPA/DHA 1.9 g + psychotherapy; PLB + psychotherapy	8	EPDS, HDRS	No difference	Psychotherapy adjunct
Krauss-Etschmann et al. (2007)	Healthy, perinatal females	270	EPA/DHA 0.65; PLB	18	HDRS	Decreased depression	Concomitant folic acid
Llorente et al. (2003)	Postpartum females	89	DHA 0.2 mg	12	EPDS	No difference	No psych history
Lucas et al. (2009)	Psychological distress ± MDE, females	120	EPA/DHA 1.2 g; PLB	8	HDRS	Decreased depression in women without MDE	No difference in MDE
Marangell et al. (2004)	MDD, perinatal females	7	EPA/DHA 2.96 g; PLB	12	HDRS, EPDS	No difference	Euthymic upon study entry
Mozurkewich et al. (2013)	Perinatal females, risk for depression or hx of depression	126	EPA-rich; DHA-rich; PLB	40–48	BDI	No difference	
Su et al. (2008)	MDD, pregnant	24	EPA/DHA 3.4 g; PLB	8	HDRS	Decreased depression	Monotherapy, no antidepressants

patients remaining on antidepressant and psychotherapy regimens. Four of the positive studies were monotherapy designs, however one included only patients with mild to moderate depression (Mozaffari-Khosravi et al. 2013), a second was conducted with pediatric subjects (Nemets et al. 2006), and the remaining two were in geriatric subjects (Sinn et al. 2012; Rizzo et al. 2012). Therefore, there is limited evidence supporting ω-3 monotherapy or augmentation therapy to standard antidepressants in adult patients with severe MDD.

CVD and depressive disorders are intimately linked, and depressive symptoms are associated with a 2–2.5-fold increase in morbidity and mortality in CVD patients (Lett et al. 2008; Whooley et al. 2008). During investigations into ω-3 effectiveness in CVD, the question arose as to the impact of ω-3 supplementation on depressive symptoms in the setting of CVD. There are four studies that have examined the impact of ω-3 treatment on depression severity in patients with established CVD or cardiac risk factors. The largest study was a retrospective data analysis by Colangelo et al. of the coronary artery risk development in young adults (CARDIA) dataset of 3,317 young adults with no prior psychiatric history (Colangelo et al. 2009). At 10-year follow-up, women who received ω-3 PUFA had lower rates of depression than women who did not receive PUFA. Haberka et al. also found significantly lower rates of depressive symptoms in patients with recent myocardial infarction but no existing psychiatric history, who received ω-3 treatment after the index cardiac event (Haberka et al. 2013). These results suggest that ω-3 supplementation may have a protective effect against the development of mood disorders; however, they do not address the impact of ω-3 treatment on existing depressive symptoms. In the OMEGA study, 2,081 patients entered a depression sub-study after 12 months of EPA/DHA or placebo supplementation (Zimmer et al. 2013). Three hundred patients had mild depression and entry into the depression sub-study. In this analysis, ω-3 supplementation was ineffective as a monotherapy but did seem to improve depressive symptoms when taken in combination with an antidepressant medication. However, Bot and colleagues found that the ω-3 plus antidepressant combination was no better than antidepressant medication alone in decreasing depressive symptoms in patients with depression and comorbid type II diabetes mellitus (Bot et al. 2010).

ω-3 use in treatment of depression is especially appealing when looking for safer alternative treatment to antidepressants in the perinatal period. Hibbeln and Davis suggested that ω-3, in particular DHA, is transferred to the fetus for intrauterine neurodevelopment, and that resultant DHA deficiency in mothers would possibly increase rates of perinatal depression (PD) (Hibbeln and Davis 2009). Nine studies examined ω-3 supplementation effects on PD. Four studies with a total of 430 patients found a decrease in overall rates of PD with ω-3 use, however their designs vary significantly. Two of the studies investigated healthy patients with no psychiatric history (Freeman et al. 2006b; Krauss-Etschmann et al. 2007), another study included only patients with PD (Su et al. 2008), while Mozurkewich studied both healthy subjects and those experiencing “psychological distress” (Lucas et al. 2009). They found significantly decreased rates of depression in response to ω-3 supplementation only in healthy subjects not currently in a major depressive episode (MDE), but no benefit to patients in an MDE. The remaining five studies included a total of 455 patients, and failed to find a benefit for ω-3 supplementation, but only three of these studies included women with diagnosed PD (Su et al. 2008; Freeman et al. 2008; Marangell et al. 2004). Taken as a whole, these studies suggest that this may be a small effect size benefit associated with ω-3 therapy for MDD. This may be due to the heterogeneity of the patients enrolled in those trials.

Bipolar Disorder

Stoll et al. reported that ω-3 supplementation markedly decreased depressive symptoms in bipolar patients (Stoll et al. 1999). Because of the public health need for developing effective treatments for bipolar depression that did not increase the risk for mania or metabolic syndrome, there was tremendous enthusiasm for this work. It led to a series of ω-3 augmentation studies for bipolar disorder (Table 18.2). Three

Table 18.2 ω-3 efficacy in bipolar disorder

Study	Population	N	Intervention	Duration (weeks)	Primary outcome measure	Results	Comments
Chiu et al. (2005)	BPAD I, acute mania	15	EPA/DHA 6.8 g; PLB	4	YMRS	No difference	
Clayton et al. (2009)	Adolescents; BPAD I & II, Bipolar NOS	18	EPA/DHA 1 g	6	YMRS, HRDS	Decreased depression	No benefit in mania
Frangou et al. (2006)	BPAD I & II	75	EPA 1 g; EPA 2 g; PLB	12	HDRS	Decreased depression	Sig. decrease in primary outcome, but not secondary
Gracious et al. (2010)	BPAD I & II	51	ALA 6.6 g; PLB	16	CDRS	No difference	
Keck et al. (2006)	BPAD I & II, Bipolar NOS	121	EPA 6 g; PLB	16	IDS-C	No difference	
Murphy et al. (2012)	BPAD I	45	CYT 2 g+ω-3 4 g; ω-3 4 g; PLB	16	Retention in study	No difference	Secondary outcomes: MADRS, YMRS, GAF—no difference
Stoll et al. (1999)	Mania or depression	44	EPA/DHA 9.6 g; PLB	16	HDRS	Decreased depression	Sig. decrease in primary outcome, but not secondary

studies found a significant decrease in depression severity, but none showed benefit in mania. Four other studies failed to substantiate these findings demonstrating no efficacy for depressive or manic symptoms with ω-3 supplementation.

Mood Disorder Meta-analyses

Ten meta-analyses evaluated available randomized controlled trials of ω-3 supplementation in mood disorders (Table 18.3). Two meta-analyses, Bloch and Hannestad, and Rogers et al., included 13 separate RCTs investigating ω-3 therapy

Table 18.3 Meta-analyses: ω-3 efficacy in treatment of all-type depressive disorders

Study	*N*	Intervention	Duration (weeks)	Results
Appleton et al. (2006)	MDD=7 Bipolar=2 Schizophrenia=1 Borderline PD=1 Healthy=1	ω-3 (EPA, DHA); PLB	4–6	No significant benefit
Bloch and Hannestad (2012)	MDD=13	ω-3 (EPA, DHA); PLB	4–16	No significant benefit
Dennis et al. (2013)	PD=6	ω-3; acupuncture, massage, bright light therapy	4–8	No significant benefit
Freeman (2006)	MDD=5 Bipolar=3 Schizophrenia=4	ω-3 (EPA, DHA); PLB	4–16	Benefit in unipolar and bipolar depression
Kraguljac et al. (2009)	MDD=8 Bipolar=5	ω-3 (EPA, DHA); PLB	4–16	Benefit in unipolar and bipolar depression; no benefit in mania
Lin and Su (2007)	MDD=3 MDE=1 Bipolar=2	ω-3 (EPA, DHA); PLB	4–16	Benefit in unipolar and bipolar depression; no benefit in mania
Montgomery and Richardson (2008)	Bipolar=5	ω-3 (EPA, DHA); PLB	4–52	Benefit in bipolar depression; no benefit in mania
Rogers et al. (2008)	MDD=13	ω-3 (EPA, DHA); PLB	4–16	No significant benefit
Sarris et al. (2012)	Bipolar=6	ω-3 (EPA, DHA); ALA; PLB	4–16	Benefit in bipolar depression; no benefit in mania
Sublette et al. (2011)	MDD=12 MDE in other=2 Bipolar=1	ω-3 (EPA); PLB	4–16	Benefit in unipolar depression with EPA between 200 and 2,200 mg; no difference with DHA

for patients with MDD and reported no significant benefit (Rogers et al. 2008; Bloch and Hannestad 2012). A Cochrane Review in perinatal depression by Dennis and Dowswell also found no significant benefit with ω-3 use in perinatal depression (Dennis et al. 2013). Montgomery and Richardson, and Sarris et al. examined RCTs comparing ω-3 supplementation for bipolar depression versus bipolar mania (Montgomery and Richardson 2008; Sarris et al. 2012). They both found overall benefit for ω-3 supplementation in bipolar depression but not in mania. Five other meta-analyses included multiple psychiatric disorders (Appleton et al. 2006; Freeman 2006; Kraguljac et al. 2009; Sublette et al. 2011; Lin and Su 2007). Freeman, Lin et al., and Kraguljac et al. determined there was significant effectiveness of ω-3 in both unipolar and bipolar depression, while Appleton et al. failed to find benefit in psychiatric disorders generally. Sublette et al. included a majority of MDD studies ($n = 12$) with only one BPAD study, and concluded there was a role for EPA augmentation, but not DHA, in unipolar depression. In general, these meta-analyses suggest: (1) ω-3 monotherapy for mood disorders is not effective for a heterogeneous group of patients with MDD or BPAD and (2) that there may be a role for ω-3 as an augmenting agent for some patients with unipolar and bipolar depression.

Schizophrenia

The pathogenesis of primary psychotic disorders, in particular schizophrenia, is postulated to include lipid changes and abnormalities in cell membranes (Fusar-Poli and Berger 2012). Low levels of ω-3 PUFA in peripheral blood cell membranes have been found during the early disease stages of antipsychotic-naïve patients with schizophrenia (Reddy et al. 2004; Yao et al. 2004). This has led to investigations into ω-3 adjunctive treatment in schizophrenia. Seven studies have gauged efficacy of ω-3 in the treatment of schizophrenia (Table 18.4). Amminger et al. reported a decrease in transition to schizophrenia in a study of ultra-high risk patients with no existing psychotic disorder, for subjects who had ω-3 versus placebo. The subjects who received ω-3 therapy had fewer psychiatric symptoms at follow-up and better psychosocial functioning. Mossaheb et al. also investigated ultra-high risk patients with no established diagnosis of schizophrenia, and found an overall decrease in all psychotic symptoms over a 12-week period of ω-3 supplementation (Mossaheb et al. 2013). Peet et al. similarly found significant improvement in psychotic symptoms in subjects with schizophrenia who received ω-3 supplements as wither monotherapy or as an adjunctive therapy (Peet et al. 2001). Yet, three studies did not find a reduction in psychosis with ω-3 treatment (Bentsen et al. 2013; Emsley et al. 2008; Fenton et al. 2001). One trial by Bentsen et al. actually reported an increase in psychosis associated with the combination of EPA and vitamin C.

Several meta-analyses evaluated the overall value of ω-3 in psychotic disorders. One meta-analysis failed to show a decrease in psychotic symptoms with EPA (Fusar-Poli and Berger 2012). Other meta-analyses focused primarily on mood

Table 18.4 ω-3 efficacy in primary psychotic disorders

Study	Population	*N*	Intervention	Duration (weeks)	Primary outcome measure	Results	Comments
Amminger et al. (2010)	Ultra-high risk adolescents, no existing psychotic disorder	81	ω-3 1.2 g; PLB	12	Transition to psychotic disorder	Decreased transition to psychotic disorder	Improved positive sx, negative sx, and functioning
Arvindakshan et al. (2003)	Schizophrenia	33	EPA/DHA 0.3 g + VitE/C	16	BPRS, PANSS	Decreased psychosis	Augmentation study; not placebo-controlled
Bentsen et al. (2013)	Schizophrenia	99	EPA2 g + VitE/C; EPA2 g; VitE/C; PLB	16	PANSS	No difference with combo; separately, both increased psychosis	Augmentation study
Emsley et al. (2006)	Schizophrenia, TD	77	EPA2 g; PLB	12	ESRS	No difference	Augmentation study; Study underpowered
Fenton et al. (2001)	Schizophrenia, Schizoaffective	87	EPA 3 g; PLB	16	Cognitive battery	No difference	Augmentation study; cognition study
Mossaheb et al. (2013)	Ultra-high risk adolescents, no existing psychotic disorder	81	EPA/DHA 1.2 g; PLB	12	PANSS	Decrease in all PANSS subscale scores	Step-wise progression of affect over 12 weeks
Peet et al. (2001)	Schizophrenia	45	EPA; DHA; PLB	12	PANSS	Decreased psychosis	Augmentation arm of study
Peet et al. (2001)	Schizophrenia	26	EPA; PLB	12	PANSS	Decreased psychosis	Monotherapy arm of study

disorders, but also included some studies in schizophrenia, and likewise did not determine benefit with ω-3 supplementation for psychosis (Appleton et al. 2006; Freeman 2006). As was the case with mood disorders, the current utility of ω-3 supplementation in the clinical management of patients with psychotic disorders is an open question. The results to date investigating ω-3 supplementation in psychotic disorders suggest a need for additional, well-powered, definitive clinical investigations.

Anxiety Disorders

There are very limited data published investigating ω-3 supplementation in treatment of anxiety disorders. Four small studies have attempted to evaluate ω-3 on anxiety symptoms (Table 18.5). Three of the studies found a decrease in anxiety symptoms, however all patients included had unspecified anxiety symptoms without a diagnosed anxiety disorder (Buydens-Branchey and Branchey 2006; Buydens-Branchey et al. 2008; Kiecolt-Glaser et al. 2011). Two of those studies sampled from a population of substance abusers, while the other utilized high-functioning medical students. The sole study investigating ω-3 treatment of a specified anxiety

Table 18.5 ω-3 Efficacy in anxiety disorders

Study	Population	*N*	Intervention	Duration (weeks)	Primary outcome measure	Results	Comments
Buydens-Branchey and Branchey (2006)	Unspecified anxiety; Substance abusers	24	EPA/DHA 0.55 g; PLB	12	POMS	Decreased anxiety	No defined anxiety disorder
Buydens-Branchey et al. (2008)	Unspecified anxiety; Substance abusers	22	EPA/DHA 3 g; PLB	12	POMS	Decreased anxiety	No defined anxiety disorder
Fux et al. (2004)	OCD	11	EPA × 6 weeks, then PLB × 6 weeks; PLB × 6 weeks, then EPA × 6 weeks	12	YBOCS	No difference	Augmentation study, concurrent SSRI use
Kiecolt-Glaser et al. (2011)	Unspecified anxiety; Medical students	68	EPA/DHA 2.5 g; PLB	12	BAI	Decreased anxiety	Effective in individuals without specified anxiety disorder

disorder, OCD, failed to demonstrate benefit for ω-3 treatment (Fux et al. 2004). At this time there is no sufficient evidence to draw any conclusion about ω-3 treatment for anxiety symptoms or disorders.

Borderline Personality Disorder

There have been three small studies investigating ω-3 treatment in borderline personality disorder. Results of two small randomized, placebo-controlled trials of subjects with borderline personality disorder suggest that ω-3 monotherapy (Zanarini and Frankenburg 2003) and augmentation of valproate (Bellino et al. 2014) significantly decrease impulsivity, aggression, and outbursts of anger. However, they failed to demonstrate improvement in depressive symptom severity or mood outcomes. The third study had similar clinical results, but suggested that erythrocyte ω-3 levels may be indicators of psychosocial functioning and psychopathology in borderline personality disorder (Amminger et al. 2013). These pilot studies require larger, more extensive follow-up studies before we will know the value of ω-3 therapy for borderline personality disorder.

Alzheimer's Disease and Cognitive Impairment

Results of epidemiologic studies suggest there is a benefit of ω-3 supplementation during the early stages of syndromes leading to cognitive decline, including age-related cognitive decline (ARCD) and mild cognitive impairment (MCI) (Mazza et al. 2007; Cunnane et al. 2009, 2013; Cole et al. 2005, 2009; Huang 2010; Mazereeuw et al. 2012; Shah 2013; Crupi et al. 2013; Dacks et al. 2013; Das 2008; Fotuhi et al. 2009; van der Beek and Kamphuis 2008). However, studies of ω-3 therapy in subjects with established Alzheimer's disease (AD) have been less promising (Table 18.6). Most studies found no benefit with EPA, DHA, or combination supplementation in AD on primary outcome measures (Boston et al. 2004; Chiu et al. 2008; Freund-Levi et al. 2006; Kotani et al. 2006; Quinn et al. 2010). Chiu et al. did note improved cognition in AD subjects on one measure, but failed to confirm this finding on the other primary outcome measure. Scheltens et al. noted improved memory and processing in patients with mild AD with Souvenaid supplementation (Scheltens et al. 2012). Souvenaid is a nutritional supplement containing DHA, EPA, vitamin E, vitamin C, phospholipids, choline, vitamin B12, vitamin B6, and folic acid. Across all of the studies of ω-3 in dementia, any potential benefit was only observed in the subjects with MCI (Chiu et al. 2008; Freund-Levi et al. 2006; Kotani et al. 2006; Yurko-Mauro et al. 2010). This suggests that ω-3 may be helpful in maintaining cognitive health and mitigating early cognitive decline in high-risk populations but has limited efficacy once a dementing process has become fully established.

Table 18.6 ω-3 Efficacy on cognition in Alzheimer's disease

Study	Population	*N*	Intervention	Duration (weeks)	Primary outcome measure	Results	Comments
Boston et al. (2004)	Mild-moderate AD	19	EPA 1 g	24	MMSE; ADAS-Cog	No difference	Monotherapy study; 1st 12 weeks no intervention
Chiu et al. (2008)	Mild-moderate AD or MCI	35	EPA/DHA 2 g; PLB	24	ADAS-Cog; CIBIC-plus	AD-improved CIBIC-plus; MCI-improved ADAS-Cog and CIBIC-plus	Monotherapy study; Secondary measures (MMSE, HRDS) no difference
Freund-Levi et al. (2006)	Mild-moderate AD; OmegAD Study	204	EPA 0.6 g+DHA 1.7 g; PLB	24	MMSE; ADAS-Cog	AD-no difference; MCI-improved cognition	Augmentation study with cholinesterase agents
Kotani et al. (2006)	AD, MCI, organic brain injury	39	AA/DHA 0.24 g; PLB	12	RBANS	AD-no difference; MCI-improved immediate memory & attention; OBI-improved immediate & delayed memory	
Quinn et al. (2010)	Mild-moderate AD	259	DHA 2 g; PLB	72	ADAS-Cog	No difference	Secondary measures CDR and MRI—no difference in brain atrophy
Scheltens et al. (2012)	Mild AD; Souvenir II Study	259	Souvenaid (DHA 1.2 g, EPC 0.3 g, vitamins); PLB	24	NTB domain- Z	Improved cognition	Drug-naïve patients
Yurko-Mauro et al. (2010)	ARCD	485	DHA 0.9 g; PLB	24	CANTAB PAL	Improved cognition	Population with mild cognitive impairment

Conclusion

In conclusion, preclinical studies suggest that ω-3 fatty acids are important in membrane stabilization and are important mediators of inflammation. The data in neuropsychiatric disorders is equivocal at this time. Studies are hampered by small sample size, and the heterogeneity of the study samples, differences in the dosage and formulation of ω-3 agents (and "placebo compounds"), differences in study outcome measures, and statistical approaches. When viewed in total, the data do suggest: (1) there may be a subgroup of patients with MDD and bipolar depression who benefit from ω-3 supplementation but they would constitute a minority of patients, (2) there may be a benefit to ω-3 supplementation for some individuals at risk of developing a psychotic disorder, and (3) ω-3 supplementation may be an as of yet unclear protective effect for some individuals at risk for developing Alzheimer's disease. Further, thoughtful research is warranted in this area of study.

References

Amminger GP, Schafer MR, Papageorgiou K, Klier CM, Cotton SM, Harrigan SM, et al. Long-chain omega-3 fatty acids for indicated prevention of psychotic disorders: a randomized, placebo-controlled trial. Arch Gen Psychiatry. 2010;67(2):146–54. PMID: 20124114. Epub 2010/02/04. eng.

Amminger GP, Chanen AM, Ohmann S, Klier CM, Mossaheb N, Bechdolf A, et al. Omega-3 fatty acid supplementation in adolescents with borderline personality disorder and ultra-high risk criteria for psychosis: a post hoc subgroup analysis of a double-blind, randomized controlled trial. Can J Psychiatry. 2013;58(7):402–8. PMID: 23870722. Epub 2013/07/23. eng.

Andreeva VA, Galan P, Torres M, Julia C, Hercberg S, Kesse-Guyot E. Supplementation with B vitamins or n-3 fatty acids and depressive symptoms in cardiovascular disease survivors: ancillary findings from the SUpplementation with FOLate, vitamins B-6 and B-12 and/or OMega-3 fatty acids (SU.FOL.OM3) randomized trial. Am J Clin Nutr. 2012;96(1):208–14. PMID: 22648722. Epub 2012/06/01. eng.

Antypa N, Smelt AH, Strengholt A, Van der Does AJ. Effects of omega-3 fatty acid supplementation on mood and emotional information processing in recovered depressed individuals. J Psychopharmacol. 2012;26(5):738–43. PMID: 22004690.

Appleton KM, Hayward RC, Gunnell D, Peters TJ, Rogers PJ, Kessler D, et al. Effects of n-3 long-chain polyunsaturated fatty acids on depressed mood: systematic review of published trials. Am J Clin Nutr. 2006;84(6):1308–16. PMID: 17158410. Epub 2006/12/13. eng.

Arvindakshan M, Ghate M, Ranjekar PK, Evans DR, Mahadik SP. Supplementation with a combination of omega-3 fatty acids and antioxidants (vitamins E and C) improves the outcome of schizophrenia. Schizophr Res. 2003;62(3):195–204. PMID: 12837515. Epub 2003/07/03. eng.

Bellino S, Bozzatello P, Rocca G, Bogetto F. Efficacy of omega-3 fatty acids in the treatment of borderline personality disorder: a study of the association with valproic acid. J Psychopharmacol. 2014;28(2):125–32. PMID: 24196948. Epub 2013/11/08. eng.

Bentsen H, Osnes K, Refsum H, Solberg DK, Bohmer T. A randomized placebo-controlled trial of an omega-3 fatty acid and vitamins E+C in schizophrenia. Transl Psychiatry. 2013;3:e335. PMID: 24346133. PMCID: PMC3906471. Epub 2013/12/19. eng.

Bloch MH, Hannestad J. Omega-3 fatty acids for the treatment of depression: systematic review and meta-analysis. Mol Psychiatry. 2012;17(12):1272–82. PMID: 21931319. PMCID: PMC3625950. Epub 2011/09/21. eng.

Boston PF, Bennett A, Horrobin DF, Bennett CN. Ethyl-EPA in Alzheimer's disease—a pilot study. Prostaglandins Leukot Essent Fatty Acids. 2004;71(5):341–6. PMID: 15380822.

Bot M, Pouwer F, Assies J, Jansen EH, Diamant M, Snoek FJ, et al. Eicosapentaenoic acid as an add-on to antidepressant medication for co-morbid major depression in patients with diabetes mellitus: a randomized, double-blind placebo-controlled study. J Affect Disord. 2010;126 (1–2):282–6. PMID: 20466431.

Buydens-Branchey L, Branchey M. n-3 polyunsaturated fatty acids decrease anxiety feelings in a population of substance abusers. J Clin Psychopharmacol. 2006;26(6):661–5. PMID: 17110827. Epub 2006/11/18. eng.

Buydens-Branchey L, Branchey M, Hibbeln JR. Associations between increases in plasma n-3 polyunsaturated fatty acids following supplementation and decreases in anger and anxiety in substance abusers. Prog Neuropsychopharmacol Biol Psychiatry. 2008;32(2):568–75. PMID: 18060675. PMCID: PMC2275606. Epub 2007/12/07. eng.

Carney RM, Freedland KE, Rubin EH, Rich MW, Steinmeyer BC, Harris WS. Omega-3 augmentation of sertraline in treatment of depression in patients with coronary heart disease: a randomized controlled trial. JAMA. 2009;302(15):1651–7. PMID: 19843899. PMCID: PMC3477788. Epub 2009/10/22. eng.

Chiu CC, Huang SY, Chen CC, Su KP. Omega-3 fatty acids are more beneficial in the depressive phase than in the manic phase in patients with bipolar I disorder. J Clin Psychiatry. 2005;66(12):1613–4. PMID: 16401167. Epub 2006/01/13. eng.

Chiu CC, Su KP, Cheng TC, Liu HC, Chang CJ, Dewey ME, et al. The effects of omega-3 fatty acids monotherapy in Alzheimer's disease and mild cognitive impairment: a preliminary randomized double-blind placebo-controlled study. Prog Neuropsychopharmacol Biol Psychiatry. 2008;32(6):1538–44. PMID: 18573585.

Clayton EH, Hanstock TL, Hirneth SJ, Kable CJ, Garg ML, Hazell PL. Reduced mania and depression in juvenile bipolar disorder associated with long-chain omega-3 polyunsaturated fatty acid supplementation. Eur J Clin Nutr. 2009;63(8):1037–40. PMID: 19156158. Epub 2009/01/22. eng.

Colangelo LA, He K, Whooley MA, Daviglus ML, Liu K. Higher dietary intake of long-chain omega-3 polyunsaturated fatty acids is inversely associated with depressive symptoms in women. Nutrition. 2009;25(10):1011–9. PMID: 19195841. PMCID: 2798585.

Cole GM, Lim GP, Yang F, Teter B, Begum A, Ma Q, et al. Prevention of Alzheimer's disease: omega-3 fatty acid and phenolic anti-oxidant interventions. Neurobiol Aging. 2005;26 Suppl 1:133–6. PMID: 16266772.

Cole GM, Ma QL, Frautschy SA. Omega-3 fatty acids and dementia. Prostaglandins Leukot Essent Fatty Acids. 2009;81(2–3):213–21. PMID: 19523795.

Crupi R, Marino A, Cuzzocrea S. n-3 fatty acids: role in neurogenesis and neuroplasticity. Curr Med Chem. 2013;20(24):2953–63. PMID: 23746276. Epub 2013/06/12. eng.

Cunnane SC, Plourde M, Pifferi F, Begin M, Feart C, Barberger-Gateau P. Fish, docosahexaenoic acid and Alzheimer's disease. Prog Lipid Res. 2009;48(5):239–56. PMID: 19362576.

Cunnane SC, Chouinard-Watkins R, Castellano CA, Barberger-Gateau P. Docosahexaenoic acid homeostasis, brain aging and Alzheimer's disease: can we reconcile the evidence? Prostaglandins Leukot Essent Fatty Acids. 2013;88(1):61–70. PMID: 22575581.

da Silva TM, Munhoz RP, Alvarez C, Naliwaiko K, Kiss A, Andreatini R, et al. Depression in Parkinson's disease: a double-blind, randomized, placebo-controlled pilot study of omega-3 fatty-acid supplementation. J Affect Disord. 2008;111(2–3):351–9. PMID: 18485485.

Dacks PA, Shineman DW, Fillit HM. Current evidence for the clinical use of long-chain polyunsaturated n-3 fatty acids to prevent age-related cognitive decline and Alzheimer's disease. J Nutr Health Aging. 2013;17(3):240–51. PMID: 23459977. Epub 2013/03/06. eng.

Das UN. Folic acid and polyunsaturated fatty acids improve cognitive function and prevent depression, dementia, and Alzheimer's disease—but how and why? Prostaglandins Leukot Essent Fatty Acids. 2008;78(1):11–9. PMID: 18054217.

Dennis CL, Dowswell T. Interventions (other than pharmacological, psychosocial or psychological) for treating antenatal depression. Cochrane Database Syst Rev. 2013;7:CD006795. PMID: 23904069. Epub 2013/08/02. eng.

Doornbos B, van Goor SA, Dijck-Brouwer DA, Schaafsma A, Korf J, Muskiet FA. Supplementation of a low dose of DHA or DHA+AA does not prevent peripartum depressive symptoms in a small population based sample. Prog Neuropsychopharmacol Biol Psychiatry. 2009;33(1): 49–52. PMID: 18955102. Epub 2008/10/29. eng.

Emsley R, Niehaus DJ, Koen L, Oosthuizen PP, Turner HJ, Carey P, et al. The effects of eicosapentaenoic acid in tardive dyskinesia: a randomized, placebo-controlled trial. Schizophr Res. 2006;84(1):112–20. PMID: 16632329. Epub 2006/04/25. eng.

Emsley R, Niehaus DJ, Oosthuizen PP, Koen L, Ascott-Evans B, Chiliza B, et al. Safety of the omega-3 fatty acid, eicosapentaenoic acid (EPA) in psychiatric patients: results from a randomized, placebo-controlled trial. Psychiatry Res. 2008;161(3):284–91. PMID: 18962989. Epub 2008/10/31. eng.

Fenton WS, Dickerson F, Boronow J, Hibbeln JR, Knable M. A placebo-controlled trial of omega-3 fatty acid (ethyl eicosapentaenoic acid) supplementation for residual symptoms and cognitive impairment in schizophrenia. Am J Psychiatry. 2001;158(12):2071–4. PMID: 11729030. Epub 2001/12/01. eng.

Fenton JI, Hord NG, Ghosh S, Gurzell EA. Immunomodulation by dietary long chain omega-3 fatty acids and the potential for adverse health outcomes. Prostaglandins Leukot Essent Fatty Acids. 2013;89(6):379–90. PMID: 24183073. PMCID: 3912985.

Ferguson JF, Mulvey CK, Patel PN, Shah RY, Doveikis J, Zhang W, et al. Omega-3 PUFA supplementation and the response to evoked endotoxemia in healthy volunteers. Mol Nutr Food Res. 2014;58:601–13. PMID: 24190860.

Fotuhi M, Mohassel P, Yaffe K. Fish consumption, long-chain omega-3 fatty acids and risk of cognitive decline or Alzheimer disease: a complex association. Nat Clin Pract Neurol. 2009;5(3):140–52. PMID: 19262590.

Frangou S, Lewis M, McCrone P. Efficacy of ethyl-eicosapentaenoic acid in bipolar depression: randomised double-blind placebo-controlled study. Br J Psychiatry. 2006;188:46–50. PMID: 16388069. Epub 2006/01/03. eng.

Freeman MP. Omega-3 fatty acids and perinatal depression: a review of the literature and recommendations for future research. Prostaglandins Leukot Essent Fatty Acids. 2006;75(4–5):291–7. PMID: 16930971. Epub 2006/08/26. eng.

Freeman MP, Rapaport MH. Omega-3 fatty acids and depression: from cellular mechanisms to clinical care. J Clin Psychiatry. 2011;72(2):258–9. PMID: 21382308. Epub 2011/03/09. eng.

Freeman MP, Hibbeln JR, Wisner KL, Davis JM, Mischoulon D, Peet M, et al. Omega-3 fatty acids: evidence basis for treatment and future research in psychiatry. J Clin Psychiatry. 2006a;67(12):1954–67. PMID: 17194275. Epub 2006/12/30. eng.

Freeman MP, Hibbeln JR, Wisner KL, Brumbach BH, Watchman M, Gelenberg AJ. Randomized dose-ranging pilot trial of omega-3 fatty acids for postpartum depression. Acta Psychiatr Scand. 2006b;113(1):31–5. PMID: 16390366. Epub 2006/01/05. eng.

Freeman MP, Davis M, Sinha P, Wisner KL, Hibbeln JR, Gelenberg AJ. Omega-3 fatty acids and supportive psychotherapy for perinatal depression: a randomized placebo-controlled study. J Affect Disord. 2008;110(1–2):142–8. PMID: 18206247. Epub 2008/01/22. eng.

Freund-Levi Y, Eriksdotter-Jonhagen M, Cederholm T, Basun H, Faxen-Irving G, Garlind A, et al. Omega-3 fatty acid treatment in 174 patients with mild to moderate Alzheimer disease: OmegAD study: a randomized double-blind trial. Arch Neurol. 2006;63(10):1402–8. PMID: 17030655. Epub 2006/10/13. eng.

Fusar-Poli P, Berger G. Eicosapentaenoic acid interventions in schizophrenia: meta-analysis of randomized, placebo-controlled studies. J Clin Psychopharmacol. 2012;32(2):179–85. PMID: 22367656. Epub 2012/03/01. eng.

Fux M, Benjamin J, Nemets B. A placebo-controlled cross-over trial of adjunctive EPA in OCD. J Psychiatr Res. 2004;38(3):323–5. PMID: 15003438. Epub 2004/03/09. eng.

Gertsik L, Poland RE, Bresee C, Rapaport MH. Omega-3 fatty acid augmentation of citalopram treatment for patients with major depressive disorder. J Clin Psychopharmacol. 2012;32(1): 61–4. PMID: 22198441. PMCID: PMC3375825. Epub 2011/12/27. eng.

Giles GE, Mahoney CR, Kanarek RB. Omega-3 fatty acids influence mood in healthy and depressed individuals. Nutr Rev. 2013;71(11):727–41. PMID: 24447198.

Giordano E, Visioli F. Long-chain omega 3 fatty acids: molecular bases of potential antioxidant actions. Prostaglandins Leukot Essent Fatty Acids. 2014;90(1):1–4. PMID: 24345866.

Goren JL, Tewksbury AT. The use of omega-3 fatty acids in mental illness. J Pharm Pract. 2011;24(5):452–71. PMID: 21940602.

Gracious BL, Chirieac MC, Costescu S, Finucane TL, Youngstrom EA, Hibbeln JR. Randomized, placebo-controlled trial of flax oil in pediatric bipolar disorder. Bipolar Disord. 2010;12(2):142–54. PMID: 20402707. PMCID: PMC3024033. Epub 2010/04/21. eng.

Grenyer BF, Crowe T, Meyer B, Owen AJ, Grigonis-Deane EM, Caputi P, et al. Fish oil supplementation in the treatment of major depression: a randomised double-blind placebo-controlled trial. Prog Neuropsychopharmacol Biol Psychiatry. 2007;31(7):1393–6. PMID: 17659823. Epub 2007/07/31. eng.

Haberka M, Mizia-Stec K, Mizia M, Gieszczyk K, Chmiel A, Sitnik-Warchulska K, et al. Effects of n-3 polyunsaturated fatty acids on depressive symptoms, anxiety and emotional state in patients with acute myocardial infarction. Pharmacol Rep. 2013;65(1):59–68. PMID: 23563024. Epub 2013/04/09. eng.

Hibbeln JR, Davis JM. Considerations regarding neuropsychiatric nutritional requirements for intakes of omega-3 highly unsaturated fatty acids. Prostaglandins Leukot Essent Fatty Acids. 2009;81(2–3):179–86. PMID: 19619995. PMCID: PMC3182570. Epub 2009/07/22. eng.

Huang TL. Omega-3 fatty acids, cognitive decline, and Alzheimer's disease: a critical review and evaluation of the literature. J Alzheimers Dis. 2010;21(3):673–90. PMID: 20634589.

Jazayeri S, Tehrani-Doost M, Keshavarz SA, Hosseini M, Djazayery A, Amini H, et al. Comparison of therapeutic effects of omega-3 fatty acid eicosapentaenoic acid and fluoxetine, separately and in combination, in major depressive disorder. Aust N Z J Psychiatry. 2008;42(3):192–8. PMID: 18247193. Epub 2008/02/06. eng.

Keck Jr PE, Mintz J, McElroy SL, Freeman MP, Suppes T, Frye MA, et al. Double-blind, randomized, placebo-controlled trials of ethyl-eicosapentanoate in the treatment of bipolar depression and rapid cycling bipolar disorder. Biol Psychiatry. 2006;60(9):1020–2. PMID: 16814257. Epub 2006/07/04. eng.

Kiecolt-Glaser JK, Belury MA, Andridge R, Malarkey WB, Glaser R. Omega-3 supplementation lowers inflammation and anxiety in medical students: a randomized controlled trial. Brain Behav Immun. 2011;25(8):1725–34. PMID: 21784145. PMCID: PMC3191260. Epub 2011/07/26. eng.

Kotani S, Sakaguchi E, Warashina S, Matsukawa N, Ishikura Y, Kiso Y, et al. Dietary supplementation of arachidonic and docosahexaenoic acids improves cognitive dysfunction. Neurosci Res. 2006;56(2):159–64. PMID: 16905216.

Kraguljac NV, Montori VM, Pavuluri M, Chai HS, Wilson BS, Unal SS. Efficacy of omega-3 fatty acids in mood disorders – a systematic review and metaanalysis. Psychopharmacol Bull. 2009;42(3):39–54. PMID: 19752840. Epub 2009/09/16. eng.

Krauss-Etschmann S, Shadid R, Campoy C, Hoster E, Demmelmair H, Jimenez M, et al. Effects of fish-oil and folate supplementation of pregnant women on maternal and fetal plasma concentrations of docosahexaenoic acid and eicosapentaenoic acid: a European randomized multicenter trial. Am J Clin Nutr. 2007;85(5):1392–400. PMID: 17490978. Epub 2007/05/11. eng.

Lesperance F, Frasure-Smith N, St-Andre E, Turecki G, Lesperance P, Wisniewski SR. The efficacy of omega-3 supplementation for major depression: a randomized controlled trial. J Clin Psychiatry. 2011;72(8):1054–62. PMID: 20584525. Epub 2010/06/30. eng.

Lett H, Ali S, Whooley M. Depression and cardiac function in patients with stable coronary heart disease: findings from the Heart and Soul Study. Psychosom Med. 2008;70(4):444–9. PMID: 18434493. PMCID: PMC2675877. Epub 2008/04/25. eng.

Lin PY, Su KP. A meta-analytic review of double-blind, placebo-controlled trials of antidepressant efficacy of omega-3 fatty acids. J Clin Psychiatry. 2007;68(7):1056–61. PMID: 17685742. Epub 2007/08/10. eng.

Llorente AM, Jensen CL, Voigt RG, Fraley JK, Berretta MC, Heird WC. Effect of maternal docosahexaenoic acid supplementation on postpartum depression and information processing. Am J Obstet Gynecol. 2003;188(5):1348–53. PMID: 12748510. Epub 2003/05/16. eng.

Lopresti AL, Maker GL, Hood SD, Drummond PD. A review of peripheral biomarkers in major depression: the potential of inflammatory and oxidative stress biomarkers. Prog Neuropsychopharmacol Biol Psychiatry. 2014;48:102–11. PMID: 24104186.

Lucas M, Asselin G, Merette C, Poulin MJ, Dodin S. Ethyl-eicosapentaenoic acid for the treatment of psychological distress and depressive symptoms in middle-aged women: a double-blind, placebo-controlled, randomized clinical trial. Am J Clin Nutr. 2009;89(2):641–51. PMID: 19116322. Epub 2009/01/01. eng.

Marangell LB, Martinez JM, Zboyan HA, Kertz B, Kim HF, Puryear LJ. A double-blind, placebo-controlled study of the omega-3 fatty acid docosahexaenoic acid in the treatment of major depression. Am J Psychiatry. 2003;160(5):996–8. PMID: 12727707. Epub 2003/05/03. eng.

Marangell LB, Martinez JM, Zboyan HA, Chong H, Puryear LJ. Omega-3 fatty acids for the prevention of postpartum depression: negative data from a preliminary, open-label pilot study. Depress Anxiety. 2004;19(1):20–3. PMID: 14978781. Epub 2004/02/24. eng.

Mazereeuw G, Lanctot KL, Chau SA, Swardfager W, Herrmann N. Effects of omega-3 fatty acids on cognitive performance: a meta-analysis. Neurobiol Aging. 2012;33(7):1482.e17–29. PMID: 22305186. Epub 2012/02/07. eng.

Mazza M, Pomponi M, Janiri L, Bria P, Mazza S. Omega-3 fatty acids and antioxidants in neurological and psychiatric diseases: an overview. Prog Neuropsychopharmacol Biol Psychiatry. 2007;31(1):12–26. PMID: 16938373.

Mischoulon D, Freeman MP. Omega-3 fatty acids in psychiatry. Psychiatr Clin North Am. 2013;36(1):15–23. PMID: 23538073. Epub 2013/03/30. eng.

Mischoulon D, Papakostas GI, Dording CM, Farabaugh AH, Sonawalla SB, Agoston AM, et al. A double-blind, randomized controlled trial of ethyl-eicosapentaenoate for major depressive disorder. J Clin Psychiatry. 2009;70(12):1636–44. PMID: 19709502. PMCID: PMC2918427. Epub 2009/08/28. eng.

Montgomery P, Richardson AJ. Omega-3 fatty acids for bipolar disorder. Cochrane Database Syst Rev. 2008;(2):CD005169. PMID: 18425912. Epub 2008/04/22. eng.

Mossaheb N, Schafer MR, Schlogelhofer M, Klier CM, Cotton SM, McGorry PD, et al. Effect of omega-3 fatty acids for indicated prevention of young patients at risk for psychosis: when do they begin to be effective? Schizophr Res. 2013;148(1–3):163–7. PMID: 23778032. Epub 2013/06/20. eng.

Mozaffari-Khosravi H, Yassini-Ardakani M, Karamati M, Shariati-Bafghi SE. Eicosapentaenoic acid versus docosahexaenoic acid in mild-to-moderate depression: a randomized, double-blind, placebo-controlled trial. Eur Neuropsychopharmacol. 2013;23(7):636–44. PMID: 22910528. Epub 2012/08/23. eng.

Mozurkewich EL, Clinton CM, Chilimigras JL, Hamilton SE, Allbaugh LJ, Berman DR, et al. The Mothers, Omega-3, and Mental Health Study: a double-blind, randomized controlled trial. Am J Obstet Gynecol. 2013;208(4):313.e1–9. PMID: 23531328. Epub 2013/03/28. eng.

Murakami K, Miyake Y, Sasaki S, Tanaka K, Arakawa M. Fish and n-3 polyunsaturated fatty acid intake and depressive symptoms: Ryukyus Child Health Study. Pediatrics. 2010; 126(3):e623–30. PMID: 20713476. Epub 2010/08/18. eng.

Murphy BL, Stoll AL, Harris PQ, Ravichandran C, Babb SM, Carlezon Jr WA, et al. Omega-3 fatty acid treatment, with or without cytidine, fails to show therapeutic properties in bipolar disorder: a double-blind, randomized add-on clinical trial. J Clin Psychopharmacol. 2012;32(5):699–703. PMID: 22926607. Epub 2012/08/29. eng.

Nemets B, Stahl Z, Belmaker RH. Addition of omega-3 fatty acid to maintenance medication treatment for recurrent unipolar depressive disorder. Am J Psychiatry. 2002;159(3):477–9. PMID: 11870016. Epub 2002/03/01. eng.

Nemets H, Nemets B, Apter A, Bracha Z, Belmaker RH. Omega-3 treatment of childhood depression: a controlled, double-blind pilot study. Am J Psychiatry. 2006;163(6):1098–100. PMID: 16741212. Epub 2006/06/03. eng.

Peet M, Horrobin DF. A dose-ranging study of the effects of ethyl-eicosapentaenoate in patients with ongoing depression despite apparently adequate treatment with standard drugs. Arch Gen Psychiatry. 2002;59(10):913–9. PMID: 12365878. Epub 2002/10/09. eng.

Peet M, Brind J, Ramchand CN, Shah S, Vankar GK. Two double-blind placebo-controlled pilot studies of eicosapentaenoic acid in the treatment of schizophrenia. Schizophr Res. 2001;49(3):243–51. PMID: 11356585. Epub 2001/05/18. eng.

Perica MM, Delas I. Essential fatty acids and psychiatric disorders. Nutr Clin Pract. 2011;26(4): 409–25. PMID: 21775637. Epub 2011/07/22. eng.

Persons JE, Robinson JG, Ammann EM, Coryell WH, Espeland MA, Harris WS, et al. Omega-3 fatty acid biomarkers and subsequent depressive symptoms. Int J Geriatr Psychiatry. 2014; 29:747–57. PMID: 24338726.

Quinn JF, Raman R, Thomas RG, Yurko-Mauro K, Nelson EB, Van Dyck C, et al. Docosahexaenoic acid supplementation and cognitive decline in Alzheimer disease: a randomized trial. JAMA. 2010;304(17):1903–11. PMID: 21045096. PMCID: PMC3259852. Epub 2010/11/04. eng.

Reddy RD, Keshavan MS, Yao JK. Reduced red blood cell membrane essential polyunsaturated fatty acids in first episode schizophrenia at neuroleptic-naive baseline. Schizophr Bull. 2004;30(4):901–11. PMID: 15957200. Epub 2005/06/16. eng.

Rizzo AM, Corsetto PA, Montorfano G, Opizzi A, Faliva M, Giacosa A, et al. Comparison between the AA/EPA ratio in depressed and non depressed elderly females: omega-3 fatty acid supplementation correlates with improved symptoms but does not change immunological parameters. Nutr J. 2012;11:82. PMID: 23046564. PMCID: PMC3499393. Epub 2012/10/11. eng.

Rogers PJ, Appleton KM, Kessler D, Peters TJ, Gunnell D, Hayward RC, et al. No effect of n-3 long-chain polyunsaturated fatty acid (EPA and DHA) supplementation on depressed mood and cognitive function: a randomised controlled trial. Br J Nutr. 2008;99(2):421–31. PMID: 17956647. Epub 2007/10/25. eng.

Rosenblat JD, Cha DS, Mansur RB, McIntyre RS. Inflamed moods: a review of the interactions between inflammation and mood disorders. Prog Neuropsychopharmacol Biol Psychiatry. 2014;53:23–34. PMID: 24468642. Epub 2014/01/29. Eng.

Sanchez-Villegas A, Henriquez P, Figueiras A, Ortuno F, Lahortiga F, Martinez-Gonzalez MA. Long chain omega-3 fatty acids intake, fish consumption and mental disorders in the SUN cohort study. Eur J Nutr. 2007;46(6):337–46. PMID: 17717628. Epub 2007/08/25. eng.

Sarris J, Mischoulon D, Schweitzer I. Omega-3 for bipolar disorder: meta-analyses of use in mania and bipolar depression. J Clin Psychiatry. 2012;73(1):81–6. PMID: 21903025. Epub 2011/09/10. eng.

Scheltens P, Twisk JW, Blesa R, Scarpini E, von Arnim CA, Bongers A, et al. Efficacy of Souvenaid in mild Alzheimer's disease: results from a randomized, controlled trial. J Alzheimers Dis. 2012;31(1):225–36. PMID: 22766770. Epub 2012/07/07. eng.

Shah R. The role of nutrition and diet in Alzheimer disease: a systematic review. J Am Med Dir Assoc. 2013;14(6):398–402. PMID: 23419980.

Shen W, Gaskins HR, McIntosh MK. Influence of dietary fat on intestinal microbes, inflammation, barrier function and metabolic outcomes. J Nutr Biochem. 2014;25(3):270–80. PMID: 24355793.

Sinn N, Milte CM, Street SJ, Buckley JD, Coates AM, Petkov J, et al. Effects of n-3 fatty acids, EPA v. DHA, on depressive symptoms, quality of life, memory and executive function in older adults with mild cognitive impairment: a 6-month randomised controlled trial. Br J Nutr. 2012;107(11):1682–93. PMID: 21929835. Epub 2011/09/21. eng.

Stoll AL, Severus WE, Freeman MP, Rueter S, Zboyan HA, Diamond E, et al. Omega 3 fatty acids in bipolar disorder: a preliminary double-blind, placebo-controlled trial. Arch Gen Psychiatry. 1999;56(5):407–12. PMID: 10232294. Epub 1999/05/08. eng.

Straub RH. Evolutionary medicine and chronic inflammatory state-known and new concepts in pathophysiology. J Mol Med (Berl). 2012;90(5):523–34. PMID: 22271169. PMCID: PMC3354326. Epub 2012/01/25. eng.

Su KP, Huang SY, Chiu TH, Huang KC, Huang CL, Chang HC, et al. Omega-3 fatty acids for major depressive disorder during pregnancy: results from a randomized, double-blind, placebo-controlled trial. J Clin Psychiatry. 2008;69(4):644–51. PMID: 18370571. Epub 2008/03/29. eng.

Sublette ME, Hibbeln JR, Galfalvy H, Oquendo MA, Mann JJ. Omega-3 polyunsaturated essential fatty acid status as a predictor of future suicide risk. Am J Psychiatry. 2006;163(6):1100–2. PMID: 16741213. Epub 2006/06/03. eng.

Sublette ME, Ellis SP, Geant AL, Mann JJ. Meta-analysis of the effects of eicosapentaenoic acid (EPA) in clinical trials in depression. J Clin Psychiatry. 2011;72(12):1577–84. PMID: 21939614. PMCID: PMC3534764. Epub 2011/09/24. eng.

Tanskanen A, Hibbeln JR, Tuomilehto J, Uutela A, Haukkala A, Viinamaki H, et al. Fish consumption and depressive symptoms in the general population in Finland. Psychiatr Serv. 2001; 52(4):529–31. PMID: 11274502. Epub 2001/03/29. eng.

Timonen M, Horrobin D, Jokelainen J, Laitinen J, Herva A, Rasanen P. Fish consumption and depression: the Northern Finland 1966 birth cohort study. J Affect Disord. 2004;82(3):447–52. PMID: 15555697. Epub 2004/11/24. eng.

van der Beek EM, Kamphuis PJ. The potential role of nutritional components in the management of Alzheimer's disease. Eur J Pharmacol. 2008;585(1):197–207. PMID: 18374332.

Whooley MA, de Jonge P, Vittinghoff E, Otte C, Moos R, Carney RM, et al. Depressive symptoms, health behaviors, and risk of cardiovascular events in patients with coronary heart disease. JAMA. 2008;300(20):2379–88. PMID: 19033588. PMCID: PMC2677371. Epub 2008/11/27. eng.

Yao JK, Magan S, Sonel AF, Gurklis JA, Sanders R, Reddy RD. Effects of omega-3 fatty acid on platelet serotonin responsivity in patients with schizophrenia. Prostaglandins Leukot Essent Fatty Acids. 2004;71(3):171–6. PMID: 15253886. Epub 2004/07/16. eng.

Yates CM, Calder PC, Ed Rainger G. Pharmacology and therapeutics of omega-3 polyunsaturated fatty acids in chronic inflammatory disease. Pharmacol Ther. 2014;141(3):272–82. PMID: 24201219.

Yurko-Mauro K, McCarthy D, Rom D, Nelson EB, Ryan AS, Blackwell A, et al. Beneficial effects of docosahexaenoic acid on cognition in age-related cognitive decline. Alzheimers Dement. 2010;6(6):456–64. PMID: 20434961.

Zanarini MC, Frankenburg FR. Omega-3 fatty acid treatment of women with borderline personality disorder: a double-blind, placebo-controlled pilot study. Am J Psychiatry. 2003;160(1):-167–9. PMID: 12505817. Epub 2002/12/31. eng.

Zimmer R, Riemer T, Rauch B, Schneider S, Schiele R, Gohlke H, et al. Effects of 1-year treatment with highly purified omega-3 fatty acids on depression after myocardial infarction: results from the OMEGA trial. J Clin Psychiatry. 2013;74(11):e1037–45. PMID: 24330904.

Index

N. Müller et al. (eds.), *Immunology and Psychiatry*, Current Topics in Neurotoxicity 8, DOI 10.1007/978-3-319-13602-8

MIX
Papier aus verantwortungsvollen Quellen
Paper from responsible sources
FSC® C105338

If you have any concerns about our products,
you can contact us on
ProductSafety@springernature.com

In case Publisher is established outside the EU,
the EU authorized representative is:
Springer Nature Customer Service Center GmbH
Europaplatz 3, 69115 Heidelberg, Germany

Printed by Libri Plureos GmbH
in Hamburg, Germany